American Institute of Chemical Engineers

Topical Conference on Refinery Processing

Held at the 2004 AIChE Spring National Meeting

April 25-29, 2004
New Orleans, Louisiana, USA

Printed from e-media with permission by:

Curran Associates, Inc.
57 Morehouse Lane
Red Hook, NY 12571
www.proceedings.com

ISBN: 978-1-60423-478-7

Some format issues inherent in the e-media version may also appear in this print version.

ISBN: 978-1-60423-478-7

American Institute of Chemical Engineers
Proceedings
Three Park Avenue
New York, NY 10016-5991
Phone: 212-591-8100

www.aiche.org

TA - Seventh Topical Conference on Refinery Processing

THE VARIOUS METHODS OF ASPHALTENE SEPARATION: ADVANTAGES AND DISADVANTAGES

James G. Speight
CD & W Inc.,
Laramie, Wyoming 82070
E-mail: JamesSp8@aol.com

Keywords: Asphaltene, separation, analytical methods, experimental parameters

Prepared for presentation at the 2004 Spring National Meeting, New Orleans, LA, April 26-30

Unpublished

ABSTRACT

Asphaltenes are the highest molecular weight and the most polar fractions in petroleum. The characteristics of the asphaltenes and the amount of them in oil depend to a greater or lesser extent on the crude source. During refining of crude oil, asphaltenes will end up in a high percentage in the residual fuels as the light fractions (gasoline, jet fuel etc) are removed from the oil through cracking and visbreaking.

The asphaltene fraction is, broadly speaking the fraction of oil that is insoluble in n-heptane or n-pentane and soluble in benzene/toluene.

The treatment of petroleum and petroleum products with solvents varies from the more simple determination of the precipitation number of lubricating oils (ASTM D-91) to the more complex fractionation methods by which several constituent fractions are produced using methods that are reproducible and repeatable. In fact, it is often the simpler methods of testing that have led to the modern complex fraction procedures.

This paper presents a critical review of the methods by which the asphaltene fraction can be separated from crude oil and the advantages or disadvantages of using the various procedures.

1.0 INTRODUCTION

Asphaltenes are the highest molecular weight and the most polar fractions in petroleum. The characteristics of the asphaltenes and the amount of them in oil depend to a greater or lesser extent on the crude source. During refining of crude oil, asphaltenes will end up in a high percentage in the residual fuels as the light fractions (gasoline, jet fuel etc) are removed from the oil through cracking and visbreaking.

Asphaltenes are very complex molecules believed to consist of associated systems of polyaromatic sheets bearing alkyl side chains. The heteroatoms O, N and S as well as the metals V, Ni and Fe are also present in asphaltenes. The exact molecular structure of asphaltenes is not known today because of the complexity of the asphaltene molecules. Therefore the definition of asphaltenes is based on their solubility. The asphaltene fraction is, broadly speaking the fraction of oil that is insoluble in n-heptane or n-pentane and soluble in benzene/toluene.

The problems with asphaltenes has increased due to the need to extract even the heaviest crude oils as well as the trend to extract large amounts of light fractions out of crude oil by amongst other methods cracking and visbreaking. Some examples of problems due to asphaltene flocculation and/or sedimentation are:

- *Oil recovery* Wellbore plugging and pipeline deposition.
- *Visbreaking and cracking processes* Degraded asphaltenes are more aromatic (loss of aliphatic chains) and less soluble.
- *Blending of fuel oils as well as blending of fuel oils with heavy crude oils* The change of medium during mixing may cause destabilization of asphaltenes.
- *Storage problems* Sedimentation and plugging can occur due to oxidation of the asphaltenes. The increased polarity may cause asphaltene aggregation. The sludge formation can furthermore be accelerated if there is bacterial infestation in the fuel.
- *Emulsion formation*
  If there is a high degree of water contamination emulsions may be formed. Asphaltenes are believed to be responsible for the undesired stabilization of emulsions, since asphaltenes are highly polar and surface active.
- *Preheating*
  Preheating fuel oils prior to their burning encourages the precipitation of asphaltenes and coking.
- *Combustion*
  A content of 6% or more of asphaltenes in the fuel results in ignition delay and poor combustion. This will cause boiler fouling, poor heat transfer, stack solid emission and corrosion amongst other problems.

The suggestion arose in 1914 that the systematic separation of petroleum be effected by treatment with solvents. If chosen carefully, solvents effect a separation between the constituents of residua, bituminous materials, and virgin petroleum according to differences in molecular weight and aromatic character. The nature and the quantity of the components separated depend on the conditions of the experiment, namely, the degree of dilution temperature and the nature of the solvent.

The treatment of petroleum and petroleum products with solvents varies from the more simple determination of the precipitation number of lubricating oils (ASTM D-91) to the more complex fractionation methods by which several constituent fractions are produced using methods that are reproducible and repeatable. In fact, it is often the simpler methods of testing that have led to the modern complex fraction procedures.

Solvent methods have also been applied to petroleum fractionation on the basis of molecular weight. The major molecular weight separation process used in the laboratory as well as in the refinery is solvent precipitation. Solvent precipitation occurs in a refinery in a deasphalting unit and is essentially an extension of the procedure for separation by molecular weight, although some separation by polarity might also be operative. The deasphalting process is usually applied to the higher molecular weight fractions of petroleum such as atmospheric and vacuum residua.

The simplest application of solvent extraction consists of mixing petroleum with another liquid, which results in the formation of two phases. This causes distribution of the petroleum constituents over the two phases; the dissolved portion is referred to as the extract, and the non-dissolved part of the petroleum is referred to as the raffinate.

The ratio of the concentration of any particular component in the two phases is known as the distribution coefficient K:

$$K = C_1 / C_2$$

where C_1 and C_2 are the concentrations in the various phases. The distribution coefficient is usually constant and may vary only slightly, if at all, with the concentration of the other components. In fact, the distribution coefficients may differ for the various components of the mixture to such an extent that the ratio of the concentrations of the various components in the solvent phase differs from that in the original petroleum; this is the basis for solvent extraction procedures.

On the basis of the solubility in a variety of solvents, it has become possible to distinguish among the various constituents of petroleum. Crude oil generally does not contain *carboids* and *carbenes*. Residua from cracking processes may contain 2% w/w by weight or more. In fact, some of the highly paraffinic crude oils may contain only small portions of asphaltenes.

2.0 ASPHALTENE SEPARATION

It is generally molecular type, not molecular size, which is responsible for the solubility of species in various solvents. Thus, solvent extraction separates petroleum fractions according to type, although within any particular series there is a separation according to molecular size. Lower molecular weight hydrocarbons of a series (the light fraction) may well be separated from their higher molecular weight homologues (the heavy fraction) by solvent extraction procedures.

In general, it is advisable that selective extraction be employed with fairly narrow boiling range fractions. However, the separation achieved after one treatment with the solvent is rarely complete and several repetitions of the treatment are required. Such repetitious treatments are normally carried out by movement of the liquids countercurrently through the extraction equipment (*countercurrent extraction*), which affords better yields of the extractable materials.

The list of compounds that have been suggested as selective solvent for the fractionation of petroleum is very large (Nelson, 1958; Gruse and Stevens, 1960; Speight, 1999). However, before any extraction process is attempted, it is necessary to consider the following criteria:

1. The differences in the solubility of the petroleum constituents in the solvent should be substantial.
2. The solvent should be significantly less or more dense than the petroleum (product) to be separated to allow easier countercurrent flow of the two phases.
3. Separation of the solvent from the extracted material should be relatively easy.

It may also be advantageous to consider other properties, such as viscosity, surface tension, and the like, as well as the optimal temperature for the extraction process. Thus, aromatics can be separated from naphthene and paraffinic hydrocarbons by the use of selective solvents. Furthermore, aromatics with differing numbers of aromatic rings that may exist in various narrow boiling fractions can also be effectively separated by solvent treatment.

2.1 INFLUENCE OF SOLVENT TYPE

The separation of crude oil into two fractions, asphaltenes and maltenes, is conveniently brought about by means of low-molecular-weight paraffinic hydrocarbons (ASTM D-2006, ASTM D-2007, ASTM D-3279, ASTM D-4124, IP 143), which were recognized to have selective solvency for hydrocarbons, and simple relatively low molecular weight hydrocarbon derivatives. One of the standard methods (ASTM D-2006) has been discontinued but still finds use in many laboratories and represents a very convenient method of asphaltene separation.

The more complex, higher molecular weight compounds are precipitated particularly well by addition of 40 volumes of *n*-pentane or *n*-heptane in the methods generally preferred at present (Speight, et al., 1982, 1984; Speight, 1994) although hexane is used on occasion (Yan et al., 1997). It is no doubt a separation of the chemical components with the most complex structures from the mixture, and this fraction, which should correctly be called *n-pentane asphaltenes* or *n-heptane asphaltenes* is qualitatively and quantitatively reproducible.

If the precipitation method (deasphalting) involves the use of a solvent and a residuum and is essentially a leaching of the heavy oil from the insoluble residue, this process may be referred to as *extraction*. However, under the prevailing conditions now in laboratory use, the term *precipitation* is perhaps more correct and descriptive of the method.

Variation in the solvent type also causes significant changes in asphaltene yield. For example, in the ease of a Western Canadian bitumen and, indeed, for conventional crude oils, branched-chain paraffins or terminal olefins do not precipitate the same amount of asphaltenes as do the corresponding normal paraffins. The solvent power of the solvents (i.e., the ability of the solvent to dissolve asphaltenes) increases in the order

2-Methylparaffin < *n*-paraffin < terminal olefin
(*iso*-paraffin)

Cycloparaffins (naphthenes) have a remarkable effect on asphaltene yield and give results totally unrelated to those from any other non-aromatic solvent. For example, when cyclopentane, cyclohexane, or their methyl derivatives are employed as precipitating media, only about 1% of the material remains insoluble.

To explain those differences it was necessary to consider the solvent power of the precipitating liquid, which can be related to molecular properties (Hildebrand et al., 1970).

Thus the solvent power of non-polar solvents has been expressed as a solubility parameter, δ, and equated to the internal pressure of the solvent, that is, the ratio between the surface tension γ and the cubic root of the molar volume V:

$$\delta_1 = {}^3\sqrt{V}$$

Alternatively, the solubility parameter of non-polar solvents can be related to the energy or vaporization ΔRv and the molar volume,

$$\delta_2 = (\Delta Ev/V)^{1/2}$$

or

$$\delta_2 = (\Delta Hv - RT/V)^{1/2}$$

where ΔHv is the heat of vaporization, R is the gas constant, and T is the absolute temperature.

Consideration of this approach shows that there is indeed a relationship between the solubility parameters for a variety of solvents and the amount of precipitate. The introduction of a polar group (heteroatom function) into the molecule of the solvent has significant effects on the quantity of precipitate. For example, treatment of a residuum with a variety of ethers or treatment of asphaltenes with a variety of solvents illustrates this point (Speight, 1979). In the latter instance it was not possible to obtain data from addition of the solvent to the whole feedstock *per se* since the majority of the non-hydrocarbon materials were not miscible with the feedstock. It is nevertheless interesting that, as with the hydrocarbons, the amount of precipitate, or asphaltene solubility, can be related to the solubility parameter.

The solubility parameter allows an explanation of certain apparent anomalies, for example, the insolubility of asphaltenes in pentane and the near complete solubility of the materials in cyclopentane. Moreover, the solvent power of various solvents is in agreement with the derivation of the solubility parameter; for any one series of solvents the relationship between amount of precipitate (or asphaltene solubility) and the solubility parameter δ is quite regular.

In any method used to isolate asphaltenes as a separate fraction, standardization of the technique is essential. For many years, the method of asphaltenes separation was not standardized, and even now it remains subject to the preferences of the standards organizations of different countries. The use of both *n*-pentane and *n*-heptane has been widely advocated, and although *n*-heptane is becoming the deasphalting liquid of choice, this is by no means a hard-and-fast rule. And it must be recognized that large volumes of solvent may be required to effect a reproducible separation, similar to amounts required for consistent asphaltene separation. It is also preferable that the solvents be of sufficiently low boiling point that complete removal of the solvent from the fraction can be effected and, most important, the solvent must not react with the feedstock. Hence, the preference for hydrocarbon liquids, although the several standard methods that have been used are not unanimous in the ratio of hydrocarbon liquid to feedstock.

Method	Deasphalting liquid	Volume ml/gm
ASTM D-893	*n*-pentane	10
ASTM D-2006	*n*-pentane	50
ASTM D-2007	*n*-pentane	10
IP 143	*n*-heptane	30
ASTM D-3279	*n*-heptane	100
ASTM D-4124	*n*-heptane	100

However, it must be recognized that some of these methods were developed for use with feedstocks other than heavy oil and adjustments are necessary.

Although, *n*-pentane and *n*-heptane are the solvents of choice in the laboratory (other solvents can be used) Girdler, 1965; Speight, 1979) and cause the separation of asphaltenes as brown-to-black powdery materials. In the refinery, supercritical low molecular weight hydrocarbons (e.g., liquid propane, liquid butane, or mixtures of both) are the solvents of choice and the product is a semisolid (tacky) to solid asphalt. The amount of asphalt that settles out of the paraffin/residuum mixture depends on the size of the paraffin, the temperature, and the paraffin-to-feedstock ratio (Girdler, 1965; Corbett and Petrossi, 1978; Speight et al., 1984).

Insofar as industrial solvents are very rarely one compound, it was also of interest to note that the physical characteristics of two different solvent types, in this case benzene and *n*-pentane, are additive on a mole-fraction basis (Mitchell and Speight,

1973) and also explain the variation of solubility with temperature. The data also show the effects of blending a solvent with the bitumen itself and allowing the resulting solvent-heavy oil blend to control the degree of bitumen solubility. Varying proportions of the hydrocarbon alter the physical characteristics of the oil to such an extent that the amount of precipitate (asphaltenes) can be varied accordingly within a certain range.

2.2 INFLUENCE OF THE DEGREE OF DILUTION

One of the asphaltene separation methods (ASTM D-893) was developed for a specific test to determine the amount of insoluble material in lubricating oil. Another (ASTM D-2007) was designed for use with rubber extending and processing oils. Both tests recommend use of 10 ml of hydrocarbon liquid per gram of feedstock. This is not recommended in the tests that are now considered standard for asphaltenes separation (AST D-3279, ASTM D-4124) (Speight et al., 1984). However, it must be recognized that some of these methods were developed for use with feedstocks other than heavy oil and bitumen; thus, adjustments are necessary.

At constant temperature, the quantity of precipitate first increases with increasing ratio of solvent to feedstock and then reaches a maximum. In fact, there are indications that when the proportion of solvent in the mix is <35% little or no asphaltenes are precipitated. If it can be assumed that the bituminous material employed for this particular work had a surface tension of approximately 30 dyne/cm, it is possible to illustrate this phenomenon.

Thus, depending on the feedstock, there is a *solvent induction volume* in which little or no asphaltenes are precipitated. Very few data have been reported that relate to this aspect of asphaltene separation. There is fragmentary evidence by the author to show that the most polar materials (not necessarily the highest molecular weight material) separate first from the feedstock. This is in keeping with the increased paraffinic character of the feedstock as the hydrocarbon is added.

Very few data have been reported that relate to this aspect of asphaltene separation. There is fragmentary evidence to show that the most polar materials (not necessarily the highest molecular weight material) separate first from the feedstock. This is in keeping with the increased paraffin character of the feedstock as the hydrocarbon is added.

2.3 INFLUENCE OF TEMPERATURE

When pentane and the lower molecular weight hydrocarbon solvents are used in large excess, the quantity of precipitate and the composition of the precipitate changes with increasing temperature (Mitchell and Speight, 1973; Andersen, 1994).

One particular example the separation of asphaltenes from using *n*-pentane. At ambient temperatures (~21°C, 70°F) the yield of asphaltenes is 17% w/w but at 35°C (95° F), 22.5% by weight *asphaltenes* are produced using the same feedstock-pentane ratio. This latter precipitate is in fact asphaltenes plus resins; similar effects have been noted with other hydrocarbon solvents at temperatures up to 70°C (160°F). These results are self-explanatory when it is realized that the heat of vaporization ΔHv and the surface tension γ, from which the solubility parameters are derived, both decrease with increasing temperature.

2.4 INFLUENCE OF CONTACT TIME

Contact time between the hydrocarbon and the feedstock also plays an important role in asphaltene separation (Speight et al., 1984). Yields of the asphaltenes reach a maximum after approximately 8 hr, which may be ascribed to the time required for the asphaltene particles to agglomerate into particles of a *filterable size* as well as the diffusion-controlled nature of the process. Heavier feedstocks also need time for the hydrocarbon to penetrate their mass.

3.0 SUMMARY

The relevant parameters for asphaltene separation are both physical and chemical in nature and include:

1. Polarity (the presence of functional groups derived from the presence of heteroatoms in the asphaltenes) (Speight, 1994);
2. Aromaticity (the presence of polynuclear aromatic systems in the asphaltenes) (Girdler, 1965; Mitchell and Speight, 1973);
3. Molecular weight (molecular size) (Speight, 1994);
4. Three-dimensional structure (the micelle) of the asphaltene constituents as they exist in relationship with the other constituents of crude oil (Speight, 1994);
5. Solvent power of the precipitating/extracting liquid used for the separation (Girdler, 1965; Mitchell and Speight, 1973);
6. Time required to allow the precipitating/extracting liquid to penetrate the micelle which is dependent upon the ability of the hydrocarbon liquid to penetrate the micelle, indicating that the process is diffusion-controlled (Speight et al., 1984);
7. Ratio of the precipitating/extracting liquid to crude oil that dictates the yield and character of the asphaltene product (Girdler, 1965; Mitchell and Speight, 1973; Speight et al., 1984);
8. Temperature, which may reduce the induction period that is a requirement of diffusion-controlled processes (Mitchell and Speight, 1973); and
9. Pressure, as employed in several refinery processes as a means of maintaining the low-boiling liquid hydrocarbon in the liquid phase (Speight, 1999, 2000).

4.0 REFERENCES

Andersen, S.I. 1994. Fuel Science and Technology International. 12: 51.

Andersen, S.I. and Speight, J.G., 1992 J.G. Preprints, Div. Fuel Chem. Am. Chem. Soc. 37(3): 1335.

ASTM. 2003. Annual Book of Standards. American Society for Testing and Materials, West Conshohocken, Pennsylvania.

Corbett. L.W., and Petrossi, U. 1978. Ind. Eng. Chem. Prod. Res. Dev. 17: 342.

Girdler, R.B. 1965. Proc. Assoc. Asphalt Paving Technologists 34: 45.

Hildebrand, J.H., Prausnitz, J.M., and Scott, R.L. 1970. Regular Solutions. Van Nostrand-Reinhold, New York.

IP. 2003. Standard No. 143. Standard Methods for Analysis and Testing of Petroleum and Related Products 1997. Institute of Petroleum, London, England.

Mitchell. D.L., and Speight. J.G. 1973. Fuel 52: 149.
Speight, J.G. 1979. Information Series No. 84. Alberta Research Council, Edmonton, Alberta, Canada.
Speight, J.G. 1994. In Asphaltenes and Asphalts. I. Developments in Petroleum Science, 40. T.F. Yen and G.V. Chilingarian (Editors). Elsevier, Amsterdam, The Netherlands. Chapter 2.
Speight, J.G. 1999. The Chemistry and Technology of Petroleum. 3rd Edition. Marcel Dekker Inc., New York.
Speight. J.G. 2000. The Desulfurization of Heavy Oils and Residua. 2nd Edition. Marcel Dekker, New York.
Speight. J.G., Long, R.B., and Trowbridge, T.D. 1982. Preprints. Div. Fuel Chem. Am. Chem. Soc. 27(3/4): 268.
Speight, J.G., Long, R.B., and Trowbridge, T.D. 1984. Fuel. 63: 616.
Yan, J., Plancher, H., and Morrow, N.R. 1997. Paper No. SPE 37232. SPE International Symposium on Oilfield Chemistry. Houston, TX.

DETECTION OF ASPHALTENE FLOCCULATION USING NMR RELAXOMETRY.

Alexandre Prunelet[1] , Marc Fleury[2] et Jean-Pierre Cohen-Addad[1]
[1]Laboratoire de spectrométrie Physique, BP 87, 38402 St Martin d'Hères
[2]Institut Français du Pétrole, 1, avenue de Bois-Préau, 92852 Rueil-Malmaison
Email: marc.fleury@ifp.fr

Key words: flocculation, NMR, asphaltene, viscosity, relaxation.

Prepared for presentation at the 2004 Spring National Meeting, New Orleans, LA, April 26-30

ABSTRACT
We present a preliminary study of the detection of flocculation by NMR relaxometry. For this purpose, various asphaltene solutions in the presence of solvents inducing aggregation and flocculation were prepared. The asphaltenes were extracted from one crude oil and various fractions of toluene and n-heptane were prepared to vary the solvent quality. We measured the transverse magnetization decays at 60 MHz using standard CPMG sequences. We show first that asphaltene aggregates have a well-defined signature typically at relaxation times of the order of 1 ms. In general, the signal can be splitted into two parts characterizing the asphaltene aggregates and the solvents. The asphaltene signature becomes bimodal near the flocculation threshold. The solvent relaxation rate varies linearly with the asphaltene concentration and this effect can be associated with viscosity variations.

INTRODUCTION
Crude oils contain asphaltenes that are of concern in the petroleum industry. These compounds can affect oil production at every stage, starting from the flow in porous media down to the transportation and the refining. Due to their tendency to aggregate and stick to solid surfaces, they can alter significantly the transport properties. Also, during various recovery processes (pressure depletion, gas injection, acidizing), these macromolecules can flocculate and generate significant economic losses. Asphaltenes are commonly defined as compounds insoluble in n-heptane and soluble in toluene. Hence there is no precise chemical definition of asphaltenes but they may exhibit quite uniform chemical properties as shown by various analyses[1]. At high dilution, the aggregate size is thought to be in the range 10-50 A[1,2], and during flocculation, they can grow up to micron-scale sizes. Due the complexity of the molecules, the prediction of asphaltene flocculation is very difficult, although there is some recent attempt to use polymerization models[3].

On the experimental side, the detection of flocculation is not an easy task at temperature and pressure. The recent development of low field NMR apparatus that are portable on site or can perform in-situ measurements suggested us to test the feasibility of flocculation detection from simple relaxometry measurements. In particular, transverse relaxation time T_2 contain a wealth of information about molecular motions and large-scale interaction between molecules. For example, viscosity can be predicted over a range covering 6 orders of magnitude. Hence, this measurement seems at first glance adequate to probe changes of molecular motions induced by increasing aggregate sizes.

THEORETICAL BACKGROUND

Nuclear relaxation times are linked to rotational and translationnal motions of the spin carrying molecules through inter and intramolecular dipole-dipole interactions. If a molecule is viewed as a rigid sphere of radius a in a medium of viscosity η, the longitudinal and transverse relaxation times are linked to the viscosity of the medium and the radius of the molecule according to[4]:

$$\frac{1}{T_1}=\frac{1}{T_2}=\frac{2\gamma^4\hbar^2\pi}{kT}\,\eta\left[\frac{a^3}{b^6}+\frac{3N\pi}{5}\right] \tag{1}$$

where b is the distance between the two interacting spins present at the surface of the sphere (i.e. a water molecule is considered), and N the density of spins per cm^3. In the above relation, the a^3/b^6 term is the rotationnal contribution, whereas the $3N\pi/5$ term is the translationnal contribution. The dependence of relaxation to viscosity and molecule size is due to the use of the Stoke's laws when considering the rotationnal and translationnal diffusion constant.

For large molecules, the rigid sphere hypothesis is obviously a crude assumption but to our knowledge, no theory is available for describing the various possible modes of relaxation. However, it should be pointed out that the rotationnal contribution is expected to be dominant for large molecules, and if the Stoke's laws are still valid, a dependence with the molecular size is expected to persist, but not necessarily with the same sensitivity. Note that the dependence of T_2 with viscosity is verified for a very large range of values [5] (1 to 10^5 cp) for crude oils (this is not true for T_1). Therefore, the measurement of the longitudinal relaxation time at low field is expected to give an information when aggregates are modified in a fluid.

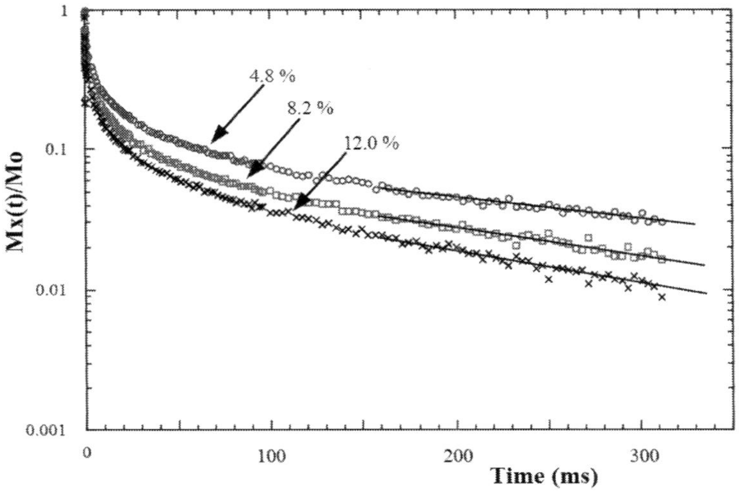

Figure 1: Example of magnetization decay signals for 3 three different asphaltene concentrations in deuterated toluene. These signals characterize the asphaltene aggregates. The strait lines indicate the relaxation time of the solvent (residual signal).

SAMPLE PREPARATION

The asphaltenes considered in this study were extracted from the Prudoe-Bay crude oil by precipitating in n-heptane according to standardized procedures (NF T60-115). The dry solid residues were reduced to powder and mixed/homogenized with toluene in a sealed container during several hours at 30°C. In parallel, 10-mm NMR tubes were modified to be easily and quickly sealed after introduction of a given solution. When ready, the solution is introduced into the NMR tube (about 100 ul) and a known quantity of n-heptane introduced before sealing the tube. Care was taken to not evaporate the solvents and modify the dilution. Using this method, we obtained 3 solutions using a deuterated toluene with different mass fractions of asphaltenes (C=4.8, 8.2 and 12.0 %) and 5 base solutions using protonated toluene with different mass fractions of asphaltenes (C=2.5, 4.7, 7.6, 8.8 and 10.7 %). To vary the solvent quality, about 8 different mass fraction of n-heptane were introduced into the above 5 base solutions. The n-heptane fraction Φ relative to the total amount of solvent (toluene + n-heptane) varied from 0 up to 50%.

NMR RELAXATION MEASUREMENTS

All the NMR experiments were performed at a proton Larmor frequency of 60 MHz (Laboratoire de Spectrométrie Physique, Grenoble). Several CPMG sequences were used to measure the transverse magnetization decays; for the short decays, echoes were recorded up to 8 ms using inter-echo times of 20, 50 100 μs, and for the long decays (up to 330 ms) characterizing the solvents, the inter-echo times were 500, 1000 and 1700 μs. For each decay, 64 scans are accumulated with a repeat delay of 10 s. Then, the data are combined to give a single curve. During this processing, we did not observe any mismatch, clearly indicating that there is no inter-echo time dependence and that the physico-chemical processes studied were stable within the measurement time (of the order of one/two hours). We subtracted also to the measured curves a parasitic constant probe signal extending up to 5 ms.

Relaxation data analysis

All the measured decay curves showed a sharp decrease at short time (10 μs – 5 ms) followed by a much slower decay representing the solvents. We show in Figure 1 the data using the deuterated toluene (without n-heptane), for which most of the signal is due to the asphaltenes (a small signal still persists from the solvent due to residual protons). The magnetization decay is clearly multi-exponential, as shown by the curvature of the signal in semi-log scale (Figure 1). For non deuterated solvents for which most of the signal is due to the solvents, the procedure is the following: first, the signal is scaled by the magnetization amplitude Mo obtained at 10 μs (first data point of the free induction decay); second, the signal is split according to:

$$\frac{Mx(t)}{Mo} = Ma + Ms = Ma + As\exp(-\frac{t}{T_{2S}}) \tag{2}$$

where Ma is the magnetization associated with asphaltenes aggregates, As and T_{2s} are respectively the amount and characteristic relaxation time of the solvent (either the toluene alone or the mixture toluene/n-heptane). It was found that a single relaxation time could well reproduce the decay curve of the solvent, as seen for example by the straight line observed in Figure 1 for times larger than 100 ms. For non deuterated solvents where more than 90% of the signal is due to the solvent, this has the advantage of better revealing the relaxation times of asphaltenes. In addition, we verified that the amplitude As was proportional to the asphaltene concentration (the curve at 4.8% is above the one at 12.0%, Figure 1), indicating that Mo represents the true total magnetization of the system. Subtracting Ms to the total signal, the magnetization Ma was then analyzed separately using two methods, a multi-exponential and a stretch exponential model:

$$Ma = \sum_i Ai\exp(-t/T2i) \qquad \text{or} \qquad Ma = \exp(-(t/\tau)^{\alpha}) \tag{3}$$

The multi-exponential model could be applied directly to the entire signal but our experience indicate that the subtraction procedure yield a numerically stable and unique distribution of the short component. The coefficients Ai were calculated using standard inverse Laplace techniques where an optimum regularization term is calculated to avoid unstable solutions. For the stretch exponential model, the coefficients α and τ were calculated by linear regression in log(log(M))-t scale. The multi-exponential model has the advantage of visualizing rapidly the relaxation time distribution present in the system.

Aggregates characterization

The comparison between deuterated and protonated toluene experiments provided convincing evidence that the asphaltene aggregates could be characterized, even though the signal contribution in the later is relatively small. This is due to the relaxation time contrast (several orders of magnitude) between the two terms Ma and Ms. It has important practical consequences. Hence, we focus here on the experiments performed in protonated toluene.

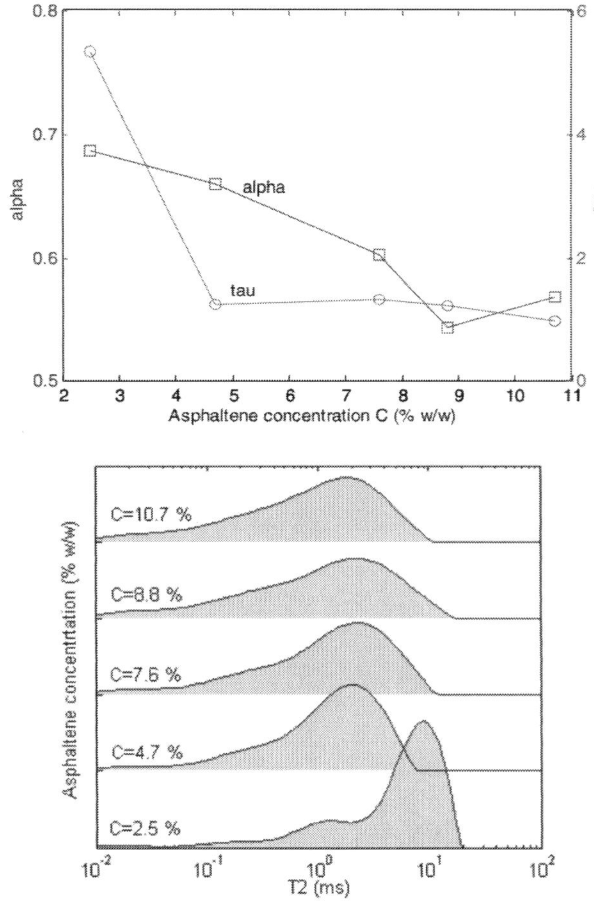

Figure 2: Characterization of asphaltene aggregates in toluene. Left: parameters α and τ of the stretch exponential model. Right: amplitude Ai of the multi-exponential model.

The measured decay curves Ma could be fitted precisely with both the multi-exponential and the stretch exponential models. In general, the aggregates are characterized by a large distribution of relaxation times (Figure 2). Using the stretch exponential model, we found α values much smaller than 1. In the multi-exponential model, the distribution spans over more than one decade. In both models, the dominant relaxation time, presumably related to the size of the aggregates, lies at about 2 ms, except for the low concentration solution in which the dominant relaxation time is about 2 times larger. It is suspected that higher dilution will produce smaller aggregates and hence larger relaxation times. Furthermore, the relaxation rate $1/T_{2S}$ of the solvent varies linearly with the asphaltene concentration (Figure 3) according to:

$$\frac{1}{T_{2S}} = 0.264\ C + 0.511 \tag{4}$$

where T_{2S} is expressed in second and C in % w/w. The interpretation of this observation is discussed later.

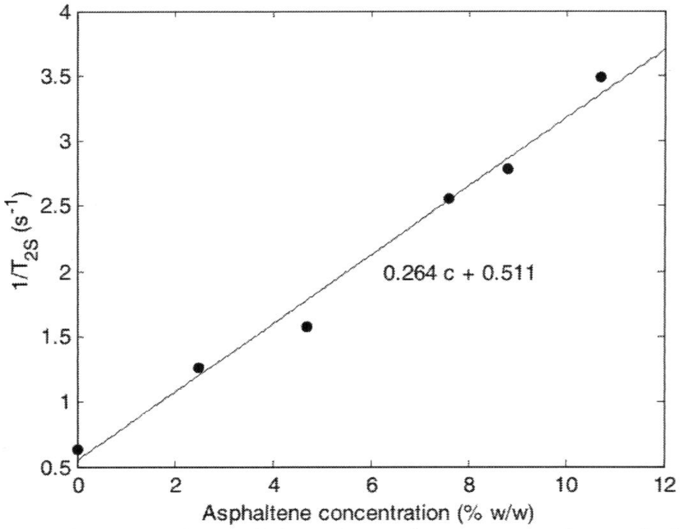

Figure 3: The relaxation rate $1/T_{2S}$ of the toluene increases as the asphaltene concentration increases.

Flocculation

Since NMR relaxation times are related to the size of the asphaltene structures, the NMR signal is expected to vary during the flocculation process. From the base solution with different initial asphaltene concentrations Ci, various quantities of n-heptane were introduced in order to induce flocculation. The data were analyzed as explained above. The relaxation time of toluene and n-heptane are respectively 1590 ms and 1670 ms. Therefore, the relaxation time of the binary mixture does not vary very much in all the experiments. It is however influenced by the asphaltenes, as will be seen later.

The first approach is to detect the changes in relaxation time when the quantity of n-heptane is increased gradually. We present the results obtained on two solutions with initial asphaltene concentration of 2.5 and 8.8 % w/w (Figure 4). For the two solutions, we see that the distribution is splitted into two distinct relaxation modes when the n-heptane concentration is close to the flocculation threshold. Far above this threshold, there is no signification change although the solution is considered to be flocculated by visual examination of the tube. However, for one solution (7.6 %) the splitting into two relaxation modes was not observed, and for the other solutions it was less pronounced.

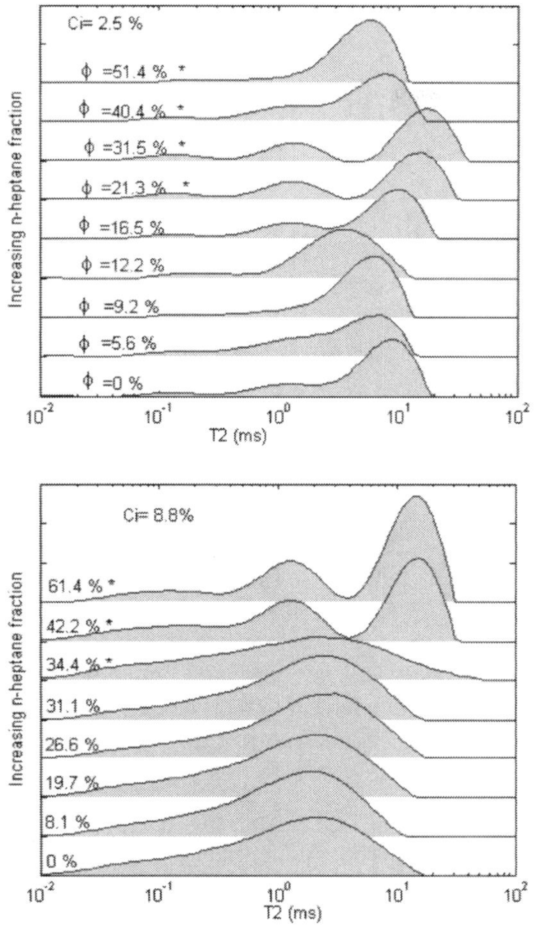

Figure 4: Relaxation time distributions of the asphaltene aggregates for two initial asphaltene concentrations and various fraction Φ of n-heptane (w/w) in the toluene/n-heptane mixture. The star indicates when the solution is considered as flocculated, as observed visually.

The second approach is to follow the quantity As (eq. 1) corresponding to the number of protons with a long relaxation time (the solvent). For the initial base solution, we verified that As decreases linearly with the asphaltene concentration. When adding n-heptane to the base solution, the total proportion of solvent increases in the mixtures and therefore, As increases (Figure 5). However, for n-heptane fraction above flocculation threshold (about 35%), As decreases although the total amount of solvent still increases. This means that protons from the solvent are relaxing faster and have shifted at smaller relaxation time. This strongly suggests the existence of trapped solvent molecules inside the aggregates as the flocculation proceeds. This phenomenon was not observed on the other solutions presumably due to a lack of measurement sensitivity of As at low concentration C.

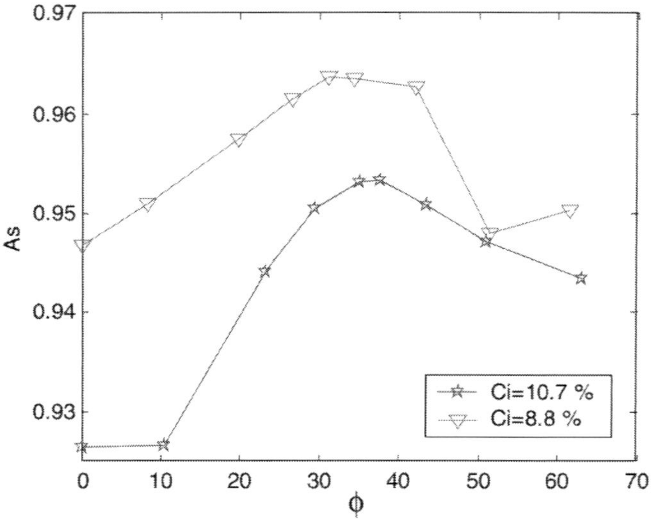

Figure 5: Magnetization amplitude As as a function of the n-heptane fraction Φ in the toluene/n-heptane mixture.

Finally, it is also of interest to analyze the influence of the asphaltenes on the solvent properties (term Ms in eq. 1). The relaxation time of a mixture of toluene and n-heptane is nearly constant and we have seen above (Figure 3) that the toluene relaxation rate depends on asphaltene concentration (eq. 4). In a similar way, the relaxation rate of the solvent $1/T_{2S}$ (eq. 2) for all the asphaltene-toluene-heptane solutions varies linearly with the asphaltene concentration (Figure 6) according to:

$$\frac{1}{T_{2S}} = 0.273\ C + 0.451 \tag{5}$$

The slope is close to the one found for asphaltene-toluene solutions. Note that the graph in Figure 6 gathers all the data points before and after flocculation.

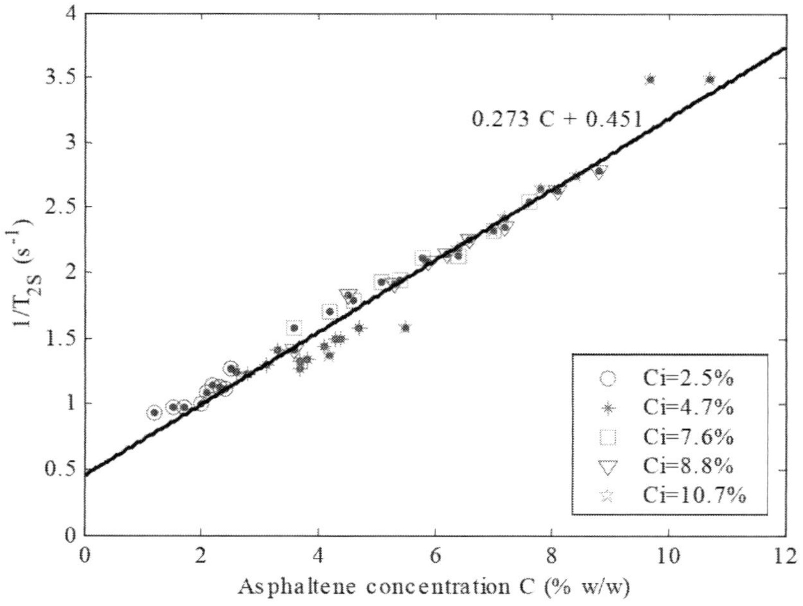

Figure 6: Relaxation rate of the solvent binary mixture as a function of the asphaltene concentration C. The different symbols indicate the inital asphaltene concentration Ci of the solutions. The unique slope indicate that there is no sharp viscosity transition during flocculation.

DISCUSSION

The NMR relaxation data collected are consistent with the flocculation process: as the quality of the solvent decreases (i.e. the n-heptane fraction increases), the aggregate size increases (the relaxation time decreases) up to the point where a solid like structure is formed (the splitting of the relaxation time distribution of the asphaltenes). The bimodal signature is observed near the flocculation threshold and has been observed for all solutions except one. Therefore, we think that it might be a good indicator of the onset of flocculation but other independent measurements should be performed. Note that the two modes are separated by a factor of ten yielding a very robust signature in terms of numerical analysis. The absence of a bimodal signature at higher n-heptane concentration above flocculation threshold is in apparent contradiction with the proposed signature. We suspect that the lower relaxation mode becomes gradually undetectable using standard CPMG sequences.

The flocculation threshold is also indicated by the decrease of the amplitude of the solvent (As) , interpreted with a trapping mechanism of the solvent molecules inside the aggregates. However, analyzing the solvent relative amplitude is a less sensitive method.

The linear relationship of the solvent relaxation rate with the asphaltene concentration can be interpreted as a viscosity effect. Unfortunately, viscosity measurements are not available and we rely on published data on similar systems. In the range of concentration considered, the solution studied can be considered in the dilute domain[6,7] for which the viscosity increases linearly with concentration. For similar solution, Fenistein et al.[6] and Henault et al.[7] found a viscosity increase of a factor of about 2 to 3 when C increases from 0 up to 10% (w/w). If we interpret the variation of the solvent relaxation rate as linked to viscosity, the measured slope (eq. 4) indicate an increase of a factor of 2.6 in agreement with these authors. Furthermore, since the dominant relaxation time of the asphaltene aggregates is not linked linearly with concentration (Figure 2) and is nearly constant above C=2.5%, the viscosity changes can therefore only be reproduced by the changes of the relation time of the solvent. However, we expect to observe a viscosity transition near the flocculation threshold [6] and this is not shown by the relaxation rate of the solvent for the flocculated solutions (Figure 6, eq. 5). Therefore, the solvent relaxation time may not indicate all viscosity fluctuations. NMR dispersion and diffusion experiments could be useful to provide a better understanding of these phenomena.

CONCLUSION

Using asphaltenes extracted from the Prudoe-Bay crude oil, we measured and analyzed the relaxation time distribution of asphaltene-toluene solutions and asphatene-toluene-heptane solutions. The asphaltene concentration varied from 2.5 up to 10.6% and are considered in the dilute regime. The n-heptane/toluene fraction varied from 0 up to about 50%, covering the flocculation threshold. The asphaltene aggregates have a signature at low relaxation time (1 ms) while the solvent is seen at about 1.5 s. During flocculation, the asphaltene signature changes from a wide unimodal to a bimodal relaxation time distribution. The relaxation rate of the solvent varies linearly with the asphaltene concentration. These variations can be interpreted as a variation of viscosity, in agreement with measurements on similar solutions published in the literature.

ACKNOWLEDGEMENTS

We thank L. Nabzar (IFP) for providing some asphaltene samples and D. Broseta and J.F. Argillier for helpful discussion.

REFERENCES

1. H Groenzin, O.C. Mullins, *J. Phys. Chem. A* 103 (1999) 11237
2. J.N. Roux, D. Broseta, B. Demé, *Langmuir* 17 (2001) 5085
3. M. Agrawala, H.W. Yarranton, *Ind. Eng. Chem. Res.* 40 (2001) 4664
4. A. Abragam A., Principles of Nuclear Magnetism, Clarendon Press, Oxford, (1961).
5. K.J. Dunn., D.J. Bergman, G.A. La Torraca, Nuclear Magnetic Resonance petrophysical and Logging Applications, Seismic Exploration Vol. 32, Pergamon, (2002)
6. D. Fenistein, L. Barré, D. Broseta, D. Espinat, J.-N. Roux, M. Scarsella., Langmuir 14 (1998) 1013
7. I. Henault, J.F. Argillier, C. Pierre, M. Moan., Offshore Technology Conference, Houston 5-8 May, Society of Petroleum Engineers (2003)

INVESTIGATION OF THE BEHAVIOR OF ASPHALTENES USING RHEOLOGICAL MEASUREMENTS - APPLICATION TO HEAVY OILS

I. Hénaut, J-F. Argillier, C. Coustet
Institut Français du Pétrole
1 et 4 avenue de Bois Préau 92852 Rueil-Malmaison, France

Key words: asphaltenes, resins, structure, viscosity.

Prepared for Presentation at the 2004 AIChE Spring National Meeting, New-Orleans, Louisiana,25-29 April

Unpublished

ABSTRACT
Heavy crude oils represent a strategic source of hydrocarbons and give rise to many experimental works. Their main chemical characteristic is their high asphaltene content. The study presented in this paper focuses on these particular molecules. We have shown that they mainly contribute to the high viscosity of heavy oils thanks to a structural organization. This was determined using SAXS and rheology, the rheological characterization enabled the determination of the intrinsic viscosity and the interaction parameter k' corresponding to the Huggins theory, which is usually applied to polymers. This last parameter allows the quantification of the interactions between asphaltenes under different conditions and in various organic solvents. Therefore, we introduced asphaltenes in increasingly more complicated solvents: aromatics, aromatics containing polar components, and deasphalted oil. It was shown that the strength of the interactions between asphaltenes is strongly dependent upon temperature and solvent quality. The experiments performed with maltenes, the natural and real environment of asphaltenes, showed that resins within heavy oils strongly diminish both the interactions between asphaltenes and their colloidal properties.

INTRODUCTION

Because of their physicochemical characteristics, asphaltenes are a challenge to almost all petroleum operations: they tend to deposit within reservoir rocks, they facilitate the formation of water-in-oil emulsions, they adsorb onto catalysts during processing etc… These issues have given rise to research programs to determine the molecular and colloidal properties of asphaltenes and to counteract their hindrance to production, transportation and refining. A recent paper reviews the fundamental and practical research performed in the 20th century [Sheu, 2002]. It shows that many methods have been employed to investigate asphaltenes: scattering techniques (SAXS [Dwiggins, 1965], SANS [Roux et al., 2001]), surface tension, viscosity, dielectric relaxation, conductivity [Sheu et al., 1991], calorimetry [Andersen and Birdi, 1991], and fluorescence depolarization [Groezin and Mullins, 2000]. Most of these experiments were performed in simple organic solvents: and therefore, there are still some questions to be resolved in the case of a real oil, which is a far more complicated environment for asphaltenes. Thus this study focuses mainly on the behavior of asphaltenes in heavy oils. It uses rheology, which has the advantages of being easily applied to both organic solvents and crude oils, and of being sensitive to structural changes. The ultimate goal of this study is to establish the link between the high viscosity of heavy oils and their structure, This is desirable to the oil industry in order to promote the production of heavy oils, the reserves of which are both strategic and huge. A particular difference between heavy oils and conventional is the high asphaltene content in the former with amounts reaching up to 20% in weight. The first part of the paper is dedicated to asphaltenes introduced into aromatics. To investigate their behavior in relation to the solvent quality of their environment, we then dissolved them in increasingly more complicated solvents: aromatics containing polar components, and deasphalted oil. Special attention was then paid to resins, which are abundant in heavy oils and appear to interfere with asphaltenes.

VISCOSITY DATA FOR ASPHALTENES IN AROMATICS

Asphaltenes are crude oil components defined according to their solvent solubility (n-alkane insoluble and toluene soluble). This classification implies that asphaltenes are not a well-defined molecule but a distribution of various molecular structures. Despite this diversity, some physicochemical properties are commonly accepted: the presence of polycondensed aromatic rings, paraffinic branches, and polar functions coexisting inside a loose structure of the archipelago type [Strausz et al., 1992] (fig.1).

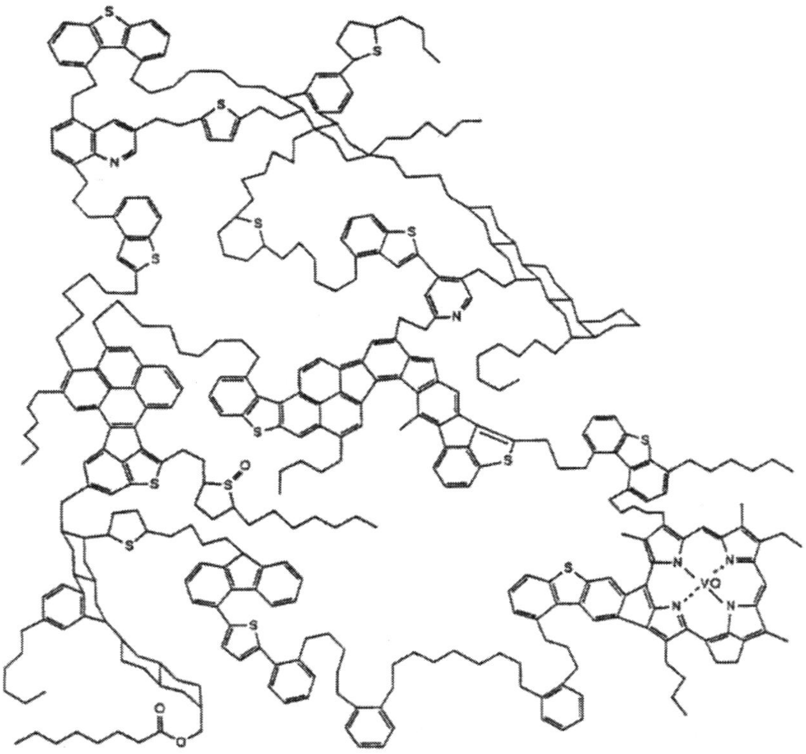

Fig.1 – Hypothetical asphaltene molecule [Strausz et al., 1992].

n-heptane insoluble asphaltenes were extracted from an Orinico Belt crude oil. After being properly milled, they were introduced into xylene under stirring within 2 hours. The concentration of asphaltenes was in the range of 2 - 16% of weight. The viscosity measurements of the samples were realized on the controlled-stress rheometer AR2000 with double concentric cylinders. Shear rate sweeps (10 to 1000s^{-1}) were performed at 20°C. Solution preparations and testing were carefully replicated for all samples. The flow curves obtained show that the asphaltenes/xylene systems are Newtonian. Using a colloidal approach, the relative viscosity of asphaltenes, η_r, can be defined as $\eta_r = \frac{\eta_{solution}}{\eta_{Solvent}}$. Figure 2 presents the plot of η_r as a function of asphaltenes concentration by weight.

Two domains can be distinguished:

- the dilute regime, where the plot is linear,
- the semi-dilute regime, where the plot deviates strongly from linearity.

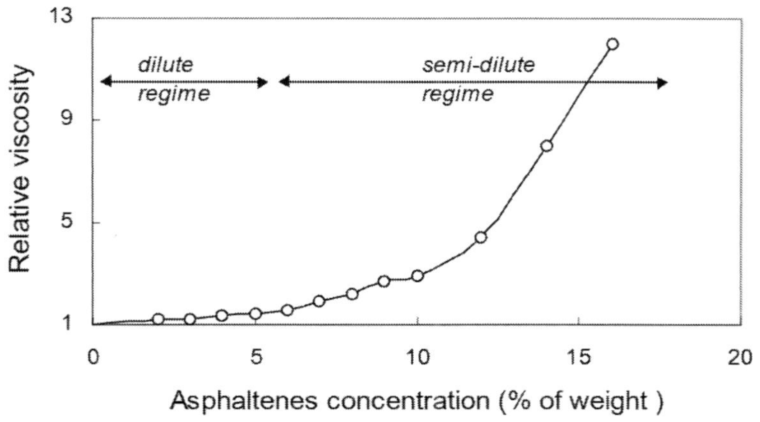

Fig.2 – Viscosity data for n-heptane insoluble asphaltenes dissolved in xylene, T=20°C.

To interpret these results, SAXS experiments were carried out on the samples [Coustet et al., 2004] (fig.3). In the dilute regime, asphaltenes form independent fractal particles with a molecular weight Mw of 145 000 g/mol and a radius of gyration equal to 85 Å. The fractal dimension was found to be close to 2. This value had already been determined by Fenistein, 2000 for diluted asphaltenes in toluene. In the semi-dilute system, interparticles interferences appear in the scattering curves. This indicates a large-scale aggregation with a fractal dimension of around 2 . This structural evolution leads to the overlapping of asphaltenes.

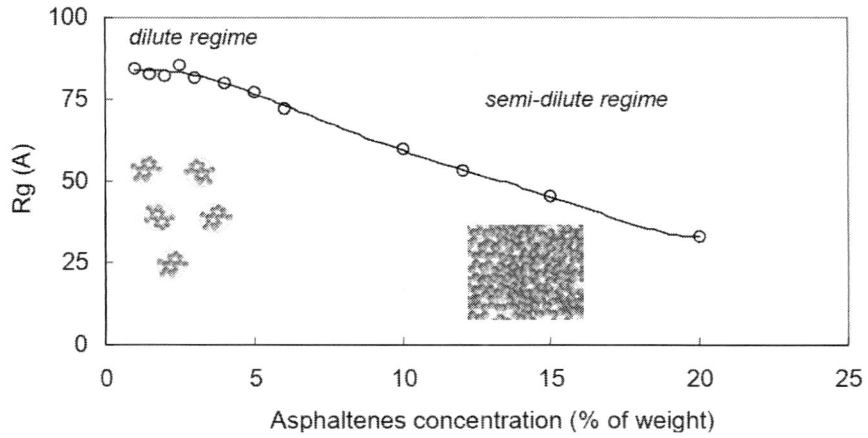

Fig.3 – Radius of gyration of n-heptane insoluble asphaltenes dissolved in xylene, T=20°C (SAXS, Coustet et al., 2004].

From these preliminary results, we deduced that the increase in viscosity in the semi-dilute regime is related to the state of aggregation of asphaltenes.
The association characteristics of asphaltenes have been extensively studied to determine the stability of the formed aggregates. Due to the chemical composition of asphaltenes, many forces play an important role, including H-bonds, □-□ interactions, ionic attraction, etc... These were recently listed by Murgish [2002]. All the interactions seem to be more or less influenced by temperature and solvent property. In any event, little converging information is available regarding the effects of these two factors on the size and shape of the aggregates of asphaltenes. Hydrodynamic properties are commonly employed to study macromolecules in solution. In the same way, we tried to explore the effects of temperature and solvent on the characteristics of the aggregates using a viscosity analysis based on Huggin's equation [1942]. This approach is explained below.

Influence of temperature and polar components on the behavior of asphaltenes.
One can write Huggins' s equation as :

$$\eta_r = 1 + [\eta]c + k'[\eta]^2 c^2 + \dots$$

where $\square_r$ is the relative viscosity, [□], the intrinsic viscosity, and k' a constant. For low concentrations, the series expansion in c is truncated and the equation is rearranged as:

$$\frac{\eta_r - 1}{c} = [\eta] + k'[\eta]^2 c$$

The intrinsic viscosity is obtained from the intercept of $\frac{\eta_r - 1}{c}$ vs. c graph. The constant k' is calculated using the slope of the plot. This method is used to provide the contribution of the shape effect through the linear term ([□]) and the contribution of the interactions with the quadratic term (k'). It had only seldom been employed for asphaltenes [Bouhadda et al., 2000] or similar system such as coal extracts [Wargadalam et al., 2001]. We propose applying the Huggins's equation to the n-heptane insoluble asphaltenes that we dissolved in three different solvents: toluene, toluene/hexanol = 90/10 (v/v) and toluene/hexanol = 70/30 (v/v). The viscosity measurements were carried out in the controlled-stress rheometer AR2000 under the same conditions described earlier. The experiments were repeated at temperatures ranging between 5 and 30°C.
As an example, the $\frac{\eta_r - 1}{c}$ vs. c plot is presented in figure 4 for pure toluene and the case of toluene containing some alcohol.

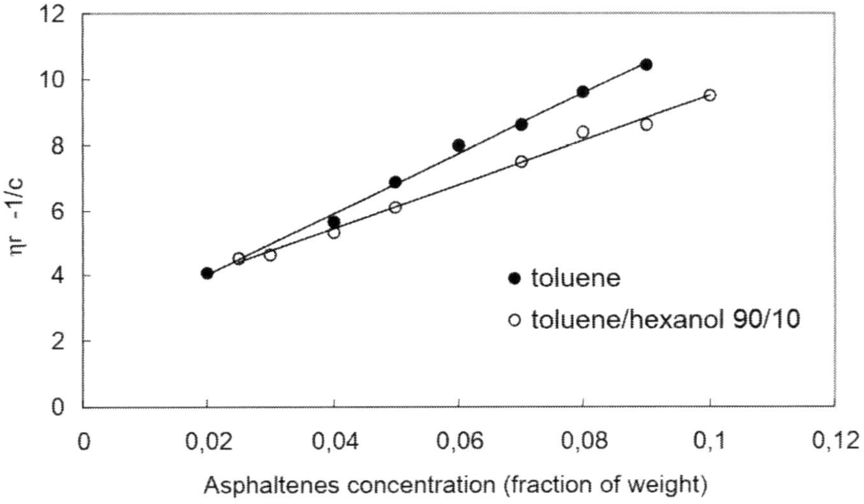

Fig4 – Application of Huggin's equation to n-heptane insoluble asphaltenes, T=20°C

The intrinsic viscosities and interaction terms derived from all the plots are shown in table 1. The most striking result is that the incorporation of the alcohol in toluene leads to a decrease in the interaction term between asphaltenes. It can also be noted that in the presence of hexanol, the interaction term does not vary with temperature. We ascribe this phenomenon to the specific interactions such as hydrogen bonds that can easily form between the asphaltenes and alcohol. Hexanol surrounds the diluted particles of asphaltenes and creates a kind of apolar layer at their periphery. The described mechanism is close to the one of peptization by resins in heavy oils [Pfeiffer and Saal, 1940]. Another way to describe the influence of polar solvents is that they decrease the attraction between polar moieties of asphaltenes. In pure toluene, the surfaces of asphaltenes remain polar and consequently the corresponding interaction term is high. We consider the increase in k' with temperature in toluene as a consequence of the dissociation of asphaltenes that liberates and increases the number of potential attractive sites. This dissociation also makes the particles more numerous and diminishes their distance [Tanaka et al., 2003]. As the asphaltenes get closer, the short-range interactions become stronger and k' increases. When alcohol is present, its screening effect makes the liberated polar surfaces resulting from the thermal dissociation less attractive and therefore k' is not affected.

	toluene		toluene/hexanol 90/10(v/v)		toluene/hexanol 70/30(v/v)	
T (°C)	[η] g/g	k'	[η] g/g	k'	[η] g/g	k'
5	3,9	5,4	3,7	5,4	5,6	2,3
10	3,3	9,3	3,9	4,5	6	1,8
20	2,2	21,7	3,2	6,2	5,9	1,4
30	2	21,5	3,5	4,5	5,6	1,3

Table.1 – [□] and k' of n-heptane insoluble asphaltenes.

As mentioned earlier, the Huggins's equation provides information about the interactions (k') and the shape of the aggregates ([□]). According to Scheraga and Mandelkern [1953], [□] can be written as

$$[\eta] = \frac{\nu N V_e}{M}$$

where □ is the shape factor, N Avogadro's number, M the molecular weight and V_e the effective volume of particles. The intrinsic viscosities of the n-heptane insoluble asphaltenes are shown in table.1. Compared to the asphaltenes/toluene solutions, the intrinsic viscosities of the asphaltenes toluene/hexanol systems seem to be higher. Presumably the addition of a polar component affects the aggregation of asphaltenes. In fact, SAXS revealed that the radius of gyration for asphaltenes varies with the introduction of hexanol in toluene [Coustet et al., 2004] (fig.5). Using SANS, Gawrys et al. , 2002 showed that asphaltenes form fractal aggregates in solution with dimensions between 1.7 and 2.1 and that when solvated by resins (polar components) this dimension increases to 3. According to the authors, the resins solvate much of the interstitial space and the size of the aggregates is diminished, which explains the resulting dimension. These various observations confirm the impact of the solvent quality on the aggregates of asphaltenes. In any case, additional structural analysis are still needed to interpret the evolution of the intrinsic viscosity.

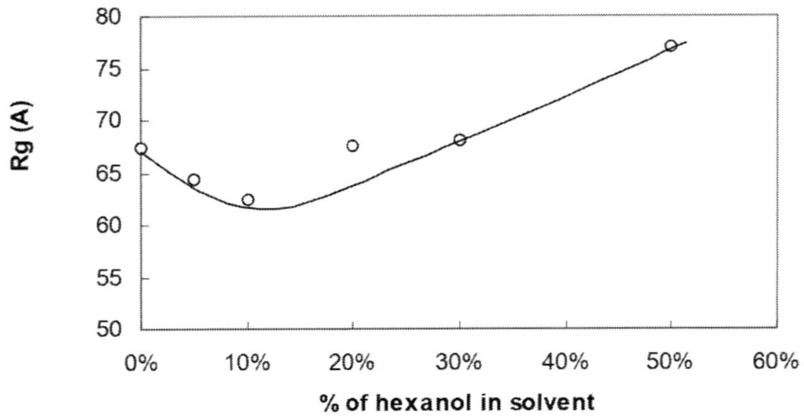

Fig5 – Influence of hexanol on the radius of gyration of n-pentane insoluble asphaltenes introduced into toluene/hexanol blends (3% in weight of asphaltenes).

In order to support the previous findings and to better mimic heavy oils, the experiments were repeated with nonylphenol instead of hexanol. In fact, according to Chang and Fogler [1994], nonylphenol offers the same characteristics as resins: an aromatic ring, an alkyl tail, and a polar head group. Both rheological and SAXS were performed on the samples. Again, the experimental results reveal that the addition of a polar component to aromatic solutions of asphaltenes increases their intrinsic viscosity, lowers their interaction, and diminishes their radius of gyration (figs 6 and 7).

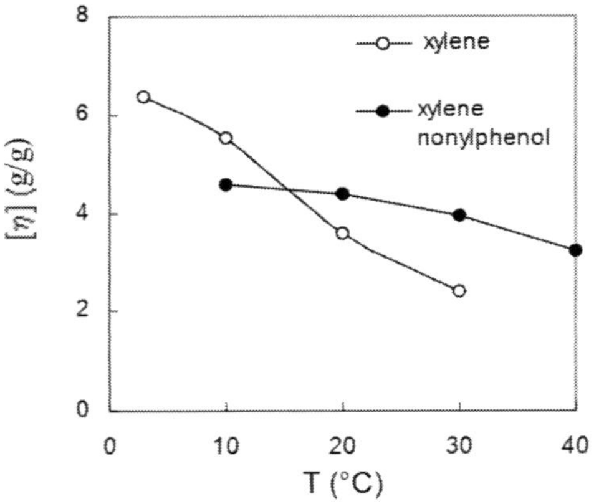

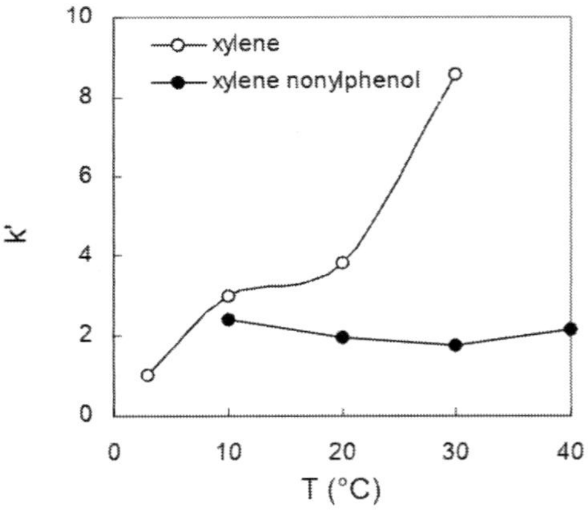

Fig.6 – Influence of nonylphenol on hydrodynamic properties of n-pentane insoluble asphaltenes introduced in xylene.

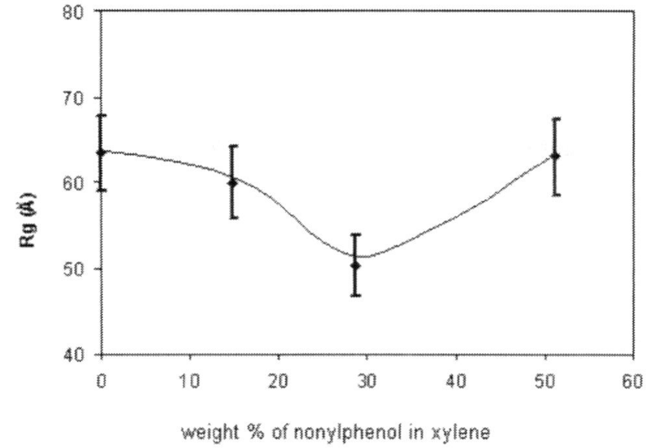

Fig.7 – Influence of nonylphenol on the radius of gyration of n-pentane insoluble asphaltenes introduced into xylene/nonylphenol blends (3% in weight of asphaltenes).

These results confirm the effect of solvent quality on the structure of asphaltenes.

Application to heavy oils.

As mentioned in introduction to this paper, rheological measurements can easily be performed in real and complex systems. Thus, a series of experiments were carried out on asphaltenes introduced into deasphalted oil (maltenes obtained after evaporation of pentane). We checked that the original crude oil and its corresponding prepared sample were equivalent [Argillier et al., 2001]. As with the organic solvents, two domains of concentrations were observed using rheological measurement and SAXS (fig. 8). Nevertheless, a difference between the experiments can be noted: the critical concentration C* separating these two domains is higher in maltenes than in xylene. We presume that this phenomenon is due to the presence of resins in maltenes which might prevent asphaltenes from overlapping. To shed some light on this, the Huggins's equation was applied in the dilute domain to the different systems. The values are shown in table.2.

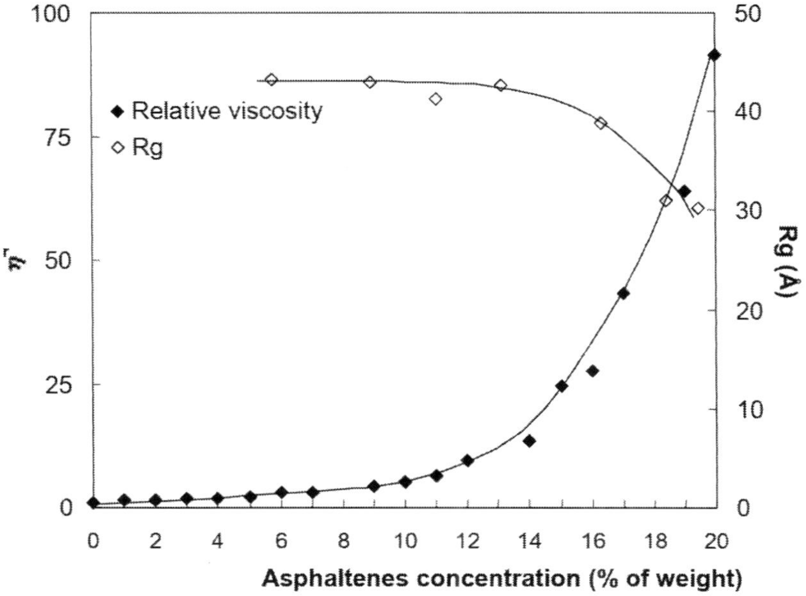

Fig.8 – Viscosity and SAXS data for n-pentane insoluble asphaltenes introduced into maltenes, T=20°C.

	xylene			
asphaltenes	[η] g/g	k'	Rg	c*
n-pentane insoluble	3,6	3,8	60	10%
n-heptane insoluble	3,7	7,2	85	6%
	maltenes			
asphaltenes	[η] g/g	k'	Rg	c*
n-pentane insoluble	12,8	1,7	42	12%
n-heptane insoluble	11,3	2,2	50	9%

Table.2 – Influence of solvent on experimental data for asphaltenes, T=20°C.

From these experimental data, it can easily be seen that the radius of gyration for asphaltenes in maltenes is lower than in xylene. This is consistent with the dissociation of the aggregates of asphaltenes resulting from the presence of resins. The Huggins's terms [□] and k' emphasize the different responses given by asphaltenes in xylene and maltenes. All these remarks clearly demonstrate that asphaltenes are very sensitive to their environment, particularly to a polar one. We therefore conclude that the colloidal state of asphaltenes in maltenes is quite different from that in simple aromatic solvents.

Conclusion

SAXS and rheological measurements have been successfully combined to study asphaltenes. They reveal how these molecules can give rise to a structural change when sufficiently concentrated. It was shown that this structural evolution leads to a pronounced increase in viscosity. This phenomenon takes place in any solvent, whether it is aromatic or polar. Despite this common tendency, we demonstrated that the properties of asphaltenes depend strongly on the solvent, particularly its polar nature. We deduced that the aggregates of asphaltenes in heavy oils are different from those in pure aromatics.

REFERENCES

Andersen S.I and K.S. Birdi. " Aggregation of asphaltenes as determined by calorimetry", Journal of colloid and Interface Science, vol 142, N°2, 1991.

Argillier J-F, L. Barré, F. Brucy, J-L Dournaux, I. Hénaut and R. Bouchard. " Influence of asphaltenes content and dilution on heavy oil rheology", SPE 69711, ITOHOS, 2001

Bouhadda Y., D. Bendedouch, E.Y. Sheu and A. Krallafa. " Some preliminary results on a physico-chemical characterization of a Hassi-Messaoud Petroleum Asphaltene", Energy & Fuels, 14, 845-853, 2000.

Chang C-L and H.S. Fogler. " Stabilization of Asphaltenes in aliphatic solvents using alkylbenzene-derived amphiphiles. 1. Effect of the chemical structure of amphiphiles on asphaltene stabilization" Langmuir, 10, 1749-1757, 1994.

Coustet. C, L. Barré, A. Pina and M. Moan. " Composition and heavy oil rheology", Journal of Oil&Gaz technology to be published in 2004.

Dwiggins C.W. " A SAXS study of the colloidal nature of petroleum", J. Phys. Chem, 69, 3500, 1965.

Fenistein, D. PhD "Structure colloïdale des produits pétroliers lourds", University Paris VI, 2000.

Gawrys .K, M. Spiecker and P. Kilpatrick. "The role of asphaltene solubility and chemistry on asphaltene aggregation" Petroleum Chemistry Division Preprints 2002, 47 (4), 332.

Groezin H. and O. Mullins. " Molecular size and structure of asphaltenes from variuos sources", Energy & Fuels, 14, 677-684, 2000.

Huggins M.L. " The viscosity of dilute solutions of long chain molecules. IV dependence on concentration", J. Am. Chem. Soc, 64, 2716-2718, 1942.

Murgish J. "Intermolecular forces in aggregates of asphaltenes and resins", Petroleum Science and Technology, vol 20, N° 9&10, 983-997, 2002.

Pfeiffer, J.P. and Saal, R.N.J.:. “Asphaltic bitumen as colloid system”, J. Phys. Chem., 44, 139-149, 1940.

Roux J.N., D. Broseta and B. Demé. " SANS study of asphaltenes aggregation : concentration and solvent quality effects", Langmuir, vol 17, 5085-5092, 2001

Scheraga H.A and L.Mandelkern. J. Am. Chem. Soc., 75, 179, 1953.

Sheu E.Y. " Petroleum Asphaltenes properties, characterization and issues", Energy & Fuels, 16, 74-82, 2002.

Shey E.Y., D.A. Storm and M. De Tar. " Asphaltenes in polar solvents", Journal of non crystalline solids, 131-133, 341-347, 1991.

Strausz .O.P, T.W. Mojelsky and E.M. Lown. " The molecular structure of asphaltenes : an unfolding story", Fuel, vol 71, December, 1355-1363, 1992.

Tanaka R, Hunt J.E., Winans R.E., Thiyagarajan P., Sato S., Takanohashi T. "Aggregates structure analysis of petroleum asphaltenes with small angle neutron scattering", Energy and Fuels, 17, 127-134, 2003.

Wargadalam V.J., K. Norinaga and M. Iino. Hydrodynamics properties of coal extracts in pyridine", Energy & Fuels, 15, 1123-1128, 2001.

ACKNOWLEDGEMENTS

The authors are grateful to L. Barré, who conducted and supervised the SAXS experiments, and to D. Espinat, M. Moan and T. Palermo for interesting discussions and useful advice.

SECONDARY EFFECTS IN SEC ANALYSIS OF OIL ASPHALTENES EVIDENCED BY CROSS SEC SEPARATIONS AND SEC-MS COUPLING

Isabelle Merdrignac[1], Bernard Desmazières[2], Olivier Laprévote[3] and Peran Terrier[2]

1. Institut Français du Pétrole, Division Physique et Analyse, BP3, F69310 Solaize
2. Université d'Évry-Val d'Essonne, Laboratoire Analyse et Environnement, CNRS UMR 8587, boulevard François Mitterrand, F91025 Évry Cedex
3. CNRS, ICSN, F91198 Gif sur Yvette

Keywords: Asphaltene, separation, analytical methods

Prepared for presentation at the 2004 Spring National Meeting, New Orleans, LA, April 26-30

Unpublished

ABSTRACT

Due to secondary effects (likely aggregation and π-interactions) the estimation of asphaltenes average molecular weights by Steric Exclusion Chromatography (SEC) is difficult.

In order to evidence these secondary effects, fractions of various asphaltenes (geochemical types I, II and III) were collected in SEC using different mobile phases (THF, dichloromethane and toluene). These fractions were reinjected using the same mobile phase in the same conditions and no modifications of elution volumes were observed. Then reinjections were performed using a different mobile phase. In most cases, an important broadening of the fraction peak was observed, as well as a change of the average elution volume.

Off-line SEC-MS coupling was then performed using combinations of different ionization sources (laser desorption, electrospray and photospray) and of different analyzers (time of flight and quadripole-time of flight).

The results of these analyses, performed in positive and negative ion modes were coherent. The main conclusions are that:

1. Fractions collected in SEC using toluene as mobile phase are the less polydispersed and their average molecular weight decreases as elution volume increases. The results were more complex with other mobile phases. Molecule elutions might not be only driven by hydrodynamic volumes but should also be affected by other factors. In these cases, the SEC mass calibration (using polystyrene) is not fairly representative of eluted molecules.

2. The average apparent molecular weights of the different fractions vary from 300 to 1000, whatever the mobile phase.
3. Fractions (F_i) collected in SEC using toluene as mobile phase and then reinjected in SEC using THF or dichloromethane were separated in several subfractions (F_{ij}). The average molecular weights of these subfractions F_{ij} measured by MS were found to be the same than the average molecular weight of the initial fraction F_i.

These results confirm clearly the existence of secondary effects in SEC. The SEC-MS measurements are consistent with other works with an average molecular weight of asphaltenes of about 700-800 Dalton (Buch et al, 2003; Hugley et al, 2002; Cunico et al, 2002).

INTRODUCTION

Conversion of heavy crude oils into more valuable products becomes an important challenge for industry. These types of feeds are rich in heteroatoms such as S, N and in metals, particularly concentrated in the asphaltene fraction.

High contents of such heteroatoms and metals cause serious problems due to the poor fuel stability during storage and to the environmental impact. Furthermore, they may cause inhibition and poisoning of catalysts used in refining processes.

Consequently it is necessary to qualify and to quantify these heteroatom-containing species in asphaltenes, with a view to understand their chemical role and their behavior in hydrotreatment processes.

The main difficulty of asphaltenes characterization resides in their high polydispersity. Asphaltenes are known to be constituted of various chemical species composed of polyaromatic and polyheteroaromatic units bearing alicyclic sites substituted and connected with aliphatic chains. In solution, asphaltenes exhibit self-assembly, micellar and colloidal behavior. Asphaltenes units may associate through π and hydrogen bonds to form micellar aggregates. Due to this property, interpretations of analytical results are often difficult and critical since they are strongly dependent on experimental conditions.

Several asphaltenes characterization approaches may be considered. Local structural analyses allow the determination of the environment (aromatic, aliphatic, S, N, metals) while colloidal structural analyses give information on the size and the molecular weight distribution of asphaltenes.

Among these methods, size exclusion chromatography (SEC) seems to provide more applicable relative information than the others. We report a study of secondary effects in SEC analysis of asphaltenes. The influences of the nature of the mobile phase and of the sample concentration were investigated. Then, the coupling of SEC and of mass spectrometry was used to estimate average molecular weights of asphaltenes fractions and to evidence the importance of secondary effects in SEC.

EXPERIMENTAL

Samples

The samples used for this study were extracted from vacuum residues using the standard heptane extraction method. We selected asphaltenes of different geochemical types:

Type I: Duri asphaltenes

Type II: Safanyia, Aramco and Oural asphaltenes

Type II hydrotreated (removal of sulfur and heavy metals): Safanyia asphaltenes

Type III: Arjuna asphaltenes

Steric Exclusion Chromatography (SEC)

Columns PL gel (250 x 8 mm, dp = 5 µm, pore diameters: 100, 1000 and 10000 Å)

Mobile phases: THF, dichloromethane and toluene

Detection: UV and refractometry

Preparative separations: injection of 20 µL

Concentration: 5 mg/mL

Dilution effects and bidimensional separations: injection of 20 µL, initial concentration from 0.01 mg/mL to 1 mg/mL

Mass spectrometry

LDI and MALDI/TOF MS analyses were performed using a Voyager DE-STR mass spectrometer (Applied Biosystems).

Reflectron and linear modes were used, in positive and negative ions mode.

Similar results were obtained with or without matrix (DHB, HCCA) and most of analyses were performed in LDI mode (no matrix).

Ionization was obtained by means of a UV laser (335 nm)
Ionspray (ISI), Atmospheric Pressure Chemical Ionization (APCI) and Atmospheric Pressure Photo Ionization (APPI) analyses were performed using a QqTOF Pulsar mass spectrometer (Applied Biosystems).

RESULTS AND DISCUSSION

SEC analyses

SEC analyses with various mobile phases

As an example, we describe the analysis of Safanyia asphaltenes. Results obtained with THF and dichloromethane were roughly similar (figure 1a). Apparent molecular weights were lower when toluene is used.

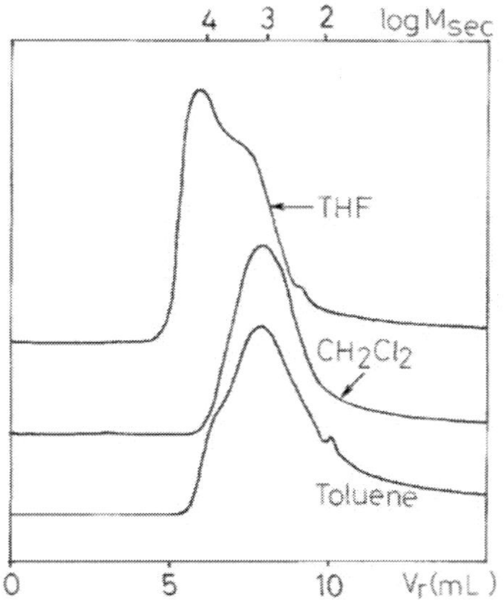

Figure 1a. SEC analysis of Safanyia asphaltenes using THF, dichloromethane and toluene (1 mL/min., 1 PL gel column, 10^3 Å, UV detection 300 nm).

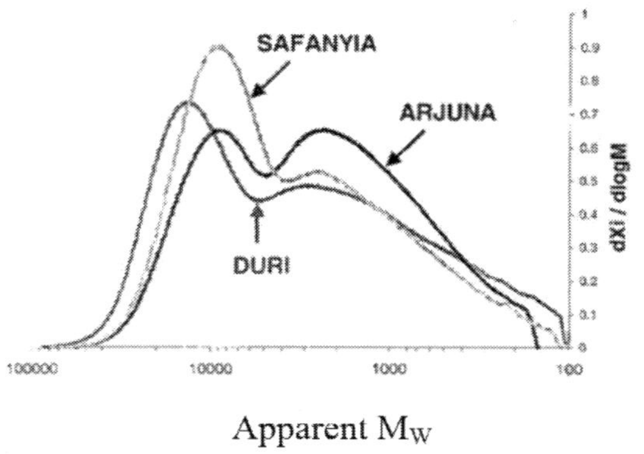

Figure 1b. SEC analysis of Duri, Safanyia and Arjuna asphaltenes in THF (1 mL/min., 3 PL gel columns, 10^2, 10^3, 10^4 Å, refractometric detection).

In similar conditions (concentration and mobile phase identical), each sample was characterized by a reproducible profile, different for each sample (figure 1b). For instance, the values of apparent molecular weights in number (M_n), in weight (M_w) and of the dispersity (I_p) are summarized below (Table 1). These values were obtained from analyses in the standard conditions (PL gel columns, mobile phase: THF, sample: 20 µL at 5 mg/mL, refractometric detection). We assumed that the refractometer response was constant for each component.

Asphaltenes	M_n	M_w	I_p
Duri	1380	7430	5.38
Safanyia	1450	6330	4.37
Aramco	1230	5180	4.21
Oural	1380	5170	3.75
Arjuna	1330	5200	3.91

Table 1. Apparent molecular weights and dispersity of different types of asphaltenes.

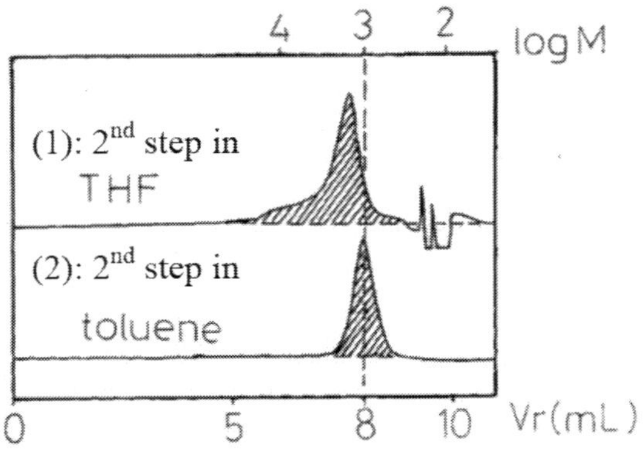

It can be concluded that, in spite of secondary effects, SEC is a good method of characterization, giving reproducible relative results, provided reproducible analysis conditions are used.

Effect of dilution in SEC

Analyses performed at different concentrations of sample resulted in a significant decrease of apparent molecular weights at low concentration (under 0.1 mg/mL). These results were obtained whatever the mobile phase and the sample. As an example, we give the evolution of the UV response at 254 nm of Duri asphaltenes (analyses at 0.01, 0.1 and 1 mg/mL) with THF as mobile phase (figure 2).

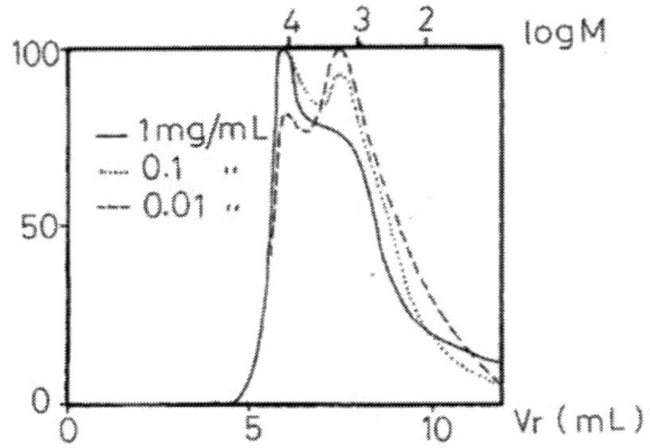

Figure 2. Example of dilution effect - Analysis of Duri asphaltenes (THF 1mL/min., 1 PL gel column 10^3 Å, UV detection 254 nm).

The decrease of apparent molecular weights is likely due to the decrease of association at low concentration.

Bidimensional SEC analyses

Bidimensional analyses of the different asphaltenes were performed using two different mobile phases in SEC. Samples were sufficiently concentrated (initially 1 to 5 mg/mL) to avoid dilution effects during the second step. In all cases, the fractions (F_i) collected at a given elution time using the first mobile phase eluted at a different elution time during the SEC analysis performed using the second mobile phase. Due to different secondary effects, the peaks were generally broad. On the opposite, elution times were not modified if the second step was performed with the first mobile phase. As secondary effects remained the same, the peaks were sharp. An example is given in figure 3 (1st step in toluene, second step in THF, Safanyia asphaltenes).

Figure 3. Example of bidimensional SEC analysis of Safanyia asphaltenes: 1st step-fraction collected at Vr = 8 mL, Msec = 1000 Dalton, mobile phase: toluene, 1 mL/min., 1 PL gel

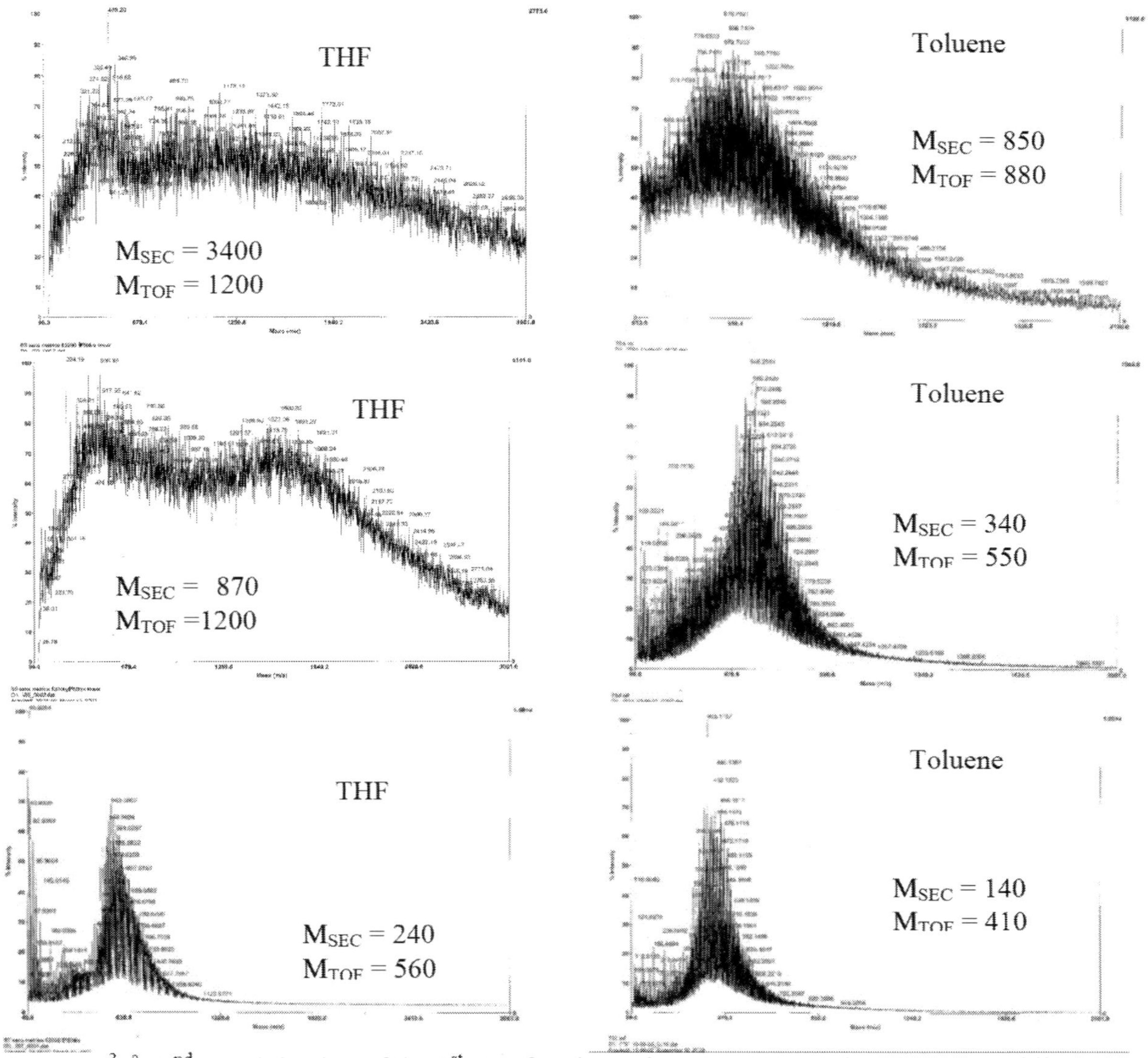

column, 10^3 Å. 2nd step-injection of the 1st step fraction using THF (1) or toluene (2) as mobile phase - UV detection 300 nm.

SEC/MS coupling

SEC/MS analyses were performed off-line. Different combinations of ions sources and of analyzers were used (see experimental):

LDI/TOF

ISI, APCI, and APPI were coupled with a Q-TOF analyzer

LDI/TOF-MS

The main results are the following:

- SEC fractions (F_i)are more polydispersed when THF and dichloromethane are used as mobile phase. On the opposite, fractions collected using toluene are the less polydispersed and elution is observed as a regular function of decreasing molecular weights (figure 4).

Figure 4. LDI/TOF analyses of SEC fractions of Safanyia asphaltenes collected using THF and toluene as mobiles phases.

- In all cases, LDI/TOF measurements are not in agreement with estimations resulting from the usual method (polystyrene calibration in SEC). This can be illustrated with Safanyia asphaltenes in table 2.

fraction	$M_{SEC\ (toluene)}$ apparent	$M_{LDI/TOF}$ positive ions
1	2110	1300
2	850	880
3	340	550
4	140	410
5	55	370

Table 2. Comparaison between apparent masses measured in SEC and LDI/TOF.

- According to LDI/TOF, the average molecular weights of SEC fractions vary between 350 and 1300. The average molecular weights of the various asphaltenes analyzed seem to be comprised between 600 and 800 Dalton.
- In the last series of experiments, the fractions (F_i) collected in SEC using toluene as mobile phase were reinjected in SEC using THF or dichloromethane as mobile phase. They were separated in subfractions (F_{ij}) and these subfractions were analyzed in LDI/TOF-MS. The average molecular weights of the (F_{ij}) fractions are the same (within ± 100 Dalton) than the average molecular weight of the initial (F_i) fraction.

It is well known that MALDI and LDI/TOF measurements can be biased due to:

- modifications of the components during ionization
- underestimation of average molecular weights if the sample is highly dispersed.

Complementary analyses were performed using different ionization sources and a different analyzer.

ISI, APCI or APPI/Q-TOF-MS

Therefore we used also ISI, APCI and APPI ions sources coupled with a Q-TOF analyzer with a view to confirm LDI/TOF results.

Due to the poor solubility of asphaltenes in media solubilizing silver salts, we could not obtain reproducible results using an ISI source (lack of sensitivity and frequent plugging were the main problems).

Very coherent results were obtained using an APCI source in positive ions mode. The average molecular weights of SEC fractions were the same than those determined in LDI/TOF (within ± 100 Dalton). For instance, a comparison of LDI and APCI analyses of a Safanyia fraction collected in SEC (mobile phase: toluene, apparent M_w in SEC: 140 from polystyrene calibration) is reported figure 5. The average molecular weight is about 410 (LDI/MS) to 450 (APCI/Q-TOF). Recently, we began to investigate the potential of APPI, using toluene as dopant. The first results confirm those obtained in LDI and APCI.

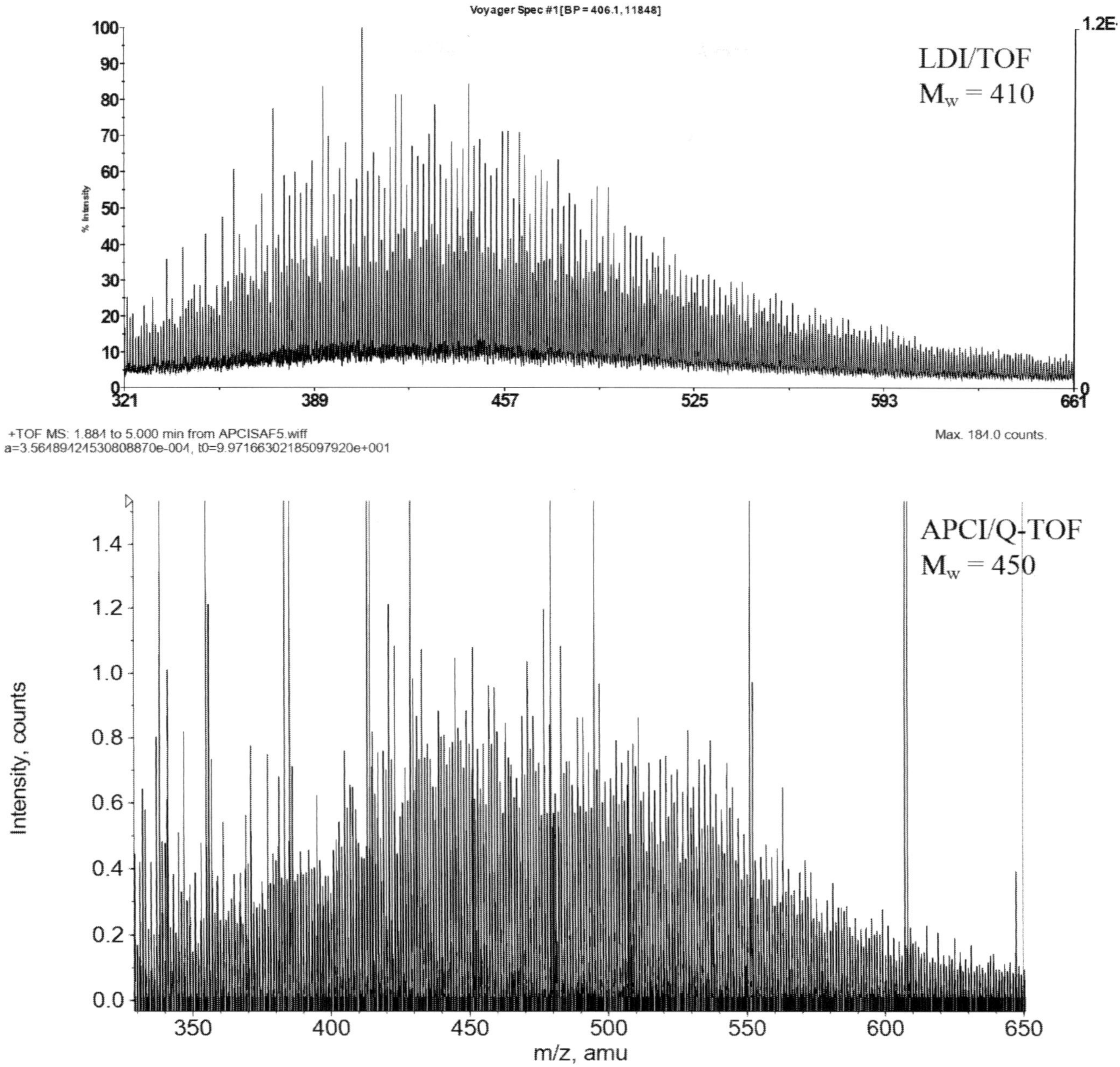

Figure 5. Comparative LDI/TOF and APCI/Q-TOF MS analyses of a SEC fraction (mobile phase toluene, apparent Mw = 140) of Safanyia asphaltenes.

CONCLUSION

The importance of secondary effects in the SEC analysis of asphaltenes was clearly evidenced by SEC and SEC/MS experiments.

The repartition of apparent molecular weights is strongly dependent of the mobile phase and of the amount of sample injected. However, reproducible characteristic profiles are obtained for each asphaltene in identical experimental conditions. Due to the secondary effects (aggregation, reversible interactions with the SEC gel), the estimations of molecular weights using polystyrene calibration is certainly false.

SEC/MS experiments performed in LDI/TOF, APCI/Q-TOF and APPI/Q-TOF MS confirmed the importance of secondary effects and gave coherent results. Whatever the sample, the average apparent molecular weights of SEC fractions were comprised between 300 and 1500 Dalton. The average molecular weights of raw asphaltenes seem to be about 600-800 Dalton, in agreement with other works (2, 3, 4).

This study will be completed by a thorough investigation of asphaltenes behavior in APPI/MS. Photo ionization appears as a very promising technique to analyze apolar polyaromatics (5).

At last, it should be particularly interesting to perform SEC-MS analyses in FT-MS at a high resolution with a view to attempt identification of the main compounds of asphaltenes (3).

REFERENCES

1. Roussis, G.S. and Proulx, R., Molecular Weight Distribution of Heavy Aromatic Petroleum Fractions by Ag^+ Electrospray Ionization Mass Spectrometry, Anal. Chem., 2002, 74, 1408-1414.
2. Cunico, R.L.; Shen, E.Y. and Mullins, O.C., Molecular Weight Measurement of UG8 asphaltene using APCI Mass Spectrometry. Proceedings of the 2002 International Conference on Heavy Organic Depositions, Puerto Vallarta, Jalisco, Mexico, Nov. 17-21, 2002
3. Hugley, C.A; Rodgers, R.P. and Marshall, A.C., Resolution of 11000 Compositionally Distinct Components in a Single Electrospray Ionization Fournier Transform Ion Cyclotron Resonance Mass Spectrum of Crude Oil, Anal. Chem., 2002, 74, 4145-4149.
4. Buch, L.; Groenzin, H.; Buenrostro-Gonzales, E.; Andersen, S.I.; Lira-Galeana, C. and Mullins, O.C., Molecular Size of Asphaltene Fractions Obtained from residuum hydrotreatment, Fuel, 2003, 92, 1075-1084.
5. Impey, G.; Kieser, B. and Alary, J-F., The Analysis of Polycyclic Aromatic Hydrocarbons by LC/MS/MS using a new Atmospheric Pressure Photo Ionization source, Proceedings of the 49th ASMS Conference, Chicago (USA), May 27-31, 2001.

Process Economics of On-Line Total Sulfur Determination

Patrick J. Moore, Rod W. Spitler, Rodney Friudenberg
Thermo Electron Corporation

Prepared for Presentation at

AIChE 2004 Spring National Meeting
April, 2004

Seventh Topical Conference on Refining Processing
Analysis/Characterization Tutorial

Unpublished

Abstract

Organic sulfur compounds are well known poisons to catalysts used in many refining processes. Typical processes of interest include isomerization and reforming. A survey of the major characteristics of these processes will be presented. To prevent catalyst poisoning by sulfur containing compounds isomerization and reforming feedstocks are typically desulfurized to sulfur (S) concentration levels less than 100 parts per billion by weight (ppb/wt) to enhance or maintain process yields and efficiencies. An accurate and reliable means to determine the trace total sulfur content on-line can deliver a rapid return on investment through an extension of catalyst life and an optimal operation of the upstream feed desulfurization process. Of the numerous sulfur detection technologies available, the pulsed ultra violet fluorescence (PUVF) and the lead acetate ($Pb(OAc)_2$) detectors are discussed in detail. The analytical performance of the PUVF detector at total sulfur concentrations of < 50 ppb/wt S to 5 ppm/wt S is reviewed. An economic analysis of a typical on-line trace total sulfur analyzer installation using the PUVF detector is presented.

Process Economics of On-Line Total Sulfur Determination

Patrick J. Moore, Rod W. Spitler, Rodney Friudenberg
Thermo Electron Corporation

Isomerization and Reforming Processes

Isomerization and reforming are important examples of processes that manufacture products used in the motor fuel pool. Each process employs catalysts tailored to significantly improve the operability and efficiency of a particular operating unit. Additional details, as well as information on other hydrocarbon processing technologies, can be found in Meyers, (1997).

The isomerization of light straight run (LSR) paraffins is believed to proceed via the formation of an olefin intermediate formed by dehydrogenating the paraffin in the presence of a noble metal, such as platinum in a hydrogen atmosphere. Hydrogen limits the formation rate of the olefin intermediate that can lead to undesirable polymerization by-products. Sufficient olefin intermediate is required to form the carbonium ion on the acid site of the catalyst. Rearrangement of the carbonium ion produces a cycloalkyl intermediate, which, in turn, creates an iso-paraffinic carbonium ion. Loss of a proton to the acid site on the catalyst creates an iso-olefin intermediate that is hydrogenated to form the analogous iso-paraffin product (Meyers, 1997). First generation Friedel-Crafts catalysts based processing of C_5 and C_6 paraffins were difficult to operate. Second generation catalysts included a metallic hydrogenation component, but operated at higher, less thermodynamically favorable, temperatures. Dual metal catalysts containing noble metal and acid functionalities provided higher catalytic activity at the more desirable lower operating temperatures.

Catalytic reforming of gasoline boiling range hydrocarbons is used to produce aromatic intermediates for the petrochemical feedstocks and high octane number components for the gasoline motor fuel pool. Feedstocks undergo a variety of mainly endothermic conversion processes in the reductive catalytic reformers such as "dehydrogenation of naphthenes to aromatics, dehydrocyclization of paraffins, isomerization of paraffins and naphthenes, dealylation of alkylaromatics, hydrocracking of paraffins to light hydrocarbons and formation of coke" (Russ, et.al., 1997). Aromatics pass through the reforming process relatively intact. Naphthalenes react with relative ease to predominantly form aromatics. Paraffins are the most difficult to convert and, because of the conversion difficulty, the severity of the reforming process is dictated by the feedstock paraffin composition. The plethora of feedstock constituents means that a large range of widely varying reaction rates occur simultaneously by both series and parallel reaction paths (Ancheyta and Villafuerte, 2000, Meyers, 1997).

The first reformers were fixed bed semiregenerative (SR) requiring periodic shut downs for regeneration of the monometallic catalyst. Bimetallic catalysts permitted reforming operating conditions characterized by lower operating pressures, higher severity, and greater paraffin conversion. Bimetallic SR reformers still had to be shut down for catalyst regeneration. In cyclic reforming, fixed bed reactors are taken out of and returned to service individually while reactors not being regenerated maintain production. In continuous regeneration, CCR, reformers spent bimetallic catalyst is removed from the last reformer reactor, is regenerated, and is returned to the first reformer reactor on a continuous basis.

Completed catalyst turnover occurs in about three days. Operating pressures of 50 psig and product octane numbers as high as 108 were obtained (Meyers, 1997). Additional information on CCR reforming and subsequent technology improvements can be found in Meyers (1997), Sutton, et.al. (1972), and Peer, et.al. (1988).

Reforming catalysts are heterogeneous. A base material, usually Al_2O_5, provides support sites for the deposited active metals. Platinum was the metal of choice for monometallic catalysts. Bimetallic catalysts use platinum and rhenium metals with high surface dispersion and a typical loading of < 1wt%. Though the platinum in the catalyst can exist in compound form, platinum in reduced form produces the best results (Russ et.al., 1997). The acid part of the dual functioning bimetallic catalyst is achieved with the addition of a 0.2 to 15 wt% halogen promoter, preferably chlorine (Meyers, 1997).

Total Sulfur Measurement

Isomerization and reforming catalysts are, to varying degrees, sensitive to feedstock sulfur concentrations. Knowing the feedstock total sulfur concentration as a function of time allows corrective actions to be taken in a timely manner. The feedstock total sulfur concentration is obtained using a on-line total sulfur analyzer. Of the available sulfur detection technologies, the lead acetate $Pb(OAc)_2$ detector and the pulsed ultra violet fluorescence (PUVF) detector will be discussed in detail. The other sulfur detection technologies reviewed include: 1) chemiluminescense, 2) flame photometric detection (FPD), 3) wavelength and energy dispersive x-ray fluorescence, and 4) titration.

Chemiluminescense is the photon emission by a chemically induced excited species as it returns to its ground state (Skoog, et.al., pg. 191, 1992). Combustion converts sample sulfur compounds to SO_2 followed by the partial reduction of SO_2 to SO. Ozone, O_3, reacts with the SO to form the excited specie, SO_2^*, which upon relaxation emits a photon that is detected and quantified with a photomultiplier tube, PMT. While amenable to laboratory installations, its use in a process environment is hampered by the necessary series of chemical reactions. Flame photometry utilizes a hydrogen/air flame to combust and reduce sulfur compounds in the sample to the excited specie S_2^* that relaxes to its ground state and emits a photon detected and quantified by a PMT. Normally, the FPD follows a gas chromatographic column. Sulfur species identification is based on the column elution time. Total sulfur is the sum of the sulfur concentrations determined for each of the chromatographic peaks.

X-ray Fluorescence is an atomic emission phenomenon, typically used for double digit ppm and higher sulfur level measurements. Absorbed x-rays excite the electrons of the sulfur atom. Relaxation proceeds by a series of electronic relaxation transitions that emit x-ray radiation. In Wavelength Dispersive X-Ray Fluorescence, (WD-XRF) the sample is targeted with a specific excitation wavelength. In Energy Dispersive X-Ray Fluorescence, (ED-XRF) all of the source wavelength target the sample. The detector and processing electronics resolve the response for a particular wavelength. Typically found in laboratory installations, titration detection technology uses constant current coulometry and is referred to as a microcoulometer. Application of microcoulometers to total sulfur measurement involves sample combustion to convert all sulfur containing compounds into a single sulfur specie, SO_2, which is captured into an iodide solution for quantification.

$Pb(OAc)_2$

Fundamentally, the $Pb(OAc)_2$ detector measures the formation of lead sulfide, PbS, and from this information infers the H_2S concentration in the sample gas. The detection phenomenon is based on the well known reaction between gaseous H_2S and $Pb(OAc)_2$, a white salt impregnated on a paper tape packaged as either a reel or a cassette (Moore, 1996). The H_2S measurement is detected inferentially by the formation of the reaction product PbS by observing the reduction of the reflected incident light intensity with a photodiode. The light source color choice is application dependent. Using the $Pb(OAc)_2$ detector for total sulfur measurements involves sample sulfur compound conversion to H_2S in two alternative ways: 1) straight hydrogenation and 2) oxi-hydrogenation (Moore, 1996). In straight hydrogenation the sample is mixed with an excess of hydrogen passes through a furnace at around 1250 °C for the quantitative thiophenic sulfur compounds conversion (Drushel, 1978 and Kissa, 1982). Sulfur compounds, such as carbonyl sulfide, COS, and methyl mercaptan, CH_3SH, are quantitatively at temperatures around 1000 °C to 1100 °C. To minimize coke formation, a two-stage conversion process called oxi-hydrogenation was introduced. Sample combustion converts hydrocarbons to CO_2 and H_2O. Sulfur compounds are converted to SO_2 and H_2O. The second step is hydrogenation, which reduces the SO_2 to H_2S. Water removal follows the conversion process. Humidification ensues after which the furnace exit gasses go to the detector for sulfur concentration quantification. The $Pb(OAc)_2$ detector has been viewed as a maintenance intensive technology. Bubbler solution replenishment, tape replacement, and hydrogen carrier gas consumption require regular attention. Because of its lead content, tape handling and disposal is subject to good hygiene practices and environmental regulations.

PUVF

The PUVF detector measures SO_2 by emission spectroscopy. A pulsed UV light source raises the ground state SO_2 molecule to an excited state. Upon relaxation, the excited SO_2 molecule releases a non-resonance photon, which is detected by a PMT, equipped with an appropriate bandpass filter. Ultraviolet light excitation of SO_2 involves raising bonding and non-bonding electrons found in pi molecular orbitals to pi* orbitals. A series of reflecting mirrors acts a bandpass wavelength selector. The pulsed UV light source, in contrast to a continuous UV light source, produces a more intense incident beam while simultaneously extending the lifetime of the lamp. A photodetector, mounted at the opposite of the spectrometer from the light source, measures the intensity of each pulse from the lamp. A feedback control circuit uses the photodetector output to maintain constant source intensity over the life of the lamp. Major components of the on-line trace total sulfur analyzer, equipped with a PUVF detector, are illustrated in Figure 1.

For trace total sulfur measurements, all the sample sulfur compounds are converted to SO_2 by oxidative combustion using air or heliox as the oxidant. Combustion products include CO_2, SO_2, and H_2O. Even though water is a combustion product, the use of a post furnace dryer is not necessarily required. Water droplet formation must be avoided due to the natural tendency of SO_2 to form sulfurous acid. Combustion gas water results from hydrocarbon oxidation and is based on the amount of sample injected. Material balance calculations, a water dew point calculation, and knowledge of the minimum post combustion ambient temperature are used to identify whether or not to include a dryer in the sample flow system.

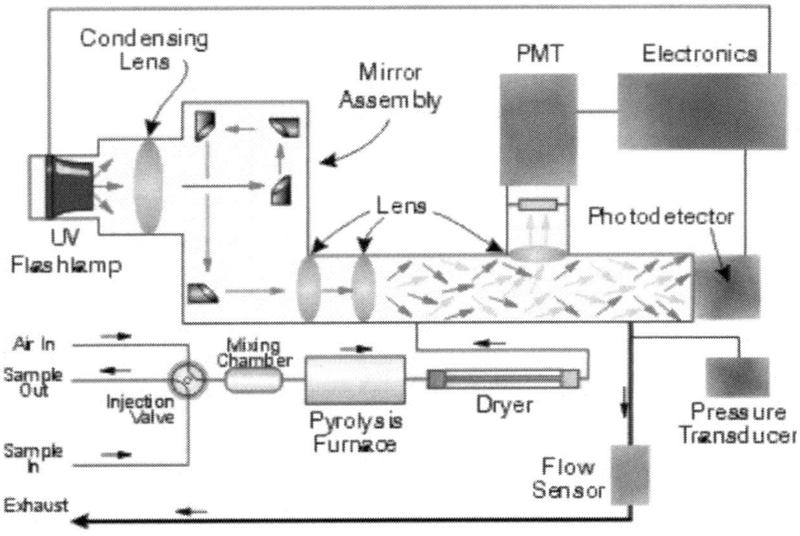

Figure 1 – Schematic of an On-Line Total Sulfur Analyzer with a PUVF Detector

Furnace carbon formation is undesirable for two reasons: 1) insufficient sulfur compound conversion to SO_2, and 2) UV fluorescent carbon compounds. Since the sample sulfur composition is small relative to the hydrocarbon concentration, carbon formation rapidly reduces the lamp intensity perceived by the photodetector in the feedback control circuit. In response to the drop, the feedback control circuit raises the lamp voltage to maintain the intensity of the lamp. The feedback control signal rate of change permits the rapid detection of carbon inside the spectrometer thus minimizing the deleterious effects of carbon deposits both in the furnace and the spectrometer by diverting the sample inlet.

The molecular population of the excited state relative to the ground state in emission spectroscopy is governed by the Boltzmann equation (Skoog et.al., 1992). Because relatively small temperature changes can result in single digit percentage changes in the quantity of excited molecules, the spectrometer is temperature controlled at about 40 °C.

PUVF Detector Performance

The determination of total sulfur in motor fuels at levels ranging from 5 ppm S (wt./wt.) to more than 1000 ppm S (wt./wt.) by sample combustion and PUVF detection with excellent analytical performance was previously reported (Spitler and Pasquale, 2003). Trace total sulfur determinations are measurements of total sulfur concentrations $\leq$ 5 ppm S (wt./wt.). In order to provide reliable total sulfur measurements at 100 ppb S (wt./wt.), a lower detectable limit, LDL, of at least 50 ppb S (wt./wt.) is required, which sets a lower limit on the definition of trace total sulfur measurements. A trace total sulfur measurement is therefore defined as a measurement between 50 ppb S (wt./wt.) and 5 ppm S (wt./wt.). Measurement linearity was conducted by preparing a series of sulfur (source: thiophene) standards in high purity, sulfur free, p-xylene. Each of these standards was introduced to the on-line trace total sulfur analyzer and the average PUVF detector frequency in kHz for approximately 30 replicates were collected. The results of the measurement linearity tests are summarized in Figure 2 below.

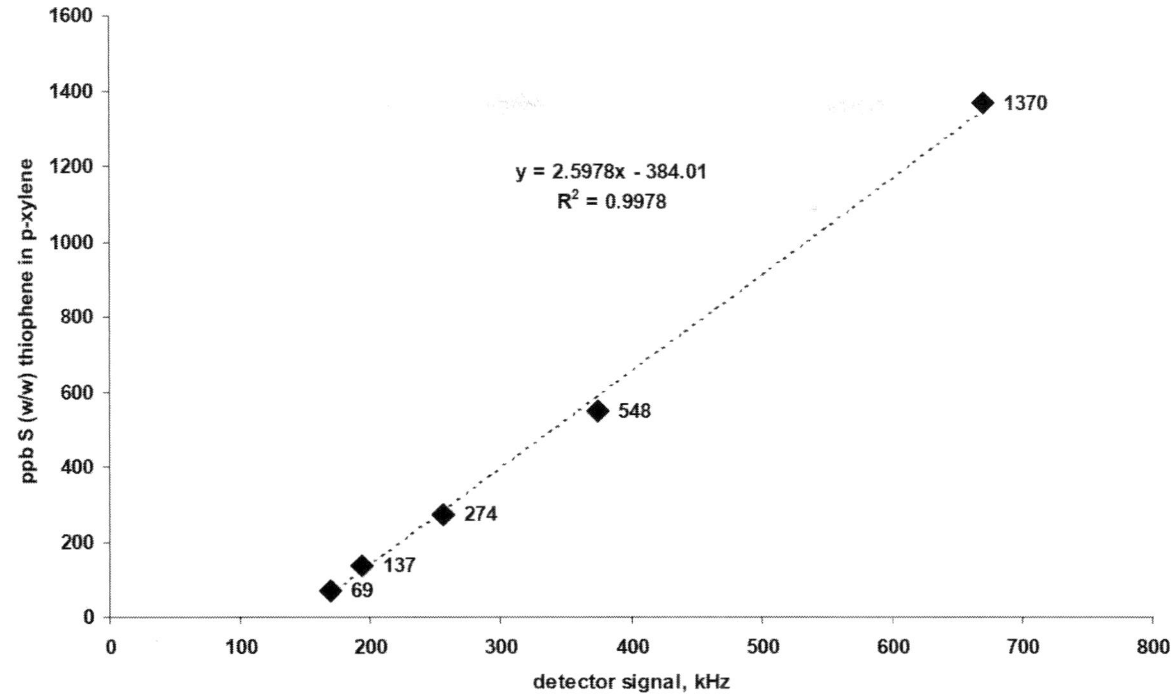

Figure 2 – PUVF Measurement Linearity

Repeatability and accuracy of trace total sulfur measurements were measured with synthetic samples prepared in high purity, sulfur free p-xylene. Prior to beginning these tests the on-line trace total sulfur analyzer was calibrated with two known standards, containing a non-zero sulfur concentration. The two point calibration is necessary to establish the detector's background reading. Calibration standards containing 137 ppb S (w/w) and 1370 ppb S (w/w) were used. Each standard was prepared from a thiophene in p-xylene stock solution. The calibration curve of y = 2.5947x – 365.07, where x is the kHz detector signal and y is the ppb S (w/w) of the calibration standard., was obtained.

Following the calibration a series of standards were introduced as unknown samples to the on-line analyzer. First, a sample of 69 ppb S (wt./wt.), thiophene in p-xylene, determined by calculation, was introduced as an unknown sample to PUVF analyzer. The results of 300 replicates yielded an Average Reported Value of 78 ppb(w/w) with a standards deviation of 5 ppb (w/w). Next a sample of 137 ppb S (wt./wt.), thiophene in p-xylene was introduced as an unknown sample. The results of 270 replicates yielded an Average Reported Value of 137 ppb(w/w) with a standards deviation of 7 ppb (w/w). A sample of 548 ppb S (w/w), thiophene in p-xylene was then introduced as an unknown. The results of120 replicates yielded an Average Reported Value of 606 ppb(w/w) with a standards deviation of 11 ppb (w/w). For each of the tests, the Calculated Value represents the gravimetrically determined sulfur concentration. The maximum difference between the average reported value and calculated value for the 69 ppb(w/w) sample is 13%. For ppb level sulfur determinations measurement, accuracies of less than 15% are generally considered to be excellent results. The standard deviation of each measurement from about 2% - 7% of the calculated value, again excellent results for ppb level sulfur measurements.

In the last test a sample of low sulfur naphtha isomerization feed was obtained from a local refinery. The sample was run continuously for a period of 10.5 hours calibrated at 1540 ppb(w/w) full scale. The statistical results of 630 replicates yielded a mean of 39.4 ppb(w/w), a median of 39.6 ppb(w/w), and a mode of 36.7 ppb(w/w). These results suggest that the observed analytical variations are due mainly to random fluctuations. The observed standard deviation was 4.3 ppb(w/w) with a maximum reading of 51.1 ppb(w/w) and a minimum reading of 27.7 ppb(w/w). The repeatability of the measurements based on two standard deviations was calculated as 0.6% of full scale.

PUVF On-Line Economic Analysis

The process economics related to catalyst deactivation due to sulfur exposure involve a large number of site specific variables. Foremost among these variables is the type of catalyst in use. As can be seen in Rhodes (1997), there are a wide range of available catalyst materials and manufacturers. While the specifics will vary between vendors and products, some general observations can provide a basis to perform a detailed sulfur economic impact statement.

Sulfur in the isomerization feedstock is undesirable because it reduces the activity of the isomerization catalyst by forming metal sulfides (Meyers, 1997). Metal sulfides reduce the amount of active metal sites needed for hydrogenation activity. Since cracking reactions compete with hydrogenation reactions, hydrogenation site reductions result in cracking dominance and the formation of light paraffins (Galperin, 1999). Feedstock sulfur concentrations are easily reduced by hydrotreating. Though the formation of metal sulfides is reversible, catalyst regeneration time represents lost production. The sensitivity of reforming catalysts to the adverse effects of sulfur exposure is well known (Bartholomew, et.al., 1982). A sulfur free feedstock is defined as having a sulfur concentration less than 20 ppb/wt, which represents the lower limit of quantification for analytical instrumentation measuring total sulfur using ASTM method 4045-87. Early reforming process feedstocks typically contained 0.1 to 1 ppm/wt sulfur (Russ, et.al., 1987). Recent reforming process feedstocks usually have sulfur levels from 0.15 to 0.5 ppm/wt (Ancheyta and Villafuerte, 2000 and Shanying, et.al., 2002).

Meyers (1997) compares the economic performance of a SR reformer with a CCR reformer assuming each is installed in a 20,000 barrels per stream day (BPSD) facility. Phansalkar, et.al. (circa 2003) identifies the effect of catalyst aging on the yields of hydrogen, fuel gas, i/n-butane, and reformate for a 20,000 BPSD facility as well as providing more recent value figures for each of these reformer products. The product values in Table 4.1.6 of Meyers (1997) were adjusted first with respect to cost and second with respect to yield using the data of Phansalkar, et.al. (circa 2003) with the following assumptions: 1) C_5^+ is reformate, 2) fuel gas is composed of methane, 3) LPG is i/n-Butane, 4) CCR C_5^+ value is \$30/bbl, 5) the ratio of SR to CCR C_5^+ value is 23/24.60, 6) the yield adjustment ratios (vol%/vol%) due to catalyst aging are applicable to SR and CCR operations and 7) the yield reductions due to catalyst aging are due to sulfur exposure. Base and adjusted product values and the assumed changes in product yields for an aged reformer catalyst are listed in Table 2. The yield changes reflect a 1.50% volume percent reduction in the reformate yield.

Table 2 – Base and Adjusted Product Values and Aged/Fresh Catalyst Yield Ratios

Product	Base Value (1997)	Adjusted Value (2003)	Aged/Fresh Catalyst Yield Ratio
C_5^+ SR	$23/bbl	$28/bbl	0.982
C_5^+ CCR	$24.60/bbl	$30/bbl	0.982
H_2	$0.34/lb	$0.48/bbl	0.837
LPG	$14/bbl	$25/bbl	1.23
Fuel Gas	$0.05/lb	$0.072/lb	1.23

Table 3 – Per Day Revenue Stream Impact of Aged Catalyst

Product	SR Adjusted (2003)	SR Adjusted (2003)	CCR Adjusted (2003)	CCR Adjusted (2003)
$/day	Fresh Catalyst	Aged Catalyst	Fresh Catalyst	Aged Catalyst
C_5^+ (SR and CCR)	$436,605	$428,746	$479,012	$470,090
H_2	$55,045	$42,073	$86,944	$72,772
LPG	$12,777	$15,716	$20,321	$24,995
Fuel Gas	$44,438	$54,659	$16,157	$19,873
Total	$548,865	$541,194	$602,434	$587,730
Daily Production Loss due to Aged Catalyst		$7,761		$14,704

Using the value and yield adjustment values indicated in Table 2, the per day revenue stream for the reformer products are calculated and are presented in Table 3. Table 3 indicates that a CCR reformer is about twice as economically sensitive compared to a SR reformer for the same reduction in the reformate yield. The differential per day revenue stream reduction can be considered as a per day profit loss. A typical on-line total sulfur analyzer installed cost is about $150,000 and represents the sum of the capital cost of the analyzer, the cost of training, sample system installation, data handling mechanisms, and site engineering. If aged catalyst yield reductions are all caused by sulfur exposure, the payback period for an analyzer in a SR reformer unit is about 20 days while in a CCR unit the payback period is about 11 days. Even if sulfur exposure accounts for only 50% of the aged catalyst yield reductions, the payback period is still very rapid. The payback is realized by controlling the sulfur exposure to the catalyst permitting extended the run times at elevated yields.

Conclusions

On-line trace total sulfur analyzers directly contribute to the efficient operation of isomerization and reforming processes through prevention of catalyst aging by sulfur poisoning. Assuming a total installed cost of $150,000 for a typical on-line PUVF total sulfur analyzer, the payback time can be as short as < 1 month.

References

Ancheyta-Juarez, Jorge and Villafuerte-Macias, Eduardo, "Kinetic Modeling of Naphtha Catalytic Reforming Reactions", Energy and Fuels, Vol. 14, pp. 1032-1037, 2000.

Bartholomew, C.H., Agrawal, P.K., and Katzer, J.R., "Sulfur Poisoning of Metals", Advances in Catalysis, Vol. 31, pp. 135-242, 1982.

Drushel, H.V., "Trace Sulfur Determinations in Petroleum Fractions", Analytical Chemistry, Vol. 50, No. 1, pp. 76-81, 1978.

Galperin, Leonid, B., US Patent 5,905,181.

Kissa, Erik, "Determination of Trace Amounts of Sulfur in Organic Solvents", Analytical chemistry, Vol. 54, No. 8, pp. 1450-1452, 1982.

Meyers, Robert A., Editor-in-Chief, Handbook of Petroleum Refining Processes, Second Edition, The McGraw-Hill Companies, Inc., New York, 1997.

Moore, Patrick J., "Hydrogen Sulfide Analyzers", Analytical Instrumentation, Instrument Society of America, pp. 343-357, 1996.

Moore, Patrick J., "Total Sulfur Analyzers", Analytical Instrumentation, Instrument Society of America, pp. 359-366, 1996.

Peer, Roger L., Bennett, Richard W., and Bakas, Steve T., "Continuous reformer catalyst regeneration technology improved", The Oil and Gas Journal, Vol. 86, no. 22, pp. 52, 54, 58-60, 1988.

Phansalkar, Sunil S., Edgar, M. Dean, Gray, Stuart, Lee, Byran, "Catalytic Reforming Optimization", Criterion Catalysts, circa 2003.

Rhodes, Anne K., "Number of catalyst formulations stable in a tough market", The Oil and Gas Journal, Vol. 95, No. 40, pp. 43-72, 1997.

Russ, Michael R., and Sechrist, Paul A., US Patent 5,614,082.

Shanying, Hu, Towler, G., and Zhu, X.X., "Combine Molecular Modeling with Optimization to Stretch Refining Operation", Ind. Engr. Chem. Res., No. 41, pp. 825-841, 2002.

Skoog, Douglas A., and Leary, James L., Principles of Instrumental Analysis, Fourth Edition, Saunders College Publishing, New York, 1992.

Spitler, Rodney, W., Pasquale, Alberto, and Friudenberg, Rodney, "Using pulsed UV fluorescence", Hydrocarbon Engineering, February 2003.

Sutton, E.A., Greenwood, A.R., and Adams, F.H., "A new processing concept: continuous Platforming", The Oil and Gas Journal, vol. 70, no. 21, pp. 52-56, 1972.

Tutorial: Upgrading Light Naphtha

David J. Shecterle, Richard R. Rosin
UOP LLC
Des Plaines, Illinois, U.S.A.

Prepared for Presentation at
Spring National Meeting
New Orleans, Louisiana
April 25-29, 2004

TUTORIAL: UPGRADING LIGHT NAPHTHA

ABSTRACT

This paper provides an overview of light (C5-C6) paraffin isomerization technology. Isomerization of light hydrocarbons continues to play an important role in refinery gasoline pool octane management as new regulations affecting gasoline composition are enacted throughout the world. Isomerization is an economic means for reducing benzene content of gasoline pools while upgrading light hydrocarbon octane values to make up for octane lost due to benzene reduction.

Original isomerization catalysts were made of chlorided alumina. An early example was UOP's I-4™ catalyst commercialized in the late 1950s. Later developments included zeolitic catalysts, such as UOP's HS-10™ catalyst commercialized in the early 1970s and higher-activity, second-generation chlorided alumina catalysts, such as UOP's I-8™ catalyst, commercialized in the early 1980s. UOP commercialized I-80™ catalyst, an even higher activity chlorided alumina catalyst, in the mid 1990s and recently improved the manufacturing process and commercialized I-8 Plus™ catalyst and I-82™ catalyst. Also in the mid 1990s, a novel sulfated zirconia catalyst, UOP's LPI-100™ catalyst, was introduced that is both robust and fully regenerable. Recently, UOP developed and commercialized a new regenerable isomerization catalyst, PI-242™ catalyst, whose performance approaches that of chlorided alumina catalysts. Each catalyst system has a unique set of operating parameters that affect both unit costs and product octane as well as potential revamps of existing units.

INTRODUCTION

The primary commercial use of the branched isomers of C_5 and C_6 paraffins is in the production of clean-burning, high-performance transportation fuels. The elimination of tetraethyl lead as a gasoline antiknock compound over the last three decades and recent regulations restricting motor fuel composition have led refiners to select alternative means of producing high quality gasoline. The choices of high-quality gasoline blending components available in a typical refinery are limited as a result of benzene-concentration restrictions, endpoint and olefin-content limitations, and potential limitations of total aromatics concentration. Consequently, isomerate, the gasoline blending component from light paraffin isomerization, is the ideal choice because of its paraffinic composition.

This paper focuses on the various process alternatives offered in isomerization of C5 and C6 paraffins. There are three main catalyst types used in the production of isomerate: zeolitic, chlorided alumina, and sulfated zirconia. The catalyst types differ significantly in their fundamental chemistry. Therefore, each of these catalyst types has specific operating or performance attributes that translate into different overall unit designs. These differences in unit designs allow refiners to choose the optimum system for their specific needs. Chlorided alumina catalysts offer the highest activity and yield. However, chlorided alumina catalysts are water sensitive and so require feed dryers which increase capital requirements. Chlorided alumina catalysts are not regenerable and require an organic chloride promoter. The need for an organic chloride promoter requires the capital and operational expense of caustic scrubbing. Zeolitic catalysts are regenerable and relatively contaminate tolerant. However, zeolitic catalysts are the lowest activity catalyst family and their higher operating temperatures require a fired heater. Sulfated zirconia catalysts have activities significantly higher than zeolitic catalysts but still are significantly less active than chlorided alumina catalysts. However, sulfated zirconia catalysts are contaminant tolerant and are regenerable. Zeolitic catalysts and sulfated zirconia catalysts require a higher hydrogen to hydrocarbon ratio and so a recycle compressor and separator are required for technologies based on these catalysts. UOP's newest isomerization catalyst, PI-242 catalyst, is contaminant tolerant, regenerable, and has activities that approach those of chlorided alumina catalysts.

PARAFFIN ISOMERIZATION REACTIONS[1]

Typical light naphtha feeds to isomerization units are comprised of many different paraffin isomers as well as naphthenes and benzene. The reactions of most interest are the isomerization reactions that yield the components with the highest blending RON values. These reactions with their blending RON values are shown below.

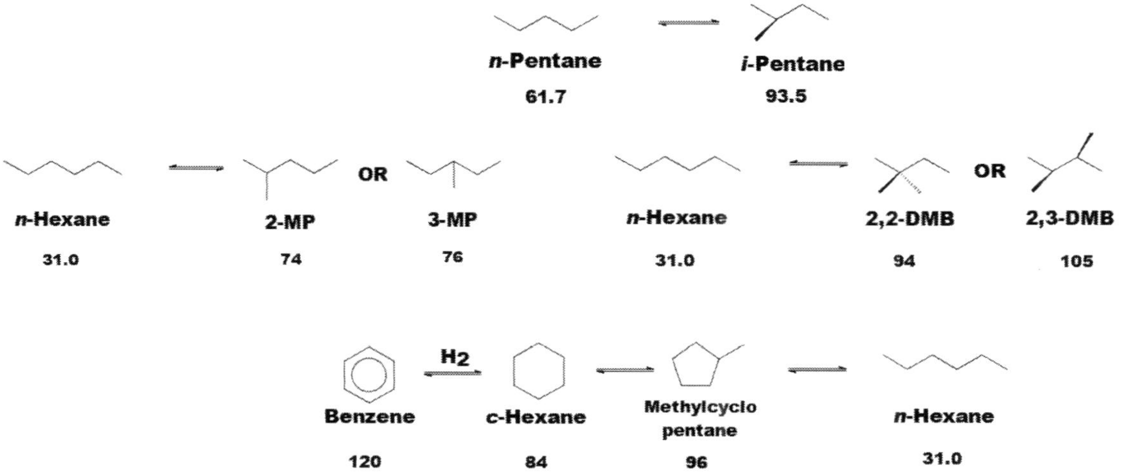

Benzene saturation in an isomerization reactor is very important to refiners as a means to control benzene in the gasoline pool. The reaction of cyclohexane to methylcyclopentane and subsequent ring opening to n-hexane is also a very desired reaction despite its loss in octane. Methylcyclopentane and, to a lesser extent, cyclohexane form very stable carbo-cations which occupy acid sites on the catalyst, preventing those sites from being utilized for paraffin isomerization. This is why the isomerization of paraffins decreases as the level of naphthenes increases. Therefore, a catalyst that promotes the rapid isomerization of cyclohexane to n-hexane is highly preferred. As refiners shift more of the benzene precursors from their reforming unit feeds to their light naphtha isomerization unit feeds, the importance of an isomerization catalyst's ring opening ability increases.

ISOMERIZATION CATALYSTS

UOP offers three different commercial catalyst systems for light naphtha isomerization: zeolitic catalyst (HS-10 catalyst), amorphous chlorided alumina catalysts (I-8 Plus catalyst and I-82 catalyst), and sulfated zirconia catalyst (LPI-100 catalyst). Each of these catalysts has a combination of Bronstead and Lewis acid sites which promote varying levels of protonation activity that is necessary for the isomerization reactions to proceed. Isomerization catalysts must also be capable of hydrogenolysis, which not only assists in the protonation step, but also serves to saturate olefin intermediates and aromatic hydrocarbons and to assist in the ring opening of cycloparaffins. This function also gives activity stability to isomerization catalysts, thereby improving the process economics.

Isomerization of C_5/C6 streams is an equilibrium limited reaction for which branched paraffin isomers are generally favored by low temperatures (see Table 1[2] and Figures 1[3] and 2). The most active catalysts, capable of operating at the lowest temperatures, will produce the highest octane products.

Zeolitic Catalysts

UOP's HS-10 catalyst is a platinum impregnated mordenite zeolite that does not contain a halide activator nor require a promoter. The zeolitic catalysts operate at higher temperatures and the product octane is thus lower due to the less favorable branched isomer equilibrium at these temperatures. Product yields are also lower for the zeolitic catalysts as a result of the higher operating temperatures and lower selectivity of these catalysts.

The main benefit of zeolitic catalysts is that they are not permanently deactivated by water or other oxygenates and are fully regenerable. The zeolitic catalysts are tolerant of sulfur but sulfur does suppress the platinum function of the catalyst which requires higher operating temperatures to maintain catalyst activity leading to lower product yields. Feed hydrotreating is not an absolute requirement with zeolitic catalysts but is recommended to get optimum catalyst performance. Because of the tolerance to feed contaminants, zeolitic catalysts are often considered in the revamp of existing hydrotreating or reforming units.

Chlorided Alumina Catalysts

UOP's amorphous chlorided alumina I-8 Plus catalyst and I-82 catalyst are the most active isomerization catalysts currently available. Their high activity means that they operate at the lowest temperatures and thus achieve the highest octane products and highest yields. These catalysts require a chloride promoter and are permanently deactivated by oxygenate compounds such as water. Units designed for chlorided alumina catalysts require continuous chloride injection and feed and makeup gas driers to remove the oxygenates. A caustic scrubber is also required to neutralize the HCL in the off gas that is formed in the reactors.

Due to its high product octanes and yields UOP's Penex™ process, which utilizes chlorided alumina catalysts, is the most popular of the isomerization unit types.

Sulfated Zirconia Catalysts

Sulfated zirconia catalysts, also known as solid super acid catalysts, form one of the newest types of isomerization catalysts. UOP's LPI-100 catalyst first commercialized in 1996 is a platinum impregnated sulfated zirconia. This catalyst exhibits an appreciable activity advantage over the zeolitic catalysts, resulting in product octanes 2 to 3 numbers higher than those achieved by zeolitic catalysts.

Sulfated zirconia catalysts are tolerant of water and other oxygenates in the feed and require no halide promoter, similar to zeolitic catalysts. These catalysts are fully regenerable using a simple oxidation procedure comparable to the one used for zeolitic isomerization catalysts. These properties make LPI-100 catalyst an ideal candidate for the conversion of existing reforming units or for increasing the capacity or performance of isomerization units using a zeolitic catalyst. This catalyst has also been chosen for new unit designs where the full performance benefit of the chlorided alumina catalysts are not required or where there is a concern about feed contaminant levels.

PI-242 Catalyst

UOP's newly commercialized PI-242 catalyst, has activities that approach those of chlorided alumina catalysts. Product yields for PI-242 catalyst are lower than those for chlorided alumina catalysts. PI-242 catalyst is tolerant of feed contaminants and is fully regenerable using a simple oxidation procedure. Due to its high activity, PI-242 catalyst, is a very attractive alternative to sulfated zirconia catalysts and is an excellent choice for conversion of existing fixed bed reforming units to isomerization service.

ISOMERIZATION PROCESS FLOWSCHEMES

UOP offers many different process flowschemes for isomerization of light hydrocarbon streams depending on the refiner's specific requirements and preferences. There are two primary types of flowschemes used for the isomerization of light hydrocarbons; hydrocarbon-once-through and hydrocarbon-recycle. Depending on feed composition and catalyst, hydrocarbon-once-through processes are capable of producing products with research octane numbers ranging from 78 to 86. Hydrocarbon-recycle flowschemes employ either molecular sieves or fractionation columns to separate normal paraffins and lower octane isomers out of the product. Next they recycle these lower octane species back to the reactor section for improved overall conversions to higher octane isomers. Hydrocarbon-recycle units are capable of producing product research octane numbers of 87 to 93 depending on feed composition and unit type chosen. See Figure 5 for a comparison of the various UOP isomerization flowscheme options.

Hydrocarbon-once-through process flowschemes are offered for each of the three main catalyst systems used for isomerization. UOP's Penex process, which utilizes chlorided alumina catalyst, is the most popular hydrocarbon-once-through process because of its high product octanes and yields. UOP's Par-Isom™ process, using PI-242 catalyst, is often used for unit conversions or where the importance of tolerance to feed contaminants or design simplicity outweighs the need for the highest product octane and yield. An example of a low cost revamp utilizing PI-242 catalyst is a Par-Isom side-car unit, in which an isomerization reactor is added to an existing semi-regeneration reforming unit. The two units share a common recycle compressor and may share product separation and stabilization facilities depending on existing equipment characteristics and product requirements. Another even lower cost scenario is the conversion of an existing, idle reforming unit to a Par-Isom process unit using PI-242 catalyst. Often only very minor changes are required to convert a reforming unit for use as a Par-Isom process unit. Hydrocarbon-once-through units employing UOP's zeolitic HS-10 catalyst are not selected very often due to their lower product octanes and similarity to unit designs that utilize PI-242 catalyst.

A variety of hydrocarbon-recycle flowschemes developed for light naphtha isomerization use either molecular sieve separation, such as the TIP™ process (Total Isomerization Process) and the Molex™ process, or fractionation, such as deisopentanizer (DIP) or deisohexanizer (DIH) columns. The advantages of the various recycle flow schemes have been discussed in detail in previous publications[4,5]. Recycle flowschemes use either adsorptive separation or fractionation to separate lower octane isomers that are recycled back to the reactor section. The selection of a separation flowscheme is dependant upon feed composition, availability of utilities, and required product octane.

PROCESS ECONOMICS

Process economics dictate the selection of the proper isomerization technology for a particular application. Of the 192 UOP isomerization units that have been commissioned, approximately 60% are hydrocarbon-once-through units. The remaining 40% utilize either molecular sieve separation or fractionation in a hydrocarbon-recycle flowscheme. Many more UOP isomerization units are currently in the design or construction phase.

A comparison of hydrocarbon-once-through unit characteristics and Estimated Erected Costs (EEC) is given in Table 2. Zeolitic catalysts are usually not considered for new unit or revamp designs due to their low conversion and relatively high capital and operating costs. Refiners who need the highest performance choose chlorided alumina catalysts. Those who can accept lower performance select Par-Isom process units with PI-242 catalyst for their low cost and higher tolerance to contaminants.

Hydrocarbon-recycle flowschemes are available for each of the major isomerization catalyst types. Hydrocarbon-recycle schemes that utilize zeolitic catalysts are not often selected for new designs due to the high capital and operating costs associated with the vapor phase adsorptive separation systems used in the TIP process. Most of the new hydrocarbon recycle units designed in the last 15 years have utilized chlorided alumina catalysts. UOP offers many different flowschemes for hydrocarbon-recycle units. The choice of a specific flowscheme depends upon the feed C5/C6 ratio, desired product octane, utility costs, and available capital. A detailed description of the various flowschemes and their economics has been previously published[5]. A Penex process unit with a DIH column (commonly referred to as a Penex/DIH unit) is the most common of the isomerization unit hydrocarbon-recycle designs.

A comparison of EECs for new unit and revamp designs for Penex process and Par-Isom process units with and without DIH columns is given in Table 3. A hydrocarbon-once-through Par-Isom unit and a Par-Isom unit with a DIH column (commonly referred to as a Par-Isom/DIH unit) have EECs that are roughly 75% of the comparable hydrocarbon-once-through Penex units and Penex/DIH units. The Penex process new unit options have higher product octanes and octane-barrels. The octane difference between Penex/DIH units and Par-Isom/DIH units is small due to the fact that the effect of the lower PI-242 catalyst activity is minimized by increasing the amount of low octane isomers recycled back to the reactors from the DIH column. The Par-Isom/DIH units have lower product octane-barrels because the strong acid function of the PI-242 catalyst causes more material to crack to lighter compounds.

The conversion EECs given in Table 3 are for the conversion of a fixed bed reforming unit of comparable size to the 10,000 BPD isomerization unit. Since the recycle compressor, reactors, reactor section heat exchangers, separator, and stabilizer can be reused for a Par-Isom unit, the cost for the conversion to a Par-Isom unit is very low. For a conversion to a Penex unit, feed and makeup gas driers as well as a net gas neutralization section have to be added, resulting in a much higher EEC. An interesting comparison is the Penex unit vs. the Par-Isom/DIH unit. Both units have similar EECs. The product octane for the converted Par-Isom/DIH unit is over 3 numbers higher than the converted Penex unit, while the product octane-barrel value for the Par-Isom/DIH process is about 2% lower than that for the Penex unit.

CONCLUSIONS

Light naphtha isomerization processes continue to be important to refiners by providing them with a low sulfur, high octane, non-aromatic blending component to increase their overall gasoline pool octane. The benzene saturation that takes place in light paraffin isomerization units is often a primary driving force behind the unit economics. UOP has continued to develop new products such as the chlorided alumina I-8 Plus catalyst and I-82 catalyst and a new class of isomerization catalyst, PI-242 catalyst.

There are many different process flowschemes available in light paraffin isomerization. The choice of the optimum flowscheme for a particular case depends on a number of factors including feed composition, desired product octane, and availability of utilities. UOP's chlorided alumina I-8 Plus catalyst and I-82 catalyst offer the highest performance available and are the best choice for refiners needing high product octanes and yields. Existing Penex unit performance can be enhanced by reloading the reactors with I-8 Plus catalyst or I-82 catalyst. UOP's new PI-242 catalyst, has activities that approach those of chlorided alumina catalysts. PI-242 catalyst is a good choice for conversion of existing assets or for new units where lower EECs outweigh the need for the highest performance.

REFERENCES

1. G.C. Anderson, M.A. Stine, R.R. Rosin and M. Hunter, "New Solutions for Light Paraffin Isomerization", Proceedings of the NPRA Conference, Mar. 21-23, 2004

2. Physical Constants of Hydrocarbons C_3-C_{10}, ASTM Committee D-2 on Petroleum Products and Lubricants and API Research Project 44 on Hydrocarbons and Related Compounds, American Society of Testing Materials, 1971.

3. D. R. Stull, and others, The Chemical Thermodynamics of Organic Compounds, Robert E. Krieger Publishing Co., Malabar, FL, 1987.

4. P. J. Kuchar, and others, "Paraffin Isomerization Innovations," Fuels Processing Technology, 35 (1993), 183-200.

5. M.J. Hunter, "Light Naphtha Isomerization to Meet 21st Century Gasoline Specifications", Oil Gas European Magazine, Feb, 2003

BIBLIOGRAPHY

L.H. Rice and others, "Tutorial: Upgrading Light Naphtha and refinery Light Ends", Proceedings of the AIChE Spring National Meeting, Mar. 14-18, 1999.

M.J. Cleveland and C.D. Gosling, "Make Light Paraffin Isomerization units Work For You," *Proceedings of the NPRA Annual Meeting*, Mar. 21-23, 1999.

M.J. Hunter, "Chances For Innovative Process at the Interface Between Refining and Petrochemistry", DGMK Conference, Berlin, 2002.

G.C. Anderson and others, "New Solutions for Light Paraffin Isomerization", Proceedings of the NPRA Conference, Mar. 21-23, 2004

Table 1
Hydrocarbon Octane Values

	Research Octane by ASTM 2699	***Motor Octane by ASTM 2700***	***(R+M)/2***
n-Butane	93.8	89.6	91.7
i-Butane	100.4	97.6	99.0
n-Pentane	61.7	62.6	62.2
i-Pentane	92.3	90.3	91.3
n-Hexane	24.8	26.0	25.4
2-Methylpentane	73.4	73.5	73.4
3-Methylpentane	74.5	74.3	74.4
2,2-Dimethylbutane	91.8	93.4	92.6
2,3-Dimethylbutane	101.0	94.3	97.6

Table 2
Isomerization Unit Comparison
Hydrocarbon-Once-Through

	Zeolite	Chlorided Alumina	Sulfated Zirconia
UOP Catalyst Type	HS-10	I-8 Plus/I-82	PI-242
Reactor Operating Temperature	High	Lowest	Low
Feed and Makeup Gas Driers	No	Yes	No
Neutralization Facilities	No	Yes	No
Charge Heater Type	Fired Heater	Steam Exchanger	Steam Exchanger
Recycle Gas Compressor	Yes	No	Yes
Product Quality, RON	78-80	82-84	80-83
[1]EEC, MM USD	~10.8	10.8	7.8

[1]EEC values are for a new 10,000 BPD unit with a heavy, 29% X-factor, feed.

Table 3
Isomerization Unit EEC

Process Unit	Catalyst	Product Quality RON	Product Yield RON*BBLS	Estimated Erected Cost	
				New Unit	Converted
Penex	I-8 Plus/I-82	83.7	821,000	\$10.8 MM	\$5.0 MM
Par-Isom	PI-242	81.4	796,000	\$7.8 MM	< \$0.25 MM
Penex-DIH	I-8 Plus/I-82	87.4	853,000	\$18.3 MM	\$11.9 MM
Par-Isom-DIH	PI-242	87.0	805,000	\$14.0 MM	\$5.1 MM

Figure 1
Pentane Equilibrium

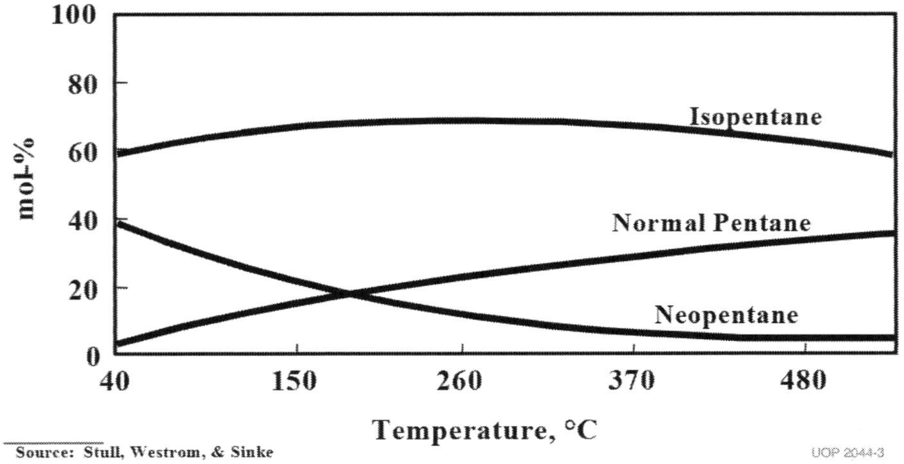

Source: Stull, Westrom, & Sinke

UOP 2044-3
UOP 2969-2

Figure 2
Hexane Equilibrium

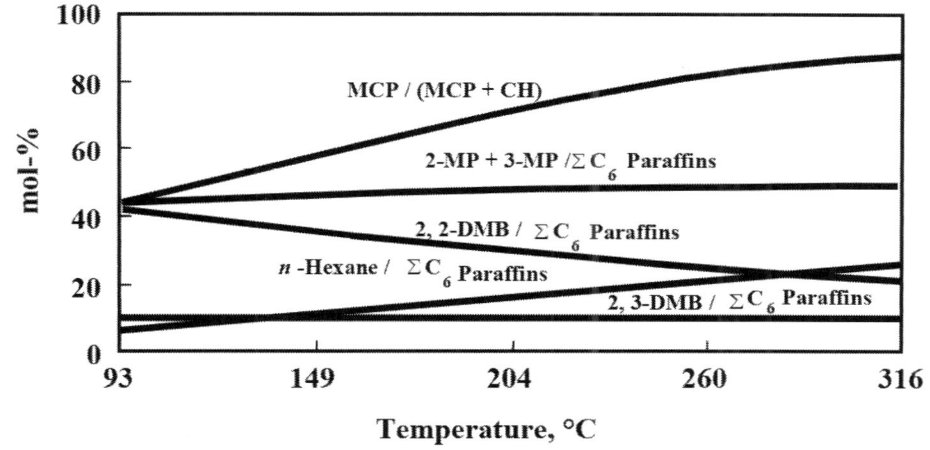

UOP 2044-4
UOP 2969-3

Figure 3
Equilibrium Octanes

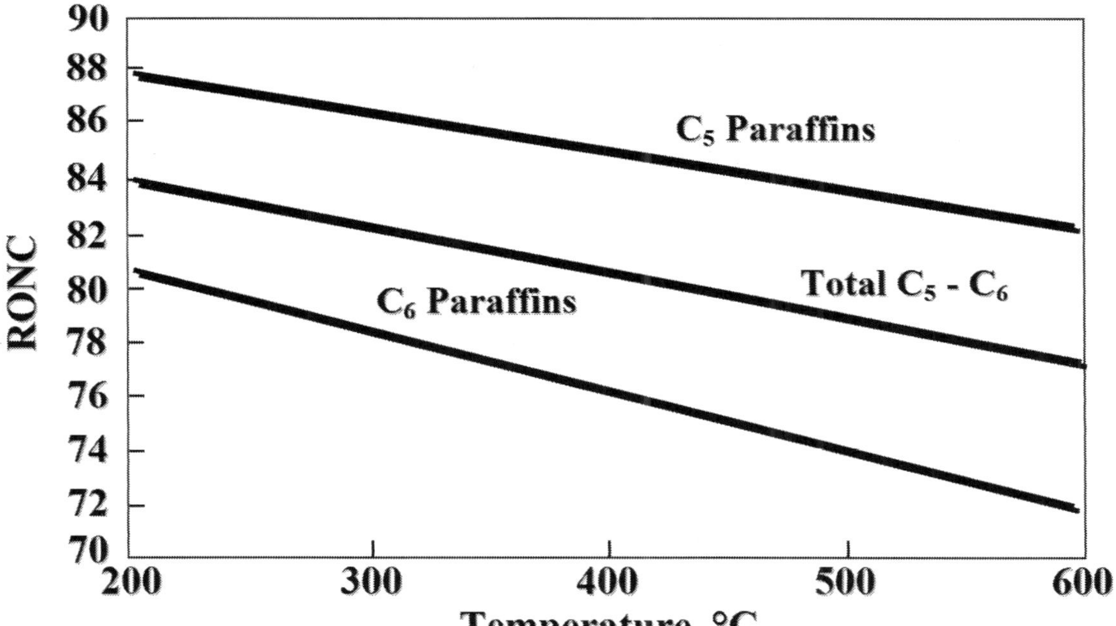

Figure 4 Isomerization Catalyst Activity Comparison

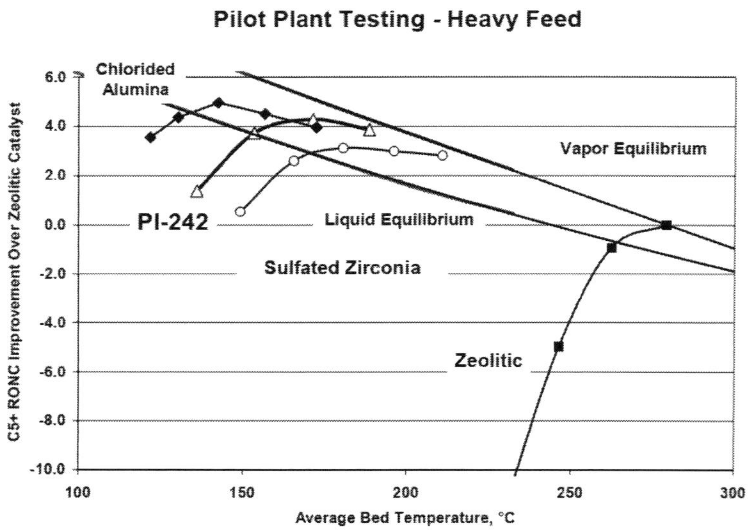

Figure 5
Isomerization Octane Portfolio Research Octane

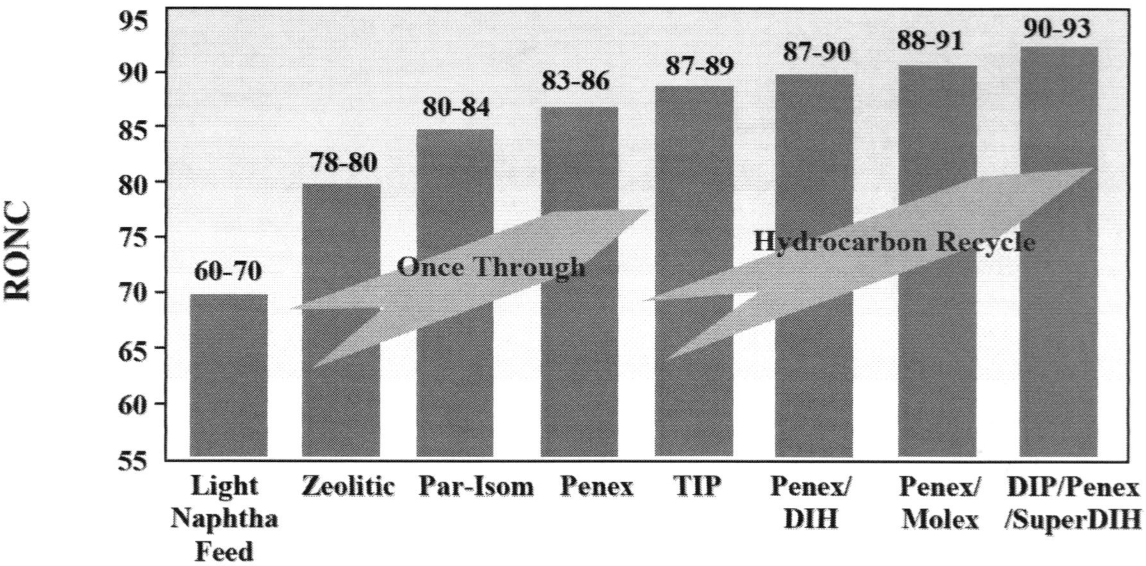

MTBE CONVERSION TO ISOOCTANE PRODUCTION: PETROBRAS CASE STORY

Marcelo Straubel Wolff
Carlos Rene Klotz Rabello
Mauro Silva
Carlos Alberto Ribeiro da Rosa
Rafael de Almeida Menegassi

PETROBRAS

Prepared for presentation at 2004 AIChE Spring National Meeting

Unpublished

ABSTRACT

Due to the ban of MTBE in California, Petrobras has decided to adapt its MTBE units to produce isooctene and/or isooctane, valuable gasoline components.

As Petrobras has 4 MTBE units and the Brazilian legislation does not allow the addition of MTBE into gasoline, the production of MTBE in Petrobras is mostly exported to the USA.

This paper shows the in-house work developed from bench scale to industrial runs in order to make it feasible to produce the isooctane that will be added to Petrobras special gasolines or exported if desired.

1. INTRODUCTION

Since the end of the 70's, MTBE has been used as a gasoline additive in the gasoline pool. Due to the approval of the Clean Air Act Amendment in the U.S. in 1990 which defined an addition of oxygenates to reformulated gasoline, the use of MTBE in the gasoline pool has increased even more in order to achieve a minimum of 2 % oxygen by weight.

In Brazil in the beginning of the 90's Petrobras decided to build and operate MTBE plants basically aiming at the external market, mainly the U.S. So, in 1996 Petrobras started up three MTBE plants and in 1997 another one that summed up a total capacity of 300,000 t / year.

In 1993 the Brazilian Congress set a legislation defining that in the internal market only ethanol could be added to the gasoline pool as an oxygenate which made exporting MTBE a must to Petrobras.

Unfortunately, MTBE has a relatively high solubility in water and due to spills from underground gasoline storage tanks a large number of contaminations occurred in drinking water systems in USA.

Even though the adverse health effect of MTBE was not definitely proved, the ban of MTBE has been proposed and approved in several states in the U.S. legislation. In California, the biggest MTBE U.S. market, the deadline for the phase-out has already set for January 1st, 2004.

As the loss of MTBE will have consequences in octane, volatility, emissions and volume of the gasoline pool, many refiners and process engineering companies that are players in the U.S. market are analyzing alternatives to replace MTBE and also making decisions about what will be done with the MTBE plants.

Petrobras as several other companies is analyzing alternatives to our MTBE plants such as:

- keeping the units running;
- shutting-down the plants;
- adapting to ETBE production;
- adapting to isooctene and isooctane production.

As it was expected, prices of MTBE in the market are going down due to the lack of demand. As this will bring low or no profit at all, keeping the units running will soon become unacceptable.

Shutting-down the plants is not as simple as it seems at first glance as costs related with the maintenance of the equipment are required. Inert gas, painting, removal of instruments and drivers are all required among others.

Also shutting-down the plants will bring another problem as the addition of the unreacted isobutene to the LPG pool will increase density of LPG to a point near or above the LPG's density specification.

Although the conversion of the MTBE plants to ETBE production is simple, as this ether is also relatively soluble in water, there is also the risk of water contamination and it was not regarded as a viable solution.

Finally the conversion to isooctane appears to be the most attractive solution as it keeps an outlet to isobutene, produces an octane booster and may reuse part of the equipment of the MTBE plant.

Isooctane production resulting from the adaptation of Petrobras MTBE plants may have different destinations such as:

- octane booster to external markets;
- octane booster to internal market;
- isooctane and isododecane cut to isoparaffinic solvents;
- additional component for special gasolines.

A final question arose about who was going to supply the technology to isooctane production. Two years ago some licensors were offering this technology but none of them had their technology proved in an industrial plant.

As the MTBE technology applied to the plants in Petrobras was developed in-house as well as their basic engineering design, Petrobras decided to evolve the isooctane technology at Cenpes, Petrobras Research and Development Center.

This paper will review the research and development work carried out at Cenpes and the current status of isooctane production in Petrobras.

2. CHEMISTRY OF THE PROCESS

MTBE synthesis:

The MTBE synthesis takes place in liquid phase at moderate temperatures and pressures using an acidic cation exchange resin as the catalyst.

As shown in figure 1, isobutene reacts with methanol in an etherification reaction. Side reactions also occur resulting in the production of tert-butyl alcohol (TBA) from the hydration of isobutene as well as di-isobutylene (DIB) from the dimerization of isobutene.

$$\text{MeOH} + \text{ISO} \rightleftharpoons \text{MTBE}$$

$$\text{ISO} + H_2O \rightleftharpoons \text{TBA}$$

$$\text{ISO} + \text{ISO} \longrightarrow \text{DIB}$$

Figure 1 – Main MTBE Process Reactions

A stoichiometric excess of methanol is employed resulting in a high selectivity in the process, near 99%. Normally conversion is around 96 – 99% depending on process configuration and catalyst activity among others.

Many catalysts are commercially available such as *Amberlyst*-15 or 35 from Rohm and Haas, K-2631 from Bayer and M-30 or M-31 from Dow Chemical.

The modification of the MTBE synthesis to isooctane production is connected with the facts that MTBE units also produce DIB and that water or methanol have an inhibiting effect on the catalyst activity.

Dimerization:

The isobutene dimerization to isooctene proceeds in liquid phase, and the same acidic resin may be employed. As shown in figure 2, the main reaction products are 2,4,4-trimethyl-1-pentene [244TMP1] and 2,4,4-trimethyl-2-pentene [244TMP2], but other isomers are also produced.

$$\text{ISO} + \text{ISO} \longrightarrow \begin{matrix}\text{244TMP1}\\ \text{244TMP2}\end{matrix}$$

Figure 2 – Main Reaction Products

When feedstocks also contain linear butenes such as FCCs feedstocks, side reactions with these components may happen resulting in other isomers and codimers. The dimethylhexenes [DMHx] are the undesired sub-products due to their low octane number. Other side reactions include the production of isobutene trimers (mainly 2,2,4,6,6-pentamethyl-3-heptene [22466PMHp3] and tetramers.

The side reactions of the principal sub-products can be seen in figure 3.

ISO + BUTENES	$\longrightarrow$	233TMP 344TMP1 344TMP2 234TMP	(TRIMETHYLPENTENES)
BUTENES + BUTENES	$\longrightarrow$	55DMHx 34DMHx	(DIMETHYLHEXENES)
n [ISO] + n [BUTENEs]	$\longrightarrow$	Trimer Tetramer Oligomer	(22466PMHp3)

Figure 3 – Side Reactions of Principal Sub-products

Another side reaction shown in figure 4 is the production of TBA. This reaction plays an important role in the isooctene production as it can control the activity of the catalyst and, consequently, the process selectivity.

$$\text{ISO} + H_2O \longrightarrow \text{TBA}$$

Figure 4 - TBA Production

In addition to the presence of this selectivity modifier, the composition of the isomeric mixture will also depend on the relative amounts of isobutene and linear butenes in the feed and also on the severity of the process.

Process conditions have to be optimized in order to make feasible a high isobutene conversion and a small linear butenes conversion in order to minimize the amount of dimethylhexenes, trimers and tetramers in the product and maximize product octane number.

It is important to note that the heat of reaction of the dimerization process (ΔH = - 19,800 kcal/kmol) is more than twice that of the MTBE (ΔH = - 9,500 kcal/kmol), and this fact may lead to an extra difficulty in the control of the reaction temperature.

In order to reduce the activity of the catalyst and minimize the oligomerization reactions, many polar compounds can be used including TBA and MTBE (which will have a similar effect to water and methanol as they balance in the process). In cases where MTBE is not desired, TBA will be the obvious choice.

Hydrogenation:

The catalytic system employed in the hydrogenation step depends on the poisons found in the feed, mainly on sulfur components.

A nobel metal or nickel-based catalyst may be employed if the feed is sulfur free or has a low level of sulfur, but in the case of a FCC C4 feed that will have a high sulfur content a hydrotreatment catalyst shall be employed as it is robust and generally insensitive to contaminants.

The main reactions that happen in the hydrogenation section are shown in figure 5, and it is important to note that they are also highly exothermic with a heat of reaction of approximately – 30,000 kcal/kmol.

ISOOCTENES + H_2	$\longrightarrow$	ISOOCTANES	HYDROGENATION
TRIMERS + H_2	$\longrightarrow$	ISODODECANES	
TETRAMERS + H_2	$\longrightarrow$	ISOBUTANE ISOOCTANES ISODODECANES	HYDROCRACKING
R-S-S-R R-SH + H_2	$\longrightarrow$	H_2S + R-H	HYDROTREATING

Figure 5 - Main Reactions Products

It is also important to note that process conditions and the catalytic system in the hydrogenation step must be carefully selected in order to minimize cracking of the isooctene stream and, consequently, a reduction in the process yield.

3. PROCESS DESCRIPTION

3.1 Dimerization section

In the dimerization section, the C4 feed is first washed with demineralized water in a washing tower in order to remove almost all acetonitrile in the feed. As this compound is a poison to the catalyst to achieve a suitable catalyst life, the acetonitrile content in the washed C4 stream shall be lower than 1 ppm wt.

The washed C4 stream from the top of the washing tower is mixed with a controlled amount of water to produce the required quantity of TBA. Unreacted C4 from downstream is recycled and mixed with fresh C4 to control the temperature rise in the reactor, and the feed to the dimerization reactor is heated in a preheater to the desired inlet reaction temperature.

The effluent of the dimerization reactor is not recycled as in a MTBE plant since this could increase the trimers and tetramers production. Also, only one reactor is applied as the reactions are not equilibrium limited as it occurs in the MTBE process.

Furthermore, as the reaction temperature is higher than that used in the MTBE production, the dimerization reactor pressure shall be increased to keep the reaction in liquid phase.

All the reactor effluent is sent to the debutenizer column under pressure control where all the unreacted C4 is fractionated from the isooctene stream and sent to storage.

The isooctene stream leaves the tower as the bottoms product and may be sent to storage to be added to the gasoline pool or to the hydrogenation section to be hydrogenated to isooctane.

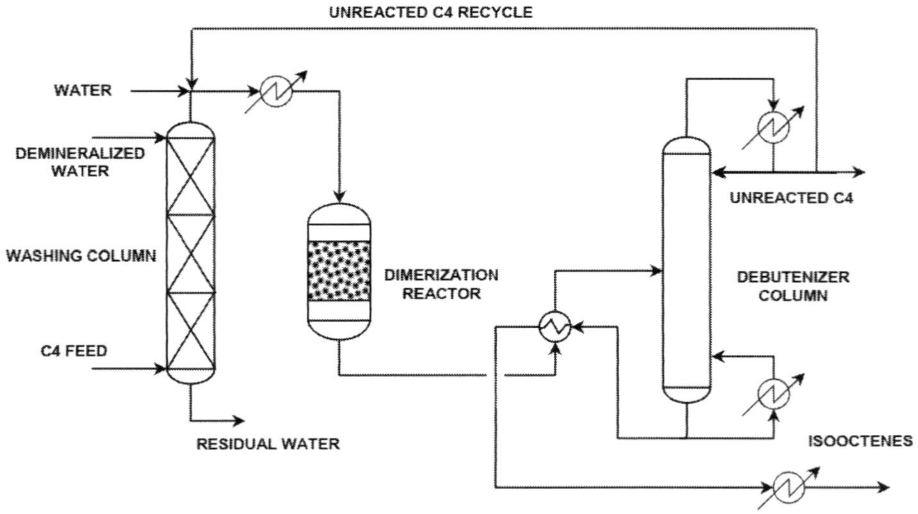

Figure 6 – Dimerization Section

3.2 Hydrogenation section

At the hydrogenation section, isooctene is pumped to the desired pressure and mixed with the recycle gas and with hydrogen make-up. Isooctane is also combined with the feed in order to control the temperature rise of the hydrogenation reactor

In addition to the hydrogenation of the olefins, almost all sulfur compounds in the feed are hydrotreated resulting in an effluent with sulfur content lower than 5 ppm.

The effluent of the reactor is cooled and after separation in a high-pressure drum the vapor phase hydrogen rich gas is recycled to be mixed with the feed and the liquid phase is split in two streams. The first one is recycled to be mixed with the feed and dilute the olefin content and the second one is sent to a stripper tower.

After being heated, the unstabilized isooctane stream is fed to the stripper where hydrogen sulfide and light components are removed from the isooctane product.

The isooctane stream from the bottoms of the stripper tower is pumped to be cooled and finally is sent to storage.

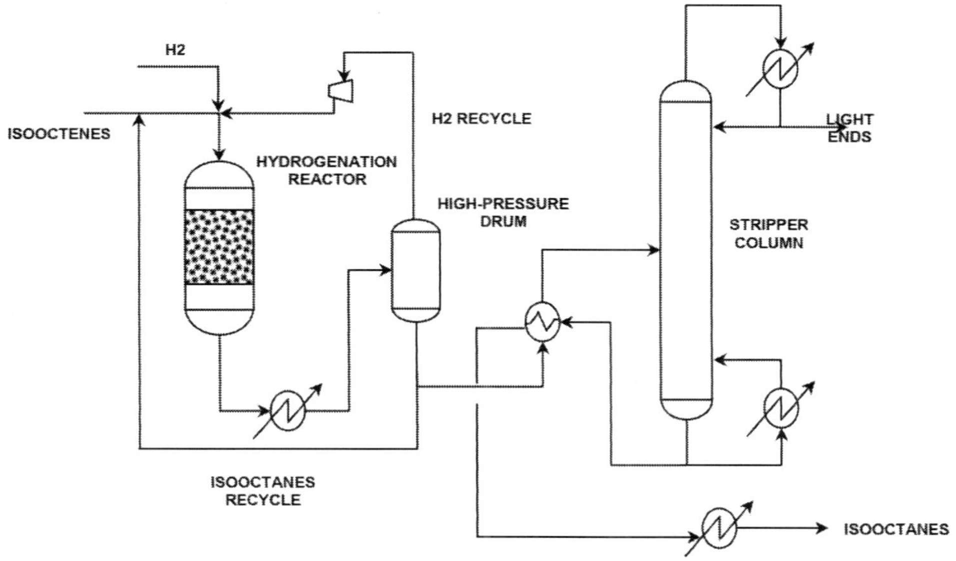

Figure 7 – Hydrogenation Section

4. PROCESS DEVELOPMENT

After a broad review in the available literature, the bench scale work began at Cenpes, Petrobras Research and Development Center, with focus on the dimerization step. The major concerns at that time were the achievable octane number and the stability of the mixture of isooctene and gasoline.

Depending on the stability of the mixture, the hydrogenation step could be disregarded or not, making the adaptation of the MTBE plants more attractive and faster.

An extensive work was developed and dimerization test runs were carried out in bench scale with a variety of isobutene concentrations, different catalysts and operating conditions. Space velocities, reaction temperatures and TBA / isobutene molar ratios were all analyzed establishing a data bank.

From the data bank, a preliminary kinetic model was developed for the dimerization reactions aiming to determine the expected yields, catalyst volume and composition of the dimerization reactor effluent.

Taking advantage of the existing configuration of one of our MTBE plants we decided to make an industrial test run and produce isooctene in that plant. The operating conditions were established based on the data bank and the kinetic model already available.

To perform this first test run previous process simulation and equipment evaluation were required in order to find where the unit could bottleneck and what was the maximum feedstock rate. Also, new operational instructions were developed and discussed with the unit operators.

Finally, minor modifications in the plant were made basically to allow the injection of reaction water for the production of the desired TBA amount and to adapt the instrumentation to the new operating conditions.

Results of the industrial test run were used to adjust the kinetic model accordingly, and the isooctene produced also provided enough feed to the hydrogenation bench scale and pilot plant located at Cenpes.

The isooctene produced in the industrial test run showed a high octane number, good selectivity and an excellent stability as can be seen in table 1. As foreseen in the previous evaluations, conversion was limited as the reaction temperature could not be increased as much as required due to equipment design pressure constraints.

Although the isooctene produced showed an excellent stability, sulfur content in the product was much higher than desired as it concentrated in isooctene which is the heavier product from the debutenizer column. So the hydrogenation section turns out to be required not only to perform the isooctene hydrogenation but also to the product hydrodesulfurizition.

Isooctene specifications	Results
RON	102 min.
MON	86 min.
Gum	1 mg/100 ml max.
RVP	15 kPa max.
Sulfur content	100 - 200 ppm wt
ASTM D-86	
10 % point	92 °C
50 % point	104 °C
90 % point	150 °C
End point	200 °C

Table 1 - Isooctene results from industrial test run

Hydrogenation also results in another advantage as it increases the Motor Octane Number while performing the saturation of the olefins in the isooctene product.

The hydrogenation work focused on the catalyst system screening and process conditions such as reaction temperatures and space velocities. The purposes were to achieve a high conversion with minimum cracking of the isooctene stream and also reduce the sulfur content to less than 5 ppm wt.

It soon became clear that the biggest challenge was how to maximize conversion without loosing yield via product cracking.

Many catalytic systems were studied and as a final result a new catalyst system was selected to perform the hydrogenation step.

From the hydrogenation data bank another kinetic model was set also aiming to determine the required catalyst space velocity, yields and conversion. Based on the kinetic model and pilot work developed the expected performance of the hydrogenation section was established. The main results are presented in Table 2.

Hydrogenation conversion	> 99% wt
Cracking conversion	< 2% wt
Sulfur content	< 5 ppm wt
RON	98-100
MON	94-96
End point	197°C

Table 2 – Results of Hydrogenation Section

Petrobras has finished a conceptual design to conclude the process configuration, the material balance, required modifications in the dimerization section, hydrogenation section equipment, utilities consumption and an estimate of the total investment to produce isooctane.

Polymer fouling and acetonitrile contamination are the main causes of shorter catalyst life in the dimerization section. As Brazilian crudes have a high nitrogen content and with the increasing processing of these crudes in Petrobras refineries, acetonitrile content in C4 FCC streams is increasing along the years making the washing of the C4 feed stream very important to the life of the catalyst.

So, in the conceptual design the replacement of the existing washing towers of the MTBE plants was considered and new high performance towers capable of treating the high acetonitrile content foreseen in the C4 feed were designed. As a consequence investment required in the dimerization section became higher than it could be in a traditional approach without the replacement of those towers.

The hydrogenation section was designed as a complete new section and existing utilities systems and storage facilities will be reused in the isooctane operation.

A new test run is being planned to be performed in another Petrobras MTBE plant with the purpose to explore a broader range of process conditions and also to further test the kinetic model developed.

As soon as the results of a detailed economic evaluation that is being carried on by Petrobras are available, the basic engineering design of the dimerization and hydrogenation sections will start. It is worth to note that much of the work has already been done in the conceptual design making basic engineering design shorter than usual.

5. CONCLUSION

Based on the previous knowledge of the MTBE technology, Petrobras felt confident to develop the in-house isooctane technology.

After a broad review in the literature, an extensive bench work was carried out and different dimerization conditions were studied. As a result, a kinetic model was evolved and an industrial test run was performed permitting the production of isooctene.

Due to the high nitrogen content in the brazilian crudes and in the FCC C4 feed, special care was given to the C4 washing aiming to maximize the acetonitrile removal and extend the dimerization catalyst life.

Special attention was also given to the catalytic system and process conditions employed in the hydrogenation section in order to maximize yield and minimize isooctene cracking.

Although isooctene produced showed an excellent stability when mixed with gasoline, hydrogenation became required in order to reduce the sulfur content in the product and allow its addition in special gasolines with ultra low sulfur content.

After conclusion of an economic evaluation the next steps will be the basic engineering design, revamp of the MTBE plants and construction of the new hydrogenation sections. The start-up is expected to happen in the second semester of 2005.

6. REFERENCES

Rabello, C. R. K. – PETROBRAS Technical Report nº 073/2000;
Rabello, C. R. K.; Wolff, M. S.; Silva, M. – PETROBRAS Internal Report nº 062/2001;
Rabello, C. R. K.; Wolff, M. S.; Silva, M. – PETROBRAS Internal Report nº 009/2002;
Rabello, C. R. K.; Wolff, M. S.; Silva, M. – PETROBRAS Internal Report nº 003/2003;
Hunszinger, P. et al. – Start-up and Operation of the First On-purpose Isooctane Unit at Alberta Envirofuels Inc., Edmonton, Canada – NPRA Annual Meeting 2003, AM-03-43;
Vandersall, M. T. – Review of Isooctane Processes as Replacements for MTBE – 12th Brazilian Congress of Catalysis.

PI-242™ Catalyst: A New, High Activity C_5-C_6 Isomerization Catalyst

S. A. Bradley, R. R. Rosin, J. G. Vassilakis, W. R. Vetter and F. Xu
UOP LLC
Des Plaines, Illinois, USA

Abstract

A new class of light paraffin isomerization catalyst has been developed and commercialized. The formulation is the result of combinatorial chemistry methodology applied to heterogeneous catalysis research. The resulting catalyst has activity significantly higher than zeolitic catalysts or sulfated-zirconia catalysts. This new product is UOP's PI-242 catalyst.

The commercial performance of the PI-242 catalyst will be presented. Possible reaction mechanisms for this higher performance will be discussed. Contaminant tolerance and regenerability will also be discussed.

Introduction

Isomerization of light naphtha streams is a significant contributor to the world's octane pools. Demand for isomerization capacity is expected to grow. Octane loss from lead phase-out or desulfurization of FCC naphtha can be offset by light paraffin isomerization. Light paraffin isomerization can also provide a cost effective solution to benzene management in motor fuels. Many catalysts and technologies currently exist for isomerization of light naphtha.

The current families of isomerization catalysts offer relative advantages and disadvantages. Chlorided alumina offers the highest activity and yield. However, chlorided alumina is water sensitive requiring feed dryers. Chlorided alumina is not regenerable and requires an organic chloride promoter. Zeolitic catalysts are regenerable and relatively contaminant tolerant. However, zeolitic catalysts are the lowest activity catalyst family. Their higher operating temperatures require a fired heater. Sulfated zirconia catalysts have activities significantly higher than zeolitic catalysts but are less active than chlorided alumina catalysts. Sulfated zirconia catalysts are contaminant tolerant and regenerable.

The activities of the three families of isomerization catalysts were measured by pilot plant testing at similar space velocities. The thermodynamics of light paraffin isomerization reactions are such that higher equilibrium product octanes are achieved at lower temperatures. Catalysts which are active at lower temperature will tend to produce higher product octanes. Figure 1 shows the relative activities of the three families of commercial isomerization catalysts. Zeolitic catalysts are active only at temperatures greater than 225°C, achieving the lowest product octane. Sulfated zirconia catalysts are active at temperatures greater than 125°C, achieving product octanes 3 RON greater than zeolitic catalysts. Chlorided alumina catalysts are active at temperatures greater than 100°C, achieving the product octanes 5 RON greater than zeolitic catalysts.

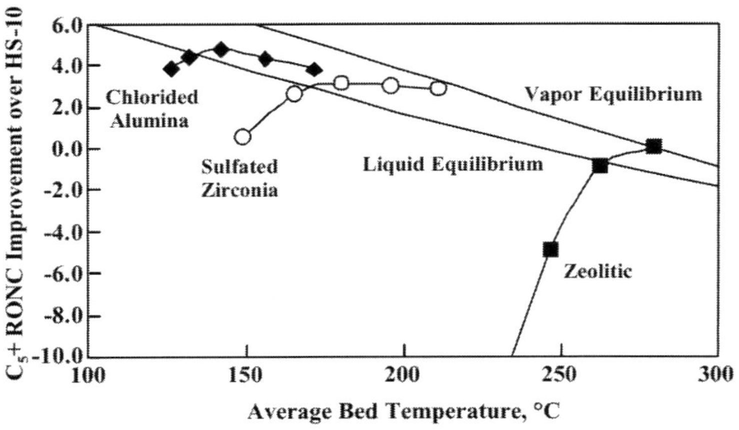

Figure 1

Isomerization Catalyst Activity Comparison

Pilot Plant Testing – Heavy Feed

The goal of this work is to discover a catalyst which combines the high activity of chlorided alumina without need for organic chloride promoter, with the regenerability and water tolerance of zeolitic catalysts. This work has developed a new class of light paraffin isomerization catalyst. This catalyst, PI-242 catalyst, has an activity comparable to chlorided alumina catalysts. The PI-242 catalyst is water tolerant and is regenerable.

Catalyst Development

The development of the PI-242 catalyst utilized a new methodology in catalyst research. The traditional approach of catalyst research has been largely based on the "Edisonian" approach of sample preparation and testing. This approach has been successful because of the similarity between catalyst preparation and testing at the bench scale, semi-works scale, and commercial application. However, the approach of "preparation and test" can be slow and expensive. Screening new catalyst formulations at the bench scale can cost thousands of dollars per formulation. Other faster, less expensive, approaches of catalyst screening have been proposed, such as computational chemistry. These approaches have not been widely successful due to the lack of correlation between the computational methodology and actual application.

Combinatorial chemistry uses the traditional approach of sample preparation and testing. However, the specific methodology of combinatorial chemistry uses a very small sample size combined with high throughput screening and integrated informatics. This allows much faster screening and ultimately smarter formulation selection.

UOP LLC has undertaken the application of combinatorial chemistry techniques to the preparation and screening of catalysts for refining and petrochemical applications. The key elements of combinatorial chemistry for isomerization catalyst research are metals and promoter impregnation, heat treatment, elemental analysis, and catalyst activity screening.

As part of the development of the PI-242 catalyst, a proprietary high throughput screening test for isomerization was developed that correlates very well with large-scale pilot plant testing formats and commercial performance. More than 500 catalyst formulations were screened in five weeks. Using the conventional catalyst development methodology, this experiment plan would have taken more than three years to complete.

CATALYST PERFORMANCE

The pilot plant test for the PI-242 catalyst was formatted to simulate a typical hydrocarbon once-through operation processing a heavy feed. The testing conditions are 450 psig and 1.6 hr-1 WHSV process a 25% x-factor feed. Pilot testing in the hydrocarbon once-through format compared the PI-242 catalyst against commercial samples of zeolitic, sulfated zirconia, and chlorided alumina catalysts. Figure 2 shows the results of the octane temperature comparison. The PI-242 catalyst has an approximate 4 RON advantage over zeolitic catalysts and a 1.5 RON advantage over sulfated zirconia catalysts compared at optimum temperature. The highest RON achieved by the PI-242 catalyst is comparable to the chlorided alumina. Figure 3 compares these same catalysts on an octane yield basis. While the PI-242 catalyst does achieves product octanes comparable to chlorided alumina, the product yield appears in this test to be 3-4 wt-% less than chlorided alumina.

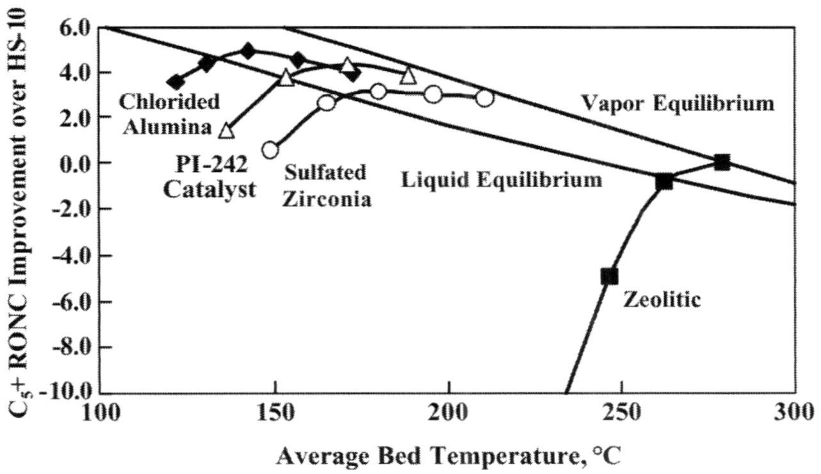

Figure 2

RON vs. Temperature Comparison

Pilot Plant Testing – Heavy Feed

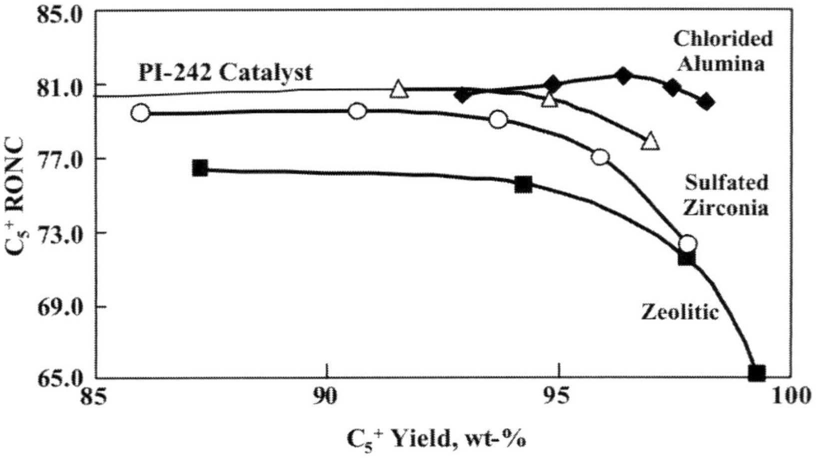

Figure 3

RON vs. C_5^+ Yield Comparison

Pilot Plant Testing – Heavy Feed

The source of the octane achieved using the PI-242 catalyst can be understood by examining mechanisms of the reactions that occur in isomerization reactors. The main reaction of interest is the isomerization of paraffins. Figure 4 shows these reactions of interest with their blending RON values. The saturation of benzene in an isomerization reactor is desired as a means to control benzene in the gasoline pool. The isomerization of cyclohexane to methyl cyclopentane and the subsequent ring opening to *n*-hexane is a very desired reaction in an isomerization reactor, despite its loss in octane. Methyl cyclopentane and, to a lesser extent, cyclohexane, form very stable carbocations which will occupy acid sites on the catalyst and prevent those sites from being utilized for paraffin isomerization. This is why the isomerization of paraffins is decreased as the level of naphthenes increases.

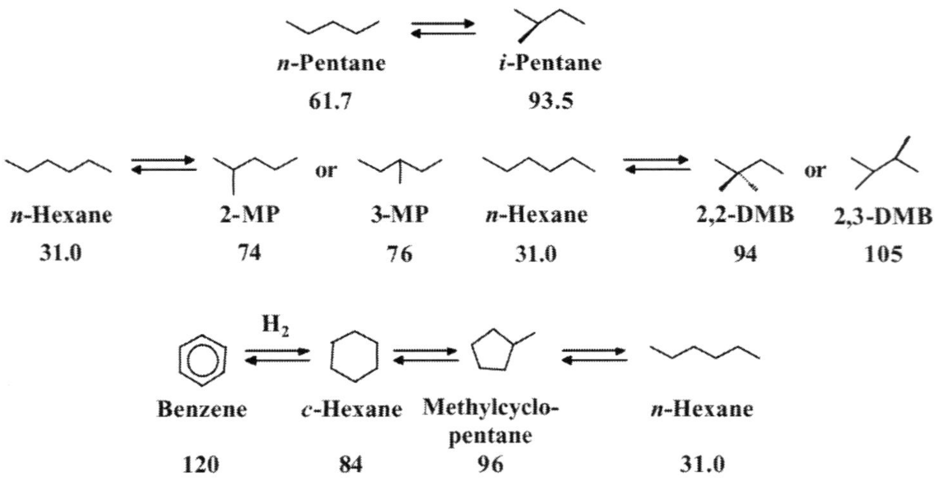

Figure 4

Isomerization Reactions

While 2,3 di-methyl butane (23DMB) has the highest RON value of the C_5 or C_6 paraffins, the equilibrium level of 23DMB is relatively low and so 23DMB is not a significant contributor to the product octane value. Isopentane (i-C_5) and 2,2 dimethyl butane (22DMB) have relatively large RON values and relatively high equilibrium levels and are the major contributors to the product octane value.

Comparisons of i-C_5 and 22DMB formation ratios are shown in Figures 5 and 6. Figure 5 shows that the i-C_5 formation ratios of chlorided alumina, sulfated zirconia, and PI-242 catalyst are very similar: 74-75%. The greatest difference is in the 22DMB formation ratio where the PI-242 catalyst achieves a 22DMB ratio 5 percentage points higher than sulfated zirconia and 3 percentage points higher than chlorided alumina.

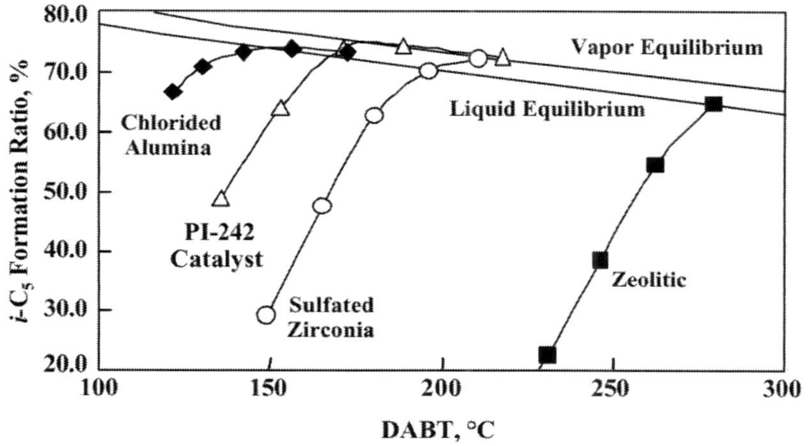

Figure 5

i-C_5^+ Formation Ratio Comparison

Pilot Plant Activities – Heavy Feed

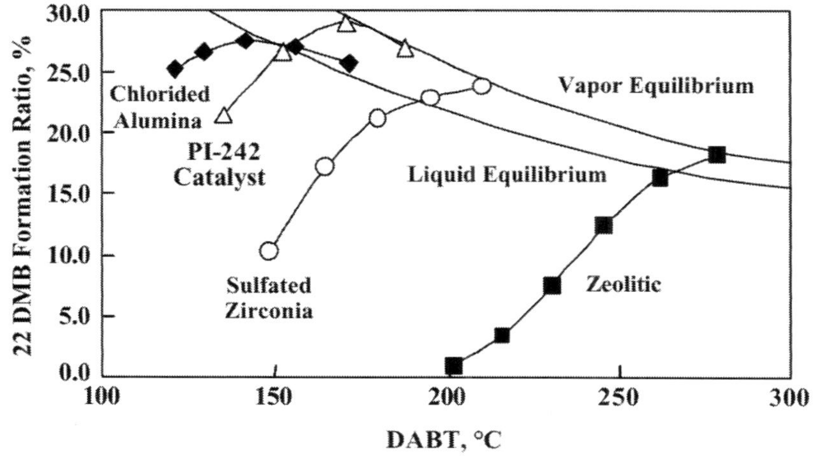

Figure 6

22 DMB Formation Ratio Comparison

Pilot Plant Activities – Heavy Feed

One source of the higher paraffin isomerization activity of the PI-242 catalyst can be understood by examining the ability of the respective catalysts to convert C_6 naphthenes. It is well known that the presence of C_6 naphthenes reduces the paraffin isomerization activity of the catalysts of this study.

Figure 7 compares the ability of chlorided alumina, sulfated zirconia, zeolitic, and the PI-242 catalyst to convert C_6 naphthenes. Due to the complex nature of the feed it is not possible to determine what fraction of the naphthenes reacts to paraffin and what fraction is cracked to C_4- light products. PI-242 catalyst has a C_6 naphthene disappearance significantly higher than chlorided alumina, sulfated zirconia or zeolitic catalysts. This higher C_6 naphthene disappearance is believed to contribute to the higher paraffin isomerization activity of the PI-242 catalyst.

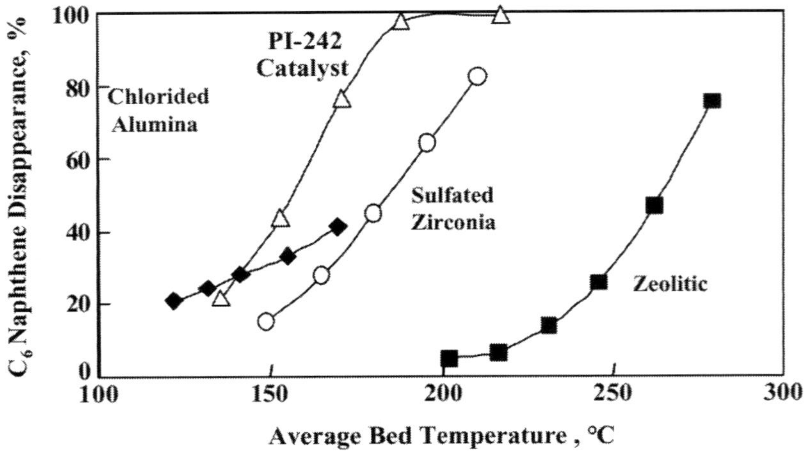

Figure 7

C_6 Naphthene Disappearance

Pilot Plant Activities – Heavy Feed

The PI-242 catalyst has been evaluated for water tolerance. Previous pilot plant testing was conducted with dry feed (~1 wt-ppm H_2O). The PI-242 catalyst was tested at the conditions in Table 1 with dry feed, 10 wt-ppm H_2O was then added to the hydrocarbon feed, the H_2O level was then increased to 30 wt-ppm H_2O, and finally the H_2O level was returned to dry levels. Figure 8 shows the product octanes as a function of the H_2O level. Increasing the H_2O from ~1 wt-ppm to 10 wt-ppm had little impact on the activity of the PI-242 catalyst. Increasing the H_2O level from 10 wt-ppm to 30 wt-ppm had significant impact on the activity at 163°C. However, when the experiment was repeated at 177°C the increase from 10 wt-ppm to 30 wt-ppm H_2O had only slight impact on the activity. For both reaction temperatures, when the feed H_2O level was returned to dry levels, the activity was fully recovered.

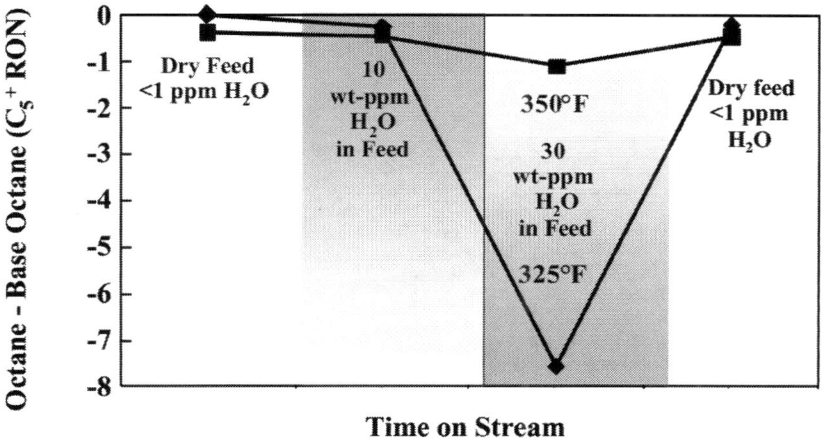

Figure 8

PI-242 Catalyst Water Tolerance

The higher activity and contaminant tolerance of PI-242 catalyst offers many interesting cost effective options for increasing the value of existing light paraffin isomerization units and many other cost effective options for new unit design. These options have been reviewed in detail elsewhere.[1,2]

PI-242 Catalyst Commercial Performance

The PI-242 catalyst has been commercialized. A North American refiner loaded the PI-242 catalyst in September 2003 replacing the LPI-100™ catalyst in an existing hydrocarbon once-through Par-Isom™ isomerization unit. The performance of the fresh PI-242 catalyst is compared against the performance of fresh LPI-100 catalyst in Figure 9. The feedstock currently run in this unit is significantly heavier than the feedstock used when the LPI-100 catalyst was first loaded. Therefore, the comparison in Figure 9 is based on the PIN number upgrade of the product over the feed. PIN number is paraffin isomerization number. It is a measure of the amount of *iso* C_5 and highest octane C_6 paraffins in a feed or product stream. It is calculated as:

***PIN* = *i*-C_5 formation ratio + 22 DMB formation ratio + 23 DMB formation ratio**

The PI-242 catalyst operation resulted in a PIN upgrade 10 PIN numbers larger than operation with the LPI-100 catalyst despite processing a heavier feed.

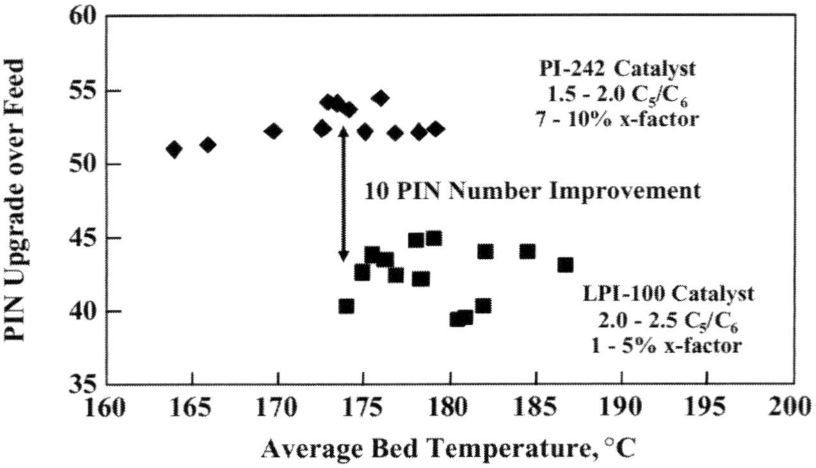

Figure 9

Par-Isom Unit Commercial Operation
LPI-100 Catalyst vs. PI-242 Catalyst

SUMMARY

A new light paraffin isomerization catalyst has been commercialized. The PI-242 catalyst has significantly higher activity than sulfated zirconia or zeolitic catalysts. The product octanes achievable with the PI-242 catalyst are comparable to those product octanes achieved with chlorided alumina catalysts. The source of this higher activity appears to be the PI-242 catalyst's ability to convert more C_6 paraffin into 22DMB and the ability to convert more C_6 naphthene. However, the PI-242 catalyst is less selective than chlorided alumina. The product yields of the PI-242 catalyst are 3-4 wt-% lower than chlorided alumina at similar product octanes.

The PI-242 catalyst is also water tolerant. Activity losses due to continuous high levels of H_2O can be recovered by raising temperature. Activity losses due to water upsets are fully recoverable when dry feed is re-established.

The PI-242 catalyst has been commercialized. Commercial operation has demonstrated a significantly higher performance over sulfated-zirconia catalysts.

REFERENCES

1. G.C. Anderson, M.A. Stine, R.R. Rosin and M. Hunter, "New Solutions for Light Paraffin Isomerization", Proceedings of the NPRA Conference, Mar. 21-23, 2004

2. N.A.Cusher, R.D.Gillespie, G.A.Marr, W. Martin, R.R.Rosin, and P.Palmer, "High Activity C5-C6 Isomerization Catalyst for Lower Capital Solutions", Proceedings of the 4th European Catalyst Technology Conference, February 18-19, 2003

Hydrocracking Chemistry and Economics In a Clean Fuels Environment

Stephen J. McGovern and CK Lee
PetroTech Consultants
3 Bedford Terrace, Mantua, New Jersey 08051

Key Words: Refining, Hydrocracking, Hydroprocessing, Clean Fuels, ULSD, Chemistry, Economics

Prepared for presentation at the AIChE Spring National Meeting 2004
Seventh International Conference on Refinery Processing
April 25-29, 2004, New Orleans, Louisiana

Abstract

Hydrocracking is a complex process that combines the benefits of hydrotreating and catalytic cracking into a single refinery process. This paper will review the reactions that determine the product selectivity and product quality, as well as the impact of process conditions ad catalyst selection on these reactions. In the past, hydrocracking has been characterized as an expensive process; however, this paper will also show why the quality demands of clean gasoline and diesel fuel can make hydrocracking an economically attractive refinery option in an era of declining crude quality

Introduction

Over the past several years, governmental agencies throughout the world have proposed and/or implemented regulations that require a significant reduction in the sulfur content of most transportation fuels. The European Union requires gasoline and diesel fuel to contain less than 50 ppm sulfur, while several European countries give significant tax credits for fuels that contain less than 10 ppm sulfur.

The US regulations for low sulfur gasoline, which begin phasing in this year, will eventually require that the average sulfur content of gasoline sold in the US be less than 30 ppm. The US ultra low sulfur diesel (ULSD) regulations will require that all on-road diesel fuel contain less than 15 ppm sulfur. (The 15 ppm US spec is essentially equal to the 10 ppm European target. The US spec is measured at the pump while the European spec is measured as the product leaves the refinery. To meet the US spec at the pump, refiners expect to produce 10 ppm product leaving the refinery.) Current US proposals for off-road diesel, if approved, will eventually require off-road diesel to also contain less than 15 ppm sulfur with railroad and marine diesel capped at 500 ppm sulfur. Fuels with less than 15 ppm sulfur are often classified as "zero sulfur fuels".

The impetus for these regulations is a reduction in the emissions from gasoline and diesel engines. The lower sulfur fuels allow the use of much more effective post engine catalyst systems to reduce tail pipe emissions. Many other countries are following the lead of the US and Europe and either have or will have similar regulations.

Although many different grades of gasoline are produced in the US, all grades are lumped together to calculate the corporate average sulfur content. The sulfur contents of the different grades of distillate fuels will each be regulated separately. Currently, there are 5 major types of distillate fuels with several sub-types. These major fuel types and their approximate market shares in the US in 2000 are given in Table 1.

Table 1

	2000 US Demand, kBPD	Percent of US Demand	Proposed Sulfur Limit, ppm
Kerosene	1,591	30.6	3000
On-Road Diesel	2,283	43.9	15
Off-Road Diesel	463	8.9	15
Marine and Railroad Diesel	336	6.5	500
Heating Oil	523	10.1	2000-5000

Over 95% of the kerosene sold in the US is used as jet fuel. There are currently no proposed regulations that would reduce the sulfur content of jet fuel. Kerosene that is sold as diesel fuel must meet the new sulfur regulations for diesel fuel. Over 85% of the heating oil sold in the US is sold in PADD 1 (northeastern US). The sulfur content of heating oil sold in the US is regulated by the individual states where it is used, not by the federal government. Current sulfur specifications vary from 2000 ppm to 5000ppm, depending on the region in which the fuel will be used. Although, no states have officially proposed reducing the sulfur specification on all heating oil, some are considering it. If one does, the rest will most likely follow.

The primary transportation fuels: gasoline, diesel and kerojet account for almost 80% of total refinery production in the US, with gasoline being almost 50%. The current growth rate for diesel fuel is higher than that for gasoline; however, because gasoline is growing from a larger base, the additional barrels of gasoline needed each year exceed the additional volume of diesel needed. In Europe, the demand for gasoline is actually declining, while the demand for diesel is growing.

Meeting these new sulfur regulations requires the construction of new refinery process units. Many articles have been published summarizing the results of various studies that attempt to predict the number and type of units that will be required. Essentially all of the sulfur in the gasoline pool comes from the FCC gasoline that is blended into the pool. The US gasoline pool is about 1/3 FCC gasoline, or about 2.5 million barrels per day of FCC gasoline. These gasoline desulfurization units will be relatively low severity, low pressure units and the total volume that will be desulfurized will be less than the total FCC gasoline volume.

Production of ULSD could require significantly more investment. The majority of current on-road diesel production at 500ppm sulfur comes from hydrodesulfurizing various refinery streams. Only a very small volume of LSD derived from very low sulfur crudes is not currently hydrotreated. These HDS units typically operate at 80-99% desulfurization, depending on feed sulfur content. At 500ppm, most of the sulfur compounds remaining in the product are the very difficult to remove dibenzothiophenes.

Desulfurizing a 500ppm diesel fuel to 10 ppm is more difficult than the production of the original 500ppm product because the most refractory sulfur compounds must now be removed.

Prior to the proposed reduction of off-road diesel sulfur content, many refiners included stream segregation as part of their strategy for ULSD production. They intended to desulfurize the easiest distillate components to 10 ppm for ULSD and relegate some of the harder to desulfurize streams such as LCO to off-road production at the current limit of 2000-5000ppm sulfur. This feed segregation strategy allowed significantly lower investment for ULSD production by revamping many existing HDS unit.

The recent proposal to lower the sulfur content of all diesel fuel is causing many refiners to rethink their ULSD strategy. The Table 2 shows the balance between distillate production in 2000 and distillate hydrotreating capacity in the US at the end of 2000.

Table 2

Kerosene & Kerojet, KBPD	Low Sulfur Diesel, kBPD	Other Distillates, kBPD	Total, kBPD	Refinery Distillate HDT Capacity, kBPD
1602	2617	1077	5296	4038 1204 Kero 2834 Dist

Note: LSD imports were less than 130 kBPD

This overall balance shows that almost 25% of the total distillate produced in the US did not go through a dedicated HDS unit. Some of the distillate blending components are by-products from vacuum gas oil or residue hydrotreaters or hydrocrackers, while others are unhydrotreated streams that might only be chemically treated for odor and corrosion control. Since essentially all current LSD is produced from diesel HDS units, it is not surprising that in mid-2003 refiners reported to USEPA that about 75% of ULSD product would come from revamped HDS units not new capacity. That implies that about 2 million barrel per day of distillate HDS capacity will be revamped and almost 1 million BPD of new capacity will be built.

LSD production currently exceeds on-road demand by over 300,000 BPD. This is because some product is downgraded in the distribution system, but also because a significant number of off-road diesels are currently using LSD because of the low price difference between low sulfur and conventional diesel fuel (less than 2 cpg). The new off-road diesel regulations will require the construction of about 500,000 BPD of additional distillate hydrotreating capacity. Feed segregation is no longer as attractive with the new regulations, so more severe HDS units will most likely be built to handle the more difficult distillate streams.

Hydrocracking is a technology option that is receiving consideration as a viable option for producing lower sulfur distillate fuels. When a refiner must install additional processing capacity, it is prudent to evaluate the entire refinery to determine if the refinery would be more profitable by shifting services of existing units and adding a different unit than originally anticipated.

Hydrocracking has been considered a rather expensive process; however, the new sulfur regulations make it a more attractive alternative. This paper will review the basic hydrocracking chemistry and its relevance in the clean fuels environment. It will also discuss the impact of process chemistry and other technology improvements on the economics of hydrocracking.

Process Chemistry

Hydrocracking, like catalytic cracking or FCC, is normally used to convert atmospheric and vacuum distillate streams with boiling points between 400 and 1100F (200-600C) into lighter products. These streams are composed of paraffins (normal and iso alkanes), naphthenes (single and multi-ring alkanes), single and condensed multi-ring aromatics as well as sulfur and nitrogen containing homologues of the hydrocarbons. Coker and FCC products can also be hydrocracked. In addition to the previously mentioned hydrocarbon types, these streams will also contain olefins. In general, the aromatic content and the number of rings in each molecule increase with the boiling range of the stream.

These various hydrocarbon types can undergo a number of reactions in a hydrocracker. Many authors have proposed reactions mechanism to explain the observed product distributions. A discussion of the detailed reaction mechanisms, however, is beyond the scope of this paper. We will limit our discussion to the overall observations of these works and their importance in the hydrocracking process.

The primary cracking catalyst in both hydrocracking and FCC is usually a Y-type zeolite although other zeolites and amorphous solid acids such as silica-alumina are also used. These are all solid acids, which contain both Bronsted and Lewis type acidity. The primary difference between hydrocracking and FCC catalysts (other than particle size) is that the hydrocracking catalyst also contains a strong hydrogenation component in addition to the cracking component. This hydrogenation component can either be a noble metal such as Platinum or Palladium or a mixed sulfide such as Nickel or Cobalt promoted Molybdenum or Tungsten.

The cracking reactions that occur in hydrocracking are essentially the same as those that occur in FCC, except that the presence of the hydrogenation component allows additional classes of reactions to occur and can affect the relative reaction rates. The different operating conditions of the two processes also cause different thermodynamic driving forces. The high temperature, low pressure FCC process favors the formation of olefins and aromatics, while the lower temperatures and higher hydrogen pressure of hydrocracking favor the saturation of olefins and aromatics.

Paraffin Conversion

The paraffins can either isomerize or crack to form lower molecular weight fragments. The rate of cracking of C7 and larger molecules is significantly higher than their rate of isomerization or the rate of cracking of C6 and lighter paraffins. The feed iso paraffins can directly crack to form smaller fragments, while the normal paraffins must first be isomerized before they can crack.

Butane is the smallest cracked fragment that is produced in appreciable quantities. Essentially no methane or ethane is produced and only a small amount of propane. The other cracked fragments are uniformly distributed among those that are possible, e.g. a C16 would produce roughly equal molar amounts of C4 through C12 products. The C7 and larger products can continue to crack. Because the rate of cracking is faster than isomerization and the normal paraffins crack very slowly, the C4, C5 and C6 paraffins that are ultimately produced have an iso/normal that significantly exceeds thermodynamic equilibrium.

In pure component studies, paraffins are very reactive, however in mixtures with aromatics or naphthenes their reactivity is suppressed by the preferential adsorption of the other species. The adsorption effect is so strong that paraffins are the least reactive species in mixtures containing aromatics or naphthenes.

Naphthene (Cycloparaffin) Conversion

Naphthenes undergo essentially the same reactions as paraffins; however, the presence of the ring structure introduces additional reactions and affects the selectivity of the cracking reactions. Dealkylation, the removal of an alkyl side chain to form a smaller paraffin and lower molecular weight naphthene is favored over opening of the naphthenic ring. Again, butane is essentially the smallest cracked fragment that will be produced.

The preference for dealkylation, coupled with the constraint of making nothing smaller than butane causes an interesting class of reactions known as the paring reactions. When a polymethyl substituted naphthene is hydrocracked, the ring does not open, as one would expect. The molecule instead isomerizes to form a side chain with at least 4 carbons. The side chain is then removed via dealkylation. The ring stability causes C9 and smaller naphthenes to have very low cracking rates, while larger naphthenes readily crack.

Naphthenes can also be dehydrogenated to form aromatics; however, hydrocracking operating conditions are usually set to favor the saturation of aromatics to naphthenes.

Multi-ring naphthenes react the same as single ring molecules with dealkylation being favored. They are more strongly adsorbed than the single ring naphthenes so they are more reactive.

Aromatic Conversion

The aromatic ring is very stable, so aromatics can only be dealkylated or saturated to form naphthenes, which then undergo the reactions discussed above. Aromatics, however, are very strongly adsorbed on the catalyst. This strong adsorption makes aromatic dealkylation the first reaction that occurs in a mixed feed. The dealkylation normally occurs between the aromatic ring and the side chain, leaving a bare ring aromatic and a paraffin with a chain length equal to the original side chain length.

As with paraffin and naphthene cracking, C3 and lighter fragments are hard to form, so C9 and lighter aromatics are very difficult to crack. The paring reaction also occurs with aromatics, so even poly-methyl aromatics can crack to form butane and benzene.

Aromatics also readily undergo transalkylation reactions at typical hydrocracking conditions. Transalkylation is the transfer of side chains among aromatic molecules. These reactions are very fast and result in very low net benzene yields from hydrocracking. When benzene is formed from dealkylation of heavier alkyl benzenes. The benzene is quickly transalkylated by other aromatics with short side chains (methyl and ethyl side chains) to primarily form toluene, xylenes and ethylbenzene.

Multi-ring aromatics are interesting reactants. Several authors have studied the reactivity of multi-ring molecules, without side chains; however, very little work has been done with multi-ring alkyl aromatics, either alone or in simple mixtures. Without side chains, the molecules undergo partial hydrogenation followed by dealkylation of naphthenic ring from the remaining aromatic core. Without alkyl side chains, the final products of the reaction of naphthalene for example are benzene and butane. The reaction of light LCO, however, which contains primarily methyl substituted naphthalenes, gives primarily C7 plus naphthenes and aromatics and C5 plus paraffins. These yields indicate the transalkylation reactions are important and also that a methyl substituted naphthenic ring on a partially saturated multi-ring aromatic, will be completely removed from the aromatic core.

Multi-ring aromatic molecules can also participate in ring growth reaction under hydrocracking conditions. Both pure component laboratory studies and commercial hydrocracking data show that aromatic molecules with a greater number of rings than are present in any feed molecule are formed in hydrocracking. The rate of formation of these molecules is low, but in recycle hydrocrackers, the recycle stream must be purged to keep them from building to a high level and causing rapid catalyst deactivation.

Shape Selectivity

The breaking of carbon-carbon bonds in the cracking reactions occurs on the acid sites of the catalyst. With zeolite-based catalysts, most of these acid sites are within the pore structure of the zeolite crystals. Essentially all paraffin molecules can fit within the pore structure of Y-zeolites. The larger multi-ring aromatics and naphthenes, however, have difficulty fitting in the pores of the Y-zeolite, so their diffusion rates are lower.

In the absence of diffusional limitations, the rate of cracking or dealkylation normally increases with the molecular weight of the molecule. With Y-zeolites, diffusional resistances become noticeable for molecules containing four or more rings. These molecules still crack, but at a lower than expected rate. Ultrastable Y-zeolites, with their smaller effective crystallite size, are usually more active than rare earth Y-zeolite.

Smaller pore zeolites, such as ZSM-5, can change the relative rates of reaction of molecular classes by taking advantage of these diffusional effects. The pore dimensions of ZSM-5 are such that normal paraffins have a much higher diffusivity than any other molecular class, including iso-paraffins. This property has been advantageous in several commercial processes. Normal paraffins in the gasoline range have very low octanes, while those in the diesel range have very poor low temperature fluidity properties. These normal paraffins can be cracked from the parent stream with very high selectivity, leaving behind a stream with improved properties at the expense of a small loss in yield.

Catalyst Deactivation

Hydrocracking catalysts can deactivate in two ways. The acid sites can adsorb basic molecules that are strongly bound to the site, eliminating its acidity and cracking activity. This is called temporary poisoning, because the basic molecule can desorb, restoring the original catalytic activity. Permanent deactivation requires catalyst replacement or regeneration.

Temporary

The most common temporary poisons are the nitrogen containing molecules in the hydrocracker feed. There are several different types of nitrogen containing molecules and poisoning strength is a function of type, size and structure of the nitrogen molecules as well as the operating conditions of the hydrocracker.

Basicity increases in the following order:

Pyrrole < Pyridine < Amine

Pyrrole and Pyridine are aromatic in nature and much less basic than the amines and are the most common types of nitrogen compounds found in petroleum. They can be converted into amines, however, by saturation of the nitrogen-containing ring.

Amines are up to 10 orders of magnitude more basic than pyrrole. At low temperatures, amines are much more strongly adsorbed by the acid catalyst and will quantitatively displace less basic compounds. At reaction temperatures, however, the nitrogen-carbon bond of the amine can be broken releasing a hydrocarbon and leaving an ammonia molecule on the acid site. Ammonia is also very basic, but because it has a low boiling point, it will desorb at temperatures around 700F. At temperatures above about 700F, ammonia is only a weak poison.

The structure and size of the nitrogen compound can also decrease its poisoning strength. Nitrogen atoms that are shielded by other rings or alkyl side chains will be much weaker poisons. Also, if a molecule is too large to fit inside the pores of the zeolite, it cannot poison the acid sites.

Permanent

Permanent deactivation primarily occurs because of coke deposition on the catalyst, although feed metals can also permanently deactivate the catalyst. Coking primarily occurs via aromatic ring growth, which is the addition of additional rings to a multi-ring aromatic to form a larger molecule. Anything that reduces aromatic ring growth will decrease the rate of catalyst deactivation. These include:

- Increasing hydrogen partial pressure

- Increasing the purge rate from the recycle liquid
- Decreasing the aromatic concentration in the reactor.
- Decreasing reactor temperature.
- Increasing the ratio of hydrogenation to cracking activity

Hydrocracking Economics

Hydrocracking has often been characterized as an expensive process with high capital and operating costs. In the past, this might have been true, because the hydrocracker products were essentially sulfur free and low in aromatics. A US refiner with large hydrocracking capacity often produced finished products that exceeded specifications.

The new fuel quality regulations, however, have eliminated this quality give away. In fact, some of the new diesel hydrotreaters that are being designed to produce ULSD are just as costly to build and operate as a hydrocracker and give lower product uplift. Hydrocrackers have often been evaluated as an alternative to catalytic cracking. In today's environment, they should be viewed more as a complement to catalytic cracking.

Outside the US, they are often added in a refinery expansion project to provide additional vacuum gas oil conversion capacity, but with a higher D/G and better quality products. Most hydrocrackers within the US were designed for a gasoline economy. The primary objective of these units was to convert distillate to gasoline.

There are three different aspects of a potential refinery project that impact the overall economics of the project. These are the capital investment that is required, the operating costs for the unit once it is built and the increase in value between the feed and products.

Capital Investment

There are two basic types of hydrocrackers: two stage, liquid recycle hydrocrackers that operate at high pressure and near 100% feed conversion and single stage, partial conversion (mild) hydrocrackers that operate at lower conversion and without product recycle. Most of the hydrocrackers that are operating in the US are the high pressure, two stage, high conversion type. Many hydrocrackers outside the US are mild hydrocrackers.

Very few grassroots high severity hydrocrackers have been built in the last 15 years. Most new hydrocrackers are mild hydrocrackers. Curve type cost estimating programs usually give a much lower capital cost for a mild hydrocracker than for a high severity hydrocracker. This cost difference usually means that the decision between hydrocracker types is made early in the project evaluation process, based on erroneous information.

The distinction between hydrocracker types is artificial. Hydrocrackers can be designed for a very wide range of conversions (10 to almost 100%), gasoline to distillate selectivities and product qualities. The equipment (and hence the capital investment that is required) is a function of the desired conversion, selectivity, product quality and feed quality, not the hydrocracker type.

The design pressure for the unit is set by the desired combination of conversion, product quality and feed quality. Higher pressures being required to maintain acceptable cycle length for higher conversions, poorer feed, or to provide additional aromatic saturation or desulfurization to improve product quality.

Liquid recycle is usually required to avoid overcracking and excess LPG production for conversions above 60% or if an unusually high D/G ratio is specified. The choice between a single stage and two-stage unit is purely an economic decision. Essentially any project targets can be met in single stage unit. A two-stage unit, however, is sometimes more cost effect for large, high conversion hydrocrackers. The second stage is essentially free of H2S and ammonia, so the effective catalyst activity is much higher. The higher activity allows the second stage to operate at lower pressure and higher space velocity. Sometimes, these cost savings offset the added investment in the additional equipment required to design and build a second reactor loop. A second stage is usually only justified in larger units, where multiple reactors, compressor and exchangers are already required.

Initial curve based, estimated capital investment for a "hydrocracker" is usually higher than for a mild hydrocracker, ever if both are designed for the same feed, conversion and product quality. Curve type estimates usually start with an historic unit total cost and than scale it for throughput and escalate it for inflation. Very few high severity hydrocrackers have been built in the last 20 years. During this time period, major advancements have been made in catalyst activity, equipment metallurgy and reactor and compressor design which have significantly lowered the cost of building a hydrocracker. Some but not all of these advancements are part of the newer mild hydrocracker designs, giving them a lower cost basis.

Hydrotreating catalyst activity has increased by almost an order of magnitude, while better metallurgy, fabrication and erection techniques have lowered the cost of building and erecting high-pressure reactors. Improvements in reactor internals have both increased the catalyst utilization and reduced the amount of treat gas and quench gas that are required for stable reactor performance. Larger, more efficient compressors are also available. The availability of high purity hydrogen has also reduced compression costs.

The cost of a hydroprocessing unit has usually been assumed to increase with the design pressure of the unit. When the units are designed to meet the same processing object, this is not necessarily true. Figure 1 shows that the total compression costs (make-up plus recycle) increase slowly with design pressure, because the volumetric flows decrease and the compression ratio increases slowly. High-pressure piping costs actually decrease because the impact of pressure on volumetric flow rate and pipe diameter is stronger than its impact on wall thickness.

Figure 1
Relative Compression and Piping Costs

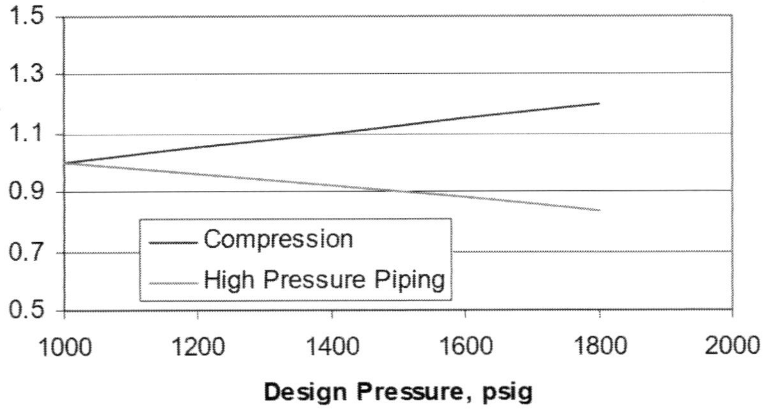

High pressure reactor costs usually scale with the overall weight of the reactors. Figure 2 shows the impact of pressure and LHSV on total reactor weights. Denitrogenation and aromatic saturation rates are strong functions of pressure. Increasing design pressure lowers the required catalyst volume, and can result in a reduction in the cost of the unit. This interaction must be evaluated in any design.

Figure 2

Effect of LHSV and Pressure on Reactor Weight

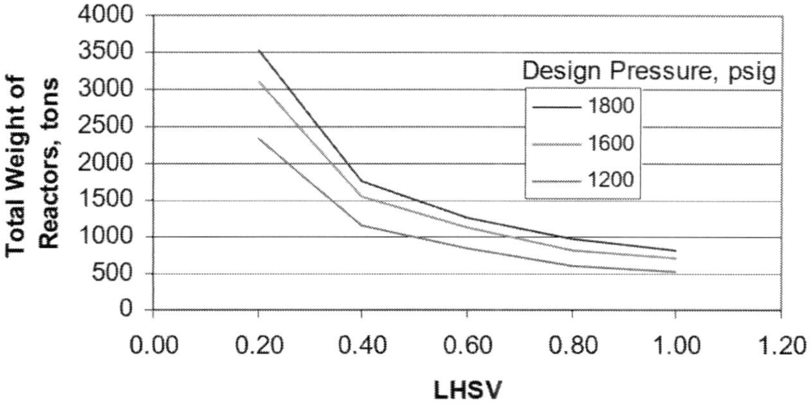

Product Value Improvement

Hydrocracking of the most difficult to desulfurize diesel components can be more profitable than just desulfurizing them. The acidic component of the hydrocracking catalyst increases the reaction rate for hindered dibenzothiophenes, while cracking of the polynuclear aromatics (PNA) gives a larger diesel cetane gain and greater volume expansion, with lower net hydrogen consumption, than saturating them.

Table 3 shows a theoretical calculation for the economic difference in potential uplift between saturating and hydrocracking a typical PNA. These economics are based on diesel at $30/bbl and hydrogen produced from natural gas a $4/MMSCF. Higher product or natural gas prices will increase the difference. The revenues also do not include differences in product quality, which would also favor hydrocracking.

Table 3

	Volume Swell, %	Hydrogen Consumption, SCFB	Revenue, $/BBL
Saturation	22	4300	-0.4
Hydrocracking	40	3440	6.4

The volume swell and hydrogen consumption values in Table 3 are net values assuming that the naphtha produced from the hydrocracker will be reformed. This revenue difference would more than justify the incremental cost of a hydrocracker over a heavy diesel hydrotreater.

Hydrocracking heavy diesel does reduce available diesel volume, however, this volume can be recovered by shifting cut points between naphtha and kerosene on the crude unit. Reducing crude unit naphtha end point provides the reformer capacity that is required to process the additional naphtha made on the hydrocracker and maintains the desired G/D. The hydrocracker naphtha is usually better reformer feed than the virgin naphtha it is replacing so the reformer yields more octane barrels. The extra kerosene that is produced is easier to desulfurize and is a better diesel blending component than the heavy diesel that was cracked. Existing diesel hydrotreaters can easily be modified to produce ULSD from this material.

Desulfurizing diesel to less than 10 ppm sulfur produces product nitrogen levels that are compatible with hydrocracking. Converting an ULSD hydrotreater to a partial conversion hydrocracker requires very little additional capital investment. The majority of the additional capital is for product fractionation to recover the naphtha that is produced. If more existing equipment can be utilized to produce ULSD by hydrocracking the worst components in a smaller new unit, then overall investment for a hydrocracker based ULSD project could be lower than one based only on hydrotreating.

HYDROCRACKING PROCESS OVERVIEW

R.K. Hoehn

UOP LLC
25 East Algonquin Road
Des Plaines, IL 60017

Key Words: Refining, Hydrocracking, Hydroprocessing, Flow Scheme

Prepared for presentation at the AIChE Spring National Meeting 2004
Seventh International Conference on Refinery Processing
April 25-29, 2004, New Orleans, Louisiana

Introduction

Hydrocracking is a versatile process that is capable of meeting a broad range of processing objectives. A recent survey included 159 refineries as containing hydrocracking units[1]. This is due to the flexibility of hydrocracking chemistry via a wide variety of catalysts which allows it to be applied to a variety of processing objectives. The hydrocracking process can utilize a broad range of feed stocks, including straight run and vacuum gas oils, coker gas oils, deasphalted oils, FCC cycle oils, thermally cracked stocks, even straight run and cracked naphthas, producing products that range from LPG, motor gasoline range material, high quality kerosene, ultra low sulfur diesel (ULSD), FCC feed, lube oil base stocks and ethylene cracker feed. This versatility is evidenced by the many different process configurations that can be applied to meet the specific processing objectives. Most major suppliers of hydrocracking technology offer several flow schemes which are suited to meeting different processing objectives. All of these configurations, however, must satisfy the same issues of feed distribution, temperature and pressure control, heat integration, product quality control and overall safety.

Basic Requirements

The flow scheme for any hydrocracking process always consists of the following basic steps[2]:

- Compression of the liquid hydrocarbon feed and hydrogen make-up gas to process operating pressure,

- Heating of the feed and recycle gas to reaction temperature, usually accomplished by heat exchange with reaction products followed by furnace heating,
- Reaction under suitable conditions of temperature and pressure,
- Cooling of the reaction products,
- Separation and recycle of hydrogen,
- Stabilization and fractionation of the liquid products.

The selection of the flow scheme is dependent on the extent of conversion, recycle of unconverted liquids, number of reactors and arrangement of reaction steps, as well as variations in reactor technology and in the gas separation and product fractionation flow schemes. This paper will focus on the most common process configurations for a hydrocracking unit in a modern refinery flow scheme, namely that of producing high quality distillates, such as kerosene and ULSD from various straight run and cracked gas oils.

Reactor Section

Once-Through Flow Scheme

The simplest flow scheme is the once-through operation shown in Figure 1[2]. Fresh feed is combined with recycle gas and the mixture is heated to reaction temperature by heat exchange and fired heat. The material passes over a number of reactor beds containing one or more different types of catalysts. Reactor effluent material is cooled first by using process exchange followed by air cooling. Water is injected into the cooled stream at the inlet of the air cooler to dissolve ammonium bisulfide which is formed as a result of removal of organic nitrogen and sulfur contained in the feedstock. The vapor and liquids are separated in the High Pressure Separator. The separator vapor is recycled to re-combine with the fresh feed. Makeup hydrogen is added to maintain the pressure, since hydrotreating and hydrocracking reactions consume hydrogen. A portion of the recycle gas is used as quench to limit the temperature rise in the reactors to the desired level. Separator liquid is let down in pressure and the vapor produced from the pressure let down is vented. Venting of this vapor enhances the recovery of LPG and light naphtha in the fractionation section, by reducing the hydrogen in the feed to the fractionation section. The liquid is then sent to the fractionation section where the various products are separated. This flow scheme does not incorporate recycle of unconverted liquid, and hence results in partial conversion of the feed to the desired products, in which at least one of the products is in the same boiling range as the feed. The unconverted oil has a higher hydrogen content than the feed and will be essentially free of sulfur and nitrogen. As such, it is an excellent material for further processing in other units. This flow scheme is therefore suitable for upgrading heavy oils for use as FCC feed, as lube oil base stocks, or ethylene cracker feed, but it does not maximize the yield of distillate products. Therefore this flow scheme would be utilized where distillate product is but one component of the overall economics setting the design of the unit. Typical conversion levels seen in once-through units range from 30% to 90% of the feedstock to distillate and lighter products.

This flow scheme is not suitable for feed stocks having a very high endpoint. This is because high endpoint feedstocks tend to have densities that are close to that of water, making the separation of reactor effluent and water difficult. In addition, feeds containing asphaltenes can cause an emulsion with the wash water in the separator which can cause operational problems. For these kinds of feeds, a flow scheme similar to that of Figure 2, but without the recycle liquid stream, would be utilized. This flow scheme incorporates a Hot Separator to remove the heavier liquid from the reactor effluent material so that the heavier components that would cause water separation problems are not contacted with water.

Table 1[3] illustrates the use of once-through hydrocracking at two different conversion levels to process a mixture of heavy vacuum gas oil and deasphalted oil to produce FCC feedstock. As the conversion is increased, the yields and properties of the diesel and lighter fractions increase. The unconverted material heavier than diesel has a high hydrogen content and essentially no sulfur or nitrogen, making it excellent FCC feedstock. The yield of FCC gasoline also increases rapidly with hydrocracking conversion.

Table 1. Once-through hydrocracking of a gas oil-deasphalted oil blend to FCC Feedstock

Feedstock		
Boiling Range, °F		
10%	763	
50%	1033	
FBP	1106	
Recovery, %	68	
API Gravity	26	
Nitrogen, wt. ppm	900	
Sulfur, wt. %	0.3	
Operating Conditions	**Moderate Conversion**	**High Conversion**
Conversion, vol.%	61	77
H_2 partial press., psia	2150	2150
Cracking Catalyst Temp., °F	Base	Base + 15°F
Product yields, vol. %		
Lt. Naphtha	13.3	24.1
Hvy Naphtha (185-320°F)	21.1	19.4
Diesel fuel (500-662°F)	9.2	6.9
FCC feed (662°F+)	39.1	23.2
Hydrogen consumption, std ft^3/bbl	1122	1380
Product Properties		
Light Naphtha		
API Gravity	85	85
RON, clear	80	81
Heavy Naphtha		
API Gravity	60	61
P/N/A	47/49/4	47/49/4
Kerosene		
API Gravity	45	46
Smoke point, mm	28	30
Diesel fuel cetane no.	65	66
FCC Feedstock		
API Gravity	32	31
Nitrogen, wt. ppm	<5	-
Aniline point, °F	284	-

Single-Stage Flow Scheme

The single-stage flow scheme illustrated in Figure 2[2] incorporates recycle of heavy unconverted liquids and hence makes possible complete conversion of the feed to distillate or lighter products. The fresh feed and recycle liquid are combined with recycle gas and heated to reaction temperature using heat exchange and fired heat. The material then passes over a number of reactor beds in one or more reactors, and each reactor may contain one or more beds of catalyst with intermediate quench and redistribution zones to control the temperature rise.

A typical arrangement would consist of a number of beds of hydrotreating catalyst followed by hydrocracking catalyst. The number of beds and quantity of each type of catalyst would be determined by the feed qualities and conversion level. Usually about 40 to 60% of the feedstock is converted into lower boiling products per pass. The reactor effluent material is partially cooled and then the vapor and liquid are separated in a Hot Separator. The Hot Separator vapor is then further cooled and the vapor and liquid are separated in the Cold Separator. The Cold Separator vapor is the recycle gas, which is compressed and re-combined with the fresh feed and recycle oil. Some of the recycle gas is also used as quench for reactor bed temperature control. The Hot Separator liquid is let down in pressure and the flashed vapors and liquid are separated in the Hot Flash Drum. The Hot Flash Drum liquid is sent to the product separation section. The Cold Separator liquid is let down in pressure, mixes with the Hot Flash Drum vapor and enters the Cold Flash Drum. Vapor from the Cold Flash Drum is vented and the Cold Flash Drum liquid is sent to the product recovery section. Venting of vapors from the flash drums enhances the recovery of light materials in the fractionation, particularly LPG and light naphtha.

Although recycle of unconverted liquid from the product separation section permits full conversion, it introduces additional problems in design and operation of the hydrocracking unit. Refractory compounds, called polynuclear aromatic compounds (PNAs) tend to accumulate in the liquid recycle and can contribute to coking and deactivation of the catalyst. Polynuclear aromatic compounds are polycyclic, condensed aromatic hydrocarbons that contain two or more fused aromatic rings[4]. These compounds are present to a certain extent in the feed, but are also formed as a byproduct of the hydrocracking reactions. PNAs in the reactor feed are preferentially adsorbed onto the catalyst as coke. PNAs present in the reactor effluent are non-volatile and have very low solubility. As their concentration in the recycle oil increases, PNAs will tend to coat the inside of equipment operating below about 350°F. This phenomenon occurs even at fairly low concentrations of PNAs. This coating becomes most evident as fouling of heat transfer surface in exchangers operating below the solubility limit, particularly the air-cooled reactor products condenser. If PNA is allowed to continue to build up, it will eventually be necessary to shutdown the unit and mechanically clean the exchanger.

The most common technique for preventing the fouling tendency due to PNA buildup is to utilize a Hot Separator to separate the reactor effluent vapor and liquid material at elevated temperature. The Hot Separator liquid contains virtually all of the PNA containing material and does not get cooled to a temperature at which it would begin coating lines and equipment. While the Hot Separator prevents the fouling action of the PNA material, it does not prevent the catalyst deactivation due to recycle of PNA compounds. The PNA concentration in the recycle oil can only be reduced by fractionation or adsorption. Minimizing the PNA concentration in the recycle oil reduces the impact of PNAs on catalyst deactivation. Adsorption using a suitable activated carbon is the most common method for eliminating PNA material from the recycle oil.

As mentioned in the description of the once-through flow scheme, a Hot Separator is also necessary for feeds that have high endpoints and contain asphaltenes. For units designed without a Hot Separator, but where the refiner desires to maximize conversion, the unit is operated at as high a level of conversion as possible and the unconverted material is tested for the presence of certain compounds known to be characteristic of PNA. Maintaining the concentration of PNA material below a certain limit sets an achievable conversion for the unit. According to the fresh feed characteristics, this may be from 90-95%. With the proper selection of catalysts and operating conditions, the single-stage hydrocracking process can produce a wide variety of products from the same feed. Table 2[4] shows how the product distribution can vary just by changing the cracking catalyst temperature to maximize naphtha, kerosene, or diesel.

Table 2. Single-Stage Hydrocracking Process Flexibility

Feedstock	Vacuum Gas Oil		
Boiling range	645-1027°F		
API gravity	21.09		
Sulfur, wt. %	2.6		
Nitrogen, wt. ppm	920		
Product Objective	**Naphtha**	**Jet Fuel**	**Diesel**
Yields, vol % FF			
C_4	11	8	7
C_5-180°F	25	18	16
Naphtha	90	29	21
Jet	--	69	15
Diesel	--	--	62
Total C_4+	126	124	121
Reactor temp, °F	Base	Base - 11°F	Base – 22°F
Product Quality			
C_5-180°F, RONC	79	79	80
Naphtha			
RONC	41	63	67
Endpoint, °F	420	250	244
Jet Fuel			
Smoke pt., mm	--	34	--
Diesel Fuel			
Cetane no.	--	55	--

Two-Stage Flow Scheme

A two-stage unit as shown in Figure 3[2] consists of a Once-Through flow scheme configuration, which comprises the first stage to process the fresh feed, and a second stage to process the unconverted oil from the first stage. There is usually a common recycle gas loop for both the first and second stage, which minimizes the capital cost, but the unit can be designed to have independent recycle gas loops. The separation of reactor effluent vapor and liquid after cooling can either utilize common facilities or can use separators and flash drums for each stage, depending on the throughput and processing objectives. Because the second stage processes only the unconverted material, only cracking catalyst is used, instead of the usual combination of hydrotreating and cracking catalysts used in the first stage. However, the second stage is also susceptible to the same problems caused by to the formation and buildup of PNA material as is the case for a single stage unit. To maximize conversion, some form of PNA control is necessary.

A two-stage unit is attractive for high throughput designs because by processing the recycle liquid in its own reactor section, the equipment size limitations are not reached until much higher fresh feed rates than would be the case in a single-stage unit. A two-stage unit would be expected to produce yields and have similar product flexibility as that of a single-stage flow scheme. However, a two-stage unit can have a higher capital cost, due to the fact that more equipment is involved.

Separate Hydrotreat Flow Scheme

A Separate Hydrotreat flow scheme as shown in Figure 4[2] is similar to a two-stage flow scheme with separate recycle gas loops for each stage, except that the fresh feed is processed over hydrotreating catalyst only. Cracking catalyst is used only in the second stage. Separate hydrotreat takes advantage of the fact that in some cases it may be desirable to use different recycle gas systems for the hydrotreating and hydrocracking reactors. The single-loop gas recycle schemes illustrated so far do not allow removal of hydrogen sulfide and ammonia between the hydrotreating and hydrocracking stages, and hence require that the hydrocracking stage operate under conditions of high concentrations of these contaminants. Hydrogen sulfide and ammonia are formed when sulfur and nitrogen compounds present in the fresh feed are removed by the treating catalyst. If two separate recycle gas loops are employed, operation of the hydrocracking reactor using catalysts that are sensitive to high levels of hydrogen sulfide and ammonia is possible. This flow scheme may also be used for feeds that contain large amounts of product-range material. This is because the product-range material is removed in the fractionation section before the material is sent to the cracking reactor. This limits over-cracking of product range material in the feed and maximizes the yield of desired products.

Fractionation Section

The design of the facilities to separate the products produced in the reactor section can have an almost limitless number of configurations. This paper will present the three most common configurations for hydrocracking units producing distillate range products.

Stripper First Flow Scheme

In this configuration, shown in Figure 5[2], liquid from the reactor section is preheated and then sent to a Stripper. This is a steam stripped column whose primary purpose is to remove the hydrogen sulfide and non-condensable material so that less expensive metallurgy can be used in the Product Fractionator.

If it is desired to recover the C_3 and C_4 material produced in the reactors, the Stripper overhead vapor is sent to an absorber to recover LPG (C_3/C_4) in this stream. The lean oil to the absorber is typically a naphtha range material, but heavier distillates can also be utilized. The rich oil from the Sponge Absorber combines with the Stripper net overhead liquid and is sent to the Debutanizer, where the LPG is produced as the overhead liquid product. Any net vapor from the Debutanizer is returned to the Sponge absorber for further recovery of LPG. The Debutanizer bottoms liquid is then sent to the Naphtha Splitter along with the overhead liquid from the Product Fractionator. The Naphtha Splitter enables separation of the feed into light and heavy naphtha fractions. Light naphtha, typically C_5 to C_6 range material, is produced overhead. This material has properties suitable for use directly as gasoline blending component. The bottoms liquid from the Naphtha Splitter is heavy naphtha.

Because hydrocracked heavy naphthas have a low octane, typically in the range of 60-70 RON, further catalytic reforming is usually necessary to be able to utilize this material as gasoline blending component. The Stripper bottoms liquid is heated, partially vaporized, and sent to the Product Fractionator. This column utilizes one or more sidecut strippers to produce the various distillate products that are desired. Sidecut strippers are used to strip the material lighter than the desired product distillation range due to the presence of this material in the vapors rising in the Product Fractionator. The sidecut strippers can either be steam stripped or reboiled to generate the stripping vapor.

The distillate products are fairly wide boiling mixtures and the degree of separation between adjacent products is usually measured by the product gap or overlap. This is defined as the difference between the 95 vol% ASTM D-86 distillation temperature of the light product and the 5 vol% ASTM D-86 distillation temperature of the adjacent heavier product. This is a measure of the degree of separation of the two products. If this difference is a positive number, it is called a gap, and if it is a negative number, it is called an overlap. The higher the gap, the better the separation and the higher the overlap, the worse the separation. Typical gaps that can be achieved for various distillate products are as follows:

	Gap
Naphtha/Kerosene	25°F
Kerosene/Diesel	10°F
Diesel/Unconverted Oil	0°F

The Product Fractionator column can either use a feed heater or can be reboiled, according to the characteristics of the endpoint of the unconverted oil. Temperatures appreciably above 700°F should be avoided in order to minimize the possibility of coking the heater tubes, and potential for thermal degradation of the bottoms product. Product Fractionator bottoms liquid is material boiling above that of the heaviest desirable distillate product. This unconverted oil can either be recycled back to the reactor section for further conversion, sent as a feedstock to other processing units, or utilized as refinery fuel oil. In high conversion hydrocracking units, most if not all of this material will be recycled back to the reactor section.

Debutanizer First Flow Scheme

In this configuration as shown in Figure 6, liquid from the reactor section is heated to a suitable temperature and fed to the Debutanizer column. The overhead products from this column are a net vapor and a LPG range material that may need further processing to remove hydrogen sulfide and possibly methane and ethane before being sold as LPG. The Debutanizer column is usually a reboiled column. Debutanizer bottoms liquid is heated and then sent to the Product Fractionator. The Product Fractionator can either be reboiled or utilize a feed heater and is similar to the Product Fractionator described above for a Stripper First flow scheme. The Product Fractionator overhead liquid is usually naphtha range material and is sent to the Naphtha Splitter for separation into a C_5/C_6 overhead product and a heavy naphtha product that usually requires further catalytic reforming to yield an octane that is suitable for gasoline blending stock.

Product Fractionator First Flow Scheme

In this flow scheme as shown in Figure 7, the reactor section liquid is heated and sent directly to the Product Fractionator. The design of the Product Fractionator column is similar to that described under the Stripper First Flow Scheme. To minimize the bottoms temperature at the bottom of the Product Fractionator, this column is operated at the lowest pressure possible. In a Stripper or Debutanizer first configuration this is not an issue, because the Product Fractionator column does not produce a net vapor, because the non-condensable material is removed before the material is fed to the Product Fractionator column.

In this configuration, the Product Fractionator feed contains non-condensable material. Due to the low operating pressure of the Product Fractionator (generally 10 psig or less), there is usually no suitable destination for the net vapor. Therefore it is necessary to utilize a net gas compressor to boost the pressure of the net vapor to that which is suitable for further processing or for use as refinery fuel gas. To improve recovery of light hydrocarbons from LPG through heavy naphtha, the compressed net vapor is recontacted with the Product Fractionator net overhead liquid to absorb some of the hydrocarbons in this range. The Recontact Drum vapor is vented and the liquid is sent to the Debutanizer column of recovery of the LPG product overhead. The Debutanizer Column bottoms is sent to the Naphtha Splitter, where light (C_5/C_6) and heavy naphtha product are produced.

This configuration has the disadvantage of requiring a rather expensive net gas compressor and because of the presence of hydrogen sulfide at elevated temperatures, the Product Fractionator and feed heater must be constructed from more expensive stainless steel metallurgy to resist corrosion.

Conclusion

This paper has attempted to describe the most common reactor section and fractionation section flow schemes seen in hydrocracking units. Due to the versatility of the hydrocracking process, a virtually limitless variety of flow scheme options may be utilized to meet the specific requirements for each application.

References

1. Stell, J., Oil & Gas Journal ,**98,**(51), 66-155 (2000).

2. Towler, G. P. and. Hoehn, R. K, "Hydrocracking: Process Technology and Equipment Engineering Issues", *Encyclopedia of Catalysis*, Horvath, I. V., (Editor-in-Chief), Wiley, 2002.

3. Ward, J. W., Fuel Processing Technology, **35,**55-85,(1993).

4. Scherzer, J., Gruia, A. J., *Hydrocracking Science and Technology*, Marcel Dekker, Inc., 1996.

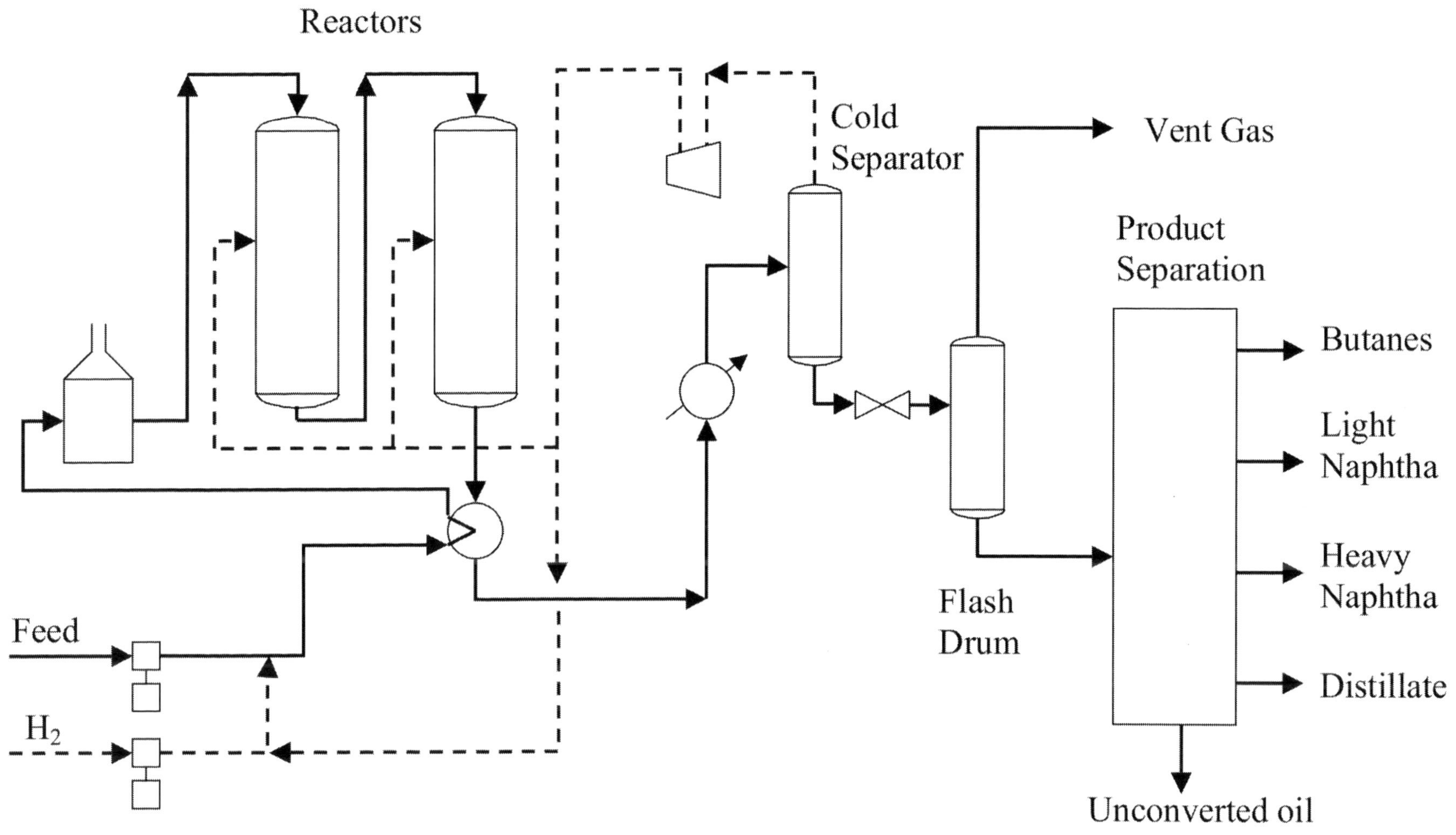

Figure 1. Once-Through Process Flow Scheme

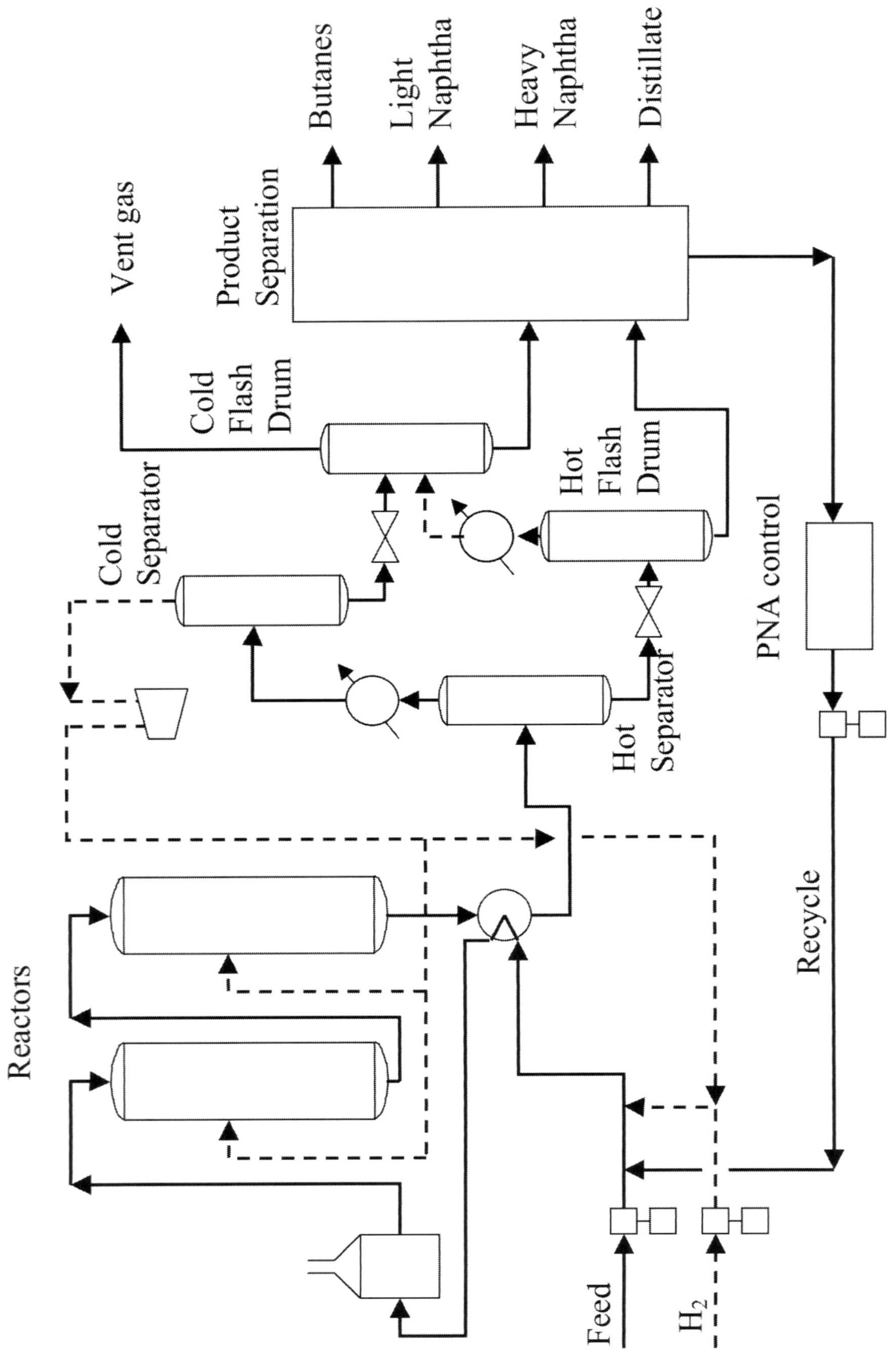

Figure 2. Single-Stage Process Flow Scheme

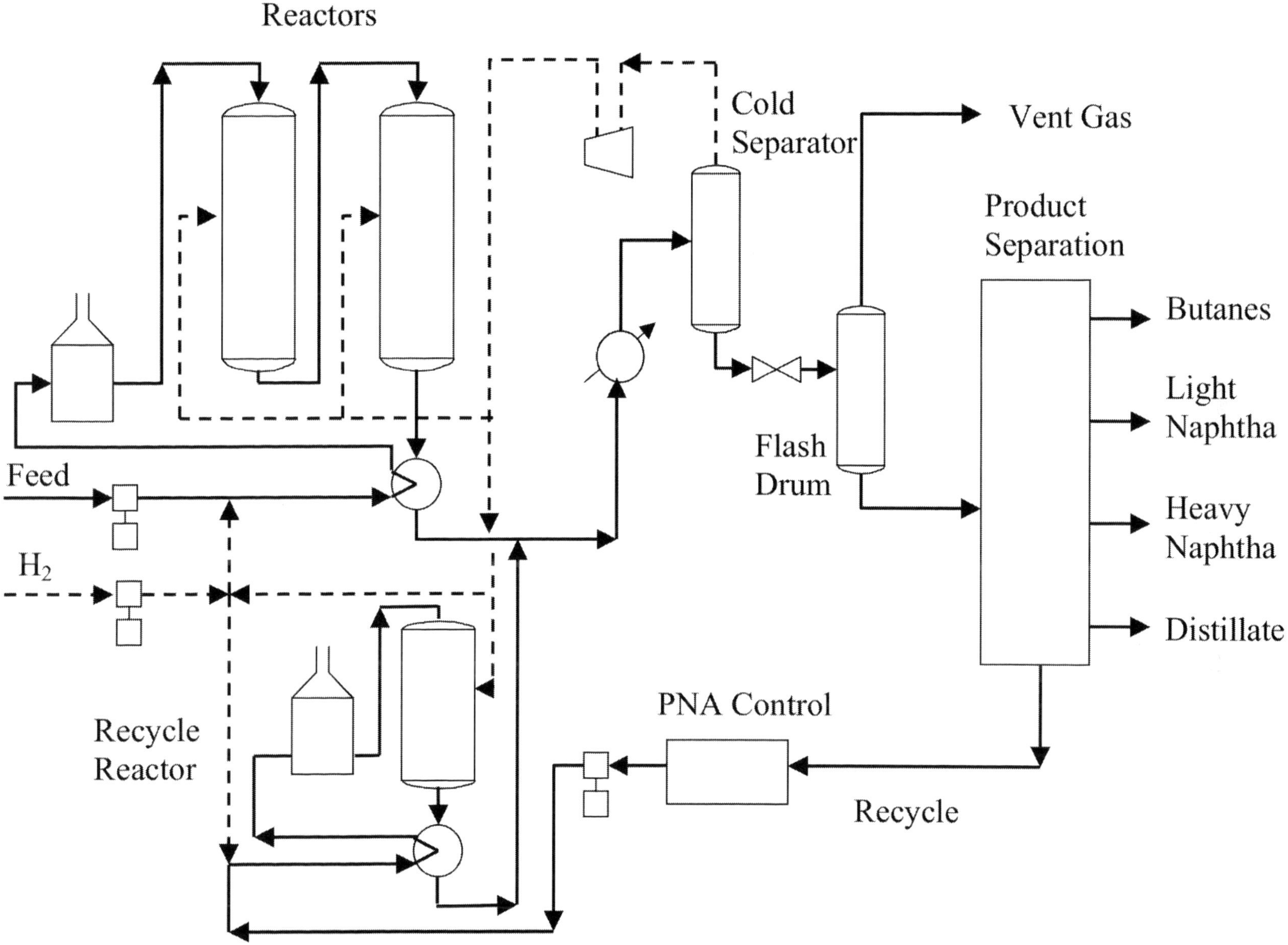

Figure 3. Two-Stage Process Flow Scheme

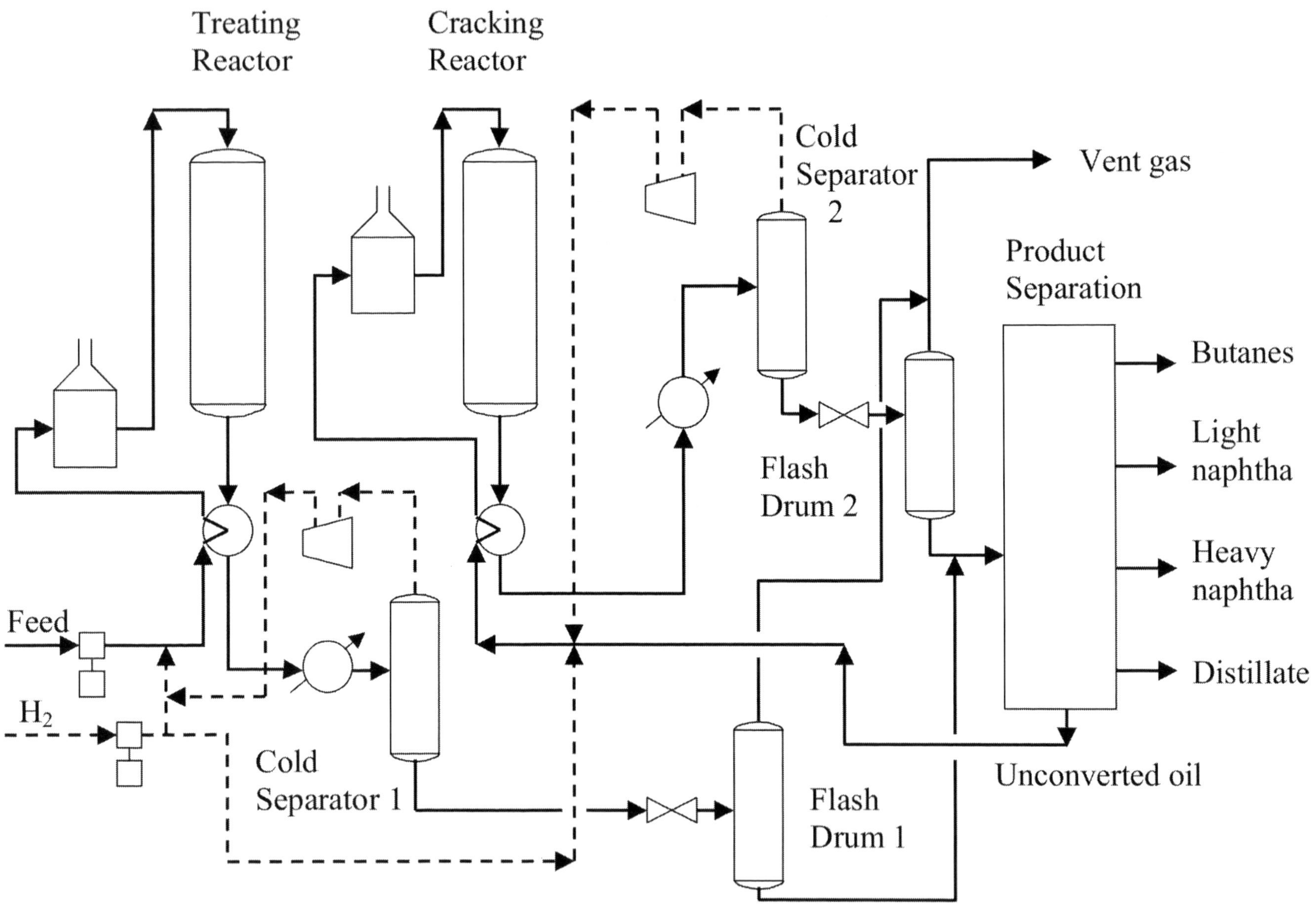

Figure 4. Separate Hydrotreat Flow Scheme

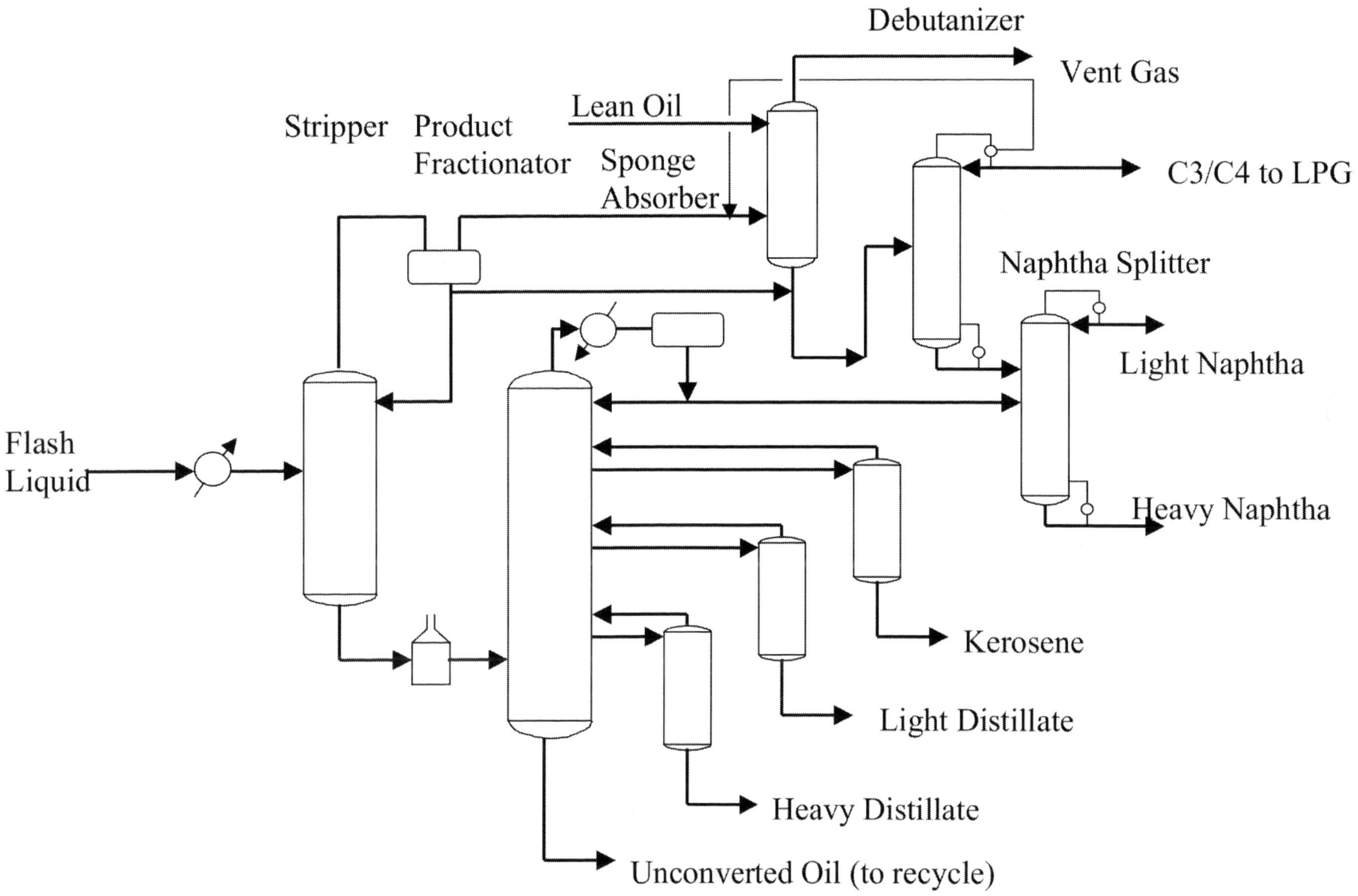

Figure 5. Typical Stripper First Product Separation Flow Scheme

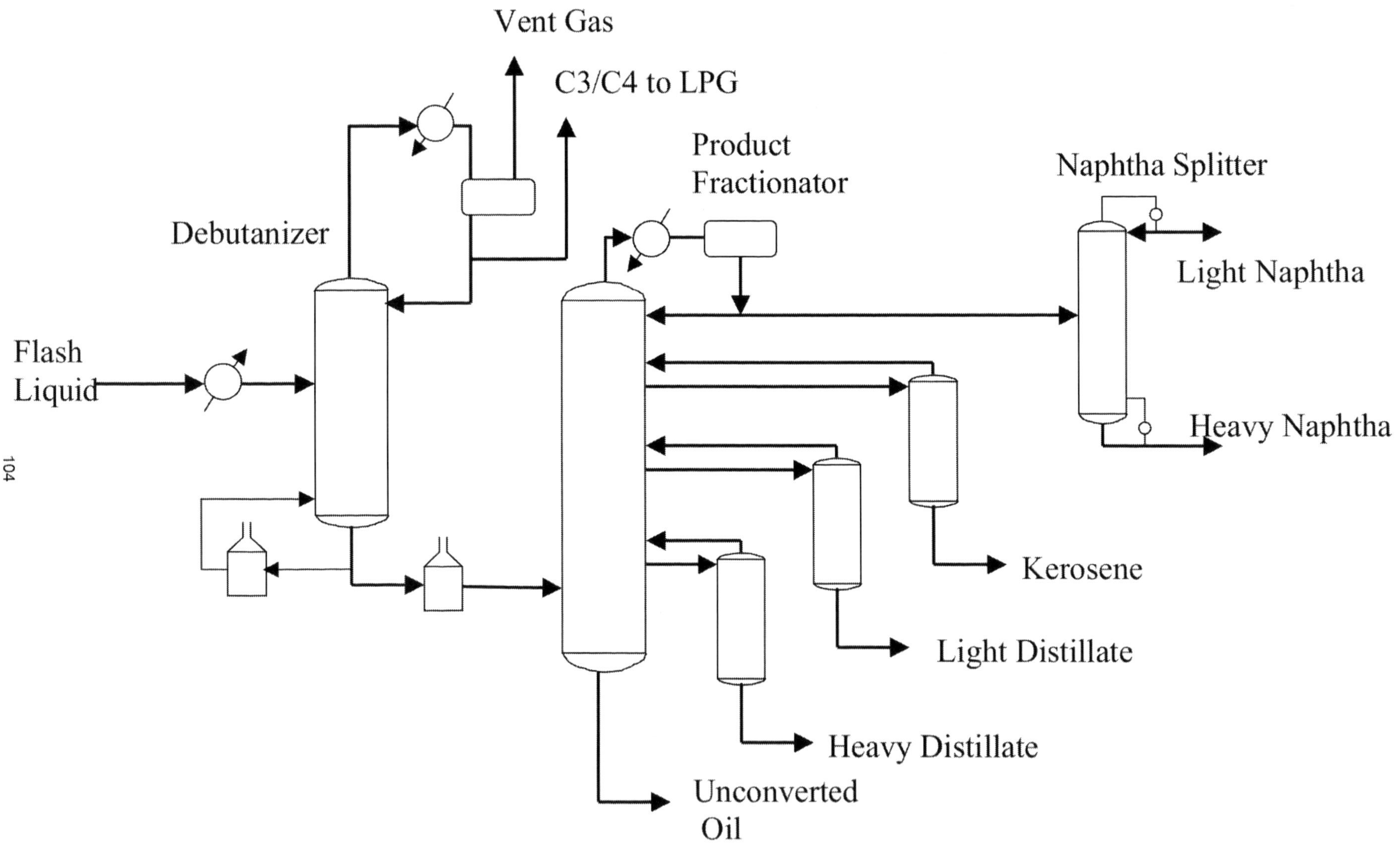

Figure 6. Typical Debutanizer First Product Separation Flow Scheme

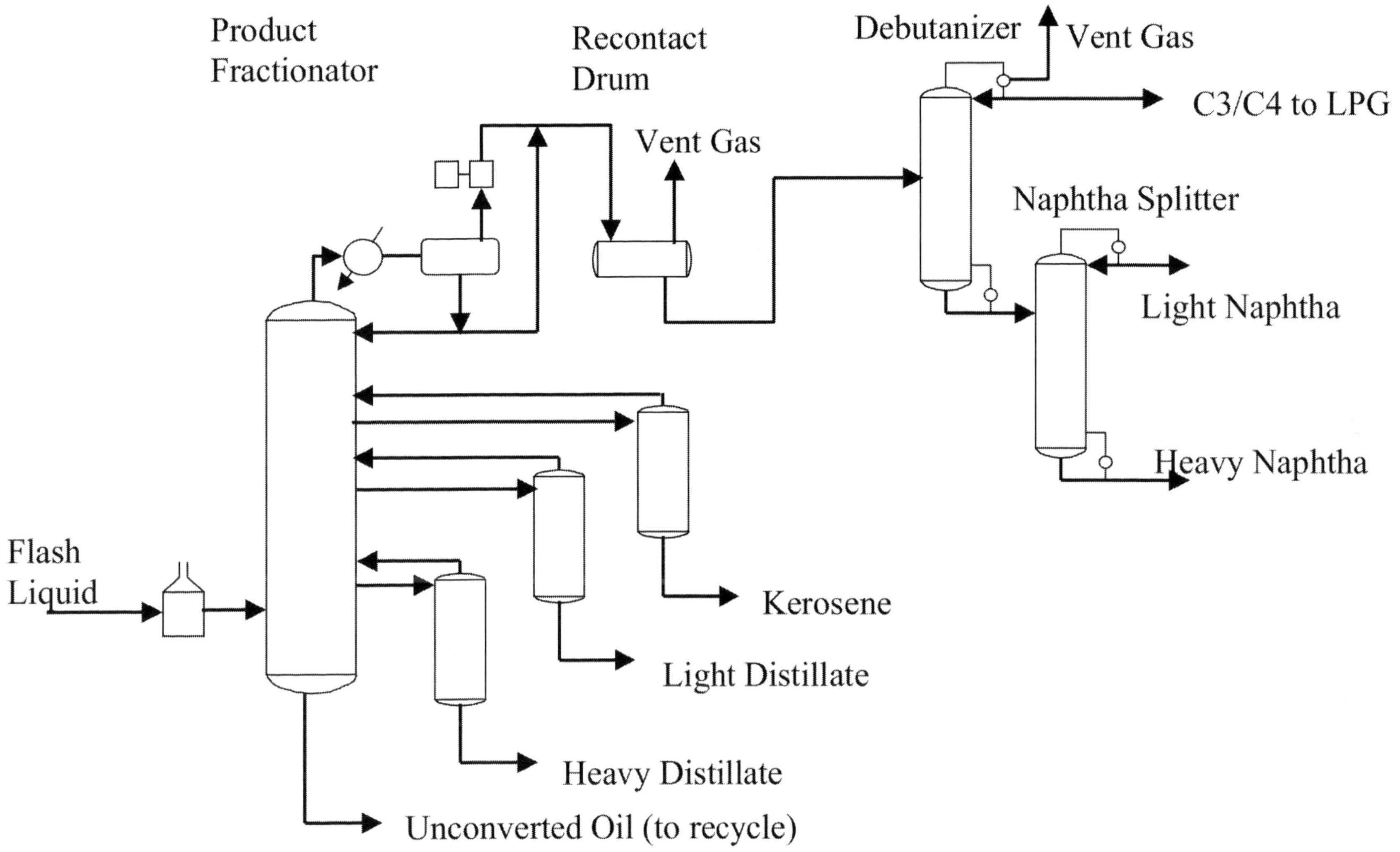

Figure 7. Typical Fractionator First Product Separation Flow Scheme

HYDROCRACKING CATALYST TECHNOLOGY

Li Wang, Hemant Gala

UOP LLC
25 East Algonquin Road
Des Plaines, IL 60017

Key Words: Refining, Hydrocracking, Hydroprocessing, Catalyst

Prepared for presentation at the AIChE Spring National Meeting 2004
Seventh International Conference on Refinery Processing
April 25-29, 2004, New Orleans, Louisiana

ABSTRACT

Hydrocracking is one the most versatile modern petroleum refining processes. Heavy feedstock of various natures and origins is cracked in the presence of high pressure hydrogen over a catalyst to make high value, high quality transportation fuels. The unconverted oil from the hydrocracking process can be used as the lube base stock or feedstock for FCC and ethylene crackers. Such diversity and flexibility also make the hydrocracking process a complex and challenging technology. Many factors affect the performance of a hydrocracking unit; however the choice of the catalyst has the greatest impact on the product stream. The dual-function hydrocracking catalyst consists of an acidic cracking component and a metal phase to carry out hydrogenation and dehydrogenation reactions. The components, their syntheses, the fabrication of catalyst, and most importantly the correlation between process output and catalyst selection will be described in this tutorial.

HYDROCRACKING CATALYST TECHNOLOGY

Hydrocracking is one the most versatile modern petroleum refining processes. Heavy feedstock of various natures and origins is cracked in the presence of high pressure hydrogen over a catalyst to high value, high quality transportation fuels. The unconverted oil from the hydrocracking process can be used as the lube base stock or feedstock for FCC and ethylene crackers. Such diversity and flexibility also make the hydrocracking process a complex and challenging technology. Many factors affect the performance of a hydrocracking unit; however the choice of the catalyst has the greatest impact on the product stream.

The modern hydrocracking process generally operates below 2500 psig. Before 1950, very high pressure hydrocracking was practiced for the purpose of converting coal to liquid fuels[1-3]. Sulfides of tungsten or molybdenum were used as catalysts in such high pressure (3000-10,000 psig), high temperature (700-1000°F) process to carry out both cracking and hydrogenation functions.

The first commercially proven modern day, low temperature, low pressure (<2500 psig) hydrocracking process was introduced in 1959[4] and quickly gained widespread acceptance and grew into a major hydrocarbon conversion process in the refinery. Many new processes[5-11] have been developed over the past forty years since the introduction of the first commercial unit. The success of the low pressure hydrocracking process is built upon the utilization of the dual-function catalysts containing both acidity and metal activity. The synergy of the two functions enabled the cracking to occur at much milder conditions and provided much improved process and product flexibilities.

HYDROCRACKING FUNDAMENTALS

To understand how catalysts perform in hydrocracking process, it is important to understand the reaction fundamentals. A simplified reaction scheme depicting the functionality of a hydrocracking catalyst and the cracking reaction path taken by a reactant, in this case, a paraffin molecule, is shown in Figure 1.[12] In this dual-function hydrocracking scheme, metallic sites are required for hydrogenation and dehydrogenation reactions, while the acid sites are necessary for isomerization and cracking activities. The choices of acidic materials and metals as well as the balance between the metal and acid have profound impact on the activity, product selectivity and stability of the catalyst.[13-17]

Since the acid function is responsible for the isomerization and the subsequent cracking reactions, the activity of the hydrocracking catalyst is primarily determined by the total acidity of the catalyst. The total acidity is determined by the acid strength, acid site population and the accessibility of the acid sites. Cracking materials frequently used for hydrocracking are amorphous silica-alumina and Y type of zeolites with various Si/Al_2 framework ratios.

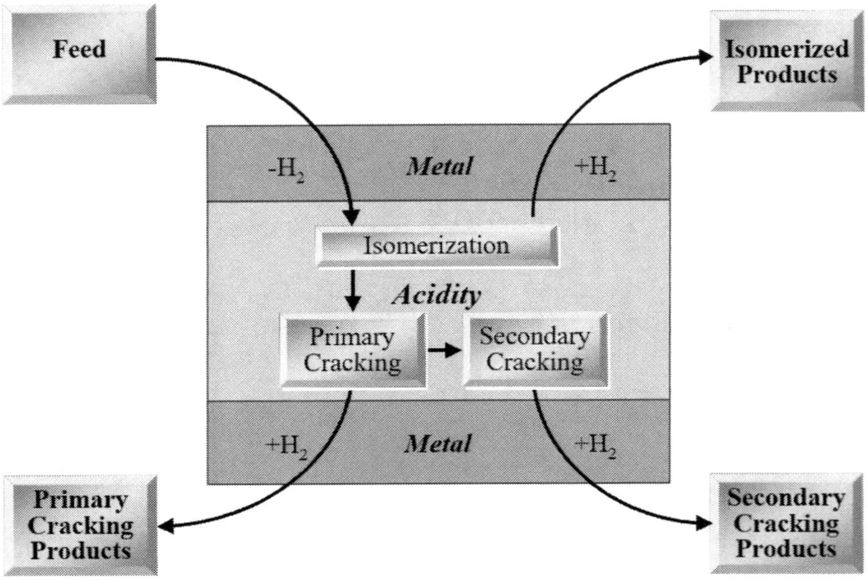

Figure 1

Hydrocracking Reaction Scheme

Metal function, typically in the form of Ni(Co)/W(Mo) pair, also affects the catalyst activity. It determines the rate of olefin formation resulting from dehydrogenation of the paraffinic components of the feed. The olefins and other unsaturates proceed to isomerize and crack over the acid sites. Therefore, it is important that the rate of olefin formation is not the rate limiting factor of a catalyst with a given acidity.

A more important function of the metal is its hydrogenation ability. A metal's hydrogenation ability is the primary factor in determining the yield and stability of the catalyst. Higher hydrogenation activity saturates the primary cracked product and reduces the probability of secondary cracking that leads to decreased liquid yield. In addition, a strong hydrogenation activity retards the coke formation rate, thus reducing rate of catalyst deactivation. Ideal cracking[18-19] is achieved when sufficient metal activity is built in the catalyst to ensure that all cracking events are considered primary cracking. Maximum yield of the desired product is obtained under the ideal cracking condition.

HYDROCRACKING CATALYSTS TECHNOLOGY

CRACKING COMPONENTS

Typical cracking materials used are amorphous mixed oxides of acidic nature and proton exchanged crystalline zeolites, most commonly the Y-type.

Amorphous mixed oxides

Many amorphous hydrocracking catalysts have been reported including silica-alumina, silica-magnesia, silica-titania, and silica-zirconia. Among them, the most frequently used material is $SiO_2 \cdot Al_2O_3$[20] because of its high acidity and low cost. The cracking activities of the amorphous catalysts are much lower than those of the zeolite containing catalysts. However, due to its high selectivity toward distillates, $SiO_2 \cdot Al_2O_3$ is used heavily in the hydrocracking processes aimed at maximizing distillates and lube base stock production.[21-22]

The preparation of amorphous $SiO_2 \cdot Al_2O_3$ is based on a sol-gel process to effect the formation of Si-O-Al chemical bonding. A physical mixture of silica and alumina does not result in an active cracking catalyst. A few commonly practiced procedures for producing $SiO_2 \cdot Al_2O_3$ are briefly described below.

One common silica-alumina producing method involves the condensation of hydrous alumina and the hydroxyls of SiO_2 hydrogel.[23-26] In this method, the diluted sodium silicate solution is acidified to induce silica to undergo polycondensation. Hydrogel particles about 30 to 50 Å in diameters are formed. The surface of the hydrogel is terminated with hydroxyl in the form of silanol groups. A solution of hydrolized aluminum is added to the silica hydrogel. At low pH (~3.0), the hydrous aluminum reacts with the silanol group to form Al-O-Si bond with elimination of water. Silica microgels decorated with surface Al in tetrahedral coordination is thus formed.

The counter ions of the silicon and aluminum salt precursors, such as sodium ions, sulfates or chlorides must be removed by washing the $SiO_2 \cdot Al_2O_3$ hydrogel repeatedly with water. Diluted ammonium solution is used to minimize sodium so that the acidity of $SiO_2 \cdot Al_2O_3$ is optimized.

The physical and chemical properties of $SiO_2 \cdot Al_2O_3$ can vary widely depending on the preparation conditions. The ratio of Si:Al of the gel, the gelling and the curing conditions are important variables in determining the final properties of the $SiO_2 \cdot Al_2O_3$.

The acidity is essentially a function of $SiO_2:Al_2O_3$ ratio. The Brönsted acidity reaches the maximum when SiO_2 content is in the range of 70 to 80%.[27-28] Therefore it is not surprising that a $SiO_2 \cdot Al_2O_3$ having about 75:25 weight ratio is typically used for producing hydrocracking catalysts. The surface areas of hydrocracking grade of $SiO_2 \cdot Al_2O_3$ in general are in the range from 300 m^2/g to 500 m^2/g.

CRYSTALLINE ZEOLITES

Many zeolites reportedly provide adequate cracking activity,[29-43] but the dealuminated ultra stable Y is most successfully applied in industrial hydrocracking. Zeolite Y is prepared by reacting sodium silicate or colloidal silica with sodium aluminate in the presence of NaOH and an amorphous sodium aluminosilicate nucleating agent.[44-46] The unit cell of Y has the formulation of $Na_x[(AlO_2)x(SiO_2)_{192-x}] \cdot yH_2O$, where y is about 260. The value x varies from 48 to 57; typical NaY zeolite has a $SiO_2:Al_2O_3$ molar ratio of 5.0 or greater. Commercially made NaY has a unit cell size of 24.65-24.70 Å, a surface area of >800 m^2/g, and a crystallite size in the range of 0.5-3 microns.

The synthesized Na-Y lacks the acidity necessary to catalyze cracking due to the presence of high concentration of Na. To generate the acidity, the sodium form zeolite has to be ion exchanged with ammonium to remove most of the Na ions. The ammonium exchanged Y is then hydrothermally treated at about 600°C in the presence of steam to reduce framework Al content and to improve its hydrothermal stability.[47-53] The unit cell size of Y decreases as a result of the increase in the framework Si:Al_2 ratio.[54] The degree of dealumination and the corresponding unit cell size reduction is controlled by the severity of the hydrothermal treatment, which is determined by the treatment temperature, time and steam partial pressure. The steam treated zeolite, referred as ultra-stable Y (USY) can be structurally stable at temperatures in excess of 1000°C.

Multiple hydrothermal treatments could be applied to further reduce the unit cell size and the Na content.[49,55] Y zeolite having a sodium level less than 0.2 wt-% is commonly used in hydrocracking catalysts. The unit cell sizes of hydrothermally treated Y vary from 24.55 to 24.20 Å. The higher unit cell size USYs are typically used in naphtha production because of their higher activity. Lower unit cell size USYs show lower activity but higher distillate product selectivity.[55-56]

Dealumination of Y zeolite can also be accomplished by various chemical treatment methods. At less than 100°C, chelating agent such as H_4EDTA can extract Al from NaY.[51,52,57] Ammonium fluorosilicate, $(NH4)_2SiF_6$, on the other hand, is able to insert Si into the location vacated by dealumination.[58]

The dealuminated Y, as described above can be prepared by many different methods at various degrees of severity. Properties of the resulting dealuminated Y including the crystallinity, surface area, framework Si:Al_2 ratio, unit cell size, the amount and the location of the non-framework alumina residue,[59-61] and finally the micro- and meso-porosities[62] can therefore be quite different. The unique combination of these properties for a given dealuminated Y defines its net performance as a hydrocracking catalyst.

The three-dimensional, wide-pore Y-zeolite produces much lower light ends compared to other zeolites with similar cracking activities. With the additional post-synthesis treatment, Y-zeolite becomes hydrothermally stable and can offer a range of acidity for considerations in various hydrocracking processes.

METALS

Two groups of metals have been used in industrial hydrocracking catalysts. Most commonly used is the base metal system, such as Ni/Mo and Ni/W combinations in their sulfided forms. Metal pairs such as Co/Mo and Co/W have limited use in hydrocracking catalysts because of their lower hydrogenation activities. The sulfided metal systems can function well in the typical hydrocracking reaction environment, where high concentrations of H_2S, NH_3 and H_2O are generated from their organic precursors present in the feedstock. The concentration of base metals in hydrocracking catalysts varies from 1 to 6 wt-% for Ni and from 8 to 20 wt-% for W.

Noble metals are also being used in hydrocracking catalysts although not as widely as the base metal catalysts. Noble metals exhibit much higher metal activities than the sulfided base metals in a clean reaction environment,[63-64] therefore, such catalysts are typically used in the second stage of a two-stage hydrocracking process. In a two-stage operation, H_2S and NH_3 generated from the first stage are mostly removed through the product separators; fresh H_2 and unconverted oil from fractionator bottom are fed into the second stage reactor with few contaminants. In the first stage operation or the single-stage process, noble metals tend to lose their advantages over the sulfided base metals due to the presence of contaminants, primarily H_2S. Without significant performance advantage, it is clearly not economical to operate with noble metal catalyst under those process conditions.

CATALYST PERFORMANCE

While the feedstock and process conditions have some effects on the catalyst performance, the catalyst selection basically dictates the output of the hydrocracking process. The catalyst in effect determines the operating temperature, the product yields, the product distribution and the product qualities.

CATALYST ACTIVITY

The activity of the most active amorphous $SiO_2{\cdot}Al_2O_3$ catalyst can easily be 100°F less active than the most active zeolitic catalyst. Only $SiO_2{\cdot}Al_2O_3$ of highest acidity is used as commercial hydrocracking catalysts. Since the introduction of zeolitic catalyst, the use of $SiO_2{\cdot}Al_2O_3$ catalyst has been limited to the production of distillates because of its inherently higher distillate selectivity.

The zeolite catalysts, on the other hand, can offer a wide range of activity for different applications. It is well accepted that the presence of tetrahedrally coordinated framework aluminum gives rise to the acidity in zeolites. When keeping the metal activity constant, the cracking activity of a zeolitic catalyst is determined by the concentration of the zeolite[65] and the framework Al content of the zeolite used. Table 1 shows the catalyst activities of a series of catalysts containing Y zeolites with various frameworks Al concentrations as indicated by the different framework $Si{:}Al_2$ ratio. The tests were conducted in once-through mode to convert light Arabian vacuum gas oil (VGO) to naphtha product. Clearly, the temperatures required to reach a constant conversion level, 65 wt-% in this test, increased with decreasing framework Al content. The activity variations were quite significant.

Table 1

Catalyst Activity As a Function of Framework $Si{:}Al_2$ Ratio

	Y-1	*Y-2*	*Y-3*
Framework $Si{:}Al_2$ Ratio	**9**	**16**	**32**
Zeolite Concentration, wt-%	**50**	**50**	**50**
Ni/W, wt-%	**3/11**	**3/11**	**3/11**
Required Rx Temp, °F	**Base**	**Base +5**	**Base +30**

Catalyst with higher activity invariably is the choice for a hydrocracking operation provided that the yield selectivity remains the same. However, it is a fact that the yield of a hydrocracking catalyst tends to decrease with increasing activity. The yield-activity correlation obtained from UOP's commercial catalysts DHC™ and HC™ illustrates (Figure 2) this point very clearly. Therefore, the decision on the catalyst selection for a hydrocracking unit must weigh both the activity and the yield selectivity.

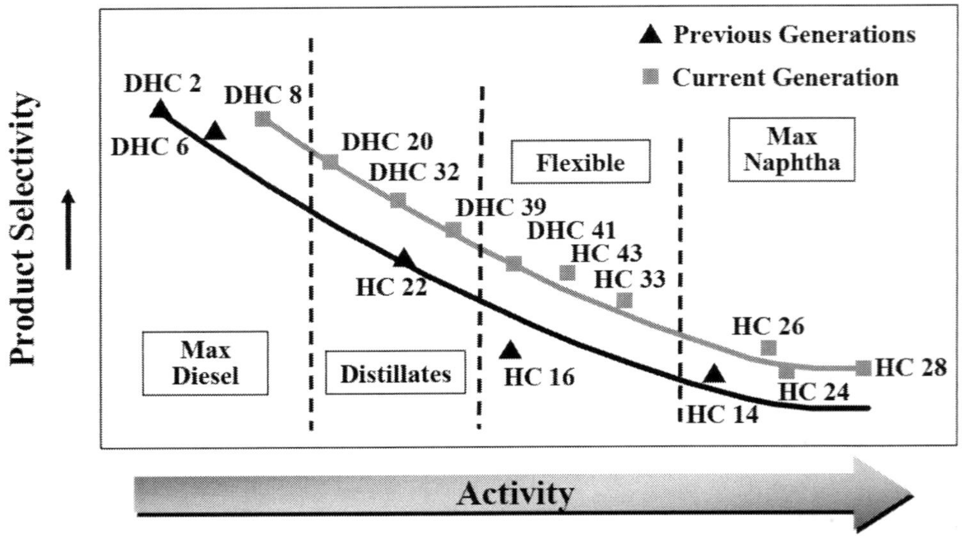

Figure 2

Product Yield and Catalyst Activity

PRODUCT YIELDS

The product yield depends on the metal:acid activity ratio of the catalyst in use. It is well acknowledged that the product yield increases with increasing metal:acid activity ratio.[18,66,67] The yield could approach maximum when a catalyst can exhibit ideal cracking behavior. The yields obtained from converting VGO to distillates as well as to naphtha for the same three catalysts discussed in Table 1 are shown in Table 2.

Table 2

Product Yields As a Function of Framework Si:Al$_2$ Ratio

	Y-1	*Y-2*	*Y-3*
Naphtha Test			
Required Rx Temp, °F	**Base**	**Base +5**	**Base +30**
Naphtha Yield, wt-%	**Base**	**Base +0.5**	**Base +1**
Distillates Test			
Required Rx Temp, °F	**Base**	**Base +3**	**Base +25**
Distillates Yield, wt-%	**Base**	**Base +1.5**	**Base +3.5**

Given that the metal contents were the same for all three catalysts, higher activity resulted in lower metal:acid activity ratio and consequently led to lower yields as evidenced by the trending shown in both distillates and naphtha tests. The magnitude of yield change was found to be more pronounced in distillates operation.

The activity of a catalyst can be readily improved by increasing the acidity of the cracking component and by increasing its concentration in a catalyst. However, to match the metal activity of a hydrocracking catalyst as its cracking activity increases is a very difficult task and, in fact, it has remained the heart of the hydrocracking catalyst research. The metal activity of the sulfided base metal catalyst has basically been optimized. Further increase of metal concentration leads to heavy agglomeration of metals and results in loss of metal activity. Noble metals can be more active than the base metal only in clean reaction environment. Until the noble metals could be made more sulfur tolerant, their use is still limited to the second stage operation.

PRODUCT DISTRIBUTION

The product distribution characteristics are quite different between amorphous catalysts and those containing Y zeolite. Figure 3 shows the incremental yield as a function of the distillation temperature of the product slates for both types of catalyst. At the same conversion level to 700°F, the amorphous $SiO_2{\cdot}Al_2O_3$ catalyst showed a much higher distillates selectivity whereas the product of Y zeolite catalyst was heavily skewed toward naphtha. The Y zeolite catalyst being 100°F more active than the amorphous catalyst lacks the proper metal/acid balance. As a result, excessive secondary cracking occurs over the Y zeolite catalyst shifting the product to lighter boiling range.

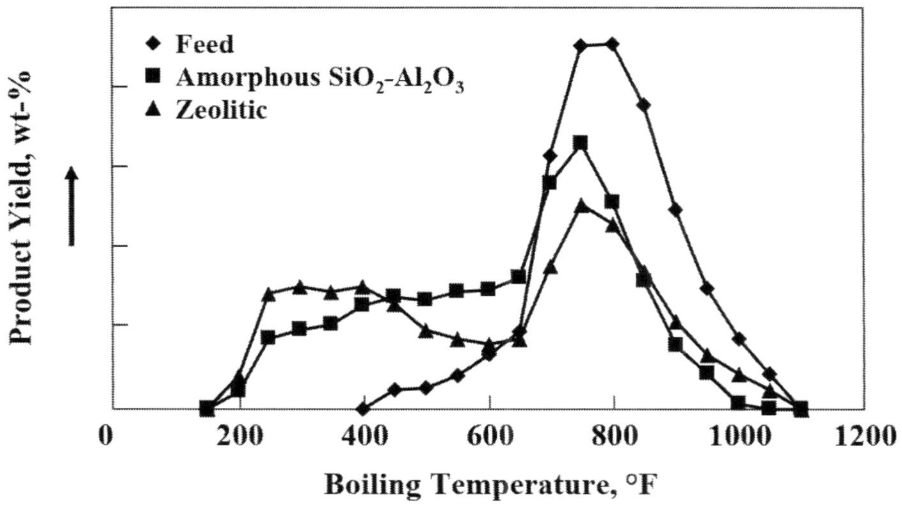

Figure 3

Product Distributions of Hydrocracking Catalysts

The different physical nature of the cracking component also contributes to the differences in the product distribution. The pore size of Y zeolite is about 7 Å compared to the pore diameter of $SiO_2 \cdot Al_2O_3$ which ranges from 40 to 200 Å. The extremely narrow pores of the Y zeolite add great difficulty to the diffusion of the reacting molecules. The intra-crystalline diffusion constraints result in higher secondary cracking and preference in cracking the lighter part of the feed, both of which contribute to making lighter products.

Hydrocracking units designed for distillate production prefer use of amorphous catalyst, despite its low activity because the zeolitic catalysts could not compete with the amorphous catalyst on distillates selectivity. On the other hand, the naphtha producing units always choose the zeolitic catalysts for their high activity and high selectivity to naphtha.

Much effort has been made to improve the distillate selectivity of the zeolitic catalyst. As shown in Table 2, the distillate selectivity was significantly improved by reducing the framework Al content. Much of the improvement was accomplished by increased metal:acid activity ratio, but the relief in diffusional constraint is also a critical factor. Pore sizes in the same range as the amorphous material are being generated during the dealumination process. The pore volume increases with increasing severity of dealumination. The presence of these meso-porosities greatly reduces the intra-crystalline diffusion constraints thus decreasing the degree of secondary cracking. This phenomenon is best illustrated by Figure 4. Two series of catalysts representing two Y zeolites of different framework alumina contents, respectively, were evaluated for their distillates selectivity as a function of zeolite concentration in the catalysts. Each series showed increased distillates selectivity with decreasing concentration of zeolite content which is attributed to increased metal:acid activity ratio. However, the distillate selectivities for the catalysts approaching zero concentration of zeolite are significantly different between the two Y zeolites. The one with higher framework Si/Al_2 showed higher distillate selectivity which presumably was the result of reduced diffusional constraints.

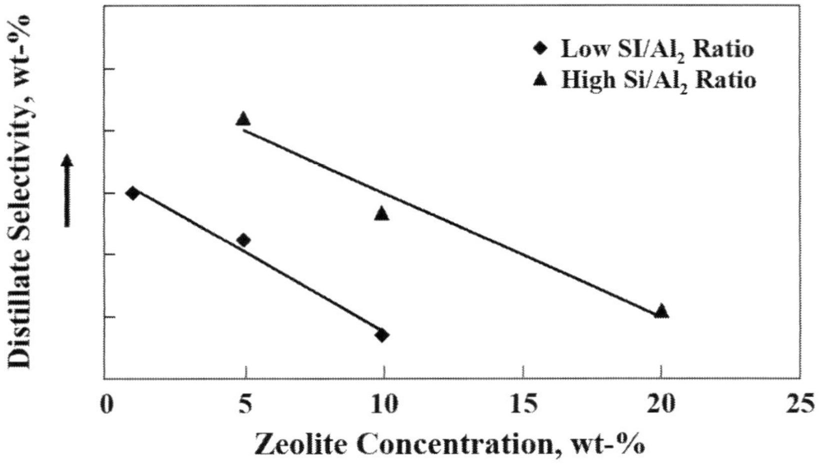

Figure 4
Distillate Selectivity of Y Zeolites

Many modern day distillate catalysts adopted the concept of combining the amorphous $SiO_2 \cdot Al_2O_3$ and the highly dealuminated Y to achieve higher activity with little or no loss in distillate selectivity relative to the amorphous only catalyst. Also the presence of amorphous cracking material complements the zeolite very well in converting favorably the heavy ends of the feedstock. The yield pattern of a composite catalyst is shown in Figure 5 to be reasonably close to that of a $SiO_2 \cdot Al_2O_3$ catalyst while being 30°F more active.

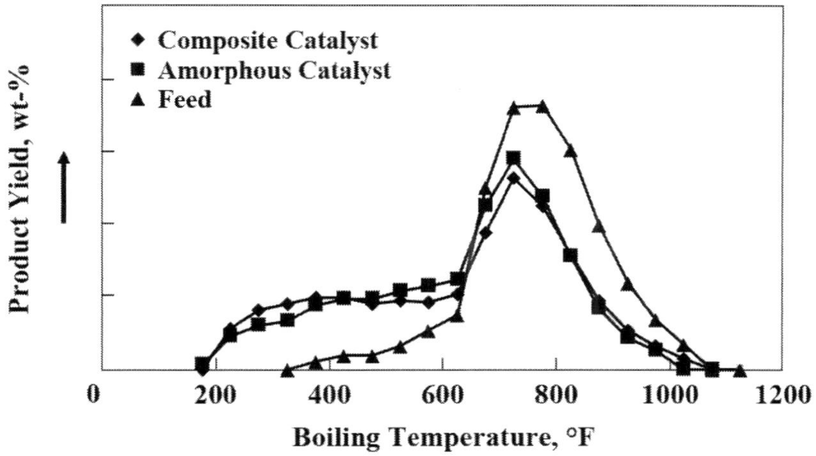

Figure 5

Comparison of Prouct Distributions between Composite and Amorphous Catalysts

PERFORMANCE STABILITY

When selecting a catalyst, one must consider the cycle length which is determined by the start-of-the-run temperature and the deactivation rate of the catalyst, and the average cycle yield which is affected by the yield loss as the cycle progresses. In general, the amorphous catalyst operating at much higher temperature deactivates faster than the zeolitic catalyst because of faster coking rate. However, the yield stability of the amorphous catalyst is known to be better than that of the zeolitic catalyst. It is hypothesized that a deteriorating metal:acid activity ratio of the zeolitic catalyst is responsible for the yield loss as the catalyst deactivates. The comparison of performance stability between amorphous and zeolitic catalysts is shown in Figure 6.

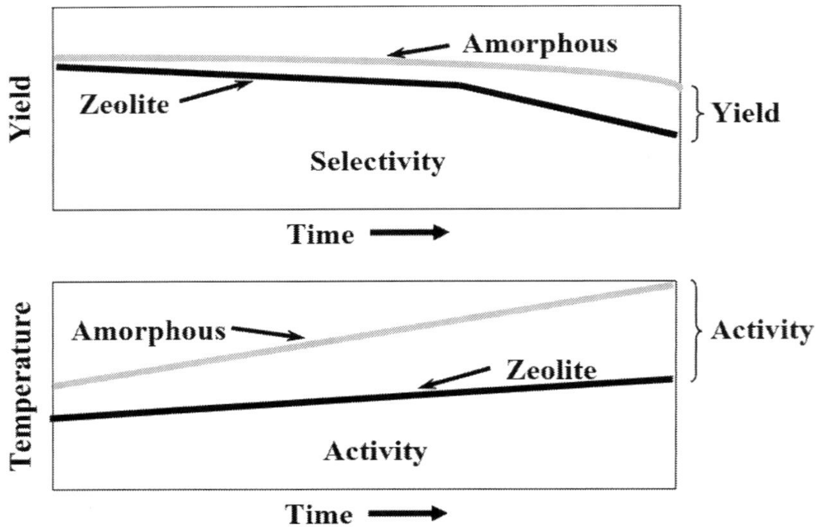

Figure 6

Performance Stability

SUMMARY

Hydrocracking catalyst technology has significantly advanced since its inception. The improvement was driven by the market needs and accomplished by better understanding of the reaction fundamentals and continued material development. From amorphous mixed oxide, to zeolite, to a combination of both, each has its unique performance niche. However there are needs to further improve the performance of hydrocracking catalyst. Materials with higher acidities than Y zeolite and having the proper geometric requirement need to be developed to increase the catalyst activity. Finding novel materials having higher hydrogenation and dehydrogenation activities than the traditional metals used in hydrocracking catalysts are also necessary in order to match the increasing catalyst acidity.

REFERENCES

1. Bergius, F., Proceedings of World Petroleum Congress, London, 2, 282 (1933).
2. Pier, M., Proceedings of World Petroleum Congress, London, 2, 290 (1933).
3. Ministry of Fuel and Power, "Report on Petroleum and Synthetic Oil Industry of Germany," B.I.O.S. Overall Report No. 1., Section C, Hydrogenation Processes, 46-73 (1947).
4. Stormont, D. H., Oil and Gas Journal, 57, (44), 48 (1959).
5. Sterba, M.J. and Watkins, C.H., Oil and Gas Journal, 58, (21) 102 (1960).
6. Oil and Gas Journal, 58, (16), 104 (1960).
7. Scott, J. W. and Patterson, N.J., Proceedings of the Seventh World Petroleum Congress, Mexico City, 4, 97 (1967).
8. Hunter, M.G. and Gentry, A.R., Annual Int. Refining Conf., May 9-12, 1994, Singapore.
9. Watkins, C.H. and Jacobs, W.L., Hydrocarbon Process, 45(5), 159 (1966).
10. Frye, C.G., Muffat, D.L. and McAninch, H.W., Oil and Gas, J., 68 (20), 69 (1970).
11. Tajbl, D.G., Handbook of Petroleum Refining Processes, Meyers, R.A. Ed., McGraw-Hill Book Co. P2-33, 1986.
12. Wang, L. and Brierley, G. R., Paper presented in 5th Annual Meeting of Asian Refinery Technology Conference, April, 2002, Bangkok, Thailand.
13. Coonradt, H. L. and Garwood, W. E., Ind. Eng. Chem. Proc. Des. Dev., Vol 3, No 1, p 38 (1964).
14. Schulz, C. , H. and Weitkamp, J., Preprints, Div. Petroleum Chem., A.C.S., Los Angeles, A-102, (1971).
15. Ward, J. W., Preparation of Catalysts III, 587 (1983).
16. Ward, J. W., Fuel Processing Technology, 35, 55 (1993).
17. Scherzer, J. and Gruia, A. J., Hydrocracking Science & Technology, Marcel Dekker, 1996.
18. Weitkamp, J, in "Hydrocracking and Hydrotreating," Ward, J. W. and Qader, S. A., Eds, A.C.S. Symposium Series, 20, Washington, D.C., 1975, p. 1.

19. Pichler, C. J., Schulz, H., Reitemeyer, H. O., and Weitkamp, J., Erdol, Kohle-Ergas-Petrochem., 25, 494 (1972).

20. Sullivan, R.F. and Scott, J.W., Heterogeneous Catalysis: Selected American Histories, ACS Symposium Series Vol 222, Washington, D.C., 1983, P293.

21. Bridge, A. G. Jaffe, J. and Powell, B. E., Proc. Mtg. Jap. Petr. Inst., Tokyo, Japan, 1982.

22. Reno, M.E., Schaefer, B. L., Penning, R. T. and Wood, B. M., NPRA Ann. Mtg., March 1987, San Antonio, Texas, AM-87-60.

23. Rylalnd, L. B., Tamele, M.W. and Wilson, J. N., in Catalysis, Vol 7 (Emmett, P. H. Ed.) Van Nostrand Reinhold, 1960, P.1.

24. Okkerse, C., in "Physical and Chemical Aspects of Adsorbents and Catalysts," (Linsen, B. C. Ed.) Academic Press, New York, 1970, p.213.

25. Jaffe, J. J., US Patent 3,637,527 (1972).

26. Lim, J. and Stamires, D., US Patent 4,142,995 (1979).

27. Sato, M, Aonuma, T and Shiba, T., Proc. 3rd Int. Cong. Catal., p.396, North-Holland, Amsterdam, (1965).

28. Tanabe, K., Solid Acids and Bases, Academic Press, New York, (1970).

29. Steinberg, K. H, Becker, K. and Nestler, K. H., Acta Phys. Cehm. 31:441 (1985).

30. Ward, J. W., US Patent 4,829,040 (1989).

31. Ward, J. W., US Patent 5,228,979 (1993).

32. Hickey Jr., K. L. and Morrison, R. A., US Patent 4,812,223 (1989).

33. Best, D. F. and Nair, V., US Patent 5,230,790 (1993).

34. Bournonville, J. P., Dufresne, P., Marcilly, C., Petit, L., Raatz, F. and Travers, C., US Patent 4,909,924 (1990).

35. Harris, T. V., Rainis, A., Santilli, D. S. and Zones, S. I., US Patent 5,106,801 (1992).

36. Zones, S. I., US Patent 4,589,977 (1986).

37. Zones, S. I., US Patent 4,589,976 (1986).

38. Apelian, M. R., Degnan Jr., T. F., Marler, D. O. and Mazzoni, D. N., US Patent 5,227,353, 1993.

39. Chu, P. and Rubin, M. K., US Patent 5,026,943 (1991).

40. Miller, S. J., US Patent 4,859,312 (1989).

41. Gortsema, F. P., Long, G. N., Pellet, R. J. and Rabo, J. A., US Patent 4,857,495 (1989).

42. Gortsema, F. P., Long, G. N., Pellet, R. J., Rabo, J. A.and Springer, A. R. US Patent 4,818,739 (1989).

43. Flanigan, E. M., Lok, B. M., Patton, R. L. and Wilson, S. T., US Patent 4,781,814 (1988).

44. Breck, D. W. and Flanigen, E. M., Molecular Sieves, Society of Chemical Industry, London, 1968, p 47.

45. Breck, D. W., US Patent 3,130,007 (1964).

46. Vaughan, D. E. W., Edward, G. C. and Barrett, M. G., US Patent 4,340,573 (1982).

47. Jacobs, P. A. and Uytterhoeven, J. Chem. Soc. Faraday Trans. 1, 69, 3359, 373 (1963).

48. Maher, P. K. and McDaniel, C. V., US Patent 3,293,129 (1966).

49. McDaniel, C. V. and Maher, P. K., in "Molecular Sieves", Soc. Chem. Ind., London, 1968, p186.

50. McDaniel, C. V. and Maher, P. K., in “Zeolite Chemistry and Catalysis,” Rabo, J. A. Ed., American Chem. Soc., Washington, 1976, p 285.

51. Kerr, G. T., J. Phys. Chem., 72, 2594 (1968).

52. Kerr, G. T., J. Phys. Chem., 73, 2780 (1969).

53. Kerr, G. T., Adv. Chem. Ser., 121, 219 (1973).

54. Ritter, R. E., Creighton, J. E., Roberie, T. G., Chin, D. S. and Wear, C. C., NPRA Annual Mtg., Los Angeles, CA, March 1986, AM-86-45.

55. Bezman, R. D. and Rabo, J. A., US Patent 4,401,556 (1983).

56. Gortsema, F. G., Pellet, R. J., Springer, A. R. and Rabo, J. A., US Patent 5,047,139 (1991).

57. Kerr, G. T., J. Phys. Chem., 71, 4155 (1967).

58. Skeels, G. W. and Breck, D. W., Proc. 6th Int. Zeol. Conf., Reno, Nevada, 1983.

59. Dwyer, J., Fritch, F. R., Qin, G. and Vickerman, J. C., J. Phys. Chem., 86, 4574 (1982).

60. Garralon, G., Fornes, V. and Corma, A., Zeolites, 8, 268(1988).

61. Dwyer, J., Fritch, F. R., Machado, F., Qin, G., Smith, S. M. and Vickerman, S. M., J. Chem. Soc., Chem. Commun., 422 (1981).

62. Patzelova, V. and Jaeger, N. I., Zeolite, 7, 240 (1987).

63. Franck, J. P. and Lepage, J. F., Proc. 7th Int. Congr. Catal., Tokyo, 1981, p792.

64. Maxwell. I. E., Catal. Today, 1, 385 (1987).

65. Howell, R. L.., Sullivan, R. F., Hung, C-W. and Laity, D. S., Paper presented at the Petroleum Refining Conference, Tokyo, Japan, Oct. 1988.

66. Sullivan, R.F. and Meyer, J. A., in "Hydrocracking and Hydrotreating," Ward, J. W. and Qader, S. A., Eds, A.C.S. Symposium Series, 20, Washington, D.C., 1975, p. 28.

67. Coonradt, H. J., Leaman, W. K. and Miale, J. N., Preprints, Div. Pet. Chem. A. C. S., Philadelphia, April 5 to 10, 1964 p 59.

uop

UOP LLC
25 East Algonquin Road
Des Plaines, IL 60017-5017

UOP 4287C

ExxonMobil Technologies for Ultra Low Sulfur Diesel Product Quality Improvement

Robert A. Bradway and Tim Hilbert

ExxonMobil Research and Engineering Co
USA

Key Words: Refining, Hydrocracking, Hydroprocessing, Clean Fuels, ULSD, Dewaxing, Arosat

Prepared for presentation at the AIChE Spring National Meeting 2004
Seventh International Conference on Refinery Processing
April 25-29, 2004, New Orleans, Louisiana

Abstract

While the sulfur specification has attracted the most attention, distillate fuels have specifications on boiling point, cetane, gravity and flow properties. Some countries have adopted new specifications on cetane and aromatics while others have proposed and discussed new specifications on these properties. A review of technologies to improve these diesel product qualities will be given with a discussion of selection process to meet various objectives. Hydrocracking, aromatic saturation and dewaxing technologies will be covered in addition to conventional hydrotreating technologies.

Corresponding States Modeling of Asphaltene Precipitation

Sang J. Park[1] *and G.Ali Mansoori*[2]

[1]Department of Chemical Engineering , Dongguk University, Pil-Dong 3-26, Joong Ku, Seoul, Korea
[2]Department of Chemical Engineering, University of Illinois at Chicago, 810 S. Clinton Street, Chicago, IL 60607-7000 USA
E-mail; [1]sjpark@dgu.edu., [2] mansoori@uic.edu

Abstract

A generalized corresponding states principle (CSP) for correlation and prediction of asphaltene precipitation data is produced and will be reported. The proposed corresponding states treatment is universal and consistent with the detailed principles of asphaltene stability and collapse from petroleum fluids. All the system parameters and properties responsible for asphaltene stability and collapse are considered in the proposed CSP treatment. Based on the literature data and the data measured by the authors the parameters and exponents of a corresponding states equation capable of describing the asphaltene precipitation behavior in various cases are reported.

Keywords: Asphaltene precipitation; Reservoir oil; Pressure depletion; CO_2-injection; Corresponding states principle.

I. Introduction

It is well known that the injection of CO_2 could result in asphaltene precipitation yielding a sticky, asphalt-like substance that clogs the reservoir and the production equipment [Vasquez and Mansoori, 2000]. This is known to be due to changes in the solubility of heavy components in the reservoir oils in presence of CO_2. In previous reports by various investigating teams [Hu et al., 2000; Hu and Guo, 2001; Yang et al. 1999, Park and Mansoori, 1988I&II, Vasquez and Mansoori 2000, Branco et al, 2001], the effects of temperature and molecular weight of n-alkane precipitants and/or CO_2 on the onset and amount of asphaltene precipitation were investigated in the case of various crude oils. Also models for prediction of onset and amount of asphaltene precipitation, flocculation and precipitation in static and dynamic cases are available [Kawanaka et al 1991, Park & Mansoori 1988 I&II, Mansoori 1997].

In this report we have introduced a generalized corresponding states principle for prediction of asphaltene precipitation due to variations of pressure, injection miscible fluid and temperature. It is also reported the predictive capability of the proposed corresponding states equation approach at various dilution ratios and the type of n-alkane precipitants, on the onset and amount of asphaltene precipitation.

II. Corresponding States Modeling of Asphaltene Precipitation

The corresponding states principle (CSP) and the conformal solution theory (CST) are used widely in development of generalized equations and charts for prediction of thermophysical properties of fluids and solids. It is extended to prediction of the behavior complex substances [Mansoori et al 1980] and prediction of, for example temperature profile in gas transmission pipelines [Edalat and Mansoori,1988]. In the case of asphaltene precipitation from an oil due to addition of a miscible solvent (like CO2) to the oil which is usually expressed in units of "weight percent of asphaltene deposited" and shown by "*W*" (for detailed derivation, see "Hu et.al., 2004").

$$W_r = f\,(D_r, R_r\, M_{r1}, M_{r2}, \mu_r, T_r, P_r, \omega_m) \qquad (1)$$

For the prediction of asphaltene precipitation, it is also defined the following variables *X* and function *Y*.

$$X = D_r . M_{r1}^{Z} . M_{r2}^{\psi} . R_r^{\zeta} . \mu_r^{\xi} \quad (z\ \psi\ \zeta\ \xi) \text{ are constants} \qquad (2)$$

$$Y = W_r . M_{r1}^{Z`} . M_{r2}^{\eta} . R_r^{\varphi} . \mu_r^{\theta} \quad (z`\ \eta\ \varphi\ \theta) \text{ are constants} \qquad (3)$$

Then the following polynomial equation can be obtained to represent the relation between *Y* and *X:*

$$Y = a(T_r P_r \omega_m) + b(T_r P_r \omega_m) X + c(T_r P_r\ \omega_m) X^2 + d(T_r P_r \omega_m) X^3 + \ldots \text{ for } X \geq X^c \qquad (4)$$

Where X^c denotes the value of *X* at the onset point of asphaltene precipitation. At constant temperature and pressure Eq. (4) will reduce to the following form:

$$Y = a + b.X + c.X^2 + d.X^3 \qquad (5)$$

The following best fit to the amount of asphaltene precipitation data of Oil A1 [Srivastava et al., 1995] is obtained as follows:

$$Y = 83.24 - 342.66X + 371.72X^2 - 31.28X^3 \quad (z = 0.1;\ z` = -2) \qquad (6)$$

In the special case of constant, molecular weight ratios, resin to asphaltene ratio and polarity ratio can be written in the following form:

$$D_r^c = X^c . M_s^{-Z} \quad \text{(z is constants)} \qquad (7)$$

The amount of asphaltene precipitation measured for CO_2-injected J2 Oil system can be well represented by the following third order polynomial:

$$Y = 2.909 - 50.51X + 82.88X^2 - 4.15X^3 \qquad (8)$$

Based on Eqs. (7) and (8), the predicted critical dilution ratio is D_r^c=48.2 mol% CO_2, which agrees well with the measured data, D_r^c=46.8 mol% CO_2. Similarly, based on Eqs. (6) and (7) the predicted onset point of asphaltene precipitation in the CO_2-injected Weyburn reservoir oil (Oil A1) is D_r^c=41 mol% CO_2, which is also in good agreement with the experimental data, D_r^c=43 mol%

CO_2 [Srivastava et al., 1995].

Figure 1 presents the phase behaviors of J2 crude oil A in contact with CO2 gas at 339 K. Observed onset of asphaltene mole fraction of CO2 entering the crude oil where the asphaltene precipitation starts to occur is about 0.506 and the observed onset pressure of asphaltene flocculation is about 20Mpa (2900 Psia). Figure 1 also shows the observed pressure/composition region of asphaltene precipitation for a mixture of carbon dioxide and J2 oil at 339 K. The region to the upper side of hollow square marks is the observed three phase region (L-V-S).

The test results are presented in Figure 2, which clearly shows that the corresponding states approach can be satisfactorily applied to correlate/ predict the amount of asphaltene deposition occurred in high-pressure CO_2-injected oil systems. In Figure 3. the effect of CO2 mole% for a given mixture of the amount of asphaltene precipitation are reported at 20 MPa. According to this figure, the amount of asphaltene deposited increases as the mole % of CO2 of the mixture increases. As described in our previous publications, it is also expected that the trend of asphaltene precipitation at different pressures decreases as pressure increases

III. Conclusions

A generalized corresponding states theory for prediction of asphaltene precipitation from petroleum fluids as a result of addition of miscible solvents at various temperatures and pressures is presented. The proposed CSP theory complements the scaling equation for asphaltene precipitation.

References

Branco, V.A.M., Mansoori, G.A., De Almeida Xavier, L.C., Park, S.J. and Manafi, H., 2001. Asphaltene flocculation and collapse from petroleum fluids" J. Petrol. Sci. & Eng., 32, 217-230.

Hirschberg, A., deJong, L. N., Schipper, B. A. and Meijer, J. G., 1984. Influence of Temperature and Pressure on Asphaltene Flocculation. SPE J, June, 283-293

Hu, Y.F., Chen, G.J., Yang, J.T., Guo, T.M., 2000. A study on the application of scaling equation for asphaltene precipitation. Fluid Phase Equilib. 171, 181–195.

Kawanaka, S., Park, S.J. and Mansoori, G.A., 1991. Organic Deposition from Reservoir Fluids: A Thermodynamic Predictive Technique" SPE Reservoir Engineering, 185-195, May.

Mansoori, G.A. and Leland, T.W., 1972. Statistical Thermodynamics of Mixtures (A New Version for the Theory of Conformal Solutions)", Faraday Transactions II, 68, 320-344.

Mansoori, G.A., 1997. Modeling of asphaltene and other heavy organic depositions", J. Petrol. Sci. & Eng., 17, 101-111.

Park, S.J. and Mansoori, G.A., 1988. Aggregation and Deposition of Heavy Organics in Petroleum Crudes, J. Energy Sources, 10, 109-125.

Park, S.J. and Mansoori, G.A., 1988. Orgainc Deposition from Heavy Petroleum Crudes (A FRACTAL Aggregation Theory Approach)", Proceed. the

UNITAR/UNDP 4th Int'l. Conf. on Heavy Crudes and Tar Sands, Edmonton, Alberta, August.
Rassamdana, H., Dabir, B., Nematy, M., Farhani, M., Sahimi, M., 1996. Asphalt flocculation and deposition: I. the onset of precipitation. AIChE J. 42, 10–22.
Srivastava, R.K., Huang, S.S., Dyer S.B., Mourits, F.M. Mourits, 1995. Quantification of asphaltene flocculation during miscible CO_2 flooding in the Weyburn reservoir. J. Can. Pet. Tech. 34, 31–42.
Vasquez, D. and Mansoori, G.A., 2000. Identification and Measurement of Petroleum Precipitates, J. Petrol. Sci. & Eng., 26, 1-4, 49-56.
Yang, Z., Ma, C.F., Lin, X.S., Yang, J.T., Guo, T.M., 1999. Experimental and modeling studies on the asphaltene precipitation in degassed and gas-injected reservoir oils. Fluid Phase Equilib. 157, 143–158.
Yelling, W.F., Metcalfe, R.S., 1980. Determination and prediction of CO_2 minimum miscibility pressure. J. Pet. Tech. 32, 160–168.
Yu-Feng Hu, Shi Li, Ning Liu, Yan-Ping Chu, Sang J. Park, G.Ali Mansoori and Tian-Min Guo, Measurement and Corresponding States Modeling ofAsphaltene Deposition in Jilin Reservoir Oils, J. Petrol. Sci. & Eng., 41, p169-182, 2004.

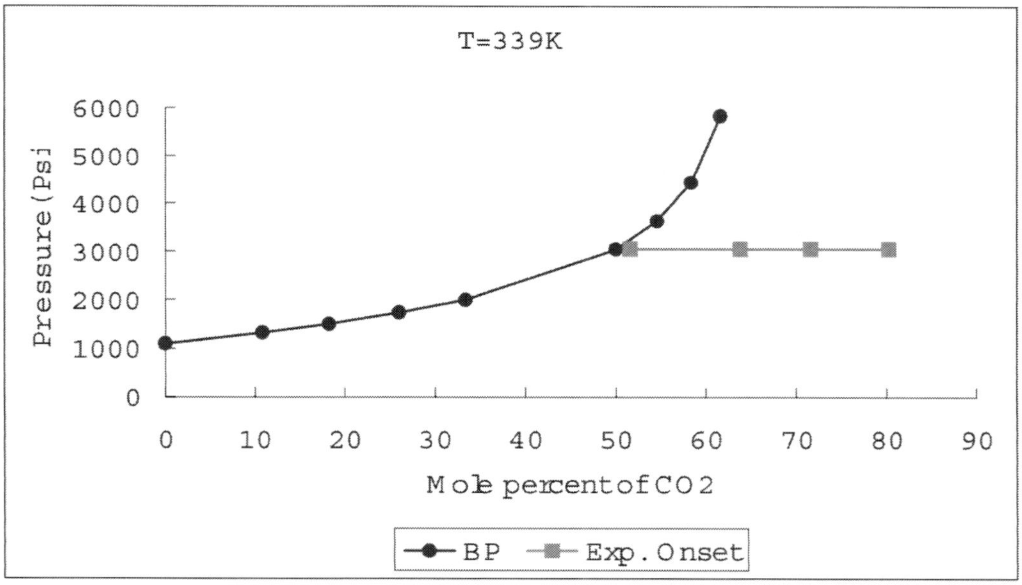

Figure 1. Phase diagram for mixtures of J2 crude oil and carbon dioxide at 339K. Region to the upper side of hollow square marks is the observed asphaltene deposition region(L-V-S).

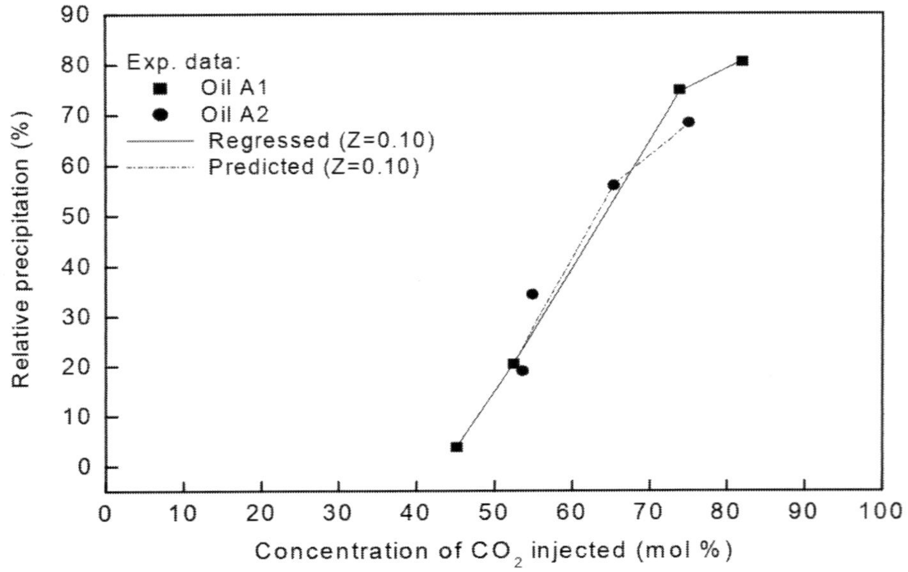

Figure 2. Effect of CO2 mole % for a given mixture of the amount of asphaltene precipitation at 20MPa.

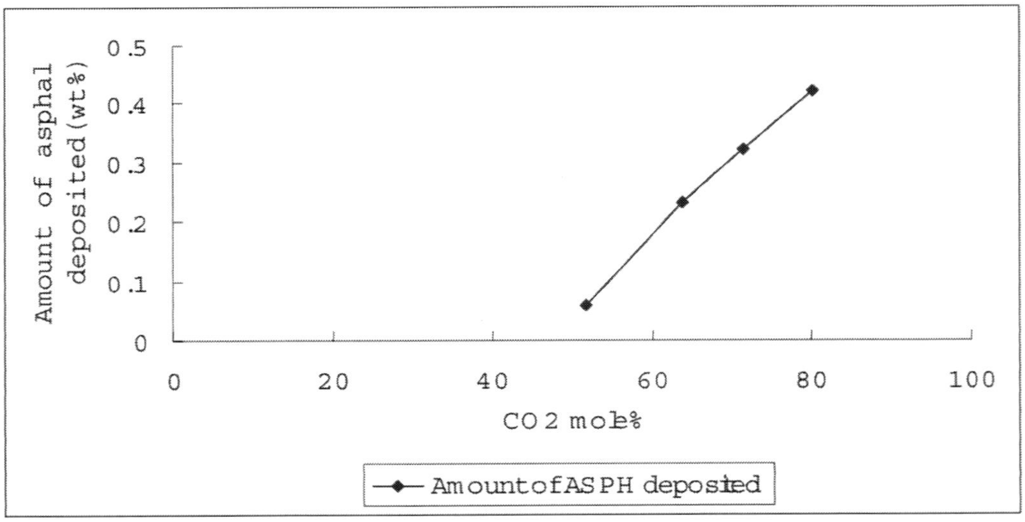

Figure 3. Comparison of calculated and experimental asphaltene precipitation in CO_2-injected Weyburn reservoir oils [Srivastava et al., 1995].

SELF-INCOMPATIBLE CRUDES AND PROCESSED RESIDS

Irwin A. Wiehe
Soluble Solutions, 3 Louise Lane, Gladstone, NJ 07934, USA

ABSTRACT

Self-incompatible oils are those containing insoluble asphaltenes that were not formed by mixing incompatible oils. Three refinery examples are given of self-incompatible oils: self-incompatible crude oils, thermally cracked oils, and oils hydrotreated with size selective catalysts that convert resins relative to asphaltenes. Methods are presented for using the Oil Compatibility Model and Tests to detect self-incompatible oils and to mitigate the fouling caused by the insoluble asphaltenes.

INTRODUCTION

Although refinery fouling is extremely costly, for a long time fouling was considered the price of doing the petroleum refining business with little hope for its mitigation. Fouling in crude preheat systems alone commonly costs United States refineries 2 to 10 cents per barrel processed in energy, maintenance, and loss of process time (1). Thus, for a refinery capacity of only 100,000 barrels per day this amounts to 0.7 to 3.6 million dollars per year. Of all the types of preheat system fouling, asphaltene deposition used to be the most mysterious to understand and the most difficult to control. However, the Oil Compatibility Model and tests now enable the prediction of what proportions and order of blending will cause asphaltenes to precipitate when one or more oils are mixed (2,3), called incompatible oils. These include both crude oils and refinery process oils (4). The Oil Compatibility Model even can be used to predict when asphaltenes will adsorb onto process surfaces from compatible, but nearly incompatible oils (5, 6). The basic parameters of the Oil Compatibility Model are two dimensionless solubility parameters, the insolubility number, I_N, and the solubility blending number, S_{BN}. These are determined by mixing individual oils with a nonpolar solvent, such as toluene, and with a nonpolar nonsolvent, such as n-heptane, and locating points of incipient asphaltene precipitation. The compatibility of any mixture of oils is predicted when the volumetric average solubility bending number is greater than the insolubility number of any oil in the mixture. The region of near incompatibility is when the mixture is predicted to be compatible but the volumetric average solubility bending number is less than 1.4 (especially less than 1.3) times the maximum insolubility number of the oils in the mixture.

There are oils that contain insoluble asphaltenes that are formed without blending oils. It is well known that the thermal conversion of petroleum resids, such as visbreaking or catalytic hydroconversion, can produce heavy oil byproducts containing insoluble asphaltene sediments. However, it is not well known that some crude oils contain insoluble asphaltenes. These will be called "self-incompatible" oils and the term will be generalized to include all oils containing insoluble asphaltenes that were not formed by blending, including those that precipitated asphaltenes as a result of thermal conversion.

SELF-INCOMPATIBLE CRUDE OILS

One point of incipient asphaltene precipitation commonly measured is to slowly titrate the oil with n-heptane and determine the maximum volume of n-heptane that can be added to the oil without precipitating asphaltenes (heptane dilution test). Surprisingly, there is a significant fraction (about 15%) of crude oils that do not precipitate asphaltenes when the added n-heptane is five times the volume of the oil. For these, we arbitrarily assign an insolubility number of zero (4). However, the solubility blending number still needs to be measured using one of the tests discussed previously (4) because such a crude oil can still be a nonsolvent for other crude oils containing asphaltenes of high insolubility number. Even more surprisingly, there are crude oils that require no n-heptane addition to precipitate asphaltenes. These contain insoluble particles, revealed by observing a drop of the oil between a glass slide and a cover slip under an optical microscope. By finding that the particles dissolve when the oil is blended with toluene but not when the oil is blended with n-heptane, one concludes that the insoluble particles are asphaltenes and the crude oil is self-incompatible.

Evaluation of Compatibility Numbers

The kinetics of dissolving insoluble asphaltenes is much slower than the kinetics of precipitation because dissolving asphaltenes in oil is both a colloidal dispersion and a solution process (7). As a result, titrating a self-incompatible oil with a solvent, like toluene, to determine points of incipient asphaltene precipitation is found to overshoot the equilibrium point. That is, a significant volume of n-heptane can be added back to the toluene-oil mixtures at these points before asphaltenes reprecipitate. Therefore, it is recommended that toluene first be added to self-incompatible oils and verified that the asphaltenes are dissolved by observing a drop of the oil-toluene mixture between a glass slide and a cover slip under an optical microscope. Then, n-heptane can be slowly titrated into this mixture to determine the incipient asphaltene precipitation point. By adding different amounts of toluene initially, one can determine a series of incipient asphaltene precipitation points. The result for the self-incompatible crude oil, Yme, is shown in Figure 1 where the volume percent toluene in the test liquid (toluene-n-heptane mixture) is plotted against 100 times the volume ratio of oil to test liquid for the points of incipient asphaltene precipitation. Previously (2), it was shown that the Oil Compatibility Model predicts that such data should be linear on this plot and that the y-axis intercept is equal to the insolubility number. While for self-compatible oils this line has a negative slope, Figure 1 shows that for self-incompatible oils the slope of the line is zero (parallel to the x-axis). The volume percent toluene in the test liquid for points of incipient asphaltene precipitation does not vary with the volume ratio of oil to test liquid within experimental error (here equal to 13). This is reasonable because the data should be linear but this line cannot intersect the x-axis for any finite value of the volume ratio of oil to test liquid. The x-axis represents test liquids containing only n-heptane and zero n-heptane (thus infinite volume ratio of oil to test liquid) is required to precipitate asphaltenes for self-incompatible oils.

The insolubility number, I_N, is equal to 13 for Yme crude oil because it is equal to the y-axis intercept. The solubility blending number, S_{BN}, is given by:

$$S_{BN} = I_N \left[1 + \frac{100}{H_D} \right]$$

where H_D is the x-axis intercept (2). Since for self-incompatible oils the x-axis intercept is infinite, the solubility blending number is equal to the insolubility number and equal to the percent toluene in the test liquid for any volume ratio of oil to test liquid (13 for Yme).

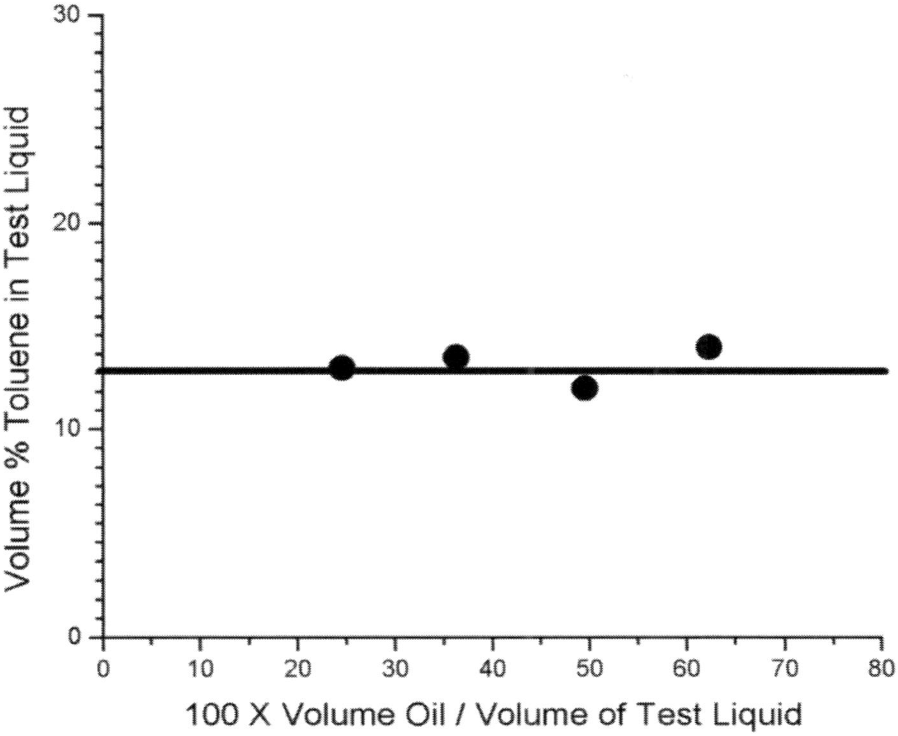

Figure 1. For the self-incompatible crude oil, Yme, the percent toluene in the test liquid at incipient asphaltene precipitation is independent of oil concentration.

Other Self-Incompatible Crude Oils

Unfortunately, the identity of self-incompatible crude oils is quite proprietary because of the competitive advantage of a refinery to avoid the high fouling caused by processing self-incompatible crudes. The identity of Yme was only permitted to be published because production of this crude was terminated. More than 10 self-incompatible crude oils have been identified by the author with insolubility numbers ranging from 12 to 58. Thus, Yme is at the low end of this range where the asphaltenes are insoluble primarily because the crude is very paraffinic and thus, a poor solvent for asphaltenes. However, at the highest end of this range the asphaltenes are insoluble because of the highly insoluble asphaltenes, the highest insolubility number measured for any unprocessed crude oil.

One might question how a self-incompatible oil can be produced through the porous media of an oil reservoir. There are at least two cases of self-incompatible crude oils. One case is that the asphaltenes come out of solution during production by the decreasing pressure as the oil flows to the surface. Soluble gases can be a good solvent for asphaltenes at high pressures but a nonsolvent at low pressures (8,9). The lowering of the temperature of the oil from reservoir conditions to surface temperatures can also promote asphaltene insolubility. However, when the oil contains considerable soluble gases, asphaltenes are more soluble at lower temperatures than higher temperatures (9). Nevertheless, some crude oils are known to plug wells with asphaltenes during production for the reasons given and many of these result in self-incompatible crude oils. A second cause of self-incompatible crude oils is that the crude oil is formed by blending incompatible crude oils near the production sites after separate production. Although in this case the term "self-incompatible" may be a misnomer, many crude oils are blends from different production sites and the term still makes sense as seen through the eyes of a refinery. Gas condensates in particular are known to be poor solvents for asphaltenes but are commonly blended into crude oils.

EFFECT OF RESID THERMAL CONVERSION

The thermal conversion of petroleum resids is known to make the asphaltenes less soluble. In terms of the Phase-Separation Kinetic Model (10) and the Pendant-Core Building Block Model (11), thermal conversion cracks pendant groups off asphaltene molecules towards a more aromatic and less soluble core. If these converted asphaltenes become insoluble in the reacting resid at thermal reaction temperatures, a liquid crystalline coke, carbonaceous mesophase, is formed (10). However, even if the converted asphaltenes are soluble at thermal reaction temperatures, they can become insoluble in the converted resid on cooling. However, the resulting sediments are insoluble asphaltenes, toluene soluble-n-heptane insoluble, rather than toluene insoluble coke (12). Thus, the resulting converted resid is self-incompatible.

Prediction of the Appearance of Asphaltene Sediments

Typically, the insolubility number is found to increase linearly with reaction time as is shown in Figure 2 for the visbreaking of Athabasca bitumen (13). Although the data in Figure 2 were reacted at various reaction temperatures, T° K, the actual reaction times, t, were transposed to an equivalent reaction time at 427°C, θ, using the Arrhenius relationship and an activation energy, E_A, of 53.1 kcal/g-mole:

$$\theta = t \ \exp[(E_A / R)(1/T - 1/700)]$$

Normally, the solubility blending number of a thermally converted resid slightly decreases linearly with reaction time. However, in this case with Athabasca bitumen the solubility blending number remained constant at 100. Nevertheless, since the Oil Compatibility Model predicts incipient asphaltene precipitation when the insolubility number equals the solubility blending number, this is predicted to occur at a reaction time of 27.2 minutes at 427°C for Athabasca bitumen. The ability to predict the reaction time at the same temperature or the temperature at the same reaction time that asphaltene sediments will appear in visbroken oil at room temperature based upon data collected before asphaltenes actually appear is particularly valuable. This has not previously been available by any technique.

Prediction of the Coke Induction Period

Since coke will not form until asphaltenes become insoluble at reaction temperatures, coke will not form until slightly longer reaction times than when insoluble asphaltene sediments begin to form at room temperature. Thus, this same method of calculating when the insolubility number equals the solubility blending number provides a conservative prediction of the coke induction period. Despite being conservative, it still is more accurate than other available methods (14) for predicting the coke induction period.

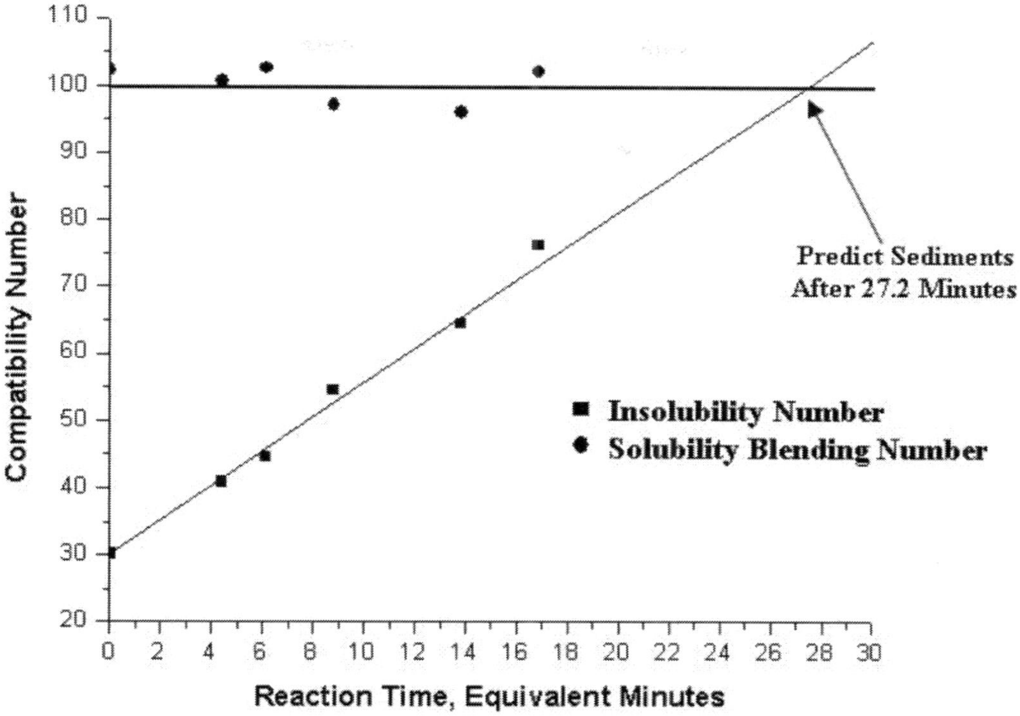

Figure 2. The Oil Compatibility Model predicts the appearance of insoluble asphaltene sediments in visbroken Athabasca bitumen at a reaction time of 27.2 minutes and a temperature of 427°C when the insolubility number and the solubility blending number become equal.

HEAT EXCHANGER FOULING AFTER RESID HYROTREATING

A fixed bed hydrotreater in a refinery is used to hydrotreat atmospheric resid before fluid catalytic cracking. The main objectives of the hydrotreater are to remove sulfur and metals and to reduce Conradson carbon residue. Large pore catalyst was put in the top of the catalyst bed to hydrotreat asphaltenes. A consistent pattern emerged during each of the first three runs that were about one year in length. No fouling was observed during most of the run but near the end of run the heat exchanger that cooled the liquid product severely fouled. As is common practice, the temperature of the hydrotreater was increased during the run to keep the sulfur level below a certain specification.

At the end of the third run before the catalyst was removed, a number of tests were done at various temperatures and space velocities. Samples of the feeds and products of these tests as well as for the start of the fourth run on fresh catalyst were collected. Microscopic examination of the oil samples revealed that all the liquid products at the end of run contained insoluble asphaltenes (self-incompatible) except one operated at low temperatures at the highest space velocity. On the other hand, the asphaltenes were soluble in the oil samples of the liquid product of the fresh catalyst and the resid feeds (self-compatible). The toluene equivalence, the minimum percent toluene required to keep asphaltenes from precipitating when mixed with n-heptane at a concentration of 2 grams of oil in 10 ml. of toluene-heptane mixture, was measured on each feed and product oil. As discussed previously, for the self-incompatible product oils the insolubility number and the solubility blending number equals the measured toluene equivalence.

Asphaltenes (n-heptane insolubles) were separated from all feeds and products and submitted for elemental analysis. The quantity of resins (polar aromatics) was measured by high performance liquid chromatography on each of the remaining deasphalted oils (15). As shown in Table 1, the quantity of asphaltenes was decreased significantly from feed to product at the start of run. The hydrogen to carbon (H/C) atomic ratio actually increased for the asphaltenes remaining in the product as compared to the feed and the sulfur significantly decreased. On the other hand, the asphaltenes from samples at the end of run barely decreased concentration, decreased in hydrogen to carbon atomic ratio, and only slightly decreased in sulfur as compared with asphaltenes in the feed. Therefore, asphaltenes were entering the large pore catalysts at start of run and being hydrotreated. At the end of run the asphaltenes were mostly by-passing the catalyst and only slightly thermally cracking to reduce sulfur, yield, and hydrogen to carbon atomic ratio (Figure 3). This was confirmed by the pore diameter of the spent large pore catalyst greatly decreasing from the deposition of vanadium and nickel. During the run when the sulfur and Conradson carbon of the product increased because asphaltenes were unconverted, the process temperature was raised. This caused the resins that could still enter the catalyst pores to be converted even more to compensate for the unconverted asphaltenes. Therefore, at end of run the resins to asphaltene ratio greatly decreased. Since the resins act as natural dispersants for asphaltenes (7), at end of run the toluene equivalence (thus, the insolubility number) increased, showing that the asphaltene solubility decreased. Figure 4 shows that below a resin to asphaltene weight ratio of 1.7 that the toluene equivalence increases approximately linearly with decreasing resin to asphaltene weight ratio.

Table 1.

Properties of Feed and Product Asphaltenes at the Start and End of Hydrotreater Run

	Asphaltene	Yield, wt%	H/C, atomic	Sulfur, wt%
Start of Run	Feed	5.4	1.25	4.54
	Product	2.8	1.36	2.76
End of Run	Feed	6.3	1.23	4.53
	Product	5.7	1.18	3.44

In conclusion, the heat exchanger was fouled by insoluble asphaltenes near the end of run because of the greater conversion of resins than asphaltenes when the large pore catalysts were partially plugged with deposited vanadium and nickel and when the hydrotreater temperature was increased for greater reactivity. Once the resin to asphaltene ratio was decreased below about 1.3, asphaltenes became insoluble on cooling without sufficient natural dispersant. While it is extremely difficult, if not impossible, to physically separate resins without separating asphaltenes, the resid hydrotreater achieved this by chemical reaction with size selective catalysts. Nevertheless, once the mechanism of fouling was known, a number of ways to interrupt the mechanism and mitigate the fouling became obvious and those that best fit into the refinery operation were successfully applied.

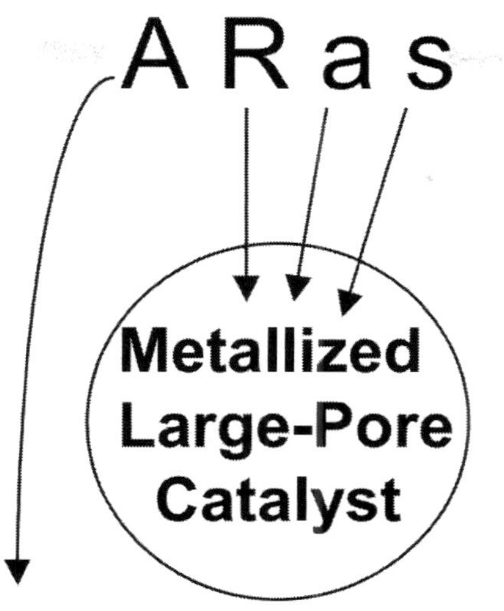

Figure 3. When the large pore catalyst partially fills with deposited nickel and vanadium near end of run, asphaltenes bypass the catalyst.

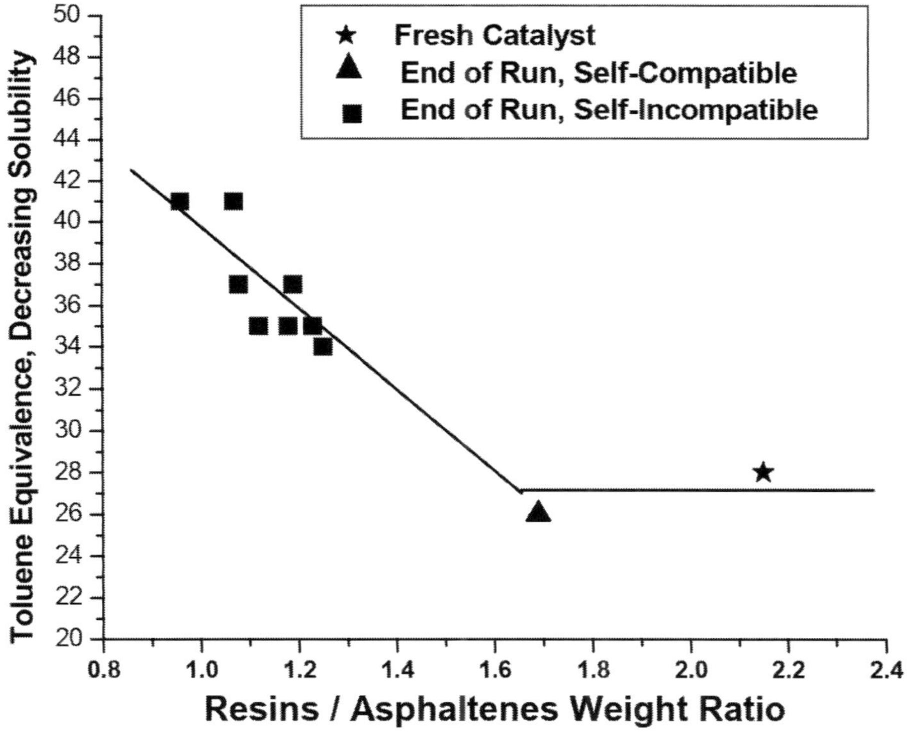

Figure 4. At the end of run for resid hydrotreating the decreasing weight ratio of resins to asphaltenes increases the toluene equivalence (thus, the insolubility number for ratios below 1.3) of the liquid product.

OVERALL CONCLUSIONS

Three refinery examples of self-incompatible oils, those with insoluble asphaltenes not formed by blending oils, were described. These examples included self-incompatible crude oils that arrive at the refinery, self-incompatible oils caused by the thermal conversion that form more aromatic and more insoluble asphaltenes, and self-incompatible oils as the result of preferentially converting resins relative to asphaltenes. It was shown that for self-incompatible oils the insolubility number and the solubility blending number are equal. As a result, only one test, rather than two, to measure incipient asphaltene precipitation is necessary to determine these numbers. The equality of the solubility blending number and the insolubility number is used as a predictor of when thermally converted oils form self-incompatible oils. Since both compatibility numbers are linear functions of thermal reaction time, one can predict this time when insoluble asphaltenes appear by measuring the compatibility numbers of the feed and of the oil after one or more thermal reaction times. Since insoluble asphaltenes are formed on cooling before coke forms at reaction temperatures, this predicted time when insoluble asphaltenes appears is a conservative estimate of the coke induction period.

Given that asphaltenes are suspended in oil by a combination of solubility and colloidal dispersion (7), asphaltenes can be precipitated by adding saturates (nonsolvent), removing small ring aromatics (solvent), and removing resins (dispersant) in addition to cracking off the asphaltene pendant groups. This paper describes the first case reported when asphaltenes were made insoluble by selectively removing resins and leaving the asphaltenes in the oil. This was accidentally accomplished by size selective catalysts after metals deposition during hydrotreating of atmospheric resid and causing heat exchanger fouling. The elucidation of this mechanism quickly led to the mitigation of the fouling.

ACKNOWLEGEMENTS

Thanks to Statoil for permission to publish the data on Yme crude oil, an example self-incompatible crude oil. The prediction of asphaltene sediments and the coke induction period was a result of consulting with the National Centre of Upgrading Technology in Devon, Canada using data of Parvis Rahimi and of Richard Parker.

REFERENCES

1. De Jong, A.R. Determining Fouling Mechanisms by Monitoring and Chemical Analysis. In *the Proceedings of the 4th International Conference on Refinery Processing*; Podder, S.K, Boock, L.T., Eds.; AIChE: New York, **2001**; 193-203.
2. Wiehe, I.A.; Kennedy, R.J. The Oil Compatibility Model and Crude Oil Incompatibility. Energy & Fuels **2000**, 14 (1), 55-59.
3. Wiehe, I.A.; Kennedy, R.J. Process for Blending Potentially Incompatible Petroleum Oils. US Patent 5,871,634, February 16, **1999**.
4. Wiehe, I.A.; Kennedy, R.J. Application of the Oil Compatibility Model to Refinery Streams. Energy & Fuels **2000**, 14 (1), 60-63.
5. Wiehe, I.A.; Kennedy, R.J; Dickakian, G. Fouling of Nearly Incompatible Oils. Energy & Fuels **2001**, 15 (5), 1057-1058.
6. Wiehe, I.A.; Kennedy, R.J. Process for Blending Petroleum Oils to Avoid Being Nearly Incompatible. US Patent 5,997,723, December 7, **1999**.
7. Wiehe, I.A.; Liang, K.S. Asphaltenes, resins, and other petroleum macromolecues. Fluid

Phase Equilibria **1996**, 117, 201-210.
8. Firoozabadi, A. Thermodynamics of Hydrocarbon Reservoirs. McGraw-Hill: New York, **1999**; 328-329.
9. Leontaritis, K.J. Asphaltene and Wax Deposition Envelopes. Fuel Sci. Technol. Int. **1996**, 14 (1&2), 13-39.
10. Wiehe, I.A. A Phase-Separation Kinetic Model for Coke Formation. Ind. Eng. Chem. Res. **1993**, 32 (11), 2447-2454.
11. Wiehe, I.A. The Pendant-Core Building Block Model of Petroleum Residua. Energy & Fuels **1994**, 8 (3), 536-544.
12. Wiehe, I.A. Mechanisms of Coke and Sediment Resulting from Resid Conversion, National Meeting of the American Chemical Society, San Francisco, **1997**.
13. Rahimi, P.M.; Parker, R.J.; Knoblauch, R.; Wiehe, I.A. Stability and Compatibility of Partially Upgraded Bitumen for Pipeline Transportation. In *Proceedings 7th International Conference on Stability and Handling Fuels*, **2000**; 177-192.
14. Schabron, J.F. Predicting Coke Formation Tendencies. Fuel, **2001**, 80, 1435-1446.
15. Robbins, W.K. Quantitative Measurement of Mass and Aromaticity Distributions for Heavy Distillates 1. Capabilities of the HPLC-2 System. J. Chrom. Sci., **1998**, 36, 457-466.

CHEMICAL CLEANING OF OIL REFINERY HEAT EXCHANGERS -- THE NEED FOR A JOINT EFFORT

EXTENDED ABSTRACT

Himanshu M. Joshi[1] **and Glen Brons**[2]

[1] 1545 Rte. 22 East, Annandale, NJ 08801, USA; E-mail: himanshu.m.joshi@exxonmobil.com
[2] 1545 Rte. 22 East, Annandale, NJ 08801, USA; E-mail: glen.b.brons@exxonmobil.com

INTRODUCTION

One component of the high cost of fouling to oil refineries is the maintenance money spent to clean fouled heat exchangers and the associated lost production while the exchanger is out of service. This presentation will highlight some specific fouling situations in oil refineries which can benefit from chemical cleaning. Cooperative efforts are needed between refinery operators, researchers, and chemical suppliers to develop the proper solvents and to find the optimum procedures to use them for cleaning.

HIGH COST OF TRADITIONAL MECHANICAL CLEANING

Large shell-and-tube heat exchangers (Figure 1) usually require disassembly of both ends and internals, removal of the tube bundle, transportation to a cleaning facility and cleaning, reassembly, and leak testing. This process can take between 3-14 days, depending on several factors like exchanger size and weight, severity of fouling, whether specialty equipment is required to extract the tube bundle etc. In the worst cases the cost for the mechanical work is as high as 40-50k$ per exchanger. A penalty of up to ten-times that cost could also be incurred, depending on the duration of outage and if production is affected while the cleaning takes place. As a result there is an obviously large incentive to develop cleaning methods which minimize or eliminate the mechanical costs and which can be carried out in short periods of time (one day or less).

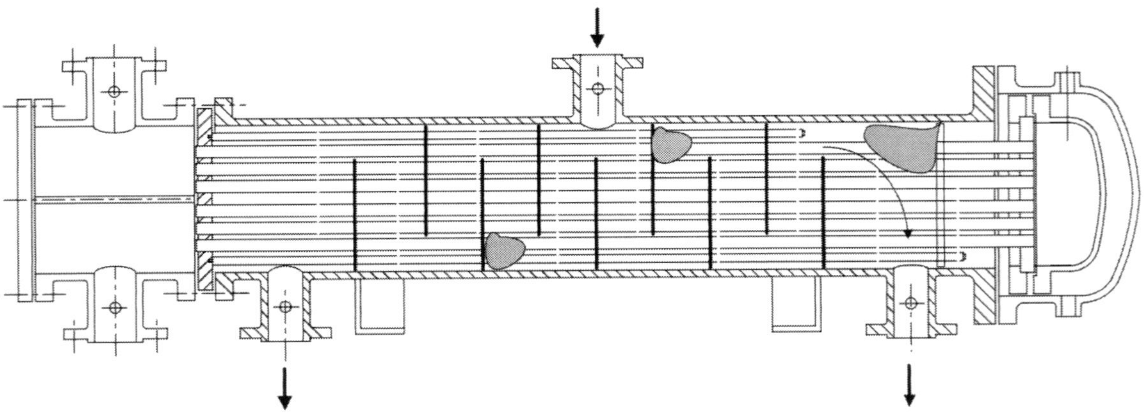

Fig. 1 Details of a typical shell-and-tube heat exchanger

CHEMICAL CLEANING AS AN ALTERNATIVE

Chemical cleaning is one technique to achieve the objectives of cheaper and quicker maintenance on fouled heat exchangers. It is broadly defined as cleaning a fouled heat exchanger using a chemical to dissolve some or all of the constituents of the solid foulant deposit. Typically, cleaning will be performed by flowing the solvent chemical through the exchanger (circulating or once-through) without the need for disassembly.

Current attempts to chemically clean heat exchange equipment have been successful in a limited number of cases, and even then, the benefits are sustained for only a relatively short amount of time compared to the alternative method of dismantling and washing with a high pressure water jet. The usual method is to circulate hydrocarbon streams like naphtha, kerosene, or gas oil, in the hope that the foulant deposit which is expected to be mostly organic can be dissolved and washed away. As seen in the results plotted below (Fig. 2), which are very typical for such methods, the chemical cleaning (referred to as LCO Soak in the plot) is not very effective.

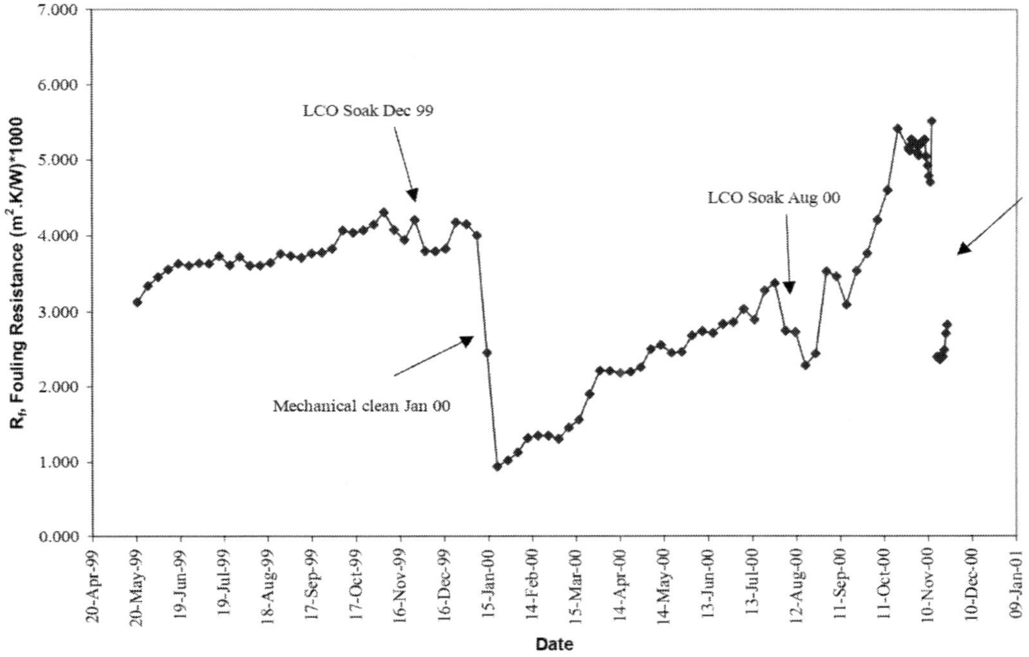

Fig. 2 Effect of Cleaning on Heat Exchanger Fouling Factors

The ineffectiveness of chemical cleaning in the above situation can be attributed to two main reasons. First, the inadequacy of the LCO as a solvent for the degraded organic material in the foulant as well as for the inorganic materials (such as iron sulfide), the latter of which may be within the deposit. Secondly, the inability to circulate a sufficiently large volumetric flow of the solvent due to the small connections used for circulation, and to get the solvent to cover the quiescent zones such as those shown on the shellside of Fig. 1

To realize the substantial benefits that effective chemical cleaning can provide, research and development is required in the following areas: (1) Determination of which types of solvents can dissolve the various types of foulant material as briefly described below, and which can be circulated safely in oil refinery equipment, (2) Procedures and equipment to ensure adequate distribution of the solvent to all areas of the fouled equipment, and (3) Understanding of the fouling process so that the cleaning can be performed at the optimum intervals considering cost, effectiveness, and time required.

Specific situations where large cost savings can be realized include: Fouling of crude oil due to organic (e.g., asphaltenes) as well as inorganic (salts, corrosion) precursors; fouling of FCC slurries due to fine catalyst particulates as well as coking; fouling of heavy hydrocrabon streams (resids) due to corrosion and coking; and fouling of mid-range hydrocrabons due to polymerization. A cooperative effort between refiners, universities, vendors, and chemical companies is needed to develop or identify chemicals that can be used for these fouling situations. The research objectives should include establishing protocols for such chemicals to be used at the right time to obtain the maximum benefit at the lowest possible cost.

CONCLUSIONS

Chemical cleaning of oil refinery heat exchangers has the potential of large economic benefits but joint research and development between the refiners and chemical companies is required before this benefit can be realized.

Paper Number <u>87e</u>

INTERPRETING THE RESULTS OF MONITORING THE PERFORMANCE OF PRE-HEAT TRAINS PROCESSING CRUDE OIL

Polley G.T.
www.pinchtechnology.com

Paper presented at the Spring National AIChE Meeting, April 25-29, 2004

Session: Fouling Mitigation

7th International Conference on Refining Processes

UNPUBLISHED

Interpreting the Results of Monitoring the Performance of Pre-Heat Trains Processing Crude Oil

Polley G.T.

The monitoring of the fouling levels of individual exchangers used in pre-heat trains processing crude oil is common practice. Such data is used to identify when individual units should be cleaned. It should also provide a guide to the effective design of such exchangers. However, in this context, the data obtained in such studies can appear confusing. Theory would suggest that the exchangers positioned at the hot end of the train would be the ones in which fouling was heaviest. Experience often yields a different picture.

In this paper we report on a study conducted by Total (reported previously by Bories & Patureaux [2002]). We then show how an analysis (incorporated in ESDU's EXPRESS program) can be used to interpret the data in a manner that yields results that are helpful in design.

Background to Study

A pre-heat train which was part of the crude distillation unit of a Total refinery was revamped in order to process 160 000 bbl/day of crude. A simplified diagram of the train is given in Figure 1.

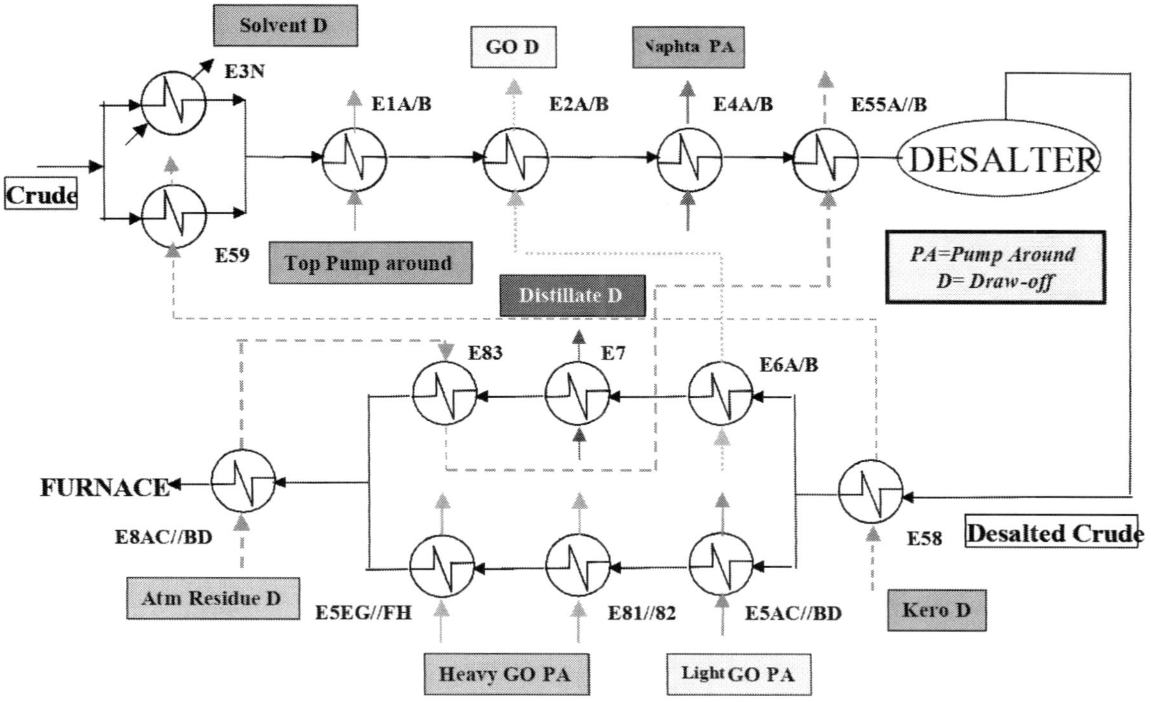

Figure 1. Simplified Pre-heat Train Structure

When the plant was re-commissioned it very rapid fouling occurred in the pre-heat train exchangers. This fouling not only led to significant decrease in heat recovery (with subsequent decrease in furnace inlet temperature and increased fuel consumption) but to a marked reduction in throughput as the furnace became bottlenecked. Rather than increasing throughput the plant modification resulted in reduced throughput. The financial losses were very high. In just three months following re-commissioning the accumulated loss was put at around 1.5M$. A solution was needed so, the individual exchangers were monitored in an attempt to understand the problem.

Data from an Exchanger Monitoring Exercise

The immediate problem encountered in analysing the data was understanding what it meant. The initial expectation was that the higher the temperature of the crude oil the worse would be the fouling. In Table 1 we rank the exchanger inlet temperatures and compare the measured fouling levels. (These number are taken from the original work of Bories & Patureaux. The units used are based on the kcal rather than on watts – as used in this paper – and the film heat transfer coefficients used to derive the values from the overall coefficient were not as accurate as those used in this paper. They are only provide a guide to the fouling levels).

Exchanger	Crude Inlet Temperature C	Level of Fouling Rate
E58	141	0.04
E6AB	148	0.47
E5AD	148	0.94
E8182	194	0.5
E7	205	0.34
E83	218	0.27
E5EH	223	1.91
E8AD	248	0.86

Table 1. Fouling Rates at Differing Crude Temperatures

There is no clear correlation between crude temperature and fouling rate.

Appreciating that each exchanger is of different design and that the velocities through the existing units had been reduced during the retrofit it can be expected that the crude velocity played a role in fouling rate. In Table 2 we rank the exchangers in order of velocity.

Exchanger	Velocity m/s	Fouling Rate
E5AD	0.9	0.94
E5EH	0.98	1.91
E6AB	1.04	0.47
E8182	1.09	0.5
E7	1.29	0.34
E8AD	1.64	0.86
E83	1.83	0.27
E58	4.05	0.04

Table 2. Fouling Rates at differing velocities

The correlation is much stronger. However, there are units in which the fouling is much higher than would be expected if velocity alone was the key. These were units operating at higher temperature.

Modelling of the Fouling Process

The results reported by Bories & Patureaux are in line with the results of recent research. Ebert & Panchal [1997] recognised that there were competing mechanisms occurring during the fouling process and where the mechanism mitigating against fouling was stronger than that promoting it fouling would not occur. They postulated a 'fouling threshold' that is a function of temperature and velocity.

Th existence of the 'fouling threshold' was subsequently demonstrated experimentally by Knudsen et al [1997].

Ebert & Panchal suggested that the mechanism promoting fouling was chemical reaction. They modelled this using very simple film theory. The volume of the reactor was taken to be proportional to the thickness of the heat transfer film (given by ratio of liquid thermal conductivity to film heat transfer coefficient) whilst the reaction rate was given by an Arrhenius Equation based on the 'film temperature'. The film heat transfer coefficient was assumed to be dependent upon Reynolds Number raised to a power 0.66.

The removal mechanism was assumed to be a wall shearing effect.

Polley et al [2002] subsequently modified this equation through the introduction of three simple changes:

- since, the Arrhenius Equation is exponential in terms of temperature, the use of wall temperature was favoured over the use of the lower value of film temperature
- for flow through tubes the Reynolds Number exponent of 0.8 appears more realistic
- rather than assume that the removal mechanism was a shearing effect, a mass transfer phenomena appeared more appropriate before a fouling layer formed.

The resultant fouling rate equation is:

$$\frac{dR_f}{dt} = \alpha \,\mathrm{Re}^{-0.8}\,\mathrm{Pr}^{-0.33}\exp\left(\frac{-E}{RT_w}\right) - \gamma\,\mathrm{Re}^{0.8}$$

Most recently, Yeap *et al.* [2003] have analysed a database of measured crude oil fouling rates obtained from both pilot plant experiments and data reconciliation on monitoring of the performance of industrial exchangers. They have compared the predictions of both the original and modified Ebert-Panchal equations and an equation based on a modification of one proposed by Epstein [1994] with this database. In the case of the Ebert-Panchal type models, found that the modified version yielded better comparisons. However, of the three models tested, the Modified Epstein Model yielded the best predictions.

Interpretation Using the EXPRESS program

Both the original and the Modified Ebert-Panchal Model have been incorporated into ESDU's EXPRESS computer program. (The Modified Epstein Model has not yet been incorporated into the program. Model development is an active research area for ESDU). The main intention has been to promote the use of such models in exchanger design (including the study of the effect of throughput on fouling propensity). Such use is demonstrated later. However, the program has recently been provide with the facility if inputting plant data (crude oil flow rate and exchanger terminal temperatures). Here the aim is the selection of model parameters which generate fouling curves (fouling resistance against time) that are in reasonable agreement with the data obtained on the plant. The results for the data obtained in this monitoring study are presented in Figures 2 to 6.

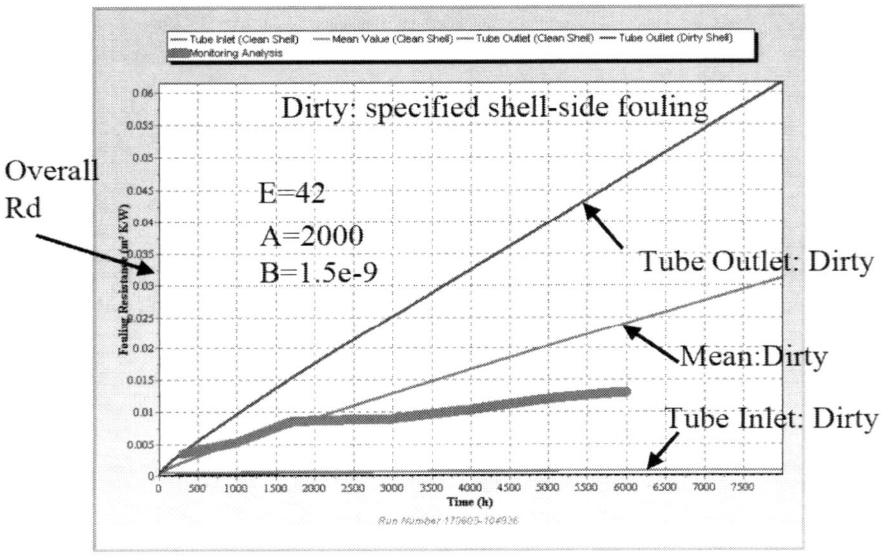

Figure 2. Predictions of Modified Ebert Panchal Model for Fouling of E6AB

The program user can obtain predictions for the situation in which there is already fouling present at the start of the exercise (by specifying a fouling resistance for the shell-side of the unit which provides the initial overall performance), or can start with a clean exchanger. In this study we opted to model the initial state by specifying a shell-side fouling resistance.

The program uses the specified fouling model to determine fouling rates at the tube inlet condition and an assumed worse condition. In a well designed exchanger using multiple passes the hot stream will leave the exchanger at the end where the crude (on the tube-side) both enters and leaves. The worse condition for fouling will be where the hot stream enters the exchanger. Unfortunately, we do not know the temperature of the crude at this point. So, we make a worse case assumption and take it to be the temperature at which the crude leaves the unit.

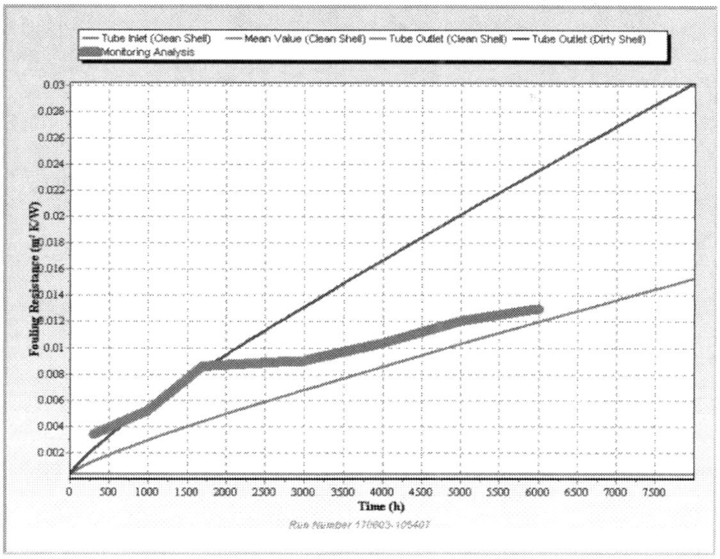

Figure 3. Fouling of Exchanger E6AB: Activation Energy 44 kJ/kmol
Crude Inlet Temp. 148 velocity 1.04

The program then generates three fouling curves: fouling at tube inlet, fouling under worse case and the mean of the two cases. At the current time we would suggest the plant data is compared with the mean fouling curve.

In Figure 2 we see a comparison between predicted and measured fouling curves for stated assumed fouling model parameters. Using the program we can quickly and easily adjust the fouling model parameters in order to get closer fit between measurement and prediction. In Figure 3 we see that by increasing the activation energy to 44 kJ/kmol we obtain closer fit.

We need to generate comparisons for each exchanger monitored. In Figure 4 we see the comparison for exchanger E5AD. The predictions here are closer to the worse case than the mean value. The conditions for this exchanger are not far removed from those in E6ab. The crude velocity is slightly lower. The predictions have been obtained for an activation energy of 42. Increasing the value to 44 would move the predicted curves down and the comparison would be poorer.

In Figure 5 we see the results for exchanger E8AD. In this unit the inlet crude temperature is 248 C and the velocity is 1.64 m/s. Again an activation energy of 42 has been used. Again raising the activation energy to 44 would make the comparison poorer.

Finally, in Figure 6 we see the predictions for exchanger E5EH which operates at a crude inlet temperature of 223 C and a velocity of 0.98 m/s. Here the predictions fall below the mean and are closer to the predictions for the exchanger inlet condition.

Increasing the activation energy from 42 to 44 would drive the predicted lower and thereby improve the comparison.

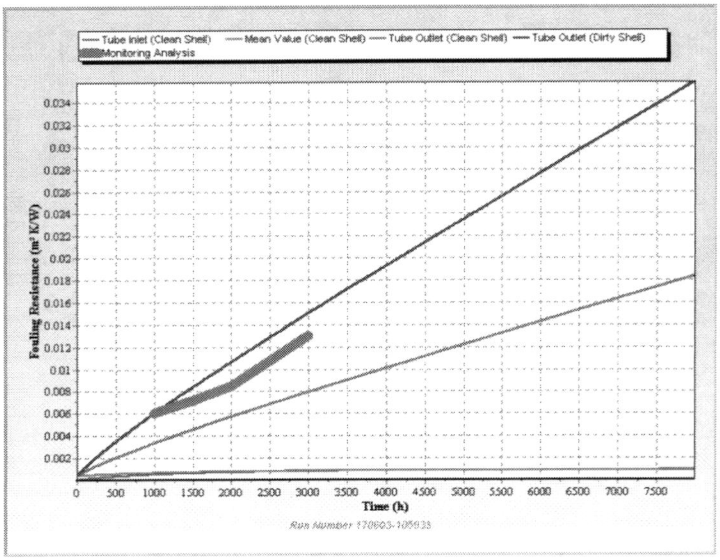

Figure 4. Fouling of Exchanger E5AD
Inlet Crude Temperature: 148, velocity 0.9 m/s

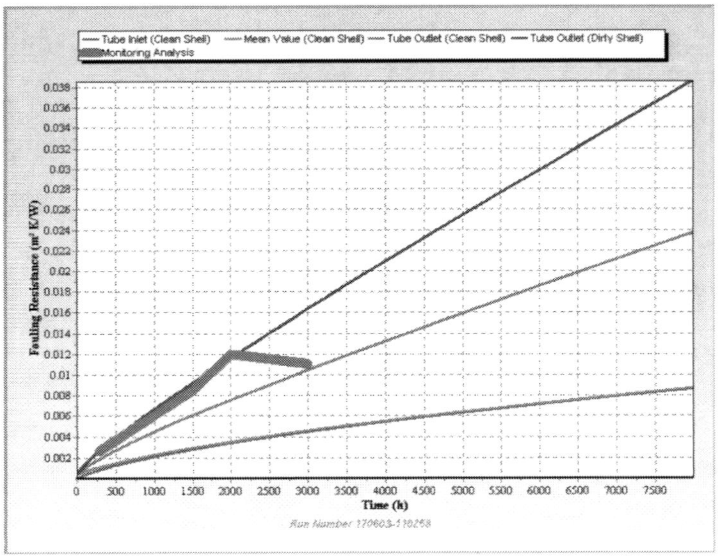

Figure 5. Fouling of Exchanger E8AD
Inlet Crude Temperature 248, velocity 1.64 m/s

On balance, we would suggest that the best activation energy to use would be 42 kJ/kmol.

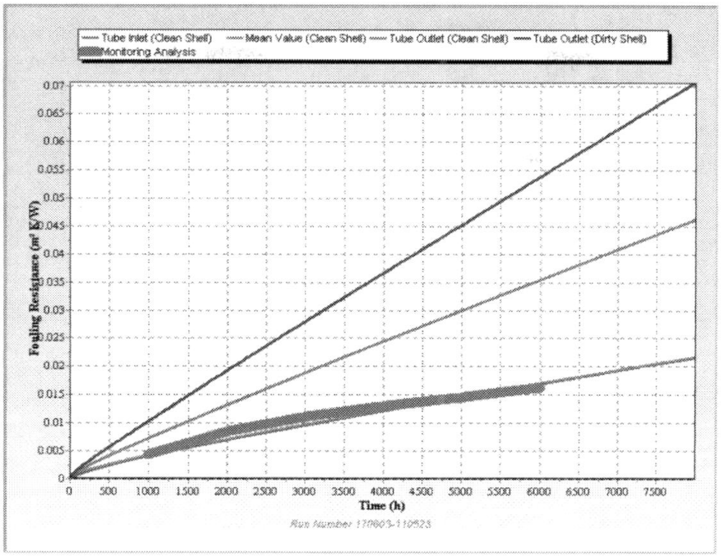

Figure 6. Fouling of Exchanger E5EH
Crude Inlet Temp. 223, velocity 0.98 m/s

Use of Fouling Models in Exchanger Design and Network Revamping

Using the above analyses we now have parameters for a model that describes the fouling process rather than just raw fouling data. We can make use of the fouling model in the design process.

First, we can use the data in exchanger design. In Figure 7 we show the 'parameter plot' (developed by Poddar and reported by Poddar & Polley [1996]) describing the exchanger 'design space' for a replacement for exchanger E5AD. In addition to showing the relationships between tube count and tube length for the constraints of allowable pressure drop and required thermal duty, the fouling model can be used to show the point on the thermal duty line at which the fouling threshold occurs. Any design positioned above this point will foul. Geometry falling below the line will be below the fouling threshold. We see that, ***if the fouling parameters identified above are correct***, the unit could be replaced with one having ***geometry that does not foul.***

We can extend our design study to include a consideration of the onset of fouling resulting from plant turndown (reduced tube velocity) for the EXPRESS program generates response curves for a variety of off design conditions. In Figure 8 (again developed for E5AD) we see how wall temperature and fouling threshold vary with exchanger throughput. As long as the threshold line is above the wall temperature line fouling will not occur. This allows us to appreciate how changes in throughput can affect exchanger fouling levels and possibly lead us to introduce safeguards into the plant operation.

Of course, the results of the exchanger analysis are subject to the values used for the fouling parameters. We have to ask the question "what are the consequences if we have got them wrong?".

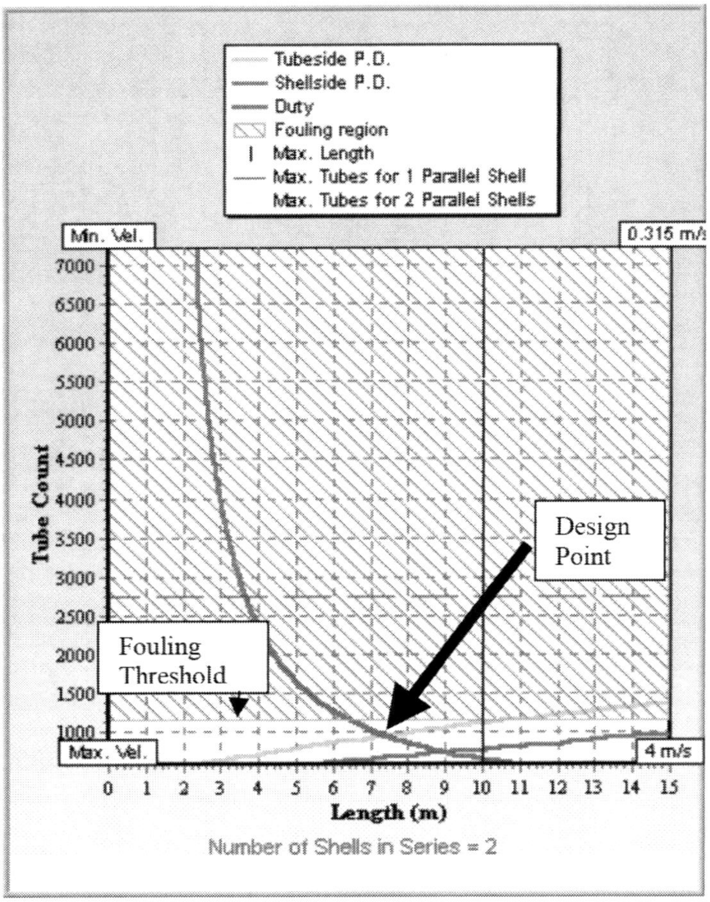

Figure 8. Design Space for Exchanger E5AD

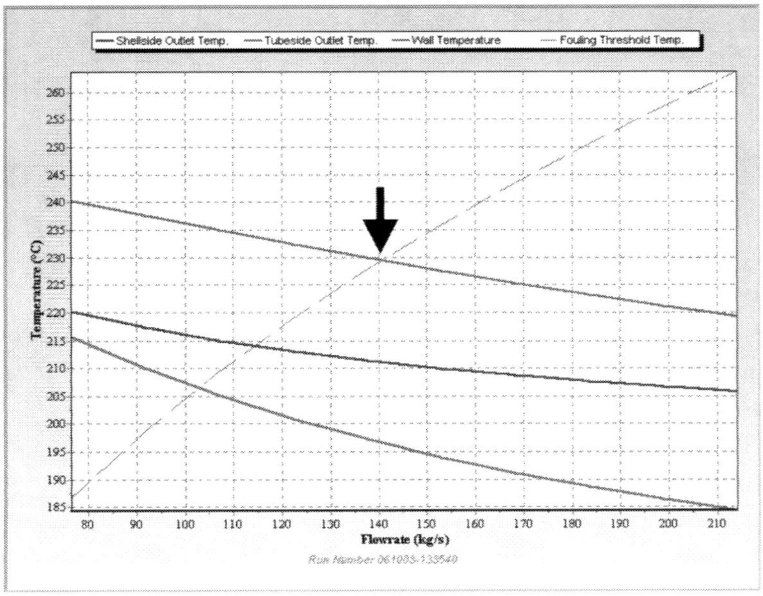

Figure 9. Onset of Fouling on Turndown

We can obtain the answer using the fouling curve facility we originally used in the monitoring analysis. Having new exchanger geometry (so changed velocity and wall temperature fields) we can generate new fouling curves. Now rather than adjust fouling parameters to fit monitoring data, we adjust them in order to examine the sensitivity of fouling levels to adjustments of parameters used.

Now fouling not only effects thermal performance, but it effects flow resistance and hence, exchanger pressure drop. The EXPRESS program allows the engineer to estimate this effect. This can be done using either a smooth layer model or a rough layer model. In Figure 10 we show a pressure drop curve for an exchanger operating above the fouling threshold.

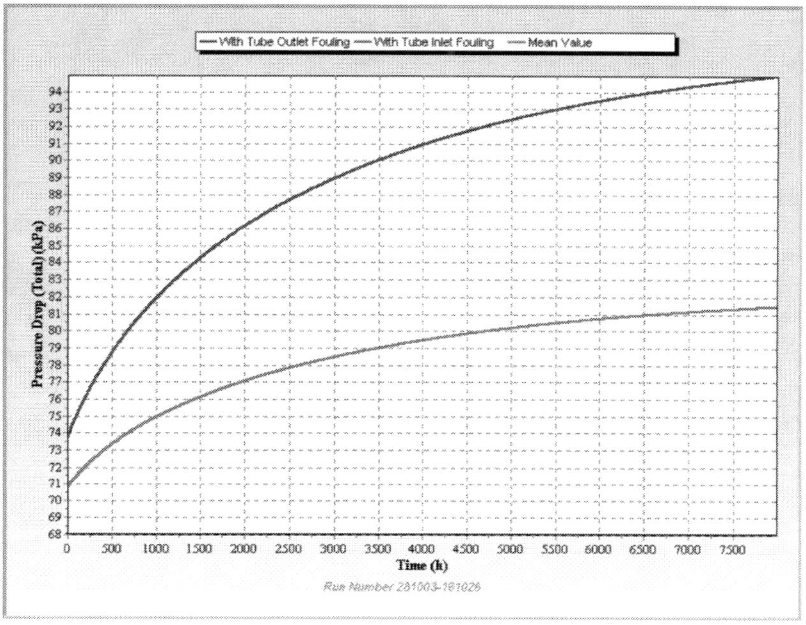

Figure 10. Pressure Drop v. Time

In this example the blue (upper) curve shows the pressure drop that would be encountered if the fouling occurring at the tube outlet was uniform across the exchanger. The absence of a third curve indicates that the unit is not fouling at the tube inlet. The second curve (red) is the pressure drop that would be encountered if the mean of the two predicted fouling levels was present across the full exchanger.

At the moment there is little information available on the roughness effects of fouling. However, if in addition to thermal performance exchanger pressure drop has been monitored, it is possible to adjust the layer roughness in order to fit the pressure drop data.

Finally, we can use the fouling model to identify changes to pre-heat train structure that mitigate fouling. This is the subject of a separate paper in this conference (Polley & Wilson [2004]).

Future Developments

The comparisons made by Yeap et al [2003] clearly show that there is a need for further research directed at understanding the development of fouling in crude oil systems and in the modelling of such systems. Some of this work will require careful experimentation in the laboratory. However, much of it will require the interpretation of data obtained on the plant.

The analysis used in this paper analysis is not very sophisticated for it involves comparison between data and the mean of two predictions developed for two point conditions. A much more meaningful result could be obtained by integrating the fouling model over both time and exchanger space. Such analysis is under development.

References

Bories, M. and Patreaux, T., 2003, Preheat train crude distillation fouling propensity evaluation by Ebert and Panchal model, *Proc.* Heat Exchanger Fouling & Cleaning: Fundamentals & Applications, Santa Fe, July 2003.

Ebert, W.A. and Panchal, C.B., 1997, Analysis of Exxon crude-oil slip stream coking data in *Fouling Mitigation of Industrial Heat Exchange Equipment*, Panchal, C.B., Bott, T.R., Somerscales, E.F.C. and Toyama, S.. (eds.), Begell House, NY, pp 451-460.

Epstein, N., 1994, A model of the initial chemical reaction fouling rate for flow within a heated tube, and its verification, *Proc. 10th Int. Heat Trans. Conf.*, Brighton, IChemE, Vol. 4, pp 225-229.

Knudsen, J.G., Lin, D. and Ebert, W.A., 1997, The determination of the threshold fouling curve for a crude oil, in: Bott, T.R., Melo, L.F., Panchal, C.B. and Somerscales, E.F.C., (eds.) *Understanding Heat Exchanger Fouling and its Mitigation*, Begell House, NY, pp 265-272.

Poddar T.K. & Polley G.T. Heat Exchanger Design Through Parameter Plotting, *Chem.Eng.Res.Des.*. 1996, 74A, 849-852

Polley G.T. & Wilson D.I. 2004, The Fouling Limit in Crude Oil Pre-heat Train Design, *AIChe Spring Meeting,* New Orleans

Yeap, B.L., Wilson D.I., Polley G.T. and Pugh S.J., 2003, Mitigation of crude oil refinery heat exchanger fouling through retrofits based on thermo-hydraulic fouling models, *Chem. Eng. Res. Des.* (*in press*).

OPTIMISING OPERATING CONDITIONS TO MITIGATE FOULING IN HEAT EXCHANGER NETWORKS

C. A. Rodriguez and R. Smith
Department of Process Integration, UMIST
Manchester, UK.

Abstract

The impact of fouling on the operation and economics of heat exchanger networks has long been recognised. In the past, however, fouling was seen as an unavoidable process and not much effort was devoted to mitigate its effects. In the last few years, the increasing global concern for reducing energy consumption has boosted the search for effective fouling mitigation strategies. In this paper, a new approach for mitigating fouling in existing heat exchanger networks is presented. It is based on the recognition that operating variables such as wall temperature and flow velocity may have a significant effect on fouling deposition rate. The approach combines the optimisation of operating conditions with the optimal management of cleaning actions in a comprehensive mitigation strategy. The application of the method is demonstrated with a case study of a refinery crude oil preheat train. Compared with the existing strategy to mitigate fouling by managing cleaning, the proposed approach leads to higher energy savings, lower operational costs and fewer disturbances to the background process.

Introduction

Fouling of heat exchanger networks (HENs) is a widespread problem. It is one of the major operational problems confronted by industries as diverse as food, textiles, pulp and paper, chemical, petroleum, glass, metals, among others [1]. The main effects of fouling on the operation of heat exchangers are to reduce the effectiveness of heat transfer and to increase the resistance to the flow of fluids. Other problems caused by fouling include corrosion, failure of equipment and contamination of products.

An important example of a system whose operation is significantly affected by fouling is the crude oil preheat train (PHT) of a refinery distillation unit (Figure 1). The distillation unit is the major energy consumer in any refinery, requiring around a third of the total energy bill of the site [2]. Between 60 - 70% of the energy required by this unit is recovered from the distillation products in the PHT [3], while the remaining heat is normally supplied in a furnace. Fouling reduces the efficiency of the PHT to recover heat, which makes it necessary to increase the consumption of fuel in the furnace or to decrease the throughput to the unit. The negative consequences on the economics of the process are obvious; it has been estimated that the cost penalty of fouling in the US refineries exceeds US$ 2 billion per year [4]. The burden caused by fouling is not only economic, but also environmental: The increase in the consumption of fuel due to fouling is responsible for the emission of hundreds of thousands of tons of CO_2 every year.

It is clear that there are important incentives to mitigate fouling in HENs. Thus, considerable attention has been paid to the development of effective fouling mitigation technologies in recent years. In this work, a new strategy for mitigating fouling in existing HENs is presented. It is based on the recognition that the propensity of some types of fouling can be reduced through the proper selection of the operating conditions of the network. The proposed method combines the optimisation of operating conditions with the scheduling of cleaning tasks in an integrated fouling mitigation framework. Although special emphasis is made on the application of the approach in PHTs, it can also be used to mitigate fouling in other kinds of HENs.

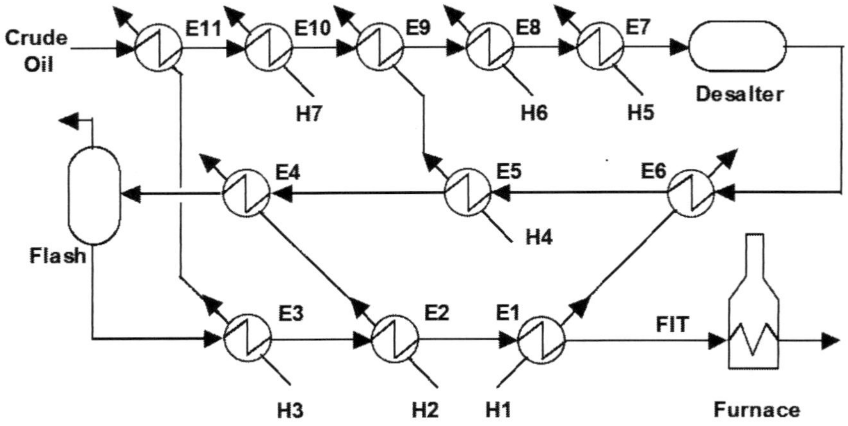

Figure 1. Simplified Flow Diagram for a Crude Oil Preheat Train [5].

Fouling Characteristics of Crude Oil Preheat Trains

In a PHT, fouling deposition mainly occurs on the crude oil side of the heat exchangers [6]. Most distillation products, which are lighter and "cleaner" than crude oil, have normally a lower fouling propensity. Nonetheless, some products such as heavy gas oil or vacuum residue may also have a high tendency to form fouling deposits.

Fouling affects most of the heat exchangers in a PHT. However, different exchangers are usually subject to different fouling mechanisms. At the cold end of the train, fouling is generally due to the deposition of salts, waxes and corrosion products. Since most the inorganic salts and solids are removed in the desalter, the deposition of these materials is not normally a problem downstream this unit. At the hot end of the network, on the other hand, the most common cause of deposition is chemical reaction fouling [2]. For exchangers at this position, fouling becomes more severe as the temperature of the crude oil increases [6].

Fouling Threshold

The concept of "fouling threshold" was first proposed by Ebert and Panchal [7] to describe a set of operating conditions, specified by wall temperature and flow velocity, below which significant fouling in crude oil heat exchangers does not occur. They suggested that chemical reaction fouling in crude oil heat exchangers is the result of two opposing mechanisms: formation and removal. Foulants are formed by chemical reaction of the crude oil near heat transfer surfaces. The removal mechanism is related to the transport of foulants by diffusion and turbulence eddies away from the surface before deposition occurs. The net fouling rate observed is the difference between the rates of formation and removal. When the rate of formation is higher than the removal rate, significant fouling may be expected. On the other hand, if the removal mechanism dominates, fouling deposition is negligible. The onset of the fouling process, or "fouling threshold", takes place when the formation and the removal mechanisms are in balance.

Ebert and Panchal presented a correlation to calculate the initial fouling rate as showed in Table 1. The first term on the right side of the equation represents the rate of formation of foulants. It relates the formation mechanism with the film temperature, an average temperature of the thermal boundary layer. The second term represents the removal mechanism, which is expressed as a function of the wall shear stress. The model is also used to calculate the fouling threshold conditions. For this, the fouling rate is set to zero and the equation is solved for either the film temperature or the wall shear stress.

Table 1. Crude Oil Fouling Threshold Models.

Model	Author(s)
$\frac{dR_f}{dt} = \alpha Re^{\beta} \, EXP(-E/RT_f) - \gamma\tau$	Ebert and Panchal [7]
$\frac{dR_f}{dt} = \alpha Re^{\beta} Pr^{-0.33} \, EXP(-E/RT_f) - \gamma\tau$	Panchal et al. [8]
$\frac{dR_f}{dt} = \alpha' Re^{-0.8} Pr^{-0.33} \, EXP(-E/RT_s) - \gamma' Re^{0.8}$	Polley et al. [9]
$\frac{dR_f}{dt} = \frac{A_{III} C_f v T_s^{2/3} \rho^{2/3} \mu^{-4/3}}{1 + B_{III} v^3 C_f^2 \rho^{-1/3} \mu^{-1/3} T_s^{2/3} \, EXP(E/RT_s)} - C_{III} v^{0.8}$	Yeap et al. [10]

The fouling threshold may be considered as the maximum film temperature, for a given flow velocity, below which significant deposition does not take place. The higher the flow velocity of the crude oil, the higher the film temperature necessary for fouling to occur. The locus of the fouling threshold conditions divides the operating space in two regions (Figure 2). For conditions of velocity and temperature below the threshold line deposition is negligible. Above the threshold line significant fouling may be expected, and the severity of the deposition increases as the conditions move away from the threshold.

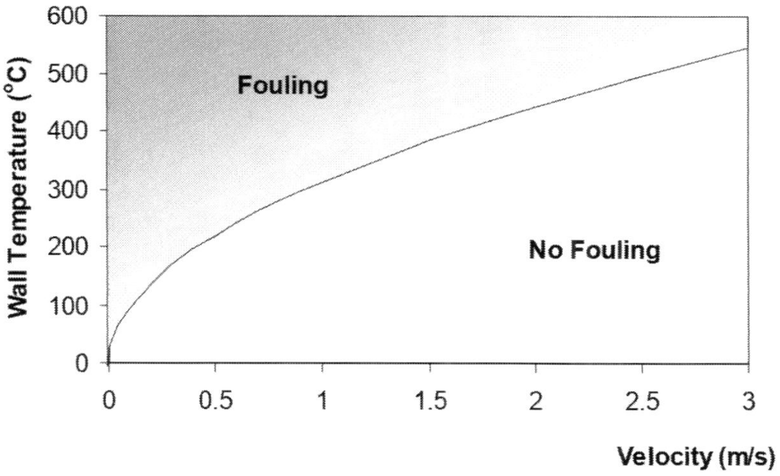

Figure 2. Threshold Film Temperature as a Function of Flow Velocity.

Several modifications to the original model of Ebert and Panchal have been proposed in order to improve its accuracy in predicting the fouling threshold and initial fouling rates. The most important changes proposed include: replacing the film temperature in the formation term by the temperature of the heat transfer surface (wall temperature); modifying the removal term so that it depends on mass transfer instead of shear-stress; and increasing the dependence of the model on the physical properties of the crude oil. Table 1 lists the improved models that have been proposed.

Changing Operating Conditions to Mitigate Fouling

The importance of the fouling threshold models is that they make it possible to identify a set of operating conditions for which fouling is very unlikely to occur. Designers can use this information to specify heat transfer equipment that operates inside the region of favourable conditions, so that significant deposition does not take place. Methods that incorporate this fouling mitigation principle for designing and retrofit of single exchangers and complete networks have recently been proposed [11-13]. In these approaches, heat exchanger geometries and network configurations are carefully selected to ensure that the operating conditions of the equipment being designed lie below the fouling threshold.

The information provided by the fouling threshold models is also the basis of the mitigation approach presented here. The aim is to minimise fouling deposition by optimising the operating conditions of the affected HEN. Knowing the location of the fouling threshold, the operating point of heat exchangers prone to fouling is moved toward the region where fouling is not expected to occur (Figure 3). The change in the operating point is accomplished without modifying the internal structure of the exchangers or the configuration of the network, which contrasts with the existing fouling mitigation techniques by retrofit.

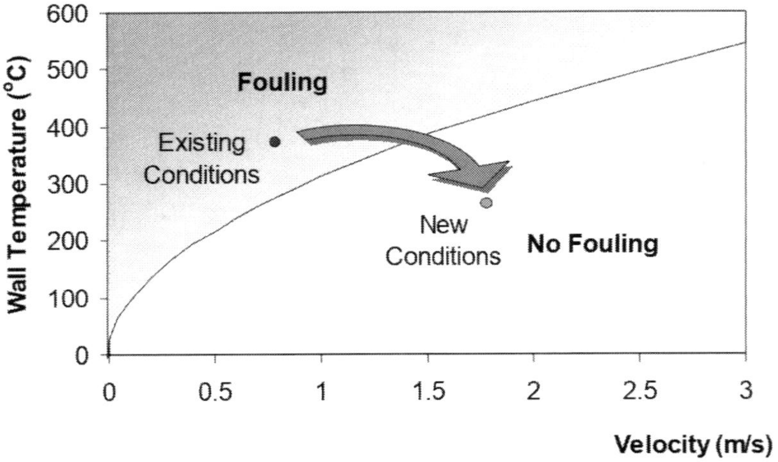

Figure 3. Changing Operating Conditions to Mitigate Fouling.

As observed in Figure 3, moving the operating conditions of a heat exchanger towards the fouling threshold implies either reducing its wall temperature, increasing the flow velocity of the crude oil passing through it or a combination of both. Wall temperature and flow velocity are not variables that can be directly controlled in a HEN. However, as shown in Table 2, they depend on other operating parameters such as flow rate and temperatures, which can be directly manipulated.

Table 2. Relationships Between Some Operational Variables in HENs.

Wall temperature	$T_s = \frac{h_{hot}}{h_{hot} + h_{cold}}(T_{hot} - T_{cold}) + T_{cold}$
Heat transfer coefficients	$h \propto v^{0.8}$ *(For the tube-side)*
	$h \propto v^{0.6}$ *(For the shell-side)*
Flow velocity	$v \propto w$

From the correlations in Table 2, it is clear that wall temperature can be reduced through any combination of the following changes in the operation of the heat exchangers: increasing the flow rate of the cold stream (crude oil in this case), decreasing the flow rate of the hot stream, reducing the hot stream temperature or reducing the crude oil temperature. Similarly, the flow velocity of the crude oil stream can be raised by increasing its flow rate, which is compatible with the reduction in wall temperature.

A direct way to improve the operation of a HEN, from the point of view of fouling, could be to change the supply conditions of the streams entering the network. For instance, the wall temperature of a troublesome exchanger may be reduced by decreasing the temperature of the hot stream at the inlet of the unit. In the same way, a desirable increase in the flow velocity of a stream across an exchanger may be achieved by increasing its flow rate. However, in most cases, stream conditions strongly depend on the background processes and cannot be changed arbitrarily. Changes in the conditions of streams should not be carried out without the careful consideration of the implications on the operation and economics of the whole process. For the sake of simplicity, the proposed approach does not consider modifications of the process, therefore flow rates and supply temperatures of all the streams entering the HEN are kept unchanged.

Given the constraints imposed on changing the configuration of the equipment and the inlet conditions of the streams, there may not seem to be any other degree of freedom available for optimising the operation of the HEN. However, most HENs exhibit features such as bypasses and stream-splits that can be used to for this purpose. Bypasses are used in many networks as a means to control the outlet temperature of the exchangers. This is particularly true in exchangers designed by following the traditional approach of oversizing to compensate for fouling effects. Bypasses are normally installed on the stream with the lower fouling propensity, which, for the case of refinery PHTs, correspond to the products of the distillation unit. Networks that have been designed or revamped by using pinch technology [14] normally contain stream splitting. Stream-splitters are also used to create parallel structures in the network, which introduce operational flexibility.

Both, bypasses and stream-splits enable operators to modify the flow distribution throughout the network without changing the total flow rate of the incoming streams. This does not only affect the flow through individual heat exchanger, but also their heat duty, and hence the temperature profiles all across the network. As discussed before, changes in stream temperatures and flow rates are sufficient to modify the wall temperature and velocity through the exchangers. Therefore, favourable operating conditions for fouling mitigation, i.e. low wall temperature and high flow velocity, can be achieved by manipulating the settings of bypasses and stream-splits in the network.

The way in which fouling can be mitigated using bypasses is illustrated with the example shown in Figure 4. The diagram represents a heat exchanger at the hot end of a PHT. Examining the fouling propensity of this unit using the Polley et al.'s model, it is found that the wall temperature at the hot end of the exchanger is higher than the fouling threshold temperature; therefore, severe fouling deposition can be expected. Figure 5 shows the trend in the FIT for a period of 3 years predicted using the same model. The extra energy consumption incurred because of the decline in FIT is equivalent to 1,300 MW·day.

Fouling in this unit can be mitigated using the bypass to reduce the flow of the hot stream passing through the exchanger. The effect of this action on the wall temperature is twofold. On one hand, the heat duty of the unit decreases, and consequently, the outlet temperature of the crude oil also decreases. On the other hand, the heat transfer coefficient of the hot stream falls as a consequence of the lower flow velocity. The resulting effect is a lower wall temperature at the hot end of the exchanger, and hence, a lower fouling propensity. However, it is important to keep in mind that the reduced fouling tendency is reached at expense of a lower heat recovery in the exchanger.

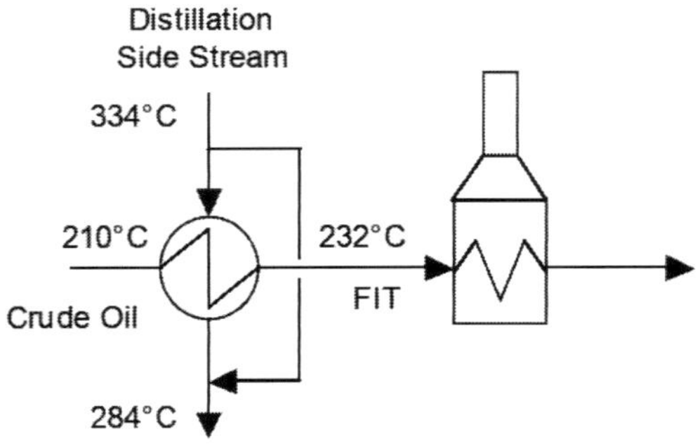

Stream Properties

	Cold Stream	Hot Stream
CP (kW / °C)	110.4	48.7
h (kW / $m^2 \cdot$°C)	1.0	1.0
Re	29900	-
Pr	12.2	-
$T_{threshold}$ = 275°C	T_s = 283°C	

Figure 4. An Example Where Fouling can be Mitigated Changing the Operating Conditions.

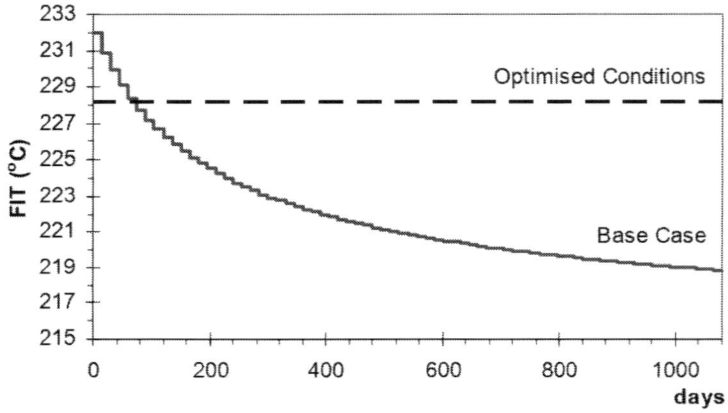

Figure 5. Profile of FIT Before and After Improving Operating Conditions.

For the example being considered, diverting 33% of the hot stream through the bypass causes the wall temperature at the hot end of the exchanger to be equal to the fouling threshold temperature. Consequently, the expected fouling deposition rate for this new set of operating conditions will be negligible. The profile of the FIT for this scenario is also presented in Figure 5. Notice that although the initial FIT is around 4°C lower than the base case, it remains constant during the whole time horizon. The total extra consumption of energy, which in this case is caused by the strategic reduction in heat recovery but not by fouling, is now 500 MW·day; this is 60% lower than the base case.

Optimising Operating Conditions

It has been demonstrated how fouling effects can be reduced by changing the operating conditions of affected HENs. Although the use of the proposed approach for mitigating fouling is normally accompanied by a reduction in the effectiveness to recover heat, in many cases this loss of heat recovery is considerable lower than the losses that fouling would cause if it were allowed to occur. Therefore, substantial energy saving can be achieved if the network is operated in the way proposed.

There may be cases, however, where the reduction in energy recovery necessary to move the operation point completely below the fouling threshold is higher than the energy penalties caused by fouling itself. Although there are numerous benefits on having fouling-free heat transfer surfaces such as lower pressure drop, lower probabilities for corrosion or equipment failure, and so forth; from the economic or energy consumption point of view, the proposed approach will be unsatisfactory in these cases.

When operating completely below the fouling threshold is not cost-effective, a better option could be to move the operating conditions of the network to a point closer to the fouling threshold but without crossing it. In this way, although fouling deposition will still be present, it will occur with a lower rate. The optimum mitigation level can be found by balancing the penalties in energy consumption caused by the induced reduction in heat recovery, against the benefits brought about by the lower fouling tendency.

The optimal set of operating conditions can be found using optimisation. Suitable objective functions may be either minimising the total energy consumption or maximising the average FIT, both evaluated over a time horizon which is normally taken as the time between planned shutdowns. The decision variables are the settings of bypasses and stream splits in the network. Given the nature of the heat transfer and fouling correlations, the resulting optimisation problem is highly non-linear.

At this point, it is worthwhile to re-examine the assumption made so far that fouling only occurs on one side of the exchangers, and therefore the reduction in the flow velocity on the other side does not accelerate deposition. Although this is true for many heat recovery systems, it is not always the case. For example, in a PHT, fouling is almost exclusively concentrated on the crude oil side of heat exchangers. The fouling tendency of light distillates, such as naphtha, kerosene and light gas-oil, is negligible even at low flow velocities. However, heavier distillation products, such as heavy gas-oil or vacuum residues, may have a significant fouling propensity. Reduction in the velocity of this kind of streams across heat exchangers may intensify deposition or even start a fouling process that was not present before.

The possibility that deposition on the product-side of the exchanger stems from the induced low flow rates can be reduced by imposing a lower boundary for the flow velocity. If the minimal flow velocity of a hot stream that allows a deposition-free operation on this side of the exchanger is known, it can be included in the optimisation model as a constraint. An even better option would be to introduce in the optimisation a correlation that relates fouling propensity of the hot streams with flow velocity. In this way, the reduction in the fouling deposition on the crude oil stream can be properly balanced against the increases in the fouling propensity of the hot stream. If there is not much information about the fouling characteristics of a hot stream, but a reduction on its flow rate is likely to increase deposition rate, the method here discussed should not be considered for the exchangers involving the stream in question.

Scheduling Cleaning Actions

As presented so far, the proposed approach is not always able to completely eliminate the occurrence of fouling throughout a HEN. The approach is only applicable to heat exchangers undergoing fouling processes strongly dependent on temperature or flow velocity. It is also necessary that the relationship between deposition rate and operating conditions is relatively well understood and can be expressed quantitatively. Many networks have a combination of exchangers in which the approach can be applied and others in which it cannot. For instance, in a PHT, fouling mechanism at the hot end of the train strongly depends on wall temperature, while at the cold end the effect of this variable on fouling is not so important; therefore, the proposed approach can only be used to mitigate fouling in the exchangers located at the hot end of the train.

To be effective, a mitigation strategy must regard fouling as a problem affecting the whole HEN rather than individual heat exchangers. No action to control fouling occurrence should be taken without considering a complete view of the network. Based on this and on the limitations previously discussed, it is clear that the optimisation of the operating conditions alone is not enough to provide an effective fouling mitigation strategy for most HENs. It is then necessary to consider an alternative method to tackle fouling in those exchangers in which the proposed approach cannot be applied. In order provide a comprehensive fouling mitigation framework, the approach here proposed is extended to include off-line cleaning of fouled heat exchanger as an alternative method to mitigate fouling.

Mechanical cleaning is perhaps the most common strategy to mitigate fouling in exiting HENs. In many cases, it is necessary to clean individual exchangers while the rest of the network is kept in operation. Taking a heat exchanger out of service for cleaning may have a great impact on the operation and economics of the system. Hence, it is necessary to carefully plan when and which unit should be cleaned. Several methods for determining the optimal cleaning schedule, mainly based on mathematical programming, have been proposed [5, 15]. The most common formulation is to discretise the time horizon into smaller time intervals and associate binary variables to each unit at each interval to indicate whether a cleaning action takes place or not.

The extended approach optimises the operating conditions and the cleaning schedule simultaneously. Fouling deposition is reduced by changing the operating conditions in the HEN, when possible. The optimisation decides whether the operating conditions should be completely moved below the threshold or a reduced fouling rate should be accepted. Cleaning is used to deal with the remains of the fouling problem. The optimisation of the cleaning schedule is incorporated using the same formulation as the existing approaches, i.e. time discretisation and use of binary variables. The main difference is that in addition to cleaning schedule, the operating variables of the network are optimised simultaneously.

Case Study

An example is used to illustrate the application of the proposed approach and to compare it with existing methods to mitigate fouling by management of cleaning. The PHT showed in Figure 1 is a simplified version of the case study presented by Smaïli et al. [5]. In the original example, the crude oil stream is split into two identical parallel branches downstream the flash separator. For the sake of simplicity, the stream splitting is here omitted without losing generality. Detailed Information about costs and design parameters of the network can be found elsewhere [5].

All the exchangers in the network are subject to fouling, although different mechanisms are responsible for it. Fouling at the cold end of the network (exchangers E5 to E11) is caused by deposition of waxes and salts, which is represented with a linear fouling model (constant fouling rate). At the hot end of the network (exchangers E1 to E4), chemical reaction fouling is considered as the dominant mechanism. For this group of exchangers, deposition rate is calculated using the Polley et al.'s threshold model. In addition, fouling deposition is considered to be mainly located on the crude oil side of the exchangers.

It is assumed that every exchanger can be taken out of service for cleaning while the rest of the train continues in operation. The cost and duration of each cleaning action is considered the same for all the units in the network, regardless of their size or condition before cleaning. It is assumed that the flow rate of the hot streams passing through the exchangers can be controlled using bypasses, so that the operating conditions of the network can be manipulated. As fouling is assumed to occur predominantly in the crude oil side, the low velocities induced by reducing the flow rate of the hot streams across the heat exchangers do not encourage fouling deposition.

The aim is to find the optimal combination of cleaning actions and operating conditions that minimise the total operational cost of the network, which comprises the cleaning and the cumulative energy cost. Note that the cumulative energy cost considered corresponds only to the extra energy consumption caused by fouling, that is, the fraction of hot utility consumption necessary to compensate for the drop in the FIT. The study is carried out over a period of 3 years, which is divided in 36 monthly intervals. The time required for cleaning each unit is15 days.

Besides the base case, in which fouling is allowed to occur without taking any action to mitigate it, two additional scenarios are considered: In the first one, only the cleaning schedule is optimised and the operating conditions of the network are kept unchanged (Case 1). This corresponds to the existing fouling mitigation strategy and is presented for comparison. In the second scenario (Case 2), both the cleaning schedule and the operating conditions are optimised simultaneously. Both cases are solved using stochastic optimisation, more specifically, simulated annealing [16]. Table 3 shows a summary of the results obtained for the cases considered.

Table 3. Results for the Different Scenarios Considered.

	Base Case No Cleaning	Case 1 Cleaning-only	Case 2 Cleaning + Opt. Conditions
Initial FIT	232.6°C	232.6°C	228.9°C
Final FIT (after 3 years)	208.0°C	219.8°C	223.4°C
Average FIT	215.4°C	222.7°C	223.9°C
Number of Cleaning Actions	-	18	10
Total Operating Cost	£ 1.38×10^6	£ 0.87×10^6	£ 0.75×10^6
Reduction in Operating Cost	-	37%	46%

The profile of the FIT and the optimal cleaning schedule obtained for Case 1 are presented in Figure 6. The sharp temporary falls in the FIT profile correspond to groups of heat exchangers being taken out of service for cleaning. The optimal cleaning schedule exhibits 18 cleaning actions distributed along the time horizon. Except for exchanger E10, all the other groups of exchangers require cleaning. As expected, units located at the hot end of the train are cleaned more frequently due to their higher fouling rates. Units E1 and E2 are required to be cleaned 5 and 3 times during the three-year period, respectively. As seen in Table 3, a result of implementing this optimal cleaning policy is the increase in the average FIT by 7°C respect to the base case, with an associated reduction in the operating cost of around 37%.

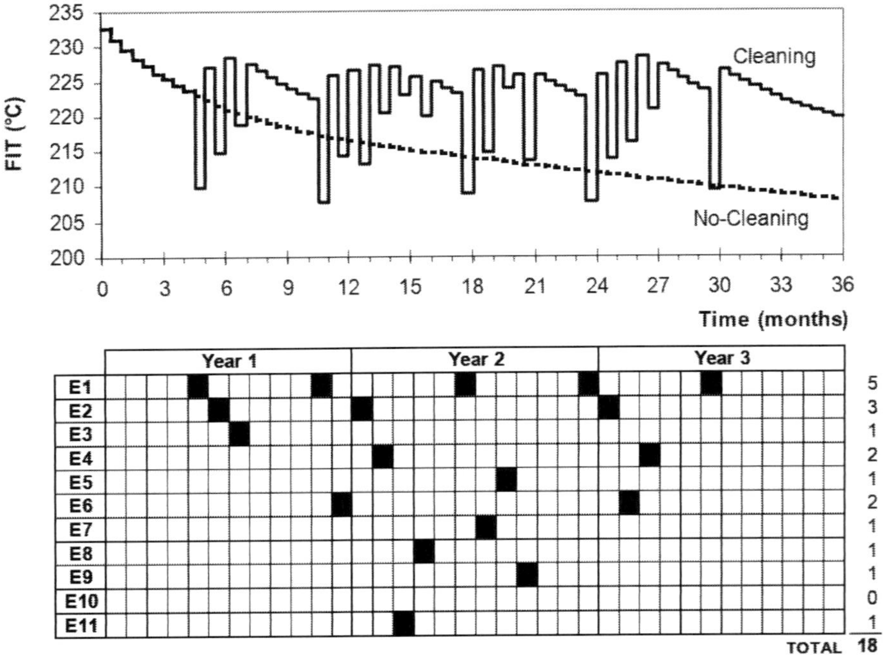

Figure 6. FIT Profile and Optimal Cleaning Schedule for the Case 1.

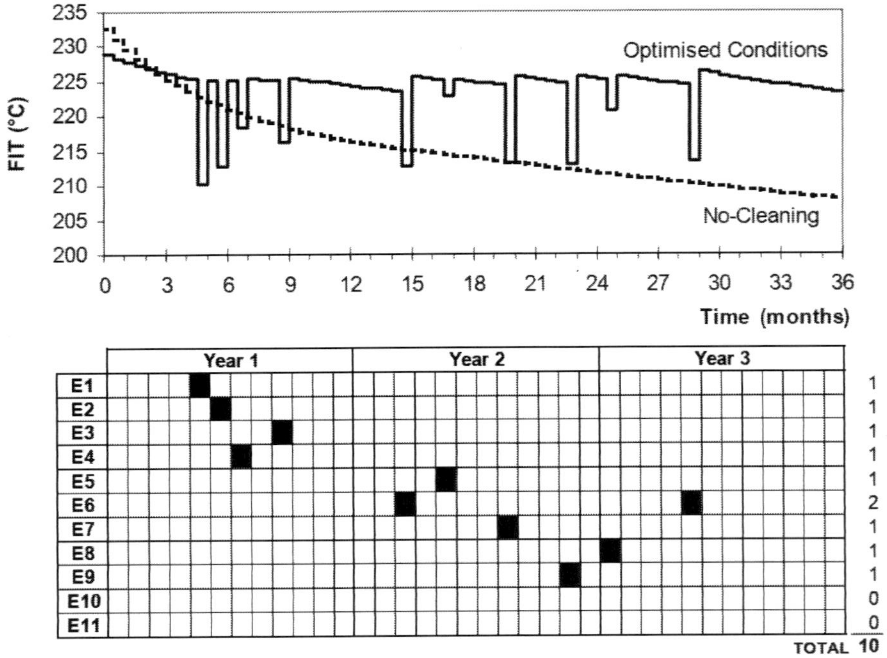

	E1	E2	E3	E4
Hot stream bypass setting	26%	21%	0%	0%

Figure 7. FIT Profile, Optimal Cleaning Schedule and Optimal Bypass Settings for the Case 2.

Figure 7 shows the FIT profile, the optimal cleaning schedule and the optimal settings of the bypasses around the exchanger at the hot end of the network for Case 2. The results indicated that the optimal operating conditions are achieved bypassing 26% and 21% of the hot stream around exchangers E1 and E2, respectively. This results in a lower fouling rate in these units, and consequently, a reduction in the required cleaning frequency. The optimal cleaning schedule requires now just 10 cleaning actions, compared with 18 required by Case 1. Noticed that although the FIT at the start of the run is lower, it declines less steeply than for the previous cases. The average FIT for this case is around 1°C higher than for the Case 1 (cleaning-only scenario), and a further 9% reduction of in the operating cost is possible.

In this example, it is assumed that the setting of the bypasses remains constant during whole time horizon. An alternative approach can be to allow these settings to vary in each time interval, so that the operating conditions of the HEN can be "finely" optimised depending on the fouling level and cleaning status of the exchangers. Although higher savings can be obtained with this approach, the number of variables and the complexity of the optimisation problem increase considerably.

Conclusions

The effect of operating variables, such as wall temperature and flow velocity, on fouling deposition in PHTs has been discussed. It has been demonstrated how fouling deposition can be mitigated by the proper selection of the operating conditions in the network. Based on this fact, a new methodology for minimising fouling in existing heat exchanger networks has been proposed. It combines the optimisation of operating conditions and cleaning schedule in a comprehensive fouling mitigation strategy.

The application of the new approach has been demonstrated with a case study. When compared with the existing mitigation strategy (i.e. optimising only the cleaning schedule), the proposed approach leads to higher savings. In addition, the lower fouling levels and lower number of cleaning actions obtained with the new approach bring about other benefits, such as lower pressure drop, lower probabilities of damage or wearing the equipment during cleaning, lower disturbances in the operation of the background process, among others.

Nomenclature

A_{III}, B_{III}, C_{III}	Constants in Yeap et al.'s model
C_f	Fanning friction factor
CP	Heat capacity
dR_f/dt	Fouling rate
E	Activation energy
FIT	Furnace inlet temperature
h	Film heat transfer coefficient
Pr	Prandtl number
R	Universal gas constant
Re	Reynolds number
T	Temperature
v	Flow velocity
w	Mass flow rate

Greek symbols

α, β, γ	Constants in Ebert-Panchal's model
α', γ'	Constants in Polley et al.'s model
μ	Viscosity
ρ	Density
τ	Wall shear stress

Subscripts

cold	Cold stream
f	Film
hot	Hot stream
s	Surface or wall

References

1. Garrett-Price, B.A., et al., Fouling of Heat Exchangers. Characteristics, Cost, Prevention, Control, and Removal. 1985, New Jersey, USA: Noyes Publications. 417.
2. ESDU, Heat Exchanger Fouling in the Pre-heat Train of a Crude Oil Distillation Unit. 2001, ESDU International: London, UK.

3. Panchal, C.B. and Huangfu, E.P., Effects of Mitigating Fouling on the Energy Efficiency of Crude-Oil distillation. Heat Transfer Engineering, 2000. 21(3): p. 3-9.
4. US Department of Energy, Fouling Minimization. Project Fact Sheet. 1999, Office of industrial Technologies, Energy Efficiency and Renewable Energy, US Department of Energy: Washington, DC.
5. Smaïli, F., Vassiliadis, V.S., and Wilson, D.I., Mitigation of Fouling in Refinery Heat Exchanger Networks by Optimal Management of Cleaning. Energy & Fuels, 2001. 15(5): p. 1038-1056.
6. Crittenden, B.D., Kolaczkowski, S.T., and Downey, I.L., Fouling of Crude Oil Preheat Exchangers. Transactions of the Institution of Chemical Engineers: Part A, 1992. 70A: p. 547-557.
7. Ebert, W. and Panchal, C.B. Analysis of Exxon Crude Oil-Slip-Stream Coking Data. in Fouling Mitigation of Industrial Heat-Exchange Equipment. 1995. San Luis Obispo; CA: Begell House, Inc.
8. Panchal, C.B., et al. Threshold Conditions For Crude Oil Fouling. in Understanding Heat Exchanger Fouling and Its Mitigation. 1997. Castelvecchio Pascoli, Italy: Begell House, inc.
9. Polley, G.T., et al., Evaluation of Laboratory Crude Oil Threshold Fouling Data for Application to Refinery Pre-heat Trains. Applied Thermal Engineering, 2002. 22(7): p. 777-788.
10. Yeap, B.L., et al., Mitigation of Crude Oil Refinery Heat Exchanger Fouling Through Retrofits Based on Thermo-hydraulic Fouling Models. Transactions of the Institution of Chemical Engineers: Part A, 2004. 82(A1): p. 53-71.
11. Butterworth, D., Design of Shell-and-Tube Heat Exchangers When the Fouling Depends on Local Temperature and Velocity. Applied Thermal Engineering, 2002. 22(7): p. 789-801.
12. Polley, G.T., et al., Use of Crude Oil Fouling Threshold Data in Heat Exchanger Design. Applied Thermal Engineering, 2002. 22(7): p. 763-776.
13. Wilson, D.I., Polley, G.T., and Pugh, S.J., Mitigation of Crude Oil Preheat Train Fouling by Design. Heat Transfer Engineering, 2002. 23(1): p. 24-37.
14. Linnhoff, B., et al., A User Guide on Process Integration for the Efficient Use of Energy. Revised First ed. 1994, Rugby, UK: Institution of Chemical Engineers. 247.
15. Georgiadis, M.C., Papageorgiou, L.G., and Macchietto, S., Optimal Cleaning Policies in Heat Exchanger Networks under Rapid Fouling. Industrial & Engineering Chemistry Research, 2000. 39(2): p. 441-454.
16. Kirkpatrick, S., Gelatt, C.D., and Vecchi, M.P., Optimization by Simulated Annealing. Science, 1983. 220(4598): p. 671-680.

The Roles of Chemical Screening and Electrostatic Field Selection in Desalting

Kenneth W. Warren & Gary W. Sams
NATCO Group, Inc.
2950 N. Loop West #700
Houston, TX 77092

Prepared for Presentation at AIChE Spring National Meeting
Recent Advances in Desalting
New Orleans, Louisiana
April 2004

Unpublished

The Roles of Chemical Screening and Electrostatic Field Selection in Desalting

Kenneth W. Warren & Gary W. Sams
NATCO Group, Inc.
2950 N. Loop West #700
Houston, TX 77092

Abstract:

Chemicals and electrostatic fields each play a significant role in the successful operation of desalters. When selecting an effective demulsification chemical it is essential to evaluate the performance in a similar processing environment. When gravity (mechanical) separation is the primary dehydration process, the demulsification chemical must provide effective water droplet flocculation as well as coalescence. When utilizing an electrostatic desalter, the high voltage field will provide flocculation and the demulsifier chemical must be capable of providing effective coagulation. In a desalting application, the final selection of a demulsifying chemical should be made to achieve effective electrostatic dehydration without adversely affecting gravity sedimentation. This paper examines the selection process for each of these critical components and describes methods to assure that they work together in an optimum fashion.

Introduction:

The desalting process may be simply described as washing crude oil containing entrained brine with fresh water or at least with water having a lower salt content than the entrained brine. The process may then be broken down into three consecutive operations:

(1) Dispersing the fresh water (usually called dilution water or wash water) into the wet crude oil stream: This is accomplished via a variety of means including pipe distributors followed by pressure drop through mixing valves perhaps followed by static mixers, and/or by the use of extremely high voltage electrostatic fields (not to be confused with the electrostatic field used for separation).

(2) Accomplishing contact between the dispersed phases of fresh water and entrained brine: Although mixing is often confused with contact, since two dispersed phases are involved, contact is really accomplished when the two types of water drops are coalesced in the desalter vessel.

(3) Achieving phase separation of the oil and water: Separation occurs in the desalter vessel as a result of drop growth followed by gravity sedimentation. Drop growth in a desalter results from the action of the demulsifier chemical in destabilizing the dispersion and electrostatically induced coalescence of the destabilized drops. Sedimentation is then controlled by the liquid flux through the vessel and the hydraulic design of the vessel and its internal components.

Two critical factors in achieving both successful and optimum desalter operation are decisions regarding selection of the type of electrostatic field to be deployed (in the design phase of the project) and the type and dosage of chemical treatment to be used (in the operation phase).

Chemical Selection:

Many different materials can contribute to stabilizing a water-in-oil emulsion – naturally occurring surfactants, precipitating organic materials such as waxes and asphaltenes, suspended solids such as mineral scales, corrosion products, and bacterial debris, residual production chemicals such as work-over fluids, corrosion inhibitors, etc. Demulsifier chemicals must be tailored to counteract whatever stabilizing influences exist in a given production stream or feedstock. However, selecting the optimum chemical blends for a given situation becomes a trial-and-error process.

Chemical selection normally begins with the classic "bottle test" in which a candidate chemical is shaken with a sample of the oil to be treated. The mixture is usually warmed in a water bath and observed for the quantity of water separated in a given time and for the "brightness" of the residual oil. Through repetition of this procedure with various chemical types, the list of candidates is narrowed and a best candidate is selected. This type of test is easily performed in the field which leads to its routine use in chemical selection at site.

While this procedure works well for processes relying upon purely gravitational separation (mechanical dehydrators), it may not identify the optimum chemical for use in an electrostatic dehydrator. The source of this discrepancy may be understood by consideration of the mechanism of drop growth and separation. The steps are as follows:

- Coagulation: Destabilization of the emulsion by chemical interaction between the demulsifier and the emulsion-stabilizing components in the oil.
- Flocculation: The gathering together of the destabilized drops into drops which are large enough to settle out of the oil in the hydraulic environment within the vessel.
- Sedimentation: The gravity-driven process by which phase separation actually occurs.

In the first and last of these steps, the responsible agents are clearly defined: The demulsifier chemical provides the coagulation and the vessel design and process conditions (flow, viscosity, etc.) along with the drop size determine the sedimentation rate. However, the flocculation process sees an overlap of responsibility. In a mechanical dehydrator, flocculating components of the chemical blend are used to assist in droplet aggregation, while in an electrostatic dehydrator this function is performed by the electrostatic field. Since the classic "bottle test" simulates a mechanical dehydrator, it favors a chemical with some flocculating properties. Chemicals with aggressive flocculating properties can tend toward sludge formation, and thus may be less effective than one designed purely for coagulation.

Comparative chemical performance between "bottle tests" and tests using an electrostatic dehydration device has routinely shown that the optimum chemical from "bottle testing" is very seldom the optimum chemical for electrostatic dehydration. The electrostatic test device is called an Electrostatic Susceptibility Tester (EST). The test procedure consists of placing a sample of heated, chemically-treated oil/water emulsion into the test cell and applying an electrostatic field. The electrical current is recorded as a function of time and at the conclusion of the test, the volume of water separated is measured and a sample of the dehydrated oil is analyzed for basic sediment and water (BS&W) by the standard ASTM D-96 method. Figure 1 shows a typical electrostatic susceptibility tester.

While the ultimate dehydration level as determined by the BS&W test is certainly of great importance, even more information can be obtained from an examination of the electrical current versus time curve. An ideal curve is shown in Figure 2. An initial rapid rise in current (**P**) results from alignment of the water dipoles with the electrostatic field followed by orientation of the polarized water drops along the lines of force of the field. This is followed by a rapid decrease of current (**F**) as the polarized drops coalesce, thereby breaking the conductive droplet chains. Finally, the current settles to a baseline (**B**) representing the overall conductivity of the oil and residual water.

Actual current vs. time curves may depart from the ideal in a number of ways. Initial response to the field may be much slower than shown. Although rare, it may be found that the particular system under test is not suitable for electrostatic treatment. More common aberrations include a poor rate of decline of the current indicating slow resolution of the emulsion or an elevated baseline indicating poor separation and high residual conductivity. These conditions may often be ameliorated by changes in chemical type or dosage or change of process conditions, usually temperature.

Chemical selection may be made based upon comparisons of both the ultimate dehydration (residual BS&W) and electrostatic response. Tables 1 – 4 illustrate the variation in chemical selection depending upon the selection method used.

Table 1: 19.5°API Brazilian Crude Oil

Chemical	Bottle Test Water in Oil % by Difference	EST Test Residual BS&W Measured %
A-1	➡ ***2.2***	2.12
B-1	4.6	2.01
C-1	5.3	1.62
C-2	5.7	➡ ***1.20***
B-2	6.0	2.35

Bold denotes best performance

Table 2: 33.3°API Japanese Refinery Feedstock

Chemical	Bottle Test Water in Oil % by Difference	EST Test Residual BS&W Measured %
D-1	➡ ***0.5***	0.30
E-1	1.0	0.31
D-2	1.5	➡ ***0.15***
F-1	2.0	0.41
F-2	2.5	0.35

Bold denotes best performance

Table 3: 25°API Kuwait Crude Oil

Chemical	Bottle Test Water in Oil % by Difference	EST Test Residual BS&W Measured %
E-2	➡ ***3.6***	1.40
E-3	4.0	1.51
C-3	4.8	1.51
C-4	5.5	➡ ***0.70***

Bold denotes best performance

Table 4: 27°API Oklahoma Crude Oil

Chemical	Bottle Test Water in Oil % by Difference	EST Test Residual BS&W Measured %
E-2	➡ ***2.1***	1.21
C-2	4.3	1.41
A-2	5.0	➡ ***0.52***
G-1	5.6	1.44

Bold denotes best performance

In each of the examples shown in the Tables, the optimum chemical for electrostatic dehydration was different than the one selected by "bottle test" methods. Such differences can become significant influences in the performance obtained by electrostatic dehydrators under actual operating circumstances. Although not shown here, the duration of the two types of test should be noted. In bottle testing, the sedimentation period allowed is traditionally from 20 to 90 minutes depending upon oil gravity. Electrostatic testing routinely allows only six minutes of field exposure regardless of oil gravity. Therefore, the results shown in the above tables illustrate not only the benefit of electrostatic fields in the selection of chemicals, but also in the rate and depth of dehydration achieved.

Selection of Electrostatic Field Type:

Because of their general high efficiency in terms of both space requirements and performance, electrostatic dehydrators are widely deployed in crude oil production operations and almost universally deployed in desalting applications. Historically, AC fields were used for crude oil applications due to their high water tolerance and low tendency toward electrolytic corrosion while DC fields were used for dehydrating aqueous wash solutions from refined products due to their increased dehydration efficiency. Today a wide array of AC, DC, combined AC/DC, and modulated fields are available to meet a much wider range of processing requirements.

The final separation step for removing water drops from oil in any electrostatic dehydrator involves gravity sedimentation. The partnership between the chemical coagulating agent and the electrostatic field is then responsible for growing the drops to as large a size as practical in order to enhance the sedimentation rate. The forces determining sedimentation rate are shown in Figure 3.

There are three major forces exerted on water drops by electrostatic fields. These are as follows:

- Dipolar Attraction: This is the attractive force produced by the positive and negative centers on the induced dipoles of the drops in an electrostatic field.
- Electrophoresis: This is the electrically induced movement of polarized drops in a uniform electrostatic field.
- Dielectrophoresis: This is the movement of polarized drops induced by a divergent electrostatic field with the movement occurring in the direction of increasing field gradient.

The mathematical definitions of these forces are shown in Figure 4. The availability of these forces depends upon the type of field selected and the electrode geometry used. In an AC field, dipolar attraction becomes the dominant force with dielectrophoresis adding assistance. Electrophoretic movement is essentially cancelled by the rapid field reversal. These forces are shown in Figure 5.

In a DC field, electrophoretic force becomes dominant with dipolar attraction providing coalescing action and dielectrophoretic movement active in the fringing areas of the field. However, as mentioned earlier, DC fields have less water tolerance than AC fields and can lead to electrolytic corrosion in conductive environments. For crude oil applications, a combination AC/DC field may be used to combine the benefits of both types of field while avoiding electrolytic corrosion. Figure 6 illustrates a combination AC/DC field. A parallel plate electrode array is used, and alternate plates are charged through reversed diodes to produce a DC field between the plates. An AC field is produced at a distance from the electrode array by virtue of overlap of the DC components. Thus, an AC field exists between the electrode array and the oil/water interface and between the array and the vessel. Bulk dehydration is accomplished in the AC field below the electrodes and removal of residual water occurs between the plates. The active forces in an AC/DC field are shown in Figures 7 and 8.

It is helpful to consider the comparative magnitudes of the three electrostatic forces available when deciding which field type to deploy. For comparison purposes, each force was calculated assuming a 750 micron drop in a 32 kV peak voltage field applied through electrodes spaced at six inches with a 0.5% BS&W drop population. It was assumed that the oil conductivity was low. The calculated forces are shown in Table 5.

Table 5: Comparison of Electrostatic Forces

Force	**Magnitude (dynes)**
Electrophoretic Force	53.00
Drag = Weight	0.010
Dipolar Force	0.004
Dielectrophoretic Force	0.002

It is apparent that electrophoresis is the strongest electrical mechanism available. Therefore, it is advisable to incorporate it into the process whenever possible.

Modulated Electrostatic Fields:

Significant performance improvements may often be gained by modulation of the electrostatic field. Although details of field modulation will be discussed in a separate paper, the basic concepts will be covered here. It should be noted that modulation, while the subject of several academic research efforts, has only been applied at the field level using the combination AC/DC field with plate electrodes.

Modulation effects may be divided into two broad categories: low frequency (~0.2 to 20 Hz), and high frequency (~ 1 kHz). The mean equilibrium drop size associated with an electrostatic field varies inversely with the magnitude of the field gradient. Therefore, it could be assumed that low field gradients would be more desirable since they relate to larger drop sizes. However, low gradient fields often lack sufficient energy to coalesce the smaller drops in a typical dispersion. Historically, this has necessitated a compromise in the selection of the transformer secondary voltage. Low frequency modulation overcomes this compromise by allowing the gradient to be continuously varied between that providing maximum mean drop size and that producing maximum coalescence of small drops. The resulting process can be envisioned as step-wise separation where small drops are coalesced at higher gradients and then grown to larger sizes as the gradient decreases. The larger drops settle a small distance during the low gradient interval, and the cycle then repeats. These relationships are shown diagrammatically in Figure 9. Modulation may occur between the limits of the lowest voltage capable of producing meaningful coalescence (threshold voltage) and the critical voltage gradient at which drops are shattered (Figure 10).

High frequency modulation, at or near the resonant frequency of the drops, results in surface perturbations which render the drops more easily coalesced. Dramatic improvements in coalescence in the laboratory have been reported in the literature. (J. Draxler and R. Marr, Int'l Chem. Engr., **33**, January, 1993) Modulation at the proper frequency energizes the drop surfaces and overcomes the stabilizing effect of interfacial tension (Figure 11) thereby causing distortion of the drop shape (J. Eow and M. Ghadiri, Chem. Engr and Processing, **42**, p. 259, 2003). Processes utilizing the benefits of high frequency modulation are now being commercialized and may be expected to significantly affect dehydrator and desalter designs in the future.

Conclusions:

Opportunities exist for optimization of dehydrator or desalter performance by improved chemical selection and by utilization of the most effective electrostatic field. Significant differences have been illustrated between the use of “bottle testing” methods versus electrostatic testing for the design of chemical treatment protocols. Only by electrostatic testing can assurance be had that the most effective chemical is being applied to an electrostatic dehydrator or desalter.

Likewise, design of the electrostatic field (gradient, power supply secondary voltage, field type, possible modulation, frequency, waveform, etc.) depends upon the characteristics of the crude oil/water stream to be treated. Proper selection of variables may require laboratory testing during the design phase if possible. When oil samples are not available, approximations can be made by relating to known oils of similar physical properties and/or oils from the same geographical region. While such approximations carry a degree of risk, they may also be key to identifying problems such as wax or asphaltene precipitation. Selection of electrostatic variables can provide powerful tools for overcoming many common difficulties in oil production and desalting.

Definition of Symbols Used in Figures

f = viscous drag force
μ = viscosity
r = drop radius
v = velocity
w = weight
g = gravitational constant
ρ = density
ρ_c = density of the continuous phase
ρ_d = density of the dispersed phase
K = constant
ε_c = dielectric constant of the continuous phase
E = electrostatic field strength
E_c = critical voltage gradient
d = distance between drops
C = constant
σ = conductivity
σ_d = conductivity of dispersed phase
σ_c = conductivity of continuous phase
t = time, seconds
γ = interfacial tension
F_d = dipolar attractive force
F_e = electrophoretic force
F_{diel} = dielectrophoretic force

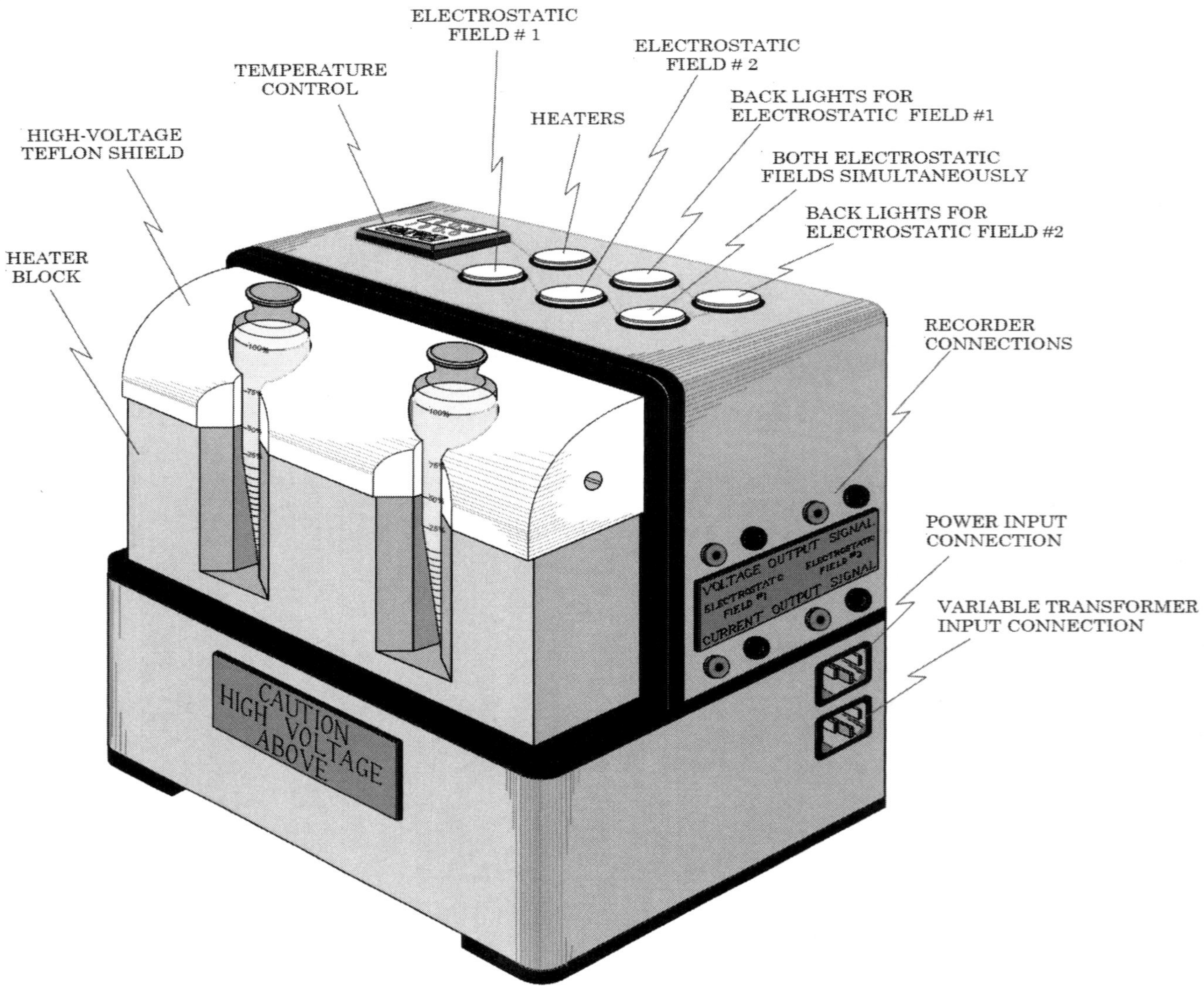

Figure 1: Electrostatic Susceptibility Tester
(courtesy Electrotech Inc.)

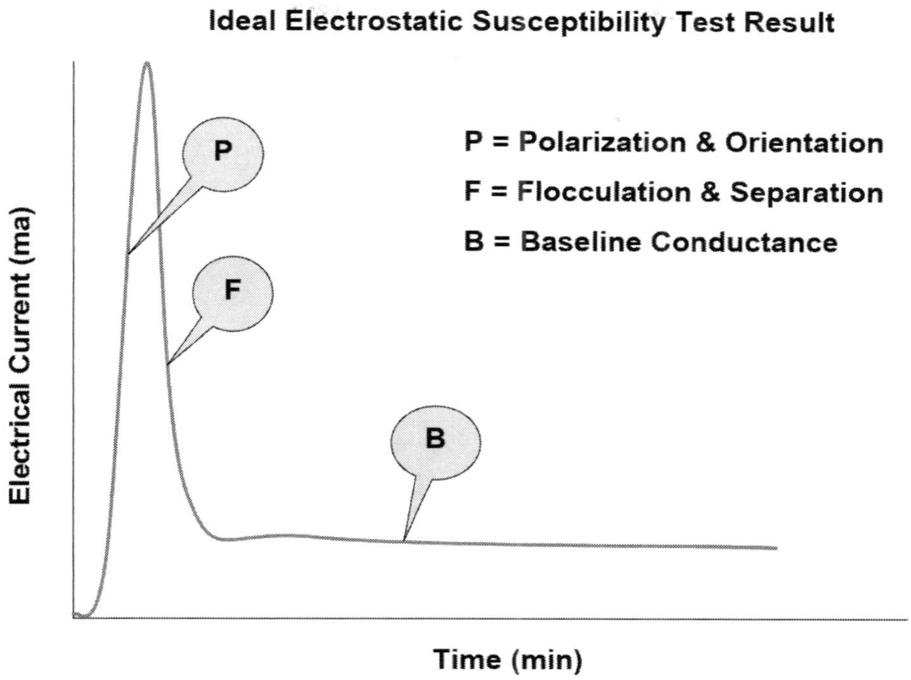

Figure 2: Ideal Electrostatic Susceptibility Curve

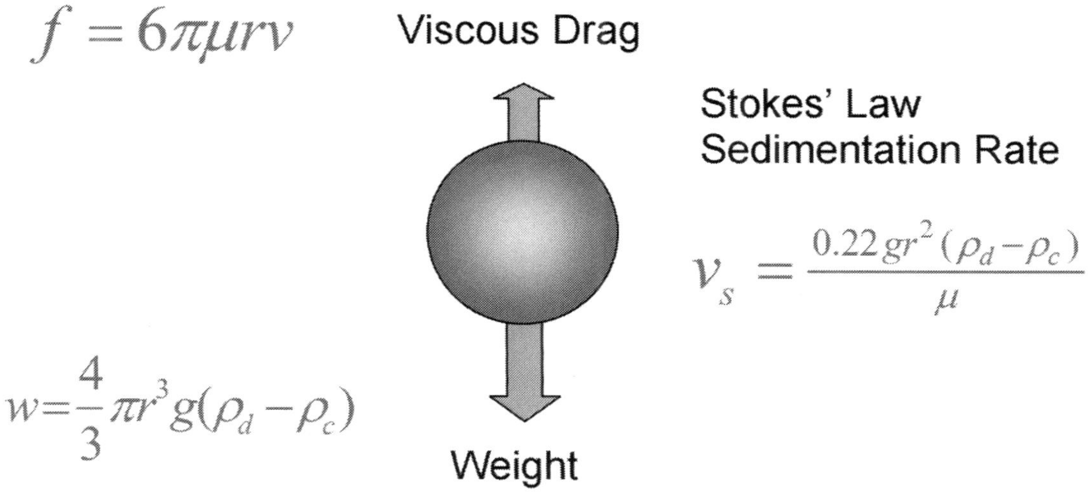

Figure 3: Forces Determining Sedimentation Rate

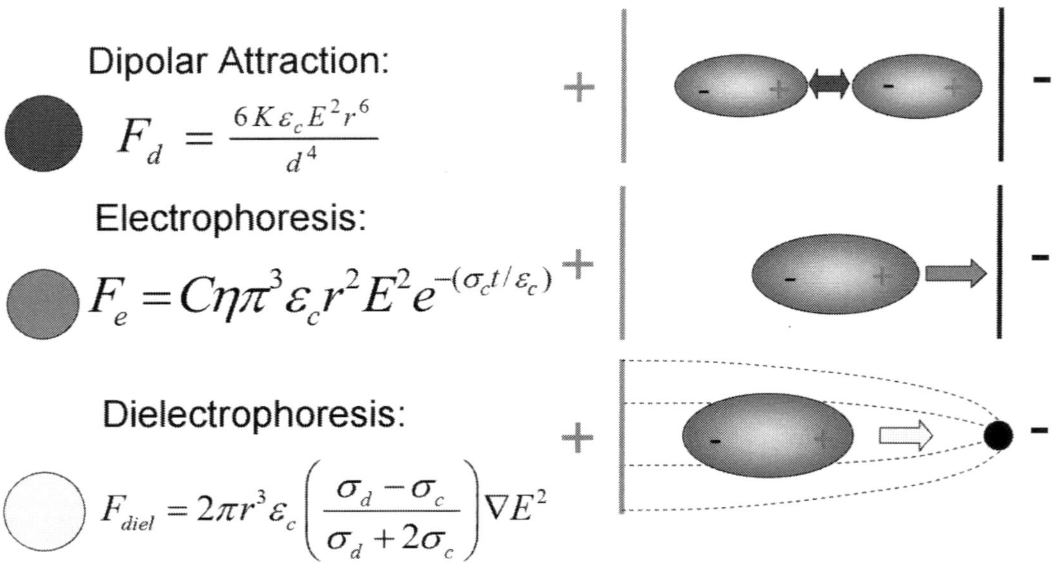

Figure 4: Definitions of Electrostatic Forces

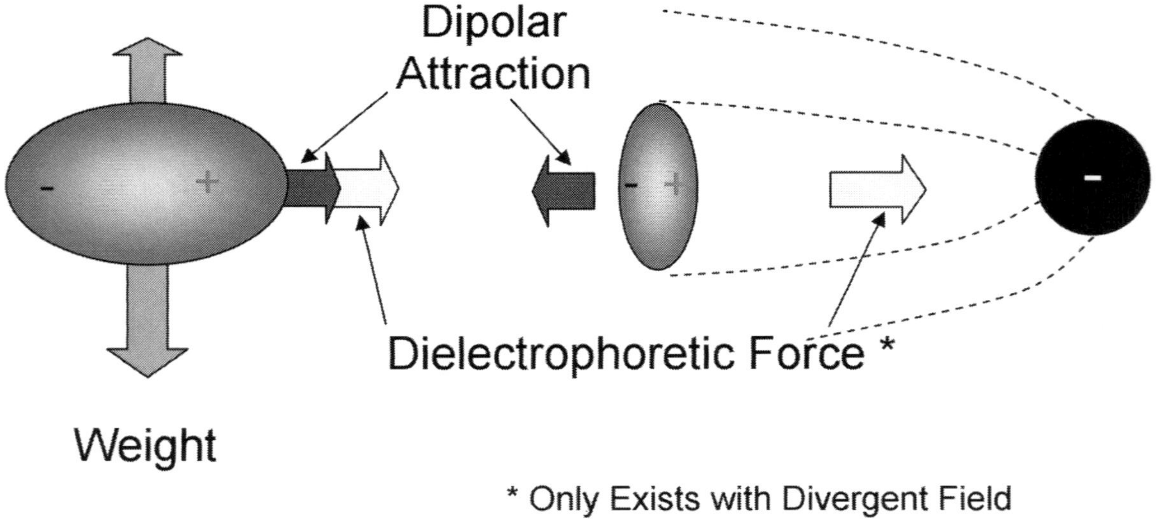

Figure 5: Electrostatic Forces in an AC Field

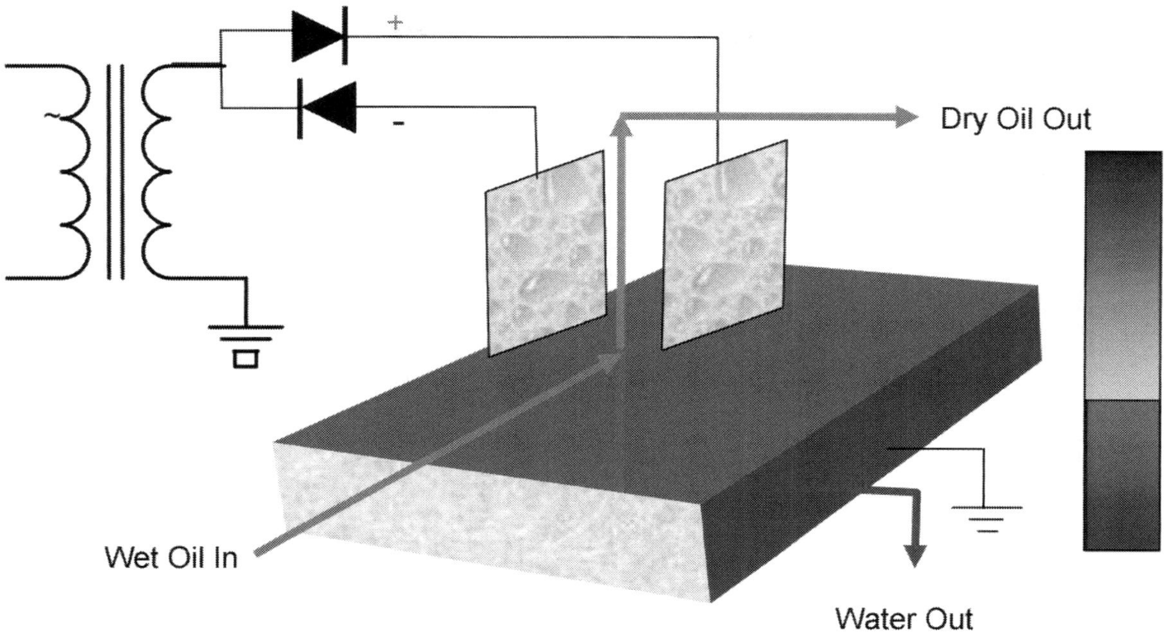

Figure 6: Combination AC/DC Dehydrator

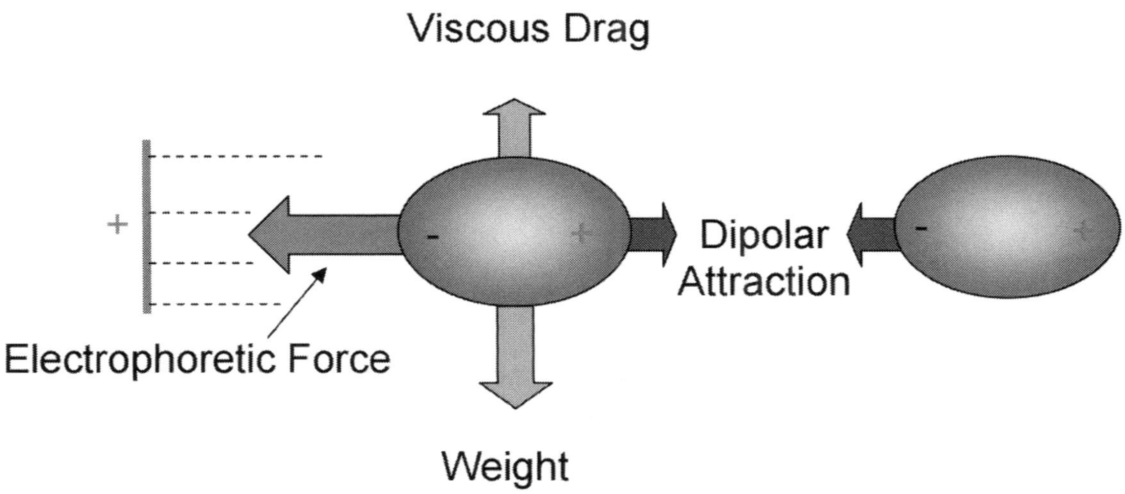

Figure 7: Electrical Forces between the Electrodes of an AC/DC Dehydrator

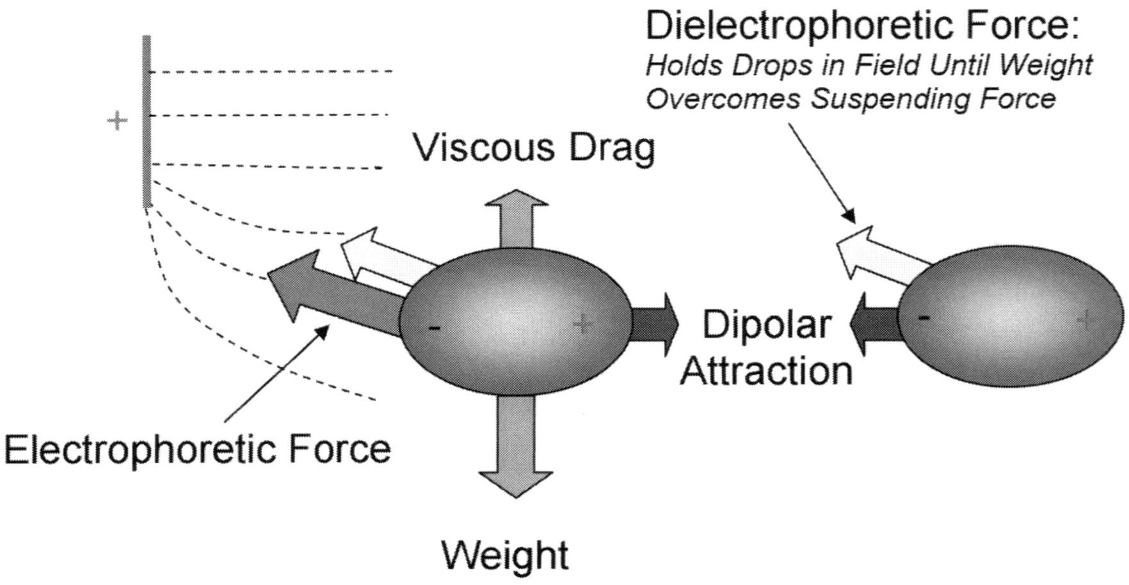

Figure 8: Electrical Forces below the Electrodes in an AC/DC Dehydrator

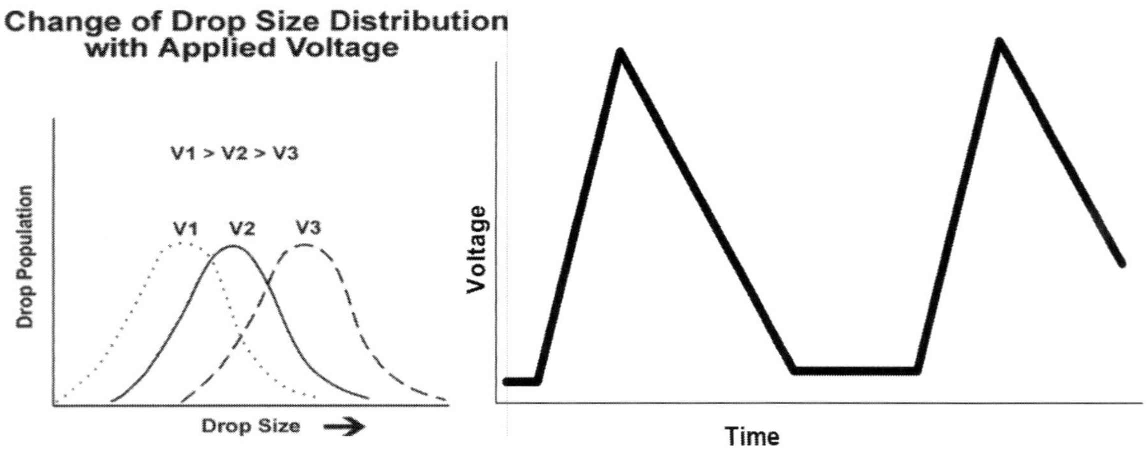

Figure 9: Low Frequency Modulation

$$E_c < \varepsilon(\gamma / d)^{1/2}$$

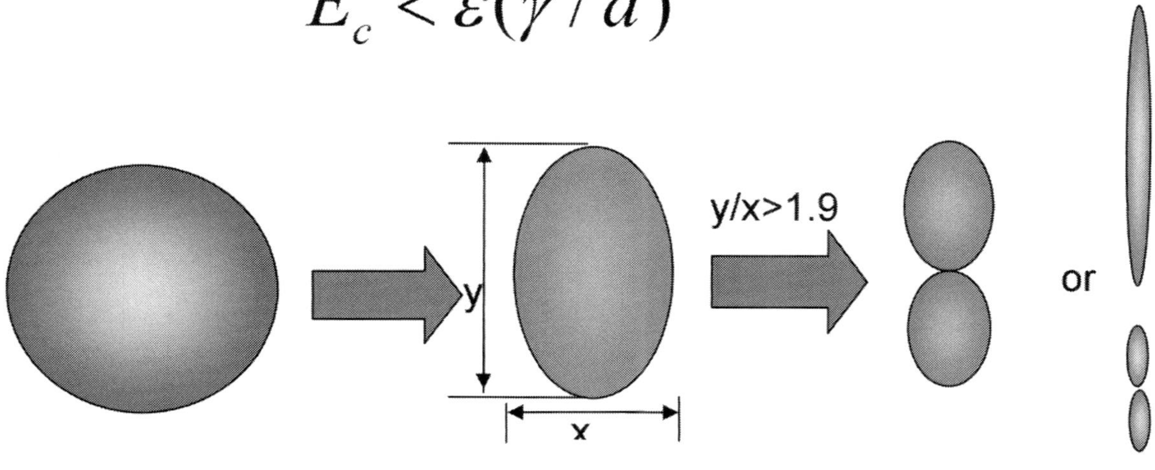

Figure 10: Critical Voltage Gradient

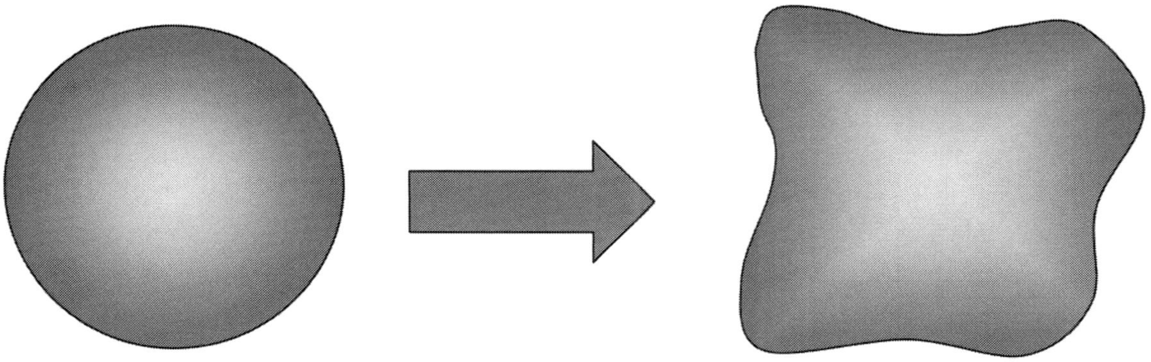

Figure 11: Drop Distortion with High Frequency Modulation

[88b] - What's Happening in Your Desalter?

Prepared for Presentation at 2004 AIChE Spring National Meeting
Session TA009 - Recent Advances in Desalting

The AIChE 2004 Spring Meeting
Session TA009 – Recent Advances in Desalting
April 25 – 29, 2004
New Orleans, Louisiana

What's Happening in Your Desalter?

Jerry Kolek, Conam Inspection & Engineering and
Lowell Pless, P.E.
QUEST TruTec, L.P.
11005 W. Fairmont Pkwy
La Porte, TX 77571, U.S.A.
Phone: 281-471-8970; Fax: 281-471-0078
E-mail: l.pless@tru-tec.com

Abstract: *Most crude oil entering refineries contains water, "salts", heavy metals, dirt, etc. Before processing the crude oil desalters are used to remove most of these undesirables. Failure to do so will cause "salts", sediments, and other solids to foul tubes in the crude unit furnace, and may foul trays and advance corrosion in the crude atmospheric column. So to prolong the operating run time of downstream equipment it is important to keep the desalter operating properly.*

Where is the oil/water interface? Is there a significant "rag" or emulsion layer? How many solids are built up in the bottom of the desalter? What is the residence time of the oil and water phases? How much water (with "salts" and other undesirables) is going downstream with the oil? There are diagnostic techniques available to answer these questions allowing refinery personnel to characterize and troubleshoot their desalter operation.

Neutron Backscatter and radioactive tracers can be used to study the desalter process. These techniques will be explained and case studies will be presented showing how these techniques may be used.

Most crude oil entering the refinery crude unit contains emulsified water, "salts", heavy metals, dirt, and who knows what else. The refinery's first "line of defense" to remove these undesirables is the crude desalter. Demulsifiers and fresh water are mixed with the crude, the mixture is heated and mixed, then sent to the desalter. In the desalter, either one or two stage settling vessels, the oil and water mixture separates into oil and water phases. The water phase contains the "salts" and other solids removed from the crude oil.

Thorough desalting prevents water and "salts" from entering the crude unit furnace, atmospheric tower, and other downstream equipment. "Salts", sediment, and other solids will foul tubes in the crude unit furnace, and have been known to foul the top trays in the atmospheric column. So to prolong the operating run time of downstream equipment it is important to keep the desalter operating properly.

There are specialized tests available to help characterize some important aspects of desalter operations available today. For example, where is the oil/water interface level? Is there a significant "rag" or emulsion layer? How many solids are built up in the bottom of the desalter? What is the residence time of the oil and water phases? How much water (with "salts" and other undesirables) is going downstream with the oil?

There are two diagnostic tests to answer these types of questions. The first test is the neutron backscatter technique. The neutron backscatter device responds to a change in the hydrogen concentration of the material inside the process. It is an excellent choice when needing to measure interface levels, whether between vapor and liquids, between liquid and solids, or between two different types of liquids. Thus a neutron backscatter survey of desalter vessels will yield a lot of information about where the various layers of material are.

Figure 1 shows a typical response seen from a desalter vessel with four different materials or phases with varying hydrogen concentration. The oil layer in the top of the desalter has the least hydrogen concentration, so that was where the minimum neutron response was. The water layer has the highest concentration of hydrogen, thus the most neutron response. In between these two layers was the "rag" or emulsion layer. In the bottom of the vessel was a neutron response indicating less hydrogen concentration than the water phase, which was interpreted to be a layer of solids or sludge.

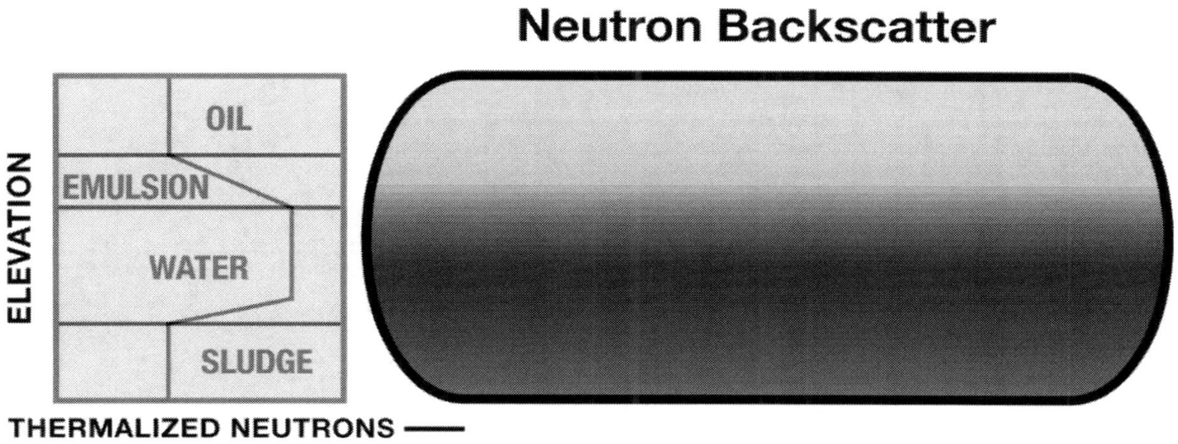

Figure 1. Neutron Backscatter is used when there is a need to measure interface levels between layers of materials with varying hydrogen concentration. This illustration shows a typical response seen from a desalter vessel with four different materials or phases.

One limitation of the neutron backscatter apparatus is that neutrons can only penetrate 4 – 6 inches (10 – 15 cm) through material. Practically speaking if a vessel wall and its insulation are thicker than 6 inches (15 cm), then neutrons cannot be used. In these cases, plus to study the residence times of vessels, radioactive tracers can be used. The next example shows a case where the desalter wall and insulation were too thick to use the neutron technique so residence times of the oil and water phases, determined by tracer tests, were used to investigate the operation.

The ultimate goal was to determine the level of sludge material built up in the desalter vessels. There was confidence that the water level instrumentation was accurate, but how much of this level was sludge in the bottom of the vessel? The water level allowed the overall volume of the water level means to be calculated. The water flow rate leaving the vessel was also being measured. Thus if the residence time of water through the vessel could be determined then the effective volume of water and sludge in the vessel could be calculated.

A test was designed using radioactive tracers to provide this critical piece of data. Performing a radioactive tracer test involves carefully selecting an appropriate liquid, solid, or gas radioisotope then injecting the tracer directly into the process and, using external radiation detectors, follow the tracer through the process. In this particular case an aqueous tracer material was used for the water phase and an organic based tracer was used for the oil phase. Figure 2 shows the layout of the desalter vessels and the detector placements used for the tracer test.

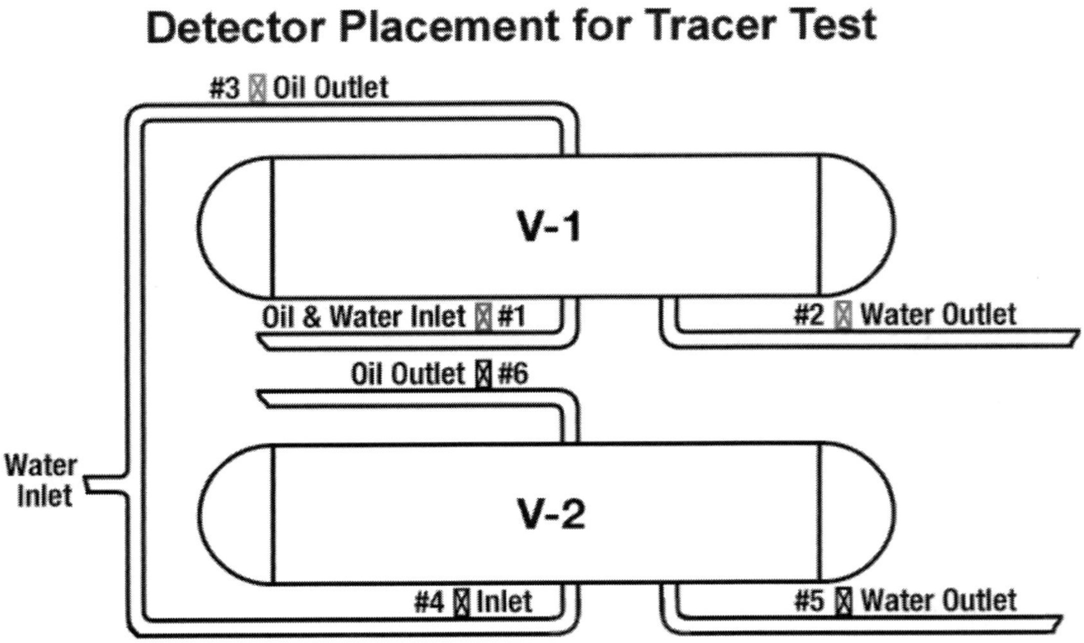

Figure 2. Radioactive tracer detector placement used to measure the oil and water residence times through crude desalter vessels.

For the first test the residence times were determined for both the oil and water phases using an organic tracer and an aqueous tracer separately. The first test was done shortly following a turnaround when the desalter vessels were known to be clean and operating well to establish baseline residence times. The organic tracer test showed that the oil phase had a mean residence time of 26 minutes through V-1 and 27 minutes through V-2. The aqueous test showed that the mean residence time for the water phase was 1.6 hours through V-1 and 1.7 hours through V-2. Using detectors 3 and 6 (refer to Figure 2) during the water test, it was calculated that approximately 3% of the inlet water left with the oil from V-1 and <2% water left with the oil from V-2. Figure 3 shows one of the data results from an aqueous tracer injection.

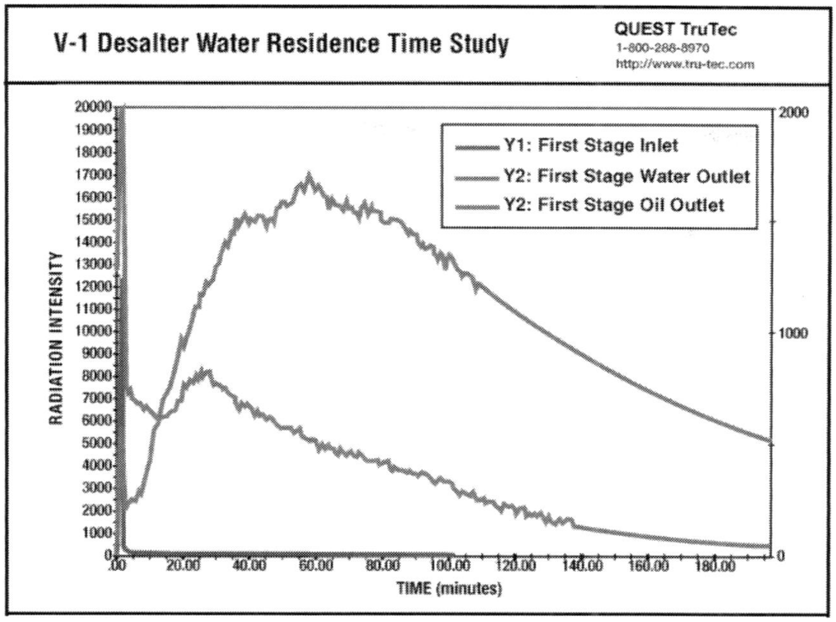

Figure 3. Typical data results from an aqueous tracer injection using the detector placements shown in Figure 2.

Six months later the test was repeated using only the aqueous tracer. Since some aqueous material leaves with the oil phase the aqueous tracer was also adequate for determining the oil phase residence time. Again the # 3 and # 6 (from Figure 2) detectors were used to measure the amount of water leaving with the oil, but also were used to measure the effective oil phase residence time. The results from this test: mean residence time of the oil phase through stage V-1 was 28 minutes, 30 minutes through V-2; mean residence time of the water phase through V-1 was 1.6 hours, 2 hours through V-2. The amount of inlet water leaving with the oil was <3% for V-1 and still <2% for V-2.

After another six months the aqueous tracer test was repeated. This time there was heightened concern about the results since downstream analysis showed the desalter operation had deteriorated. The question needing an answer was, "Had solids or sludge built up in one of the vessels, decreasing the residence time needed for good phase separation?" The results from this test showed that the concern was well placed: mean residence time of the oil phase was 24 minutes through both V-1 and V-2; mean residence time of the water phase through V-1 was 1.5 hours, and 1.1 hours through V-2. The amount of inlet water leaving with the oil was <2% for V-1, but approximately 5½ % for V-2. The water residence time through V-2 had been decreased by almost one-third, and this result showed in the amount of water leaving with the oil. Refinery engineers calculated a 2-foot sludge level had built up in V-2, which reduced the effective volume for the water phase. A sample calculation for the sludge level is shown below.

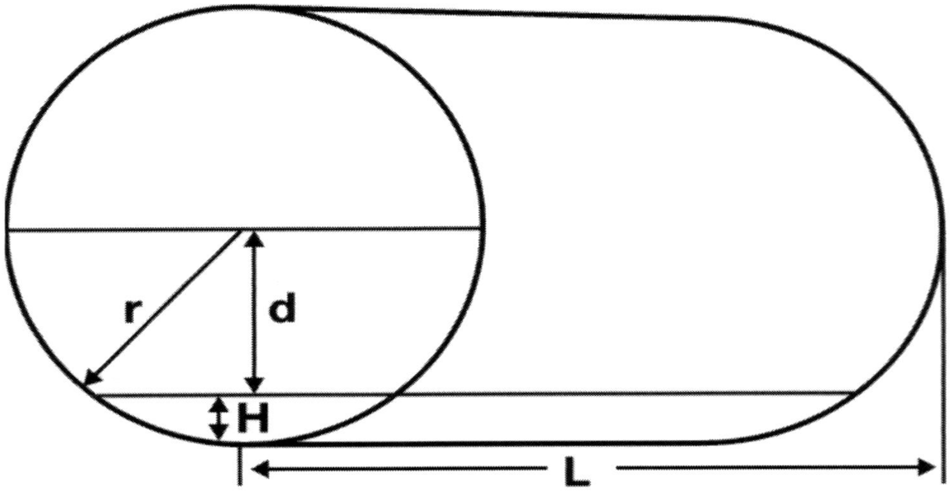

Volume=π * r^2 * L

Volume (Sector)=½ * r^2 * 2 $\cos^{-1}(\frac{d}{r})$ * L

Assume that vessel V-2 is 100 feet long, with a diameter of 25 feet. The water level was being controlled at 50%. The baseline mean residence time for the water phase was 1.7 hours. The measured water outlet flow rate was 1800 gpm. The total volume of water at 50% level in V-2 was ½ * 100 * (π r^2) = 24,500 ft^3. On the third test the mean residence time had dropped to 1.1 hours. The water flow being measured remained at 1800 gpm and the water level was still at 50%. For 1800 gpm water flow to remain in V-2 for 1.1 hours, it would take a volume of 1800 gpm * 60 min * 1.1 = 119,000 gal or 15,900 ft^3. Sludge or built up solids now occupied 24,500 ft^3 – 15,900 ft^3 = 8,600 ft^3. Based on the geometry equation for volume of a cylinder (see diagram) the height of sludge or solids calculates to: 8,600 ft^3 = $(12.5)^2$ * 100 * $\cos^{-1}$(d/12.5), with d = 10.6, $\therefore$ the equivalent height (H) of sludge was approximately 2 feet.

The only alternative at this point, short of a shutdown, was to reduce the water added to V-2 in order to increase the water phase residence time.

It was unfortunate that the equipment size made direct measurements with neutrons unsuitable in this particular case. But when one method may not work, there are usually other options available to help get that one piece of critical data needed to help solve a troublesome problem.

NEW CONTROL SYSTEM DETECTS DESALTER PROBLEMS BEFORE UPSETS OCCUR

Gary Fransen
Agar Corporation
Prepared for presentation at The AICHE 2004 Spring National Meeting

April 2004
Unpublished

ABSTRACT

The Desalter Control System commonly referred to as System # 3, uses multiple probes to control the deslater and warn the operators of possible problems---giving operations time to take corrective actions before upsets occur. This paper reviews two different types of Early Warning Instruments.

One of the major causes of desalter upsets is grid over-load. The first indications of Emulsion Build-Up in the desalter are the decrease in Volts and increase in Amps on the Grids. By detecting the presence of emulsions building upwards before they enter the grids will allow operators to take corrections before the grids begin to overload.

Another major causes of desalter upsets is high water content crude being delivered to the desalter from the tank farm. The detection of high water content crude before it arrives at the desalter allows operations to make adjustments to minimize the adverse effects of the wet crude.

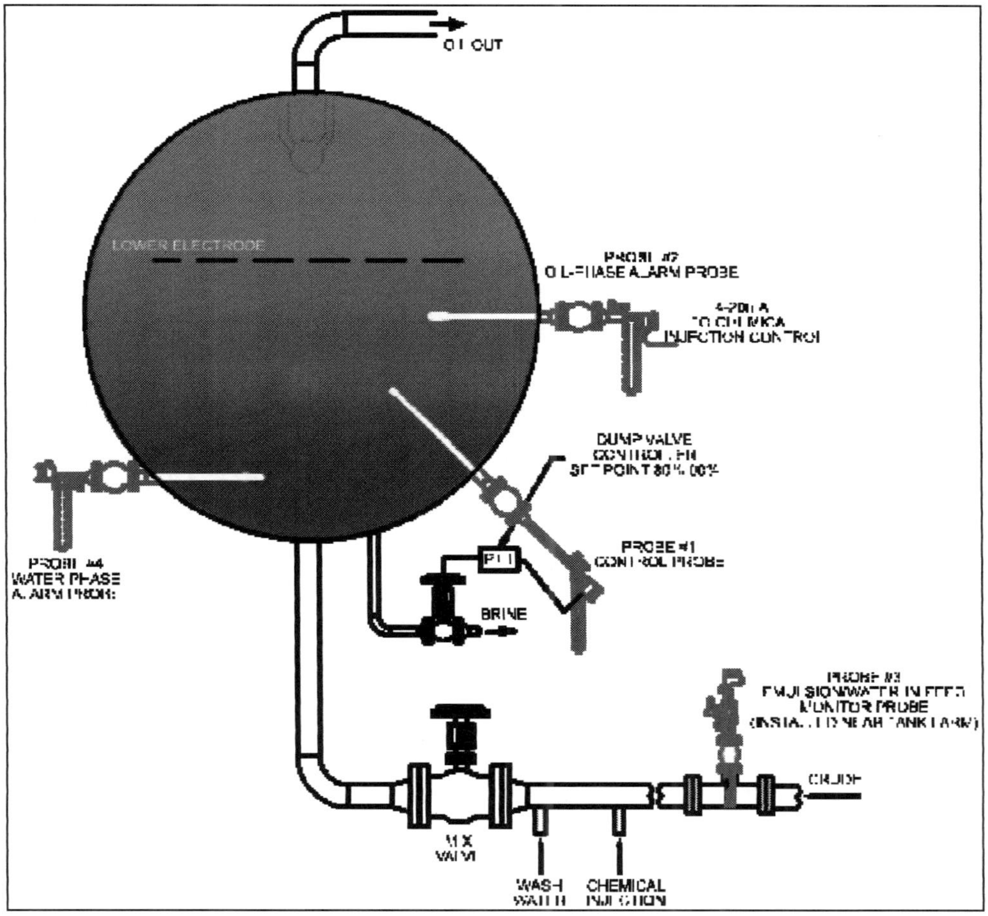

Desalter Control – System 3

It is all too easy to forget that the primary function of the desalting system is the removal of the inorganic chlorides and other water-soluble compounds from crude oil. One need not be a corrosion specialist to realize that the acids that form from these compounds can do tremendous, long term damage in the downstream processes of the refinery (as the inspection of crude tower overhead condensers can prove). However, the desalting process has become the focus of attention in areas that are no longer limited to the simple removal of salts and water. Most critical of these recent areas of concern has been the degree to which the desalting system contributes to the load of volatile organic contaminants (VOCs) at the wastewater system. In fact, the condition of the brine from a modern desalter is frequently under greater scrutiny than the condition of the desalted crude.

There can be no doubt that the operation of the desalting system is an exercise in compromise. A constant balance must be maintained between mixing intensity, wash water quality, chemical demulsifier feed and control of other parameters that can provide optimal salt removal without forming an emulsion so tenacious that it compromises the system's dehydration capabilities. Adding to this balancing act are the new legislative demands placed on effluent water quality. This presents the operator with a difficult challenge.

Optimizing the desalting process is a matter of optimizing the individual components and maximizing the efficiency of the electrostatic dehydration stage. By "pushing" the electrostatic process, one seeks to obtain the greatest amount of electrical work possible. The work performed near the grids can provide the dual benefits of enhanced salt removal (secondary mixing) and optimal coalescence.

The question therefore becomes how to maximize this electrical work.

In the vessel, the structural parameters such as vessel size, grid elevations, crude feed discharge points are all fixed. The most critical remaining variables then become interface condition and position. In fact, optimal interface control has been proven to have a significant impact on both the oil and water quality resulting from the dehydration process. Yet, in spite of the obvious need for such control, the traditional methods of control have operated on a fundamentally flawed assumption: That a Level Exists. The very term "level control" indicates the presumption that the interface between oil and water in the desalter exists at a single point (such as that observed between gasoline and water). Any review of the internal conditions in the desalter vessel via the try-lines or swing-arm will dispel this notion. There is no clear-cut level; rather the interface consists of an emulsion transition zone from oil to water with a continuously changing water percentage by volume. Understanding the true nature of the interface leads to the conclusion that efficient control comes from controlling these water percentages and not an imaginary level.

The System #3 (Desalter Control System) represents the first and only interface control system to utilize this understanding of oil/water interface as the basis for control. The system consists of the minimum of two or a maximum of four energy absorption instruments: three designated for service in the vessel and one installed well upstream of the unit in the crude feed line. In the System #3 control scheme, these probes provide continuous 4-20mA output signals that are proportional to the water concentrations (% water) at their physical locations.

Probe # 1 controls the brine outlet valve, using its ability to measure small amounts of oil in water to maintain a very high (and unstable) percentage of water several feet above the bottom of the vessel. This allows suspended oil in the water phase to separate, thus inhibiting oil undercarry as a primary control function. While Probe # 1 establishes this lower limit or bottom for the emulsion layer, Probe # 2 monitors the water content from its position in the oil phase just below the lower grid. This provides real time detection of the rate and extent of the emulsion growth (which must, by Probe # 1's control, occur in the upward direction). Probe # 2's monitoring function allows the operator to detect emulsions building towards the grids, avoid downstream upsets by providing an advance warning of such growth, and allows time for corrective measures to prevent grid over-load (transformer "trip"). Probe # 3 is an in-line monitor of the water content of the incoming crude feed and is typically located as far as possible upstream of the desalter (generally near the tank farm). This probe provides the advance warning of contaminated crude feed resulting from "tank switching" and/or the introduction of slop oil. The final element of the System # 3 is the water phase alarm probe designated # 4. It monitors the condition of the water phase below the control probe (12"-18" off bottom), alarming on the presence of suspended oil that does not readily separate or solids build-up that threatens the condition of the brine effluent. This is of particular value when low-quality sources of wash water (e.g. stripped or straight sour water) are utilized that can upset the separation process and form stable oil-in-water mixtures (reverse emulsions).

The benefits of this control scheme can be summarized as follows:
-1-Routine oil undercarry is either significantly reduced or completely eliminated, providing both environmental and economic benefits.
-2-The operator is provided with a real time "barometer" of the conditions in both the oil and water phases of the vessel as well as the crude feed line.
-3-Upsets are detected long before their effects can be felt in the unit, and the alarm outputs offer the option of automating upset response assures such as increasing or initiating chemical feeds.
-4-Control action that constantly forces emulsions growth toward the electrical grids maximizes electrical work, allowing for reduction or, in come cases, the elimination of chemical feed.
-5-By considering location of an alarm signal, more information is provided as to the source of the upset (wash water quality, wet feed, etc.).

The System 3 provides extensive information and control capabilities far beyond simple level control. Through measurement of percent water at specific points in the vessel and feed line, the overall behavior of the desalting process can be observed and the results optimized.

Below are examples of outputs of the System 3's early warning capabilities.

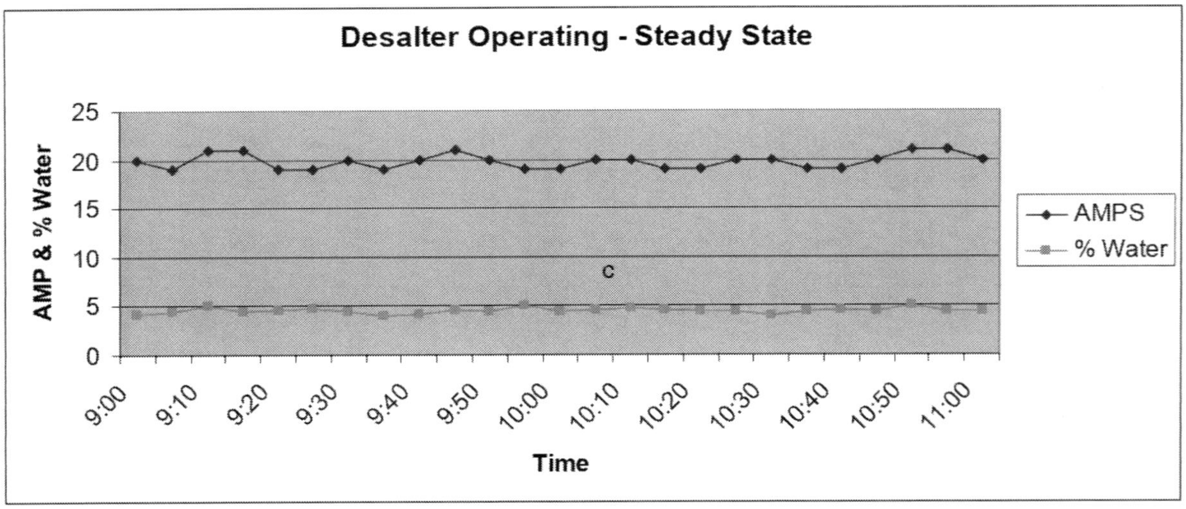

THE UPPER/EMULSION PROBE & AMPS OF THE LOWER GRID

The Desalter is operating "steady state". The Output of the Upper/Emulsion Probe is between 4-5% water, indicating that the Emulsion Layer is stable. This is confirmed by the measurement of the AMPs from the Lower Grid.

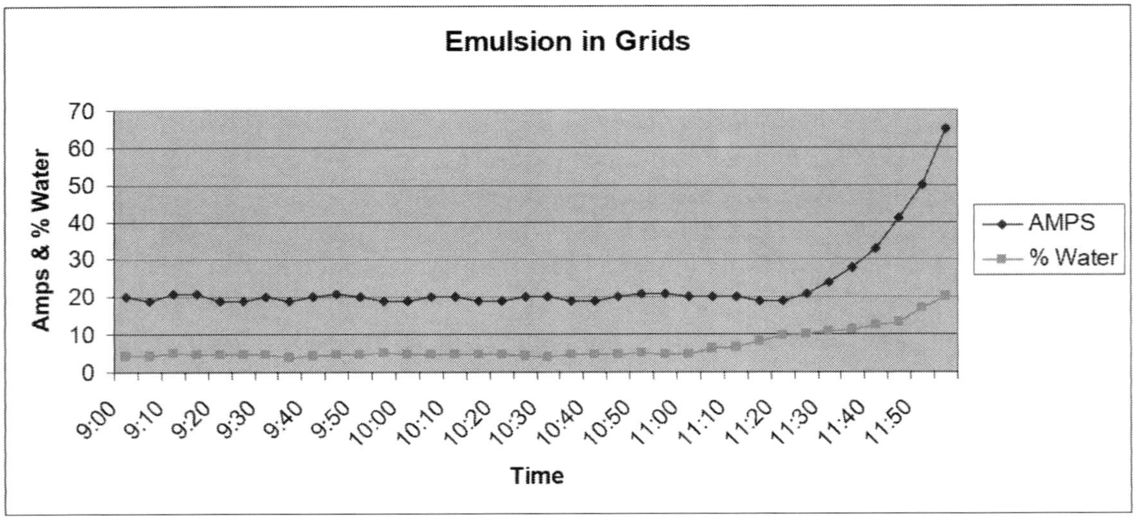

THE UPPER/EMULSION PROBE & AMPS OF THE LOWER GRID

The Upper/Emulsion Probe began to detect a growth in the Emulsion Layer (increase water content) at approximately 11:10AM. Twenty minutes later at approximately 11:30AM, the AMPs of the Lower Grid began to react (increase in AMPs) as the wet emulsion began to enter the area around the Lower Grid.

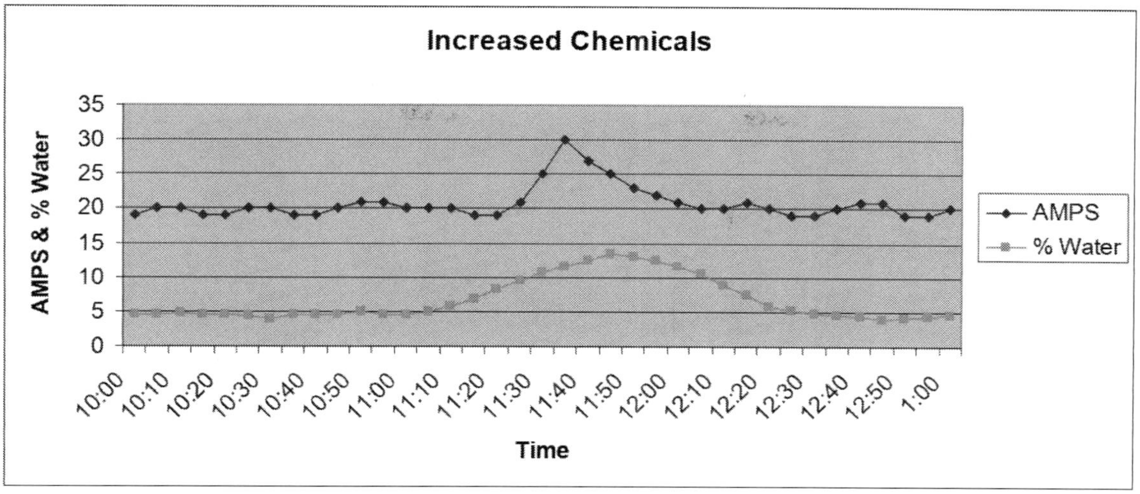

THE UPPER/EMULSION PROBE & AMPS OF THE LOWER GRID

The Upper/Emulsion Probe began to detect a growth in the Emulsion Layer (increase water content) at approximately 11:10AM. Twenty minutes later at approximately 11:30AM, the AMPs of the Lower Grid began to react (increase in AMPs) as the wet emulsion began to enter the area around the Lower Grid. If additional Emulsion Breaking Chemicals had been added at 11:35AM (when the Lower Grids 1st reacted to the Emulsion growth) the response might have looked like that shown above in the graph.

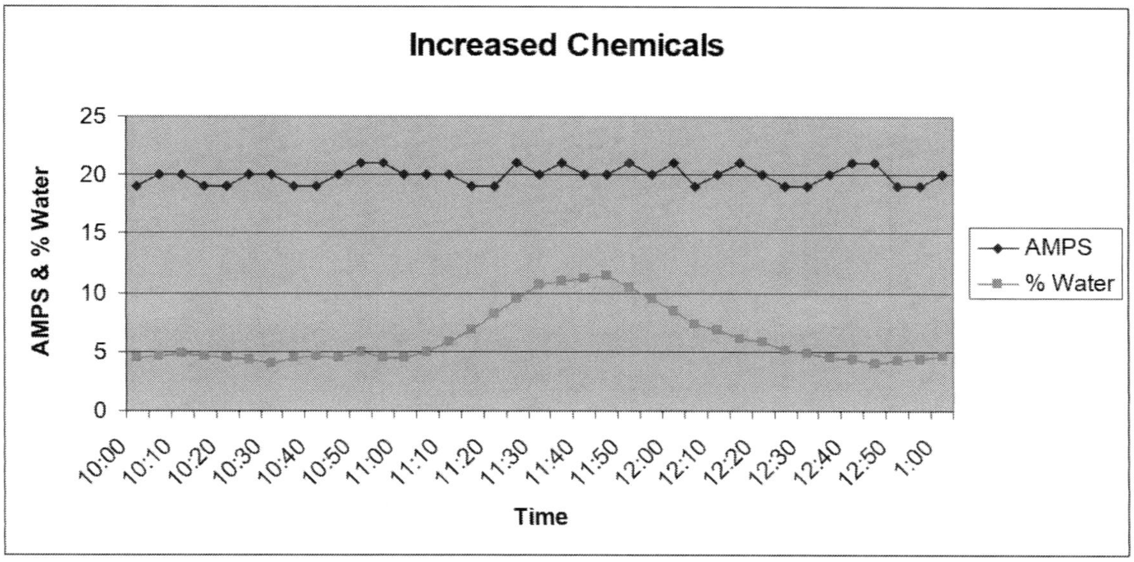

THE UPPER/EMULSION PROBE & AMPS OF THE LOWER GRID

The Upper/Emulsion Probe began to detect a growth in the Emulsion Layer (increase water content) at approximately 11:10AM. If additional Emulsion Breaking Chemicals were added immediately, the problem could have been resolved before the AMPs of the Lower Grid increased in reaction to the Emulsions.

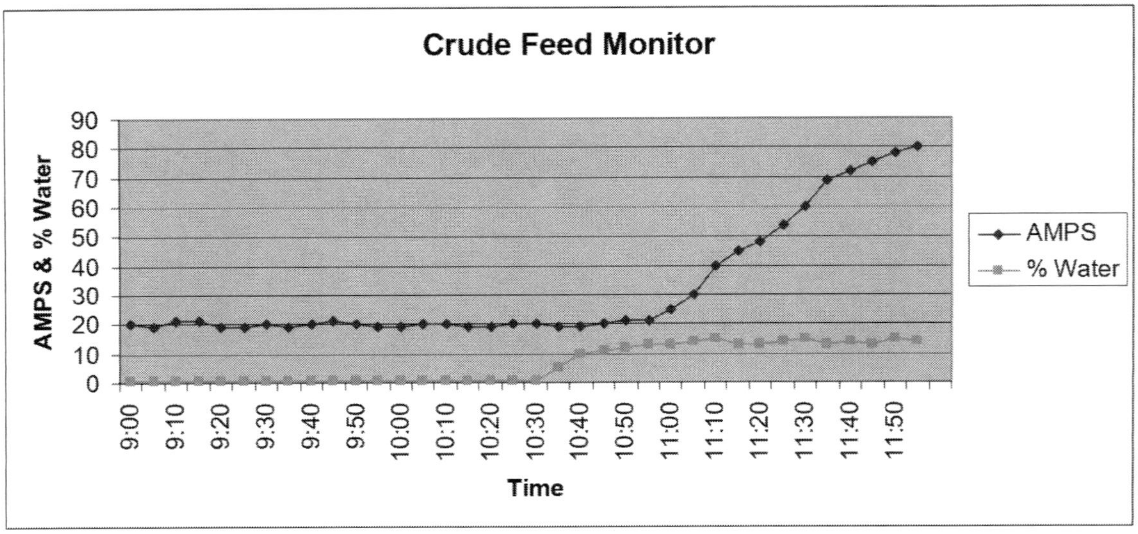

THE CRUDE FEED MONITOR & AMPS OF THE LOWER GRID

The Crude Feed Monitor began to detect Water/Emulsion in the Crude Charge Line from the Tank Farm at approximately 10:35AM. Approximately 30 minutes later at 11:05AM, the AMPS of the Lower Grid began to react to the Wet Crude Feed/Emulsions.

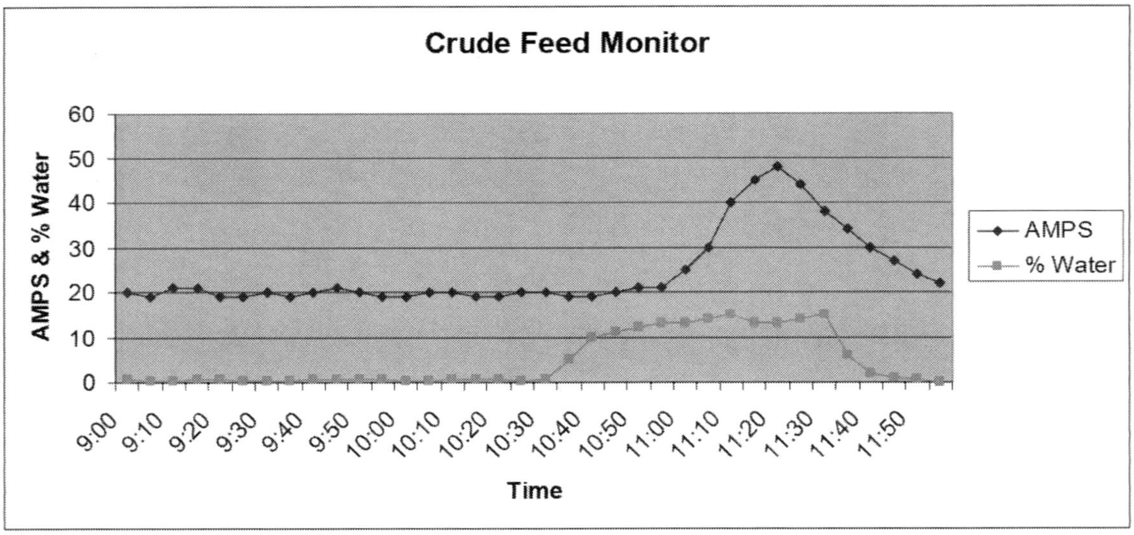

THE CRUDE FEED MONITOR & AMPS OF THE LOWER GRID

The Crude Feed Monitor began to detect Water/Emulsion in the Crude Charge Line from the Tank Farm at approximately 10:35AM. Approximately 30 minutes later at 11:05AM, the AMPS of the Lower Grid began to react to the Wet Crude Feed/Emulsions. With a 30 minute advance warning, the operators can take corrective actions---both in the Tank Farm and Desalter Operations.

Removal of Calcium and Other Metal Species from Crude Oil in the Desalting Process

By Lawrence Kremer
Joe Nguyen
Jerry Weers

Baker Petrolite Corporation
12645 West Airport Blvd.
Sugar Land, TX 77478

Prepared for Presentation at the
AIChE Spring National Meeting 2004, New Orleans, LA, April 25-29, 2004
Seventh Topical Conference on Refinery Processing
Recent Advances in Desalting

Unpublished

ABSTRACT

Some crude oils from West Africa, the North Sea, China and other locations have sufficiently high levels of calcium naphthenate to cause concern to some refineries. High calcium naphthenate content has been suspected of causing desalter problems, fouling and downgrading of coke quality. Other naturally occurring contaminants include iron compounds, nickel, and vanadium. Typically these are difficult to remove from the crude oil during normal desalter operation. Specialized additives have been developed that can significantly reduce the concentration of undesirable species in the desalting process.

INTRODUCTION

There are many metal contaminants of crude oil that cause processing problems in the refinery. Sodium chloride, calcium chloride and magnesium chloride can hydrolyze at high temperatures to form hydrogen chloride HCl gas. The HCl distills up the unit until it is condensed with water to form hydrochloric acid. Acid corrosion caused by the dissolved HCl is one of the principal forms of corrosion occurring in the refinery. The sodium and to a lesser extent calcium and magnesium are poisons to important catalysts in the refinery. In one case a refinery had to completely replace its FCC catalyst inventory when the catalyst became coated with a glass like material containing calcium. Other metal contaminants such as iron, nickel, vanadium, and copper are also poisons of FCC catalysts. Many of these contaminants can adversely affect the quality of refinery products such as petroleum coke and #6 oil. For example calcium and iron are undesirable contaminants of coke for anode manufacture and can significantly affect the premium price that can be obtained for the coke.

Some of these metal species, such as calcium naphthenate and iron sulfide, can stabilize emulsions in desalter causing many processing problems. Stable emulsions can lead to oil under carry that adversely impacts the waste water treatment facility and higher than normal water in the desalted crude oil that can force reductions in crude charge rate and adversely affect the electrical grids.[1] Recently there have been many reports of calcium naphthenates forming stable emulsions in oil field operations in the North Sea[2], West Africa[3,4], Indonesia[5], and other locations. The calcium naphthenate has been identified in stable emulsions in oil water separators, excessive sludge formation in tankage, and rapid fouling of equipment. To avoid problems associated with these metals, it is desirable to remove them from the crude oil, and the desalter is the logical unit with which to accomplish that goal. For example Chang-Ming Lee and John P. Wagner report[6] a laboratory study using the desalting process to remove metals from waste motor oil.

Refineries typically use electrical desalters to remove as much as these contaminants as possible. In the desalting process crude oil is intimately mixed with clean wash water. Water soluble and water dispersible materials are transferred to the water phase. The water and oil are separated resulting in a cleaned crude oil stream and a dirty water effluent that contains the contaminants that have been removed from the crude oil. This process has been described in detail over many decades so will not be discussed further here.

The desalter was originally introduced to remove water soluble salts such as sodium, magnesium, and calcium chloride, which are associated with corrosion in the refinery. These metal species also can poison down stream catalytic process and negatively impact finished product quality. For example sodium limits are often specified for #6 oil. Sodium, iron, and calcium often limit the quality of petroleum coke. Removal of sodium in the desalter is straight forward as most sodium species are very water soluble. Other metals such as iron and calcium can be present as both water soluble and water insoluble species. For example iron can be present as an ion in aqueous solution, an inorganic solid such as iron oxide or iron sulfide, an organic solid such as an iron imbedded in the porphyrin ring of an asphaltene, or as an organo metallic compound such as an iron naphthenate. In the desalting process, the crude oil is washed with water to extract the water soluble and water dispersible species and remove them form the crude oil. Simple water washing can remove the dissolved iron but not necessarily iron in the last three forms described above.

It has also been known for decades that the desalting process can remove inorganic solids, sometimes known as filterable solids. The typical desalter can remove 40 to 60% of the filterable solids in the original crude oil. One of the factors that is believed to limit the removal of the solids is that they tend to be coated with a hydrophobic hydrocarbon coating that prevents them from easily dispersing in water. It has been found in some cases that addition of the correct wetting agent can remove enough of this hydrophobic coating to enable the solid particles to transfer into the water phase. This is termed water wetting the solids. With proper selection of wetting agents the solids removal in the desalter can be increased to 80 to 90%.

IRON REMOVAL

Iron sulfide can cause problems both in the desalter and down stream. Iron sulfide particles are known to stabilize emulsions in the desalter. When iron sulfide stabilized emulsions occur, the rag or interface layer can build and eventually short out the grids. This can also lead to increased BS&W in the desalted crude oil and increased oil and grease in the effluent water. The iron is also a poison of FCC catalyst and a critical contaminant of premium coke. In many cases the high iron content limits the price which can be obtained for the coke. The majority of the iron is believed to come in with the crude oil and is frequently in the form of iron sulfide. Considerable effort has been expended over the years to remove the iron by washing it out in the desalter.

One approach is to water wet the surface of the iron sulfide so that it is removed from the oil and dispersed into the water and thus removed from the crude oil. Recently specialized chemistries have been developed to optimize the removal of targeted undesirable solids. For example a Midwest refinery processing a West Texas intermediate crude oil blend wanted to produce premium-grade coke in their delayed coking unit. Unfortunately, their composite crude slate contained 35 to 40 ppm of inorganic iron that produced 150 to 175 ppm of iron in the residual oil feedstock to the coker. It was determined that most of the iron was coming in with the crude oil. Attempts had been made to optimize the desalter but the iron removal efficiency was still too low to meet the iron targets in the crude oil. One problem is that iron sulfide particles tend to be much smaller than other filterable solids (such as clay and sand). It is more difficult to remove small particles than large particles from crude oil. We developed a specialized wetting agent to facilitate the removal of iron in the form of inorganic solids.

Baker Petrolite's demulsifier and iron removal chemistry was added to the crude charge to the desalter in place of the standard emulsion breaker that was being used. It performed well as a demulsifier, providing excellent dehydration of the washed oil and good effluent water, as shown in Table 1 and Figure 1. In addition, the program effectively transferred particulate inorganic iron and other metals into the water phase as shown by Figure 2. With the Baker Petrolite demulsifier and iron removal product, the refinery was able to reduce iron content in the desalted crude oil by more than half.

	% Salt Removal	% BS&W Desalter Crude	Desalter Effluent % Oil
Previous Treatment	95.8	0.1	5.0
Baker Petrolite	97.8	0.1	1.5

Table 1. In addition to effective removal of iron, BPR29821desalting additive also was more effective in removing salt and in clarifying effluent water than was the previous treatment program.

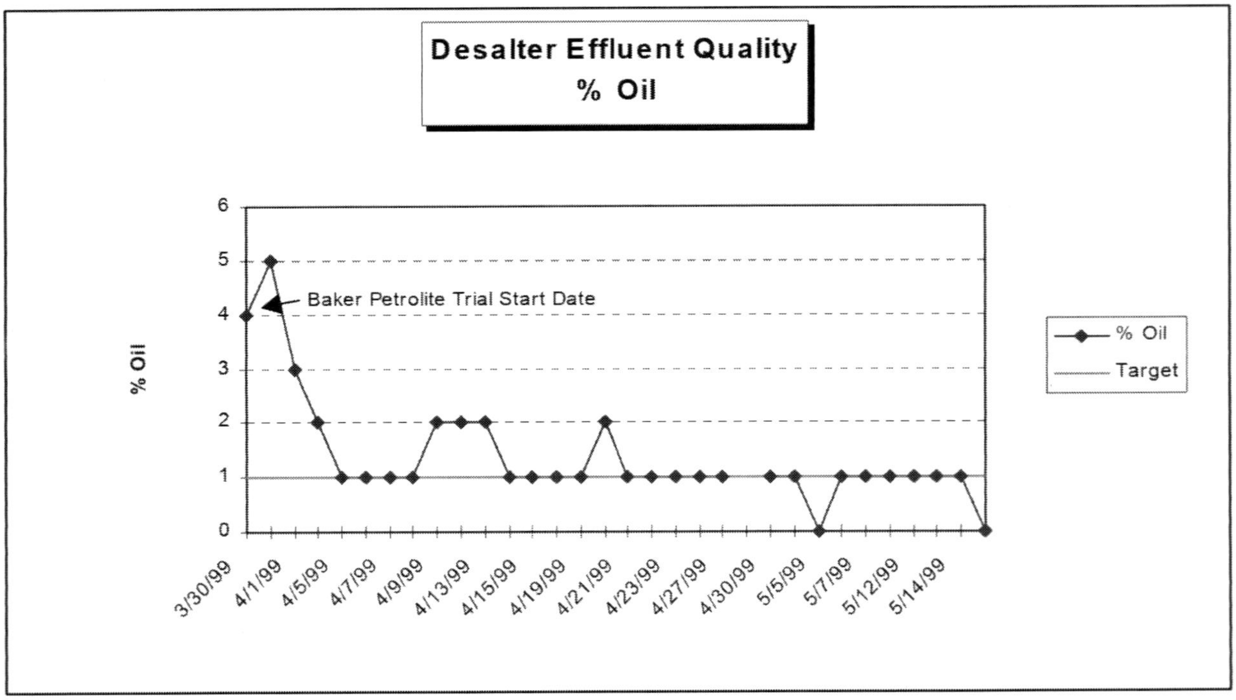

The following figure shows the improvement in iron removal when the specialized chemistry was introduced. In this case most of the iron was present as inorganic solids. If the iron was present either as an oil soluble or water insoluble organo metallic compound, then the wetting agent approach would not work.

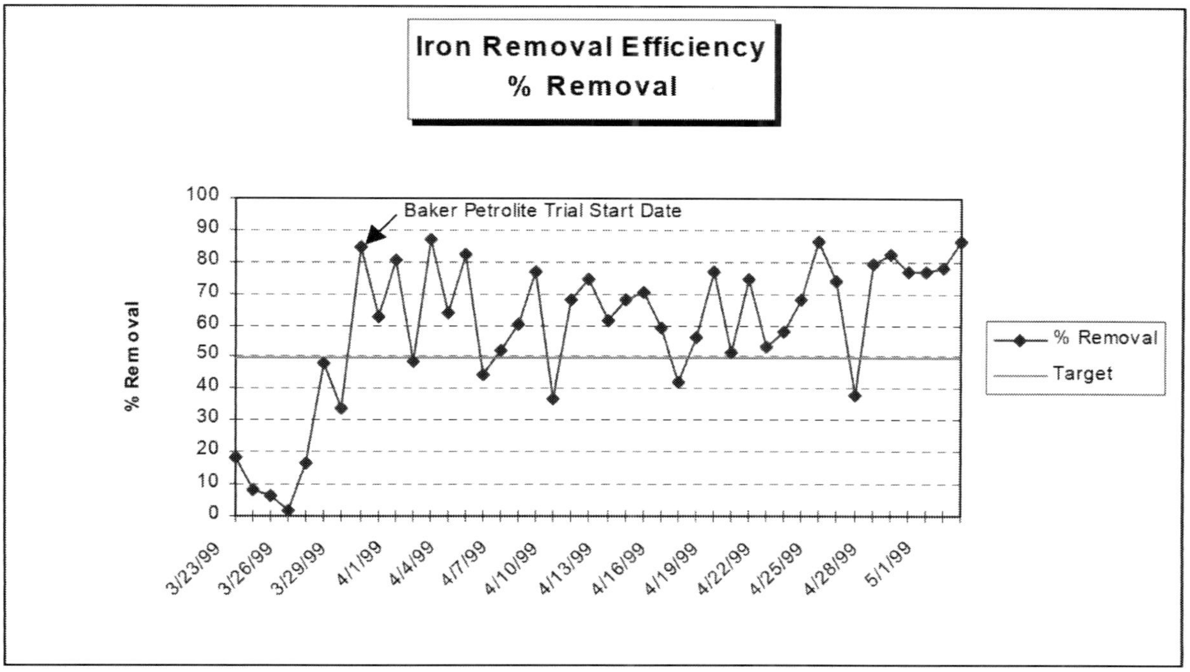

A second approach for iron removal is to chemically react the iron sulfide by converting the iron to water soluble species such that the iron is removed from the crude oil when the water phase is separated and discharged. One very effective method to chemically convert the iron sulfide to water soluble species is the use of acrolein[7].

Acrolein in water reacts in a 2:1 molar ratio with the sulfide entity, producing nontoxic, soluble, low molecular weight products. Acrolein is non-corrosive and is not an oxidizer and the reaction of acrolein with iron sulfide is irreversible. This acrolein reactivity has been used in a wide variety of applications in the oilfield including dissolution of iron sulfide from the near well bore region to stimulate injectivity of water injection well and production from oil producing wells. Acrolein has also been used to resolve oil - water emulsions that have been stabilized by iron sulfide particles. While the acrolein treatment has been successfully used to dissolve iron sulfide in the oil field production, it has not yet been used in refinery applications such as desalting.

CALCIUM REMOVAL

It has long been reported that calcium, as well as sodium and magnesium, when in the chloride form are readily removed in the desalting process. In some crude oils that have high levels of naphthenic acid, the calcium is present in the form of calcium naphthenate. For example, several crude oils from the North Sea, West Africa, and China have been reported to contain calcium naphthenates. Calcium naphthenate is not soluble in either water or oil and its density is reported to be less than water but heavier than many crude oils. Thus the calcium naphthenates tend to carry out of the desalter in the desalted crude and process with the bottom fractions. Calcium has been associated with fouling, poisoning of FCC catalyst and degrading the quality of coke. The calcium naphthenate can also stabilize emulsions in the desalter causing upsets. Under normal desalting conditions the calcium naphthenates are not removed into the water. Several additives have been tried to solubilize the calcium in water so it can be removed from the crude oil. Baker Petrolite has conducted several research projects to investigate the removal of calcium naphthenates form crude oil and to mitigate the negative affects of this material.

Baker Petrolite has developed an Electrostatic Demulsification and Dehydration Apparatus (EDDA) which ahs been described elsewhere. This laboratory test rig simulates the desalting process and has been used to screen various treatment programs that might be useful in commercial desalters. Several additives that might have efficacy in removing calcium naphthenates from crude oil were evaluated using this device.[9] Samples of the desalted and undesalted crude oil were digested to extract the calcium which was analyzed by ICP. The water samples were analyzed by ICP directly. The results for a single stage simulation are given in the following table:

Sample	Ca by ICP
Raw crude oil	**370 ppm**
Desalted crude oil	**30 ppm**
Calcium Removal Efficiency	**92%**

In this case the additive produced a 92% calcium removal in a single stage desalting process. In tests without any calcium removal additive virtually none of the calcium is removed from the crude oil. by the desalting process.

There are many important considerations that must be evaluated when implementing such a treatment program in a refinery. It is not a simple matter of changing chemicals. It is important to understand the interaction of ionic species in the water when substantial amounts of calcium are extracted into the desalter effluent. Increased scale forming tendency is a possibility. Changes in the effluent water quality may also affect the waste water treatment facility. These programs will also affect the fouling and corrosion control programs on the crude unit. Operation of the desalter and resolution of the emulsion in the desalter may be significantly different than previous experience.

For example, at one refinery[8] processing North Sea crude oil with high calcium naphthenate, an additive program was implemented to reduce the calcium in the desalted crude oil. While there were impacts on the corrosion control program, the impact on the waste treatment plant turned out to be minimal. In this case the calcium in the raw crude oil runs between 25 and 50 ppm. Chemical treatment reduces the calcium to 2 to 10 ppm in the desalted crude oil. The cost of the program is more than offset by the improved profitability of the down stream operations which had been adversely affected by the calcium.

REFERENCES

1. Phillips, C.; "Treatment of refinery emulsions and chemical wastes" *J. Ind. and Eng. Chem.* **46** (1954) pp 300-303
2. Havre, Trond Erik; "Formation of calcium naphthenate in water / oil systems, naphthenic acid chemistry and emulsion stability" *Doctoral Thesis* Department of Chemical Engineering, Norwegian University of Science and Technology, October 2002
3. Rousseau, G.; Zhou, H.; Hurtevent, C.; :Calcium carbonate and naphthenate mixed scale in deep-offshore fields" *SPE Oilfield Scale Symposium*, Aberdeen, UK, 2001; SPE 68307
4. Poggesi, G.; Hurtevent, C.; Buchart, D.; "Multifunctional chemicals for West African deep offshore fields" *SPE Oilfield Scale Symposium*, Aberdeen, UK,2002; SPE 74649
5. Gallup, D.L.;Smith, P.C.; Chipponeri ,J.; Abuyazid, A.; Mulyono, D. "Formation and mitigation of "metallic soap" sludge", Attaka, Indonesia field" *SPE International Conference on Health, Safety, and Environment in Oil and Gas Exploration and Production*, Kuala Lumpur, Malaysia, 2002; SPE 73960
6. Chang-Ming Lee and John P. Wagner "Use of Desalting Methods as a Pre-Treatment Step for Removal of Metallic Species from Oil-Continuous Emulsion Systems" *5th International Conference on Refinery Processing at AIChE 2002 Spring National Meeting* New Orleans, March 11-14, 2002, pp 653-663
7. Salma, T.; "Cost Effective Removal of Iron Sulfide and Hydrogen Sulfide from Water Using Acrolein" SPE 59708 presented at the 2000 SPE Permian Basin Oil and Gas Recovery Conference Midland TX 21-23 March 2000
8. Unpublished data, Baker Petrolite
9. Patent applied

New Methods of Application of Electrostatic Fields

Gary W. Sams & Kenneth W. Warren
NATCO Group, Inc.
2950 N. Loop W. #700
Houston, Texas 77092

Prepared for Presentation at AIChE Spring National Meeting
Recent Advances in Desalting
New Orleans, Louisiana
April 2004

Unpublished

New Methods of Application of Electrostatic Fields

Gary W. Sams & Kenneth W. Warren
NATCO Group, Inc.
2950 N. Loop W. #700

Abstract:

Desalting combines three key steps to achieve efficient salt removal. Fresh water mixing provides dilution, coalescence to promote droplet growth and phase separation to remove the contaminants. Typically, the coalescence step is accomplished by an electrostatic field to maximize the droplet growth. However, as oils become more difficult, traditional electrostatic processes fail to achieve adequate droplet coalescence for efficient desalting. Recent developments in electrostatic power supplies provide important tools for dealing with difficult crude oils. These oils include those with high conductivity, low API gravity, and high viscosity. While improvements in the hydraulic performance of an electrostatic dehydrator or desalter have successfully improved capacities, only the application of aggressive and conformable electrostatic fields has the potential to obtain the droplet growth necessary to achieve efficient dehydration at these increased oil fluxes. This paper describes the generation and application of multi-frequency electrostatic fields and the advantages realized.

Introduction:

Efficient desalting in both field and refinery applications is an essential process to reduce corrosion, control metal content and improve wastewater treatment. Electrostatic dehydrators are used almost exclusively in these desalting applications because of their high efficiency. Electrostatic desalting of crude oils is a process of three steps consisting sequentially of fresh wash water dispersion, coalescence between the connate brine and wash water and finally dehydration of the oil by phase separation. Fresh water dispersion is typically accomplished in the oil inlet piping ahead of the desalter and utilizes pipe distributors, static mixers, and mixing valves. Coalescence of the brine water and wash water is accomplished inside the desalter vessel by establishing an environment to induce collisions (coalescence) between these dispersed waters. Effective phase separation is achieved when the water droplets achieve sufficient diameter to separate rapidly. (Warren, Sams and Nakayama, 1998)

The optimum desalter environment includes an iso-kinetic oil flow to reduce short-circuiting and eliminate flow recycling and an aggressive electrostatic field capable of energizing the smallest dispersed water droplets and promoting water droplet growth. Establishing iso-kinetic flow requires reduction of the oil momentum as it exits the oil distributor orifices, redirection of the oil to achieve a uniform flow field throughout the vessel and uniform oil collection to achieve maximum utilization of the desalter volume. (Sams and Wallace, 2001) An aggressive electrostatic field consists of several key components to maximize the electrostatic forces, induce rapid droplet coalescence and enhance droplet growth to maximize phase separation. (Urdahl, Nordstad, Berry, Wayth, Williams, Bailey and Thew, 2001)

Electrostatic Fields:

Dehydrator manufacturers rely on three fundamental types of electrostatic fields to enhance coalescence of the dispersed water droplets, namely direct current (DC) fields, alternating current (AC) fields and combined AC/DC fields. DC fields are highly efficient but can promote electrolytic corrosion. Therefore, they are not used in crude oil desalting applications, but only to dehydrate refined oils having low conductivity. On the other hand, AC fields are used by all manufacturers due to their tolerance of high water cuts and non-electrolytic nature. Finally, combined AC/DC fields provide the high water tolerance of the AC field with the high efficiency of the DC field. (Burris, 1977)

AC dehydrators range from electrostatic fields utilizing one AC transformer energizing a single, horizontal electrode suspended below a grounded (earth) electrode as shown in Figure 1. A weak AC

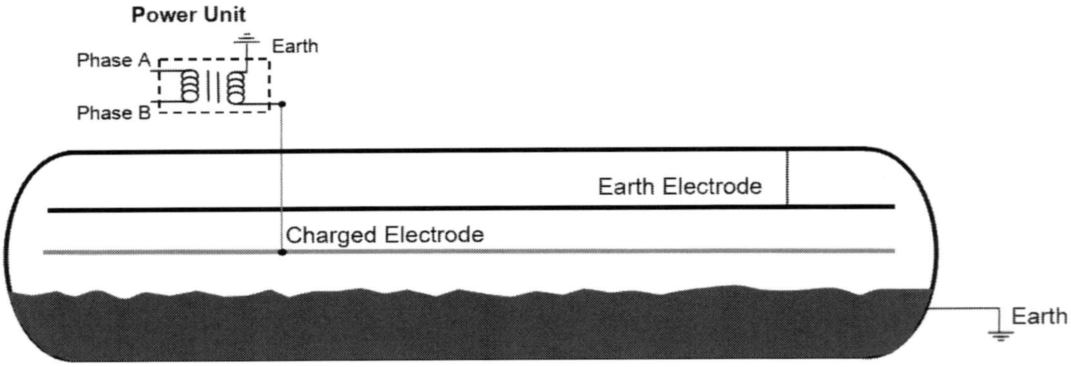

Figure 1: Conventional AC Dehydrator / Desalter

gradient is established between the energized electrode and the oil / water interface and a strong AC gradient is established between the energized and earth electrode. Wet oil entering the vessel just above the water interface is rapidly coalesced by the lower AC field and is then further coalesced and dehydrated by the upper field. Once the oil is above the earth electrode additional electrostatic coalescence is impossible since an electrostatic field can not exist above it.

A more efficient AC desalter utilizing three AC transformers and three electrodes is shown in Figure 2. These are commonly referred to as Deep-field AC processes because they establish an AC field between the oil / water interface and the oil collector. The wet oil enters just above the interface where the low AC field promotes initial coalescence and separation. Higher AC field gradients are established between the three electrodes where additional coalescence and separation achieve the desired performance.

An aggressive electrostatic process utilizes a combination AC / DC fields. These desalters consist of an array of vertical, parallel electrodes generally positioned diametrically across the vessel just above the

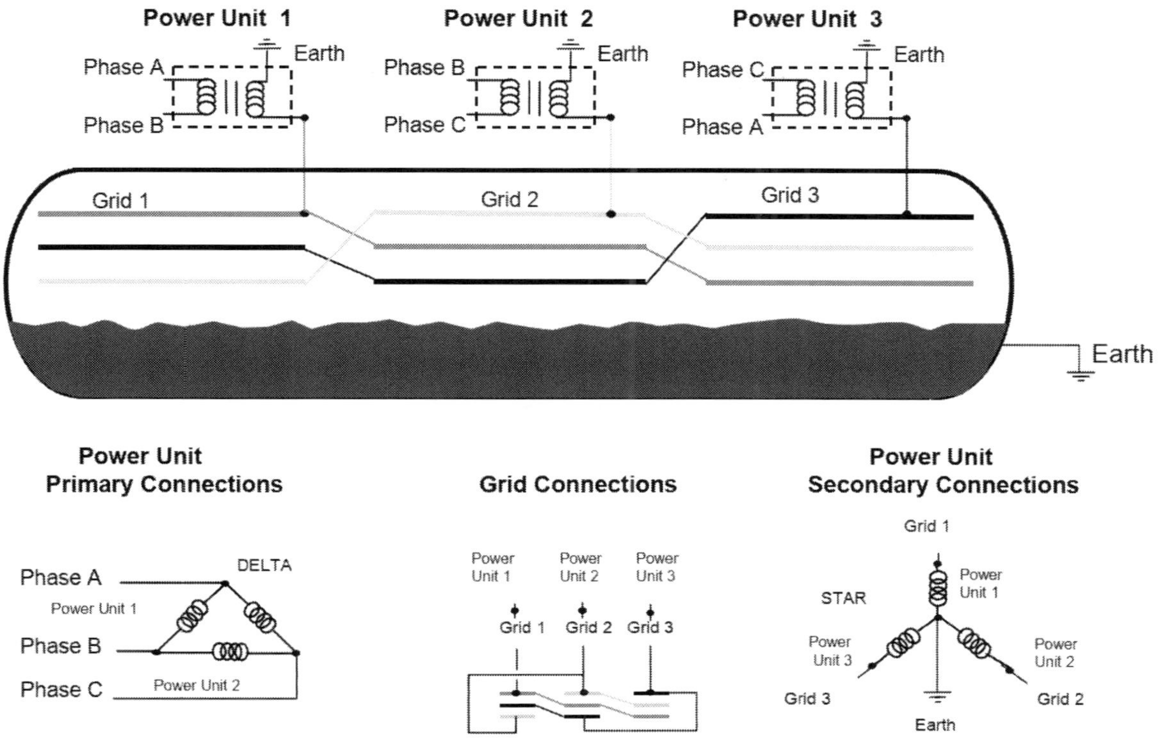

Figure 2: Deep Field AC Dehydrator / Desalter

centerline as shown in Figure 3. These desalters utilize from one to three transformers containing a pair of reversed diodes to establish a DC field between adjacent electrodes as shown in Figure 4. For a combination AC/DC desalter, an AC field is established between the bottom of the electrodes and the oil/water interface. Just like the AC desalter, this lower AC field gradient promotes initial droplet coalescence in the high water cut environment just above the interface.

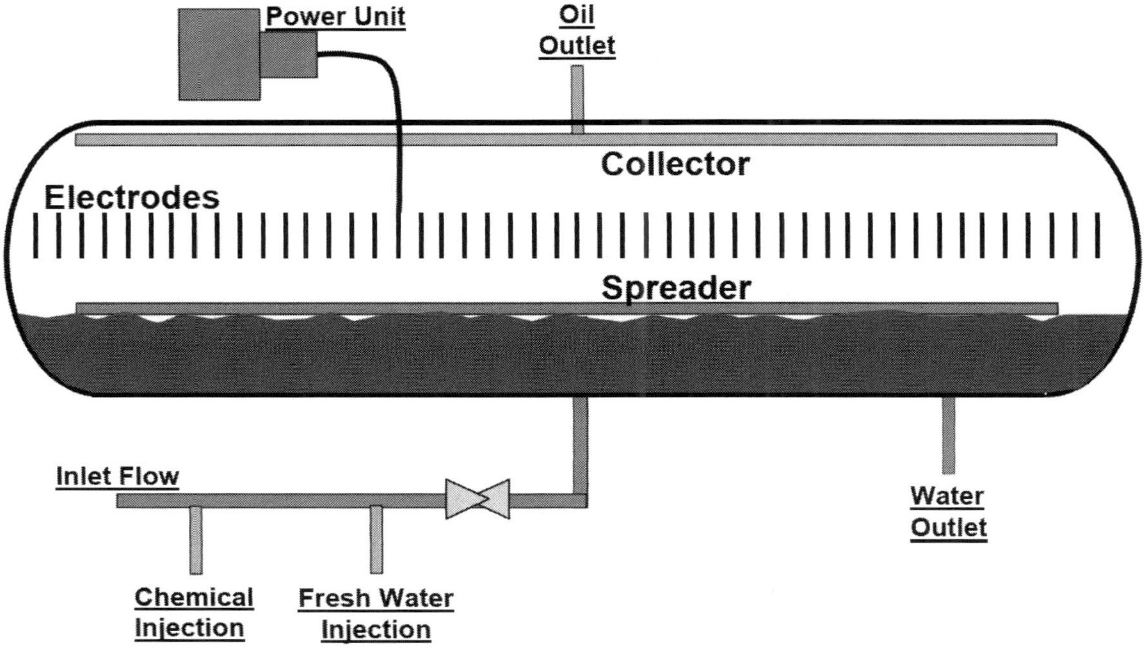

Figure 3: Combined AC/DC Desalter

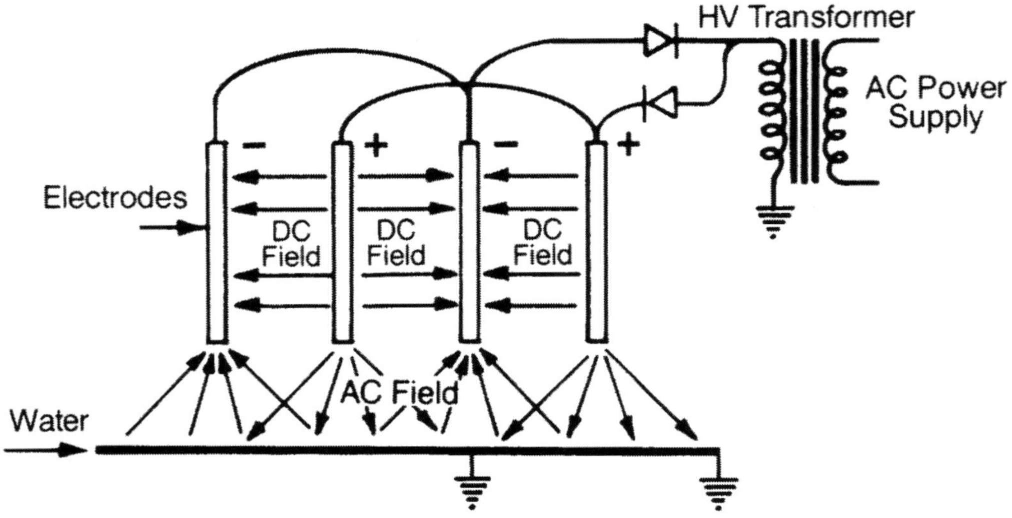

Figure 4: Combined AC/DC Wiring

Electrostatic Forces:

A water droplet suspended between a pair of electrodes is acted upon by five forces as indicated in Figure 5. (Draxler and Marrs, 1993) Two of these five forces are gravitational and hydraulic. Gravitational forces equal to the weight of the droplet act to move the water droplet toward the bottom of the vessel. Drag forces imposed by the rising oil moving pass the water droplet act to lift it toward the oil outlet. If the water droplet is larger than the Stokes' droplet diameter as calculated by the equation in Figure 6, the weight is greater than the drag and the water droplet will be separated from the oil.

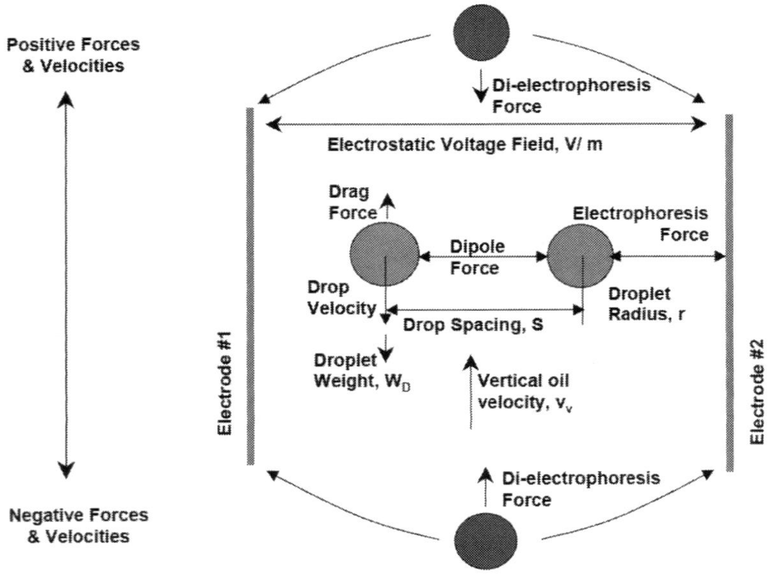

Figure 5: Coalescence Forces

To maximize the desalting process performance, the electrostatic forces must be capable of promoting droplet coalescence to diameters greater than the Stokes' diameter. The three electrostatic forces are dipolar, electrophoretic and di-electrophoretic.

$$d_{Stoke} = \left(\frac{18\, \mu_o\, v_v}{g(\rho_w - \rho_o)} \right)^{0.5}$$

Figure 6: Stokes' Droplet Diameter

1) Dipole forces are established by the alignment of the polar water molecules in the droplet and are proportional to the electric field gradient, the water droplet diameter and the spacing between droplets as shown in Figure 7.

$$F_d = \frac{6\, K\, E^2\, r^6}{s^4}$$

Figure 7: Dipole Force

2) Electrophoretic forces are both attractive and repulsive forces established in a uniform voltage field between charged droplets and the electrodes. They are proportional to the field strength, droplet diameter and oil conductivity as shown in Figure 8.

$$F_e = C\pi^3 \mu\, \varepsilon_c\, r^2\, E^2\, e^{(-\sigma_c t/\varepsilon_c)}$$

Figure 8: Electrophoretic Force

3) Di-electrophoretic forces are attractive forces established in a non-uniform field. These forces pull the droplet toward the highest voltage gradient and are proportional to the droplet diameter and the oil conductivity as shown in Figure 9.

$$F_{diel} = 2\pi r^3\, \varepsilon_c \left(\frac{\varepsilon_d^* - \varepsilon_c^*}{\varepsilon_d^* + 2\varepsilon_c^*} \right) \nabla E^2$$

Figure 9: Di-electrophoretic Force

These three electrostatic forces are present in all desalters but they can be manipulated to achieve enhanced coalescence and separation by altering the electrostatic voltage fields.

As Figure 7 shows the dipole force is dependent on the water droplet size and the spacing between them. Assuming the water droplets are uniformly sized and homogeneously distributed it is easy to see the spacing is inversely proportional to the dispersed water volume as shown in Figure 10. Therefore, as the dispersed water is coalesced and separated from the oil, the spacing between droplets increases and the dipole forces decline rapidly. Figure 10 also shows that when the droplet radius increases the spacing increases. Therefore, the dipole force weakens rapidly as droplets are coalesced and separated from the oil.

$$s = r\left(\frac{1.333\pi}{X}\right)^{0.333}$$

Figure 10: Uniform Droplet Spacing

As Figure 8 shows the electrophoretic force is independent of the droplet spacing, but is dependent on both the oil viscosity and conductivity. The electrophoretic force has a time constant equal to the ratio of the oil's dielectric constant and conductivity. As the equation in Figure 8 shows the electrophoretic force will decay most rapidly in highly conductive oil. Therefore, the force can only be sustained by replenishing the droplet charge frequently.

As Figure 9 shows the di-electrophoretic force is independent of the droplet spacing, but is dependent on the change in field strength which pulls the largest droplets into the highest field gradient. This force acts to accumulate water in areas of the electrostatic field where its divergence is greatest.

Dipolar and di-electrophoretic forces are predominating in these AC desalting processes. The drop to drop dipolar forces are greatest in the bottom of the desalter where the water content is highest and the droplets are closely spaced. The di-electrophoretic forces will pull droplets to the rods used to construct the electrode arrays and thereby increase the droplet population and the dipolar forces. Since the electrical polarity on the AC electrodes reverses every few milliseconds, the electrophoretic force also reverses direction and has little influence in the coalescence process.

The uniform DC field established between the electrodes relies on the electrophoretic force to push and pull droplets in a horizontal plane between electrodes. Once a water droplet approaches one energized electrode it is charged to the same polarity. Once the droplet is charged, the electrophoretic force then pushes the droplet toward the adjacent and oppositely charged electrode. (Urdahl, Williams, Bailey, and Thew, 1996) As the droplet approaches this electrode, the electrophoretic force pulls the droplet toward it until the droplet charge is reversed. Therefore, the electrophoretic forces provide a motive force to move the water droplet population in opposite directions between electrodes. The resulting collisions achieve efficient coalescence, large droplet sizes and rapid separation. These actions are shown schematically in Figure 11.

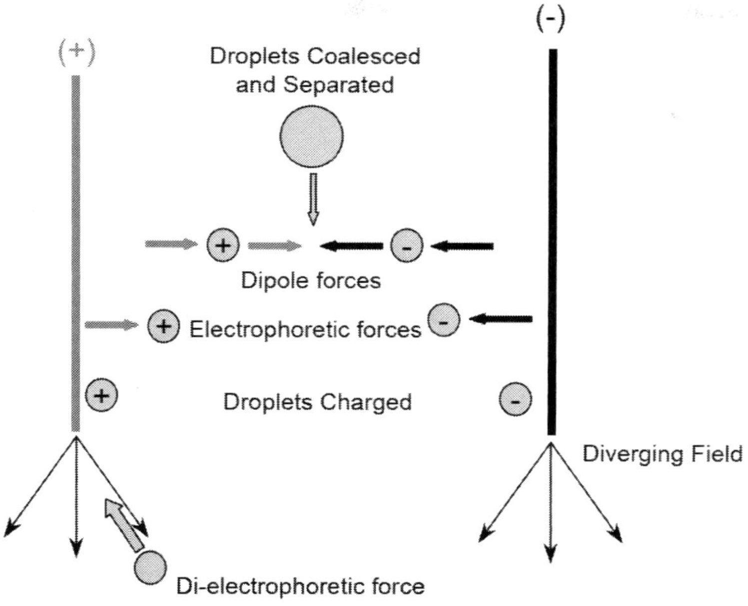

Figure 11: Combined AC/DC Droplet Forces

Oil Properties:

When determining the proper size of a desalter, the most important physical properties include the oil viscosity, oil density and water density. When these properties are used in Stokes' law along with the vertical oil velocity as shown in Figure 2, the neutrally buoyant droplet (Stokes') diameter can be calculated. All larger water droplets will have sufficient weight to overcome the viscous drag force and will settle to the interface. Smaller droplets will continue to be lifted by the flowing oil to the desalter outlet.

Certain design parameters can be adjusted to achieve the desired outlet BS&W and salt content. The oil velocity or flux is directly proportional to the desalter size and the oil flowrate. The oil viscosity is inversely proportional to the process temperature. Increasing the temperature will reduce the oil viscosity permitting smaller water droplets to settle. However, increasing the temperature will also change the density difference and affect the droplet separation rate. As Figure 12 shows the maximum density differential is around 90 to 100°C and decreases if the process temperature is increased or decreased.

A proper balance of the oil viscosity, density difference and oil flowrate is essential to ensure proper performance of a desalter. While these design parameters are certainly the primary variables that influence the desalter behavior, two other parameters play a key role in the efficiency of the electrostatic process, namely interfacial tension and oil conductivity.

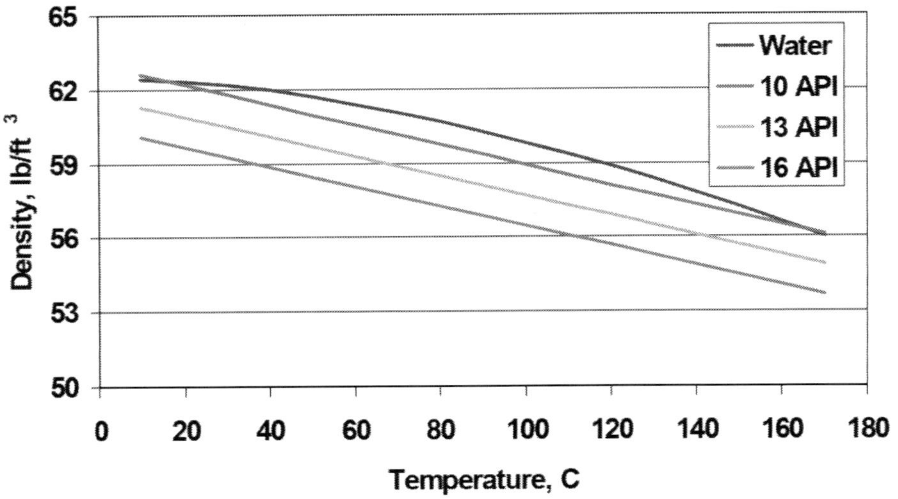

Figure 12: Oil / Water Density

Interfacial tension is usually measured in dynes/cm and ranges from 15 to 25 dynes/cm for typical oils. High interfacial tensions make it difficult for droplets to coalesce by requiring higher coalescence forces to rupture the droplet interface. Low interfacial tension makes droplet coalescence easy, but makes the coalesced droplets very unstable and easy to re-disperse. Demulsifying chemicals and electrostatic fields lower the interfacial tension to promote coalescence. (Byers and Amarnath, 1995) Excessive electrostatic droplet charge or excessive chemical treatment will reduce the interfacial tension and may promote droplet breakup and dispersion.

Oil conductivity, usually measured in nanoSiemens/meter, ranges from 40 to 80 nS/m for typical crude oils. However, for blended oils such as bitumen and SAGD (steam assisted gravity drainage) oils, conductivity of 250 nS/m has been measured. The oil conductivity is a result of excessive water in the electrode zone, polar organic and inorganic compounds and the presence of conductive solids in the oil. High oil conductivity adversely affects the electrostatic process in three ways. First, increasing oil conductivity results in an increase in resistive heating, which is not contributory to the electrostatic coalescence process. This may require the manufacturer to use a larger power unit in order to maintain the efficiency of the electrostatic process. Second, high oil conductivity will reduce the electrophoretic forces which reduces droplet mobility in the combined AC/DC electrostatic desalter. Finally, the di-electrophoretic forces also decrease to further reduce the coalescence efficiency in both the AC and combined AC/DC processes.

Understanding the roles played by oil conductivity and interfacial tension it is possible to establish an electrostatic field that aggressively promotes droplet coalescence and achieves deeper dehydration. (Eow and Ghadiri, 2003)

Electrostatic Voltages:

The combined AC/DC field utilizes these electrostatic forces in such a fashion that manipulation of the electrostatic voltage can benefit from the oil conductivity and interfacial tension. For clarity, the remainder of the paper will consider the combined AC/DC electrostatic field only.

In nearly all desalter applications a single voltage level is applied to the electrodes to achieve a beneficial level of dehydration and desalting. However, as the force equations show the smaller droplets require higher voltages to develop sufficient force to overcome the interfacial tension and promote coalescence. However, if the voltage is too high, the electrostatic forces may exceed the interfacial forces resulting in droplet breakup and dispersion.

Two voltages define the limits of an efficient dehydration process. The first voltage can be thought of as a 'threshold' voltage. Figure 13 shows the results of a lab experiment conducted with a combined

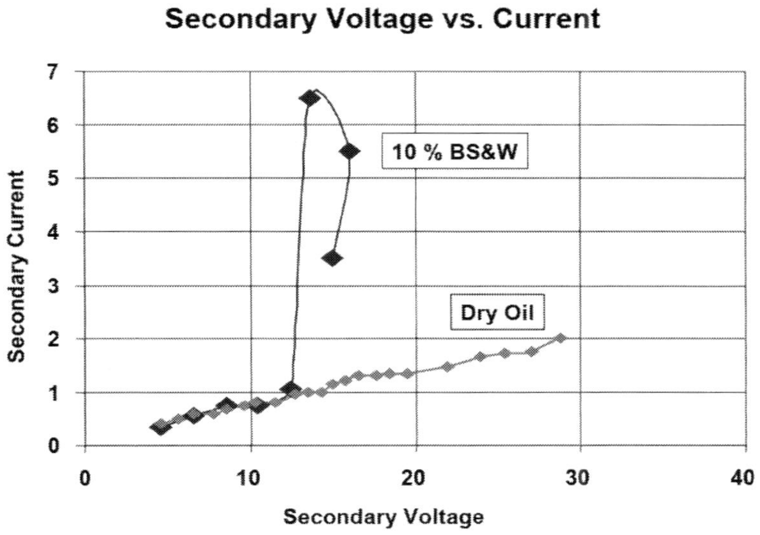

Figure 13: Threshold Voltage

AC/DC field. This experiment was used to determine the threshold voltage for a 20°API oil containing 10% water flowing through a pilot treater. The applied voltage was increased slowly starting at a low voltage and while measuring the secondary current. At 12.5 the secondary current increased sharply suggesting the dispersed water was being electrically charged. Once the water was energized the current began to decrease rapidly indicating the water was coalescing. A second experiment was conducted with 'dry' oil. As Figure 13 shows the secondary current failed to spike but continued to increase linearly with the voltage. The placement and slope of the 'dry' oil line is consistent with the oil conductivity.

As the experiment clearly indicates operating below the threshold voltage fails to provide sufficient electrostatic energy to initiate coalescence. There is a process benefit in operating at a voltage level near the threshold voltage because it maximizes the droplet diameters. However, it can be shown that operating at such a low voltage fails to reach the smallest water droplets necessary to achieve the effluent water and salt specifications.

For perfect dehydration, all of the entrained water would be coalesced to a droplet diameter larger than the Stokes' diameter so they will separate from the rising oil. Coalescing the smallest water droplets requires an increased voltage capable of developing significant electrostatic forces. However, increasing the voltage also increases the electrostatic forces on the larger droplets which may cause them to fracture. Therefore, the applied voltage should not be increased above a level than would fracture water droplets of the Stokes' diameter. This voltage level can be considered the 'critical' process voltage. Sustained operation above this 'critical' voltage results in a reduction in the water droplet diameter and a significant decline in dehydration performance.

Electrostatic Frequency:

Just as there are two voltage limits that define the boundaries for efficient dehydration there are also two frequencies. These two frequencies are dependent on the rate of droplet charge and discharge and the fundamental oscillation frequency of a water droplet of Stokes' diameter.

As Figure 14 shows, when using 50 or 60 Hz power, the rate of voltage decay on the electrodes

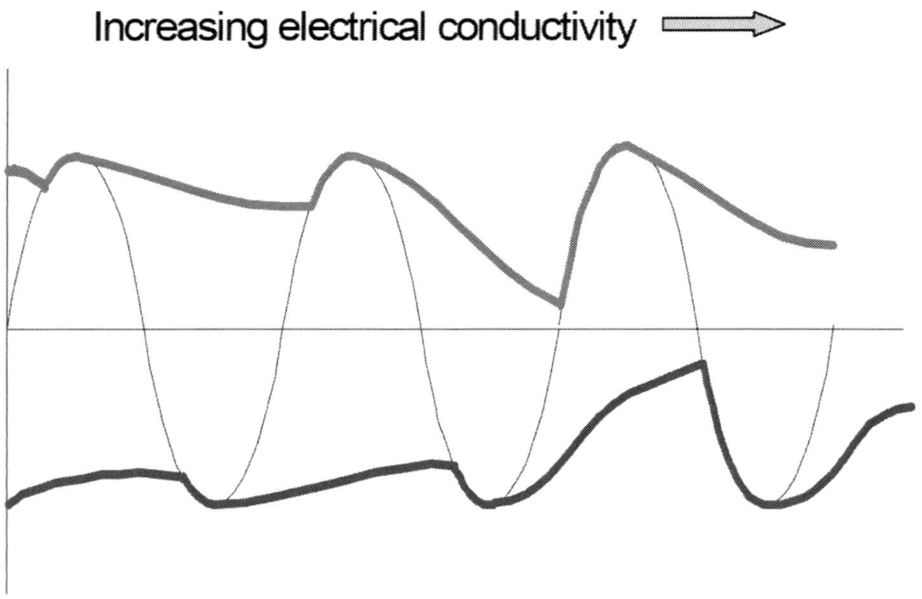

Figure 14: Applied Voltage vs. Oil Conductivity

increases as the oil conductivity increases which permits the charge on the water droplets to decrease. Furthermore, as the oil conductivity increases the electrode voltage is below the threshold level for a significant part of the voltage cycle. Therefore, to sustain the droplet charge and maintain the electrostatic forces for coalescence, the frequency of the applied voltage must be increased. For oil with a conductivity of 100 nS/m, the frequency must be near 1600 Hz. Operating above this 'base' frequency ensures the highest level of electrostatic forces.

The fundamental oscillation frequency of a water droplet can be determined using a simple mass-spring analogy. (Ivanitskii, 1998) For a Stokes' droplet diameter of 750 microns with an interfacial tension of 15 dynes/cm the fundamental frequency is 20 Hz. Therefore, when operating with a 60 Hz power unit, the Stokes' droplet is oscillating at its third harmonic. The increased droplet oscillation amplitude may result in droplet shatter due to increased electrostatic forces. However, by operating at a 'modulation' frequency below 20 Hz the destruction due to harmonic droplet oscillation can be avoided and maximum droplet growth can be promoted. (Bailes, Freestone and Sams, 1997)

Modulated Electrostatic Results:

Combining the 'base' and 'modulation' frequencies with the 'threshold' and 'critical' voltages it is possible to construct an electrostatic field that can derive a significant dehydration benefit from the oil conductivity and interfacial tension.

The dehydration benefits of these electrostatic fields have been confirmed in numerous laboratory tests conducted over the past 4 years. The results from one test are presented here showing the efficiency improvements that can be achieved by this enhanced dehydration method.

Since field data was available to confirm then lab results, Venezuelan oil was used for these comparison tests. The physical properties were determined to be:

- Oil Density @ 60°F = 0.9370 g/cc
- API Gravity = 25.1°API
- Viscosity @ 35°C = 37.31 cp
- Viscosity @ 50°C = 20.15 cp
- Interfacial Tension @ 85°F = 22.5 dyne/cm
- Conductivity @ 77°F = 42 nS/m

Lab tests were conducted in a pilot dehydrator consisting of a vertical 5 inch ID x 8 foot high column. The oil and water was premixed in a 70 gallon charge tank and held at 90°F and 5 psig pressure for the duration of the testing. The oil / water mixture was initially recycled for 45 minutes across a 50 psi pressure drop to create a tight emulsion. The oil had been previously treated with a demulsifier chemical so no additional chemical was added. To avoid additional shear, the mixture was pumped with a progressive cavity pump to a pressure of 40 psig. A shell and tube heat exchanger increased the mixture temperature to 140°F prior to entering the dehydration vessel. The vertical column has the same geometry as an 8 foot diameter dehydrator vessel and achieves results that are consistent with field results. The electrostatic configuration uses a combined AC/DC field with a single pair of oppositely charged electrodes.

For comparison, a series of tests were conducted at several oil rates and a constant temperature of 140°F using both combined AC/DC and modulated, high frequency power supplies. The AC/DC power supply was operated at a steady voltage of 23 kV (rms). The voltage limits and frequencies for a modulated, high frequency electrostatic power supply were optimized for these tests.

The commercial AC/DC vessels operated at 90 bopd/ sq.ft. and averaged 0.38% BS&W over a 6 week test period. The effluent BS&W results from the lab tests are shown in Table 1 below.

Table 1: Outlet BS&W Comparison

Oil Flowrate, Bopd/Sq. Ft.	Combined AC/DC	Modulated, High Frequency
90	0.4 %	- - -
135	0.4 %	0.1%
180	0.5 %	0.2 %
250	0.6 %	0.4 %
300	- - -	0.5 %

As these results clearly show the modulated, high frequency power supply consistently achieved improved dehydration results and was capable of achieving the same performance as the commercial units at nearly twice the oil flow rate. Even more dramatic results would have been achieved if the parameters of the modulated, high frequency power unit were optimized for the increased flows.

Figures 15 and 16 show the benefits of shifting the base and modulation frequencies on 30°API oil tested

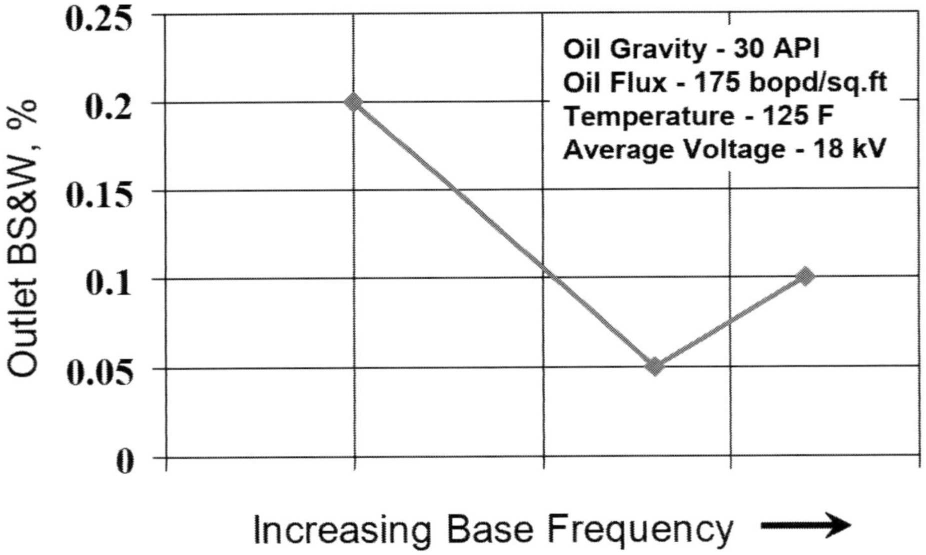

Figure 15: Base Frequency Performance

at 125°F. As the base frequency increases the electrostatic forces also increase and the dehydration performance improves significantly. However, high base frequencies can develop an excessive charge on the droplets which decreases the interfacial tension resulting in a loss of performance as shown for this oil.

Figure 16 shows the benefits obtained by varying the modulation frequency. Slow droplet oscillation stretches the droplet surface destabilizing the film to promote coalescence. It also sweeps the voltage to

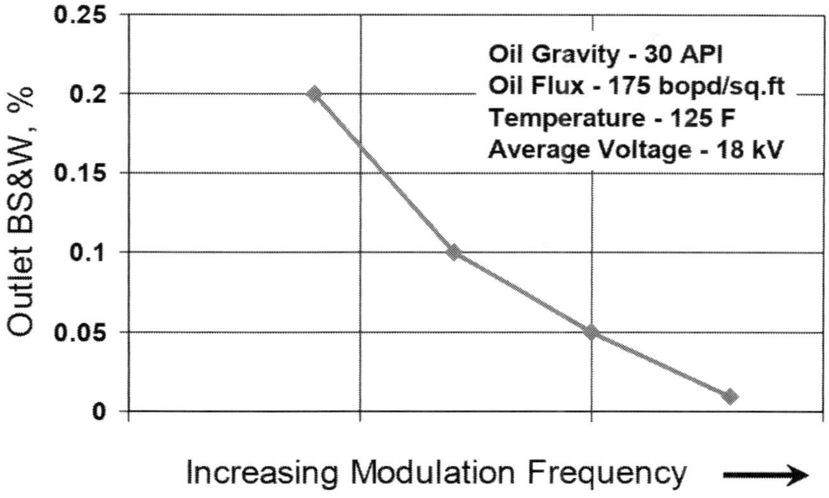

Figure 16: Modulation Frequency Performance

the maximum voltage level to energize the smallest water droplets and promote their coalescence. It then slowly sweeps to the minimum voltage level where the droplet size is maximized.

Dehydration tests conducted on a SAGD oil further demonstrated the benefit of the modulated, high frequency power unit. All three power unit configurations were used on a pilot dehydration designed to dehydrate a 6.5°API oil at a temperature of 430°F. The 'dry' oil conductivity was determined to be around 400 nS/m at the process temperature. An AC power unit failed to develop sufficient voltage to initiate and sustain dehydration. A combined AC/DC power unit permitted the applied voltage to be increased but also failed to promote coalescence. Only the modulated, high frequency power unit was able to achieve and maintain sufficient voltage for dehydration.

Conclusions:

As more difficult oils are produced and used by refiners to increase production and maximize profits, the electrostatic desalter must be capable of handling these highly conductive oils and maintain the desalter specifications. Understanding the role of oil conductivity and interfacial tension in the development of electrostatic forces makes it possible to design a customizable electrostatic field to handle these oils.

The technology is available to produce an electrostatic power unit capable of operating at a 'base' frequency and cycling at the 'modulation' frequency between the 'threshold' and 'critical' voltages. Experimental results have clearly demonstrated the process benefits that can be achieved by this technology. Field trials are currently underway to demonstrate the benefits on a commercial installation for a field dehydrator. (Sams and Wallace, 2003)

Nomenclature:

C = constant
X = BS&W, Inlet or Outlet
V_v = Vertical Oil Velocity
ρ_c = Continuous phase density
ρ_d = Dispersed phase density
w_d = Weight dispersed droplet
μ_c = Continuous phase viscosity
μ_d = Dispersed phase viscosity
σ_c = Continuous phase conductivity
σ_d = Dispersed phase conductivity
ω = frequency
ε_c = Continuous phase permittivity
ε_d = Dispersed phase permittivity
ε_c^* = Complex Continuous phase permittivity
$= \varepsilon_c - j\,\sigma_c / \omega$
ε_d^* = Complex Dispersed phase permittivity
$= \varepsilon_d - j\,\sigma_d / \omega$
F_d = Dipole Force
F_e = Electrophoretic Force
F_{diel} = Di-Electrophoretic Force
∇E = Voltage field gradient
E = Voltage
r = droplet radius
γ = Interfacial tension
d_{Stoke} = Stokes' droplet diameter

References:

Bailes, Philip J., Freestone, D. and Sams, G. W., 1997. Pulsed DC Fields for Electrostatic Coalescence of Water-in-Oil Emulsions, *The Chemical Engineer*, 23 October, 34-39.

Burris, Donald R., 1977. Dual Polarity oil dehydration, *Petroleum Engineer*, 30 – 41.

Byers, Charles H. and Amarnath, Ammi, 1995. Understand the potential of electro-separations, *Chemical Engineering Progress*, 63-69

Draxler, J. and Marr, R., 1993, Design Criteria for Electrostatic Deemulsifiers, *International Chemical Engineering*, Austria, **33**, 1.

Eow, John S. and Ghadiri, Mojtaba, 2003. Motion, deformation and break-up of aqueous drops in oils under high electric field strengths, *Chemical Engineering and Processing*, **42**, 259 – 272.

Ivanitskii, K., 1998, Modeling of deformation and breakup of drops moving in liquid, *Heat Transfer Research*, **29** (4 & 5), 225 – 234.

Sams, Gary W. and Wallace, Harry G., 2003. Field Trials Scheduled for New Dehydration Technology, *Offshore Technology Conference*, 15353.

Sams, Gary W. and Wallace, Harry G., 2001. Improving Process Efficiencies by Optimizing Fluid Hydraulics in Electrostatic Oil Dehydrators, *Offshore Technology Conference*, 13216.

Urdahl, O., Nordstad, K., Berry, P., Wayth, N., Williams, T., Bailey, A. and Thew, A., 2001. Development of a new, compact electrostatic coalescer concept, *SPE Production & Facilities*, 69196.

Urdahl, O., Williams, T. J., Bailey, A. G., and Thew, M. T., 1996. Electrostatic destabilization of water-in-oil emulsions under conditions of turbulent flow, *Transactions of the Institute of Chemical Engineers*, **74** (A), 158 – 165.

Warren, Kenneth W., Sams, Gary W. and Nakayama, Toshio, 1998. Electrostatic Fields: Essential Tools For Desalting, *AIChE 1998 Spring Meeting*, 9-11 March.

Real Time Energy Management Solutions

Joseph L. Davis

KBC Advanced Technologies, Inc
14701 St. Mary's Lane, Suite 300
Houston, Texas 77079

Prepared for presentation at the AICHE 2004 Spring Meeting, New Orleans, LA
April 28, 2004
Energy Conservation

Abstract

Over the last several years, the processing industries have made significant attempts to improving their plant energy performance. For example, many plants have made significant improvements to fired heater performance by paying careful attention to excess O2 levels. Others have installed utility system optimization programs that model the plants' steam, fuel, and power systems and identify the lowest cost mode of operation

While these efforts represent great steps forward, plant energy managers continue to face challenges in sustaining energy improvements.

This presentation will explore the use of a spreadsheet based, advanced utility optimization tool coupled with proven energy management techniques to identify, and sustain energy performance improvement.

The following components of the energy management tool will be described in some detail:

- Fuel, Steam, and Power balance development
- Employment of marginal fuel and power prices
- Energy performance benchmarking by process unit
- Energy KPI development and target setting
- Energy lost opportunity
- Plant Data Historian connection
- Data Reconciliation
- Utility cost optimization schemes
- Graphical User Interfaces
- Customized reporting
- Organizational roles and responsibilities

The following benefits of the energy management tool will be described:

- Easy to use, spreadsheet based system with graphical user interfaces
- Point and click energy stewardship report generation
- 2 – 5% annual utility cost savings
- Better understanding of process/utility interactions
- Increased visibility at the management levels
- Simplified maintenance and capital cost justification of energy initiatives

Road Blocks to Energy Improvement

- Energy has traditionally taken a backseat to other business drivers
- Fundamentals of energy efficiency often not understood
- Energy lost opportunity dollars typically not stewarded
- Energy Accounting typically dedicated full-time to BTU reconciliations and reporting and not in developing concrete energy savings initiatives
- Regionalization of processing areas inhibits plant-wide implementation of energy improvement initiatives
- Energy-analysis tools not available
- No plant-wide energy leadership team to pull together comprehensive improvement program
- Companies reluctant to engage in comprehensive "studies"

Key Success Factors

- Strong Management commitment demonstrated with appointment of Management sponsor
- Full-time Team Leader with proven leadership skills
- Energy efficiency goals ($'s), connected to EII targets, included in annual company plan process
- Developed set of meaningful Energy Influencing Variables (EIVs)
- Concise, easy to understand stewardship reports that are updated daily and reviewed with Operations
- Energy maintenance budget for funding energy-related items
- Process organization buy-in, accomplished through:
  - Regular interface with Process management
  - Involvement in developing energy EIVs and their targets

Energy Metrics - KPIs and EIVs

- Successful energy management is built around measurement and stewardship of energy-related metrics
- Key Performance Indicators (KPIs)
  - Variables that define overall performance at the management and operating unit levels
  - Limited to a few; too many leads to confusion, disinterest, and lack of credibility
  - Management level examples include refinery/unit BTs, EII, BTU/Bbl
  - Unit level examples include CITs, cycle efficiencies, etc
- Energy-Influencing Variables (EIVs)
  - Process variables at the console and engineering level that impact energy consumption
  - Targets are set and lost opportunity dollars calculated from differences in target and actual values
  - Examples include heater excess O2, stripping steam rates, exchanger fouling factors

Site-Wide Energy Balancing

- Successful energy programs must have steam, fuel, and power balances
- Fuel Balance
  - Reconcile supply vs. demand imbalance
  - Make adjustments for upcoming projects (e.g. clean fuels)
- Steam Balance
  - Each steam header is mass and energy balanced
  - Energy values for each steam level determined
- Power Balance
  - Developed from motor list
  - Switchable pump drivers identified
  - Turbine vs. Motor economics quantified
- Steam, Fuel and Power balances modeled in ProSteam software

Best Technology (BT) Benchmarking

- BT benchmarking compares energy usage to best performance unit models
- BT analysis identifies primary areas for improvement
- BT accounts for actual marginal fuel and power prices
- Provides a more practical bases for determining energy cost reduction potential than Solomon EII

BT Percent	Performance	Rank	Comments
100	Ideal results	Not practical	No refinery has a 100% BT
130-149	Pacesetter	90th percentile plus	Some Japanese and European refineries
150-179	Top performer	75th percentile	Good energy management
180-199	Average performer	50th percentile	Little attention to energy
200-250	Poor performer	Lower 25th percentile	Poor design/energy management
250+	Lowest performer	Below 25th percentile	Energy intensive facility — costly strategy

Energy Manager

- Energy Manager offers a convenient platform to monitor and steward site energy performance while acting as an enabler for sustainable energy improvement
  - Customized Tool that includes:
    - Data acquisition, validation and reconciliation
    - Benchmarking and gap analysis
    - Application of rigorous energy economics
    - Utility balancing, simulation and optimization
    - Identification of opportunities
    - Implementation guide
    - Monitoring tools
    - Multi-level reporting

Work Process

- **Daily**
  - Energy Team Leader insures that the Energy Manager is updated
  - Team Leader reviews at morning operations meeting
    - Significant EIV deviations are discussed directly with applicable console operator, either by phone or personal visit to control center
    - Comments are entered in the appropriate spaces on the stewardship pages
- **Weekly**
  - Energy Team Leader provides a 10-15 minute energy update to Management during morning operations meeting – using Energy Manager stewardship pages, but going into a little more detail on current progress of energy initiatives
  - Energy Team meets to discuss items as necessary

Data Flow in Energy Manager

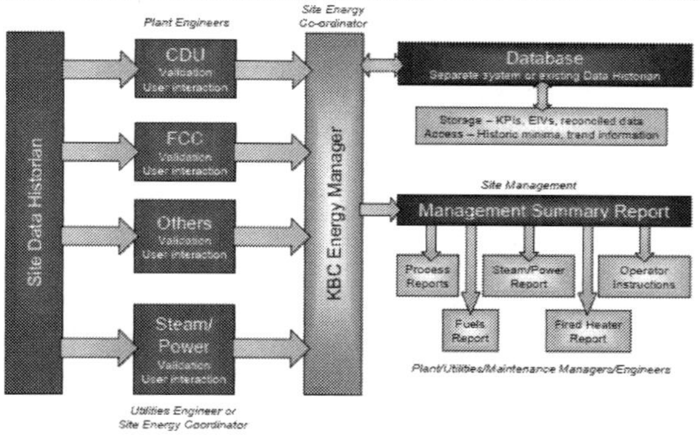

Main Menu

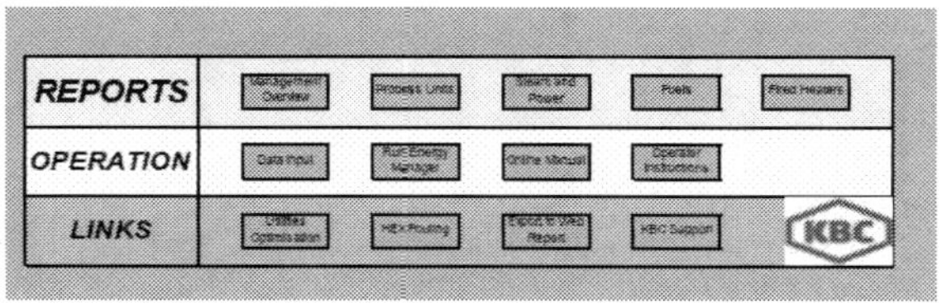

Management Report

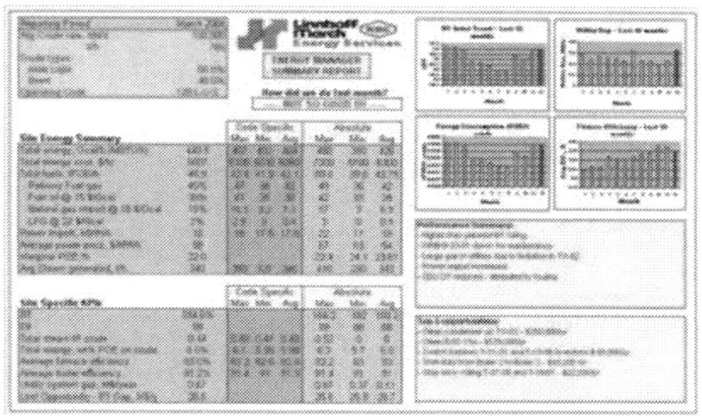

Steam and Power System

Section 1 - Steam and Power system overview

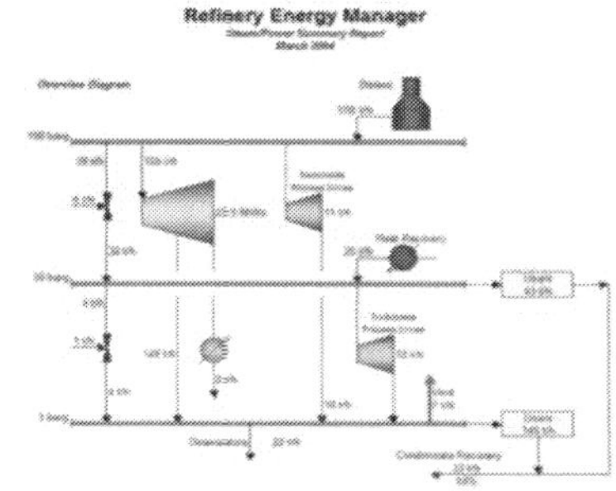

Steam and Power System

Section 2 – Data Summaries

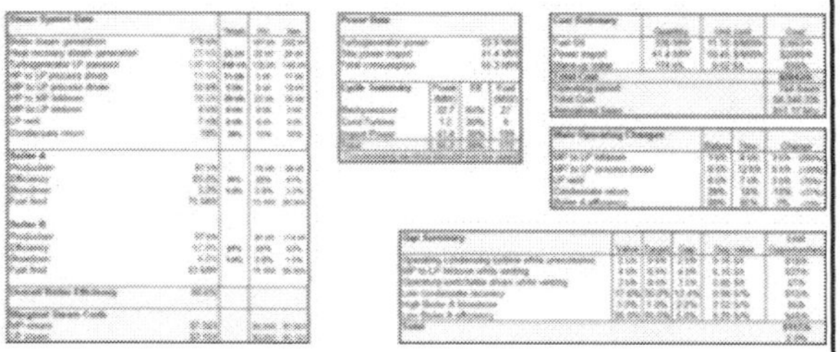

OptiSteam Option

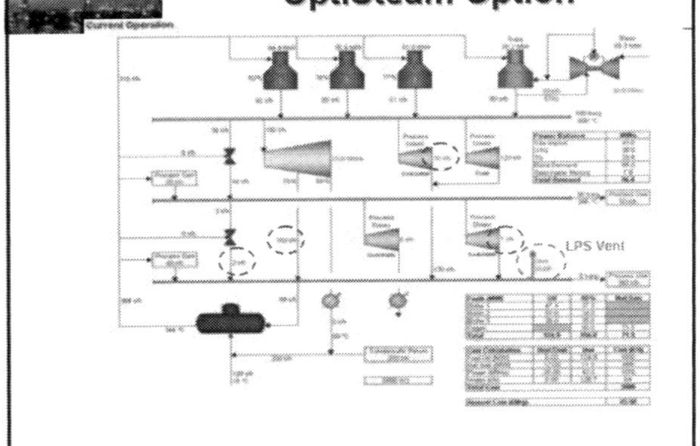

OptiSteam Option

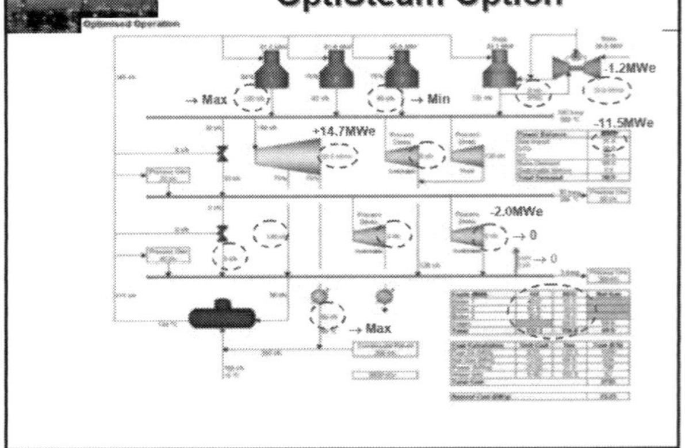

Optimizer Changes

- Turn off all switchable steam turbines exhausting at LP
- Eliminate small MP to LP letdown
- Use passout flow from TG to balance LP demand
- Operate condensing section in TG at maximum flow
- HP-MP letdown remains to balance demand but flow reduces
- Some MP to Condensing turbine drives continue to operate
- Boiler 1 and HRSG at maximum steam generation
- Boiler 3 at minimum steam generation
- Steam injection (STIG) into the GTG eliminated
- Net site power import reduces
- RFG supply to the GTG/HRSG maximized to offset natural gas import
- Boilers 2 and 3 using minimum allowable amounts of RFG
- Boiler 1 (expensive Fuel Oil) increases RFG use

Optimizer Changes

- Site operating cost
  - Before Optimization $ 4500 per hour
  - After Optimization $ 4100 per hour
  - Saving $ 400 per hour (8.9%) ($3.5 MM/year)
- Amount of changes typical of first use of the optimizer
- Number of changes will reduce as operating methods improve with time

Steam and Power System

Section 3 - Graphs and trends

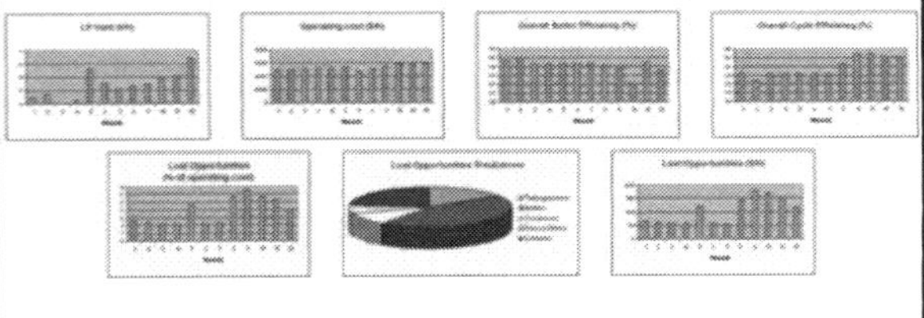

Refinery BT Report

Unit Name	Fuel	Electric	Steam	FCC Coke	Unit Total	Allow.	BET	Source	EFOM/KW	MMBTU/HR
Crude Unit 1	678	38	69		785	478	166%	Produced Fuel	279,556	2,305
Crude Unit 2	420	12	47		479	316	151%	Vaporizer	20,000	205
Hydrocracker	55	81	(64)		72	54	134%	Fuel Oil	15,000	125
Diesel Hydrotreater	26	18	6		50	20	247%	Purch Nat Gas	11,868	96
Naphtha Hydrotreater	30	3	8		41	25	165%	Cracker Coke	65,000	500
SRU	11	2	(2)		11	18	63%			
ISOM	-	3	24		27	9	294%	Purch Power KW	51,200	454
FCC Unit	35	5	(35)	500	505	207	244%	@37.9% PEE		
Gas Oil Hydrotreater	125	59	(113)		71	50	141%			
Alky Unit	-	41	87		128	94	136%	Total Supplied Energy		3,685
Coker	170	32	(30)		172	128	134%			
Hydrogen Plant	287	8	(29)		267	117	228%			
Light Ends	68	6	31		102	62	163%			
Reformer	250	6	82		338	155	218%			
Furfural	60	5	30		95	78	122%			
Offsites Area	-	145	400		545	147	372%		Supply BET	
Facility Total	2,212	464	512	500	3,689	1,957	188%	Energy Balance	188%	100.1%

Fuel Gas Report

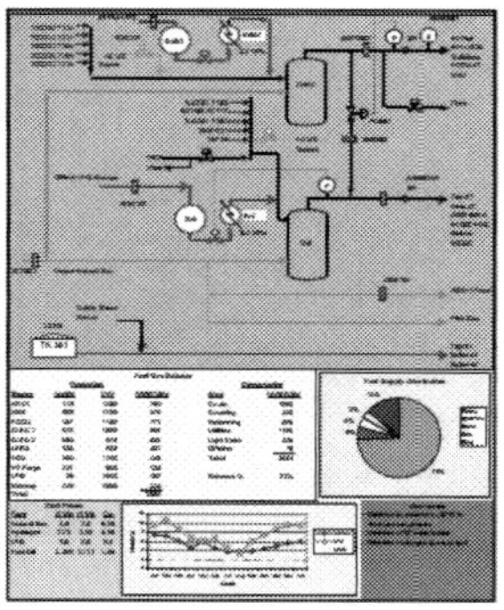

- Balance diagram
  - Fuel gas, nat gas, LPG
- Fuel oil types, sources, temp, quality /blending/cutterstock
- Pie chart showing fuel use by type
- Price of marginal fuel, flaring, hydrogen letdown, heating value
- Emissions
  - CO_2, SOx, (NOx)
- Operating Signals

Fired Heater Report

Oil = $2.51/MMBTU
Gas = $3.50/MMBTU

Macro Execution Area

Ambient Air °F = 75
# of Heaters = 21
Humidity = 0.011

Fuel Gas Mix Drums					
1	2	3	4	5	
#2 Mix D					

AREA	Equip Number	Original Fuel Oil Ton/hr	Original FG MScfh	Gas Mix Drum	Actual Stack Temp °F	Actual Stack O2%	Actual Firing EFF%	Actual Firing MMBtu/Hr	Absorbed Duty MMBtu/Hr	Target Stack Temp °F	Target Stack O2%	Target Firing EFF%	Target Firing MMBtu/Hr	FG Mol Wt	LHV Btu/Scf	Lost Opp. $/Yr
Boiler Plant	F-4501	0.5	77.7	1	363	2.4%	93%	98	85.9	400	3.2%	92.5%	89.0	15.3	904	(34,623)
Boiler Plant	F-14502		155.6	1	364	2.2%	94%	141	131.6	400	2.9%	92.1%	142.9	15.3	904	(69,028)
Boiler Plant	F-14501		160.9	1	360	2.1%	93%	148	135.9	400	2.9%	92.1%	147.6	15.3	904	(65,131)
HRSG	F-44501A		217.3	1	385	16.0%	74%	198	146.6	400	2.9%	92.1%	158.2	16.8	904	1,174,088
HRSG	F-44502A		195.4	1	382	16.0%	74%	177	131.6	400	2.9%	92.1%	142.9	15.3	904	1,033,368
CDU 3	F-22001	1.4	165.3	1	573	4.1%	87%	204	178.9	400	2.9%	92.2%	191.9	15.3	904	348,594
CDU 3	F-22002	1.3	156.1	1	549	5.2%	87%	193	168.9	400	2.9%	92.2%	182.9	15.3	904	336,578
NHT	F-13101		51.5	1	910	2.7%	78%	56	43.1	400	2.9%	92.1%	46.8	15.3	904	288,544
NHT	F-13201		55.5	1	901	2.8%	78%	59	46.1	400	2.9%	92.1%	50.1	15.3	904	278,487
Reformer 1	F-13401		144.4	1	460	3.9%	90%	131	117.3	400	2.9%	92.1%	127.5	15.3	904	94,423
Reformer 1	F-13402		112.4	1	460	3.9%	90%	102	91.4	400	2.9%	92.1%	99.2	15.3	904	72,811
HDS	F-13501		33.9	1	933	6.2%	72%	31	22.0	400	2.9%	92.1%	23.9	15.3	904	207,495
SRU A	F-14304		5.2	1	1,092	4.0%	70%	5	3.3	400	2.9%	92.1%	3.6	15.3	904	34,837
SRU B	F-14204		5.6	1	1,112	4.0%	59%	6	3.5	400	2.9%	92.1%	2.8	15.3	904	35,846
HCU	F-12502		115.9	1	711	2.7%	94%	105	87.5	400	2.9%	92.1%	95.1	15.3	904	296,759
HCU	F-12501A		0.4	1	400	5.0%	91%	0	0.3	400	2.9%	92.1%	0.3	15.3	904	106
H2 Plant	F-12601E		160.5	1	504	4.7%	88%	146	127.8	400	2.9%	92.1%	138.8	15.3	904	192,546
H2 Plant	F-12601W		174.8	1	517	3.7%	88%	159	139.6	400	2.9%	92.1%	151.5	15.3	904	196,825
RFCC	F-42533A		170.5	1	500	2.1%	90%	154	138.3	400	2.9%	92.1%	150.3	15.3	904	119,396
CDU 4	F-42001		172.1	1	385	2.7%	93%	156	145.2	400	2.9%	92.1%	157.5	15.3	904	(65,259)
CDU 4	F-42002		172.4	1	356	2.5%	93%	156	145.5	400	2.9%	92.1%	158.1	15.3	904	(69,021)
																4,392,056

Crude Unit Report

Energy Targets		
Total Energy Consumed	144	Gcal/h
BT Target Energy Consumed	74.3	Gcal/h
% of BT Target	194%	%
Delta Cost Actual to BT	580	$/h
Operating Target	105	Gcal/h
% of OP Target	137%	%
Delta Cost Operating to BT	380	$/h

Process Data		
Feed Rate	151	kBSD
sg	0.9087	
Crude Type	Sour	
Mode	Gasoil	
%Overflash	3.7	%LV
Kero Draw	16.7	T/h
LGO Draw	144.4	T/H
HGO Draw	75.2	T/h
Atm Resid yield	36	LV%

Furnace Data	Actual	Target	Delta Act to Op	Action	
Fuel Oil Fired	4.3	5	-0.7		T/h
Fuel Gas Fired	5.7	5.5	0.2		T/h
Total Fuel Fired	104	108.5	4.5		Gcal/h
CIT	267	295	-29		C
Stack Temperature	350	350	0		C
Stack O2	6%	3%	3%	Reduce	%
Furnace Efficiency	80%	83%	-3%		
Lost Opp			-3.1		Gcal/h

Comments

O2 analyser unreliable - maintenance due 5/3

Crude Unit Report (con't)

Utility Consumption	Actual	Target	Delta Act to Op	Action	
Kero Strip Steam	1	1	0		kg/h
LGO Strip Steam	4	3.5	0.5	Increase	kg/h
HGO Strip Steam	4	4	0		kg/h
Residue Strip Steam	15	16	-1	Reduce	kg/h
Total LP Steam Consumed	44	24.5	19.5		T/h
Atm Resid MP Steam Produced	3.5	5	-1.5		T/h

Comments

Charge pump turbine in operation - motor due back in service 10/3
Check Bypass around E-106

Heat Exchanger Network	Actual U kcal/hm2C	Target U kcal/hm2C	Delta Act to Op	Action
E-101 (HGO V Crude)	365	350	4%	
E-104 (LGO V Crude)	287	280	3%	
E-105 (Atm Resid V CDU)	150	200	-25%	
E-106 (Atm Resid rundown cooler)	115	190	-39%	

Comments

E-105 due for clean June, review timing of E-106 clean

Operator Instructions

Operator Instructions

Unit	Opportunity Description	Action by	Benefit $/y	Concerns
01	Increase MPA flow to 87 m3/h	MST	18,000	None
01	Clean exchangers E01-11b	MST	125,000	Check isolation valves, take out when th'put low
01	Stop slow rolling T-07-06 and T-08/01	MST	22,000	Follow procedure for hot stand-by
02	Switch turbines T-11-01 and T-13-06 to motors-$	ALD	65,000	None
06	Reduce supplementary firing of FCC COB	DFR	67,000	flue gas temperature not to drop below 900 degC
06	Increse condensing on FCC air blower turbine	DFR	35,000	Ensure that motor/generator always in export mode
06	Reduce O2 on F-06-01 from 4.5% to 3.5%	SEP	10,000	None, last month operation nat 3.5% O2
39	Clean condenser on TG-02	LOT	250,000	Include in maintenance schedule
39	Shift duty from Boiler 2 to Boiler 1	LOT	45,000	None, target for minimum duty

Advantages of Energy Manager

- Jump starts an energy program without requiring a comprehensive energy "study"
- Easy to use, spreadsheet based system with graphical user interfaces
- Point and click energy stewardship report generation
- Identifies benefits while building the system
- 2 – 5% annual utility cost savings
- Better understanding of process/utility interactions
- Increased visibility at the management levels
- Simplified maintenance and capital cost justification of energy initiatives
- Provides consistency across multiple sites
- Training tool for plant engineers and operators

A Strategy for the Successful Harvesting of "Low-Hanging Fruit"

by

Bruce L. Pretty
V.P. – Technology

and

Alan Karp
Senior Consultant

Veritech, Inc.
Sterling, VA

Abstract

In the current economic environment, energy conservation efforts are being directed almost exclusively toward "low-hanging fruit" – i.e., savings achievable in the very near term with little or no capital investment. "Low-hanging fruit" historically has been synonymous with housekeeping efforts and improved operational/maintenance practices to reduce losses. However, the "yield" of "low-hanging fruit" can be significantly increased through monitoring & targeting (M&T) programs. M&T involves more rigorous and systematic monitoring of individual energy users, coupled with an organizational commitment and pro-active program to adjust operational practices and controls to minimize energy use as plant operation varies. Key to successful M&T is the development of meaningful relationships (correlations) between dependent and independent variables that will allow energy use targets to be dynamic. Such targets provide plant staff with rational performance goals for energy use as plant operation varies, while also facilitating the quantification of M&T's benefits – or, where M&T is not yet in place, of the resulting lost opportunities. Equally important is the effective configuration and application of a plant-wide steam system model to help identify the needed correlations and to confirm and quantify net savings in the steam system, which generally represents the major cost-saving opportunity.

Introduction

Among the practical aspects of steam system model utilization for monitoring and pro-active improvement programs are:

- the availability, accessibility and quality of current and historical information on utility system operation;

- the use of the model's information gathering and database features to support historical trend analyses and, ultimately, to promote increased understanding of system performance variability and interactions;
- the ability to distill model outputs into concise, readily accessible reports that will meet the varied informational needs of plant staff.
- the interpretation of steam system model outputs and the identification of "opportunity areas" that merit the closer and ongoing attention of management, technical, and operating staff;
- the "translation" of model simulation and optimization results into "implementable" actions that are both practical and cost-effective; and
- the tracking, in economic terms, of opportunities that have been both "lost" and "captured" in conjunction with monitoring and related process improvement efforts.

Veritech's experience in steam system analysis, which encompasses more than 30 site-wide steam system models using our Excel-based *ESteam*© software, is that dealing successfully with these "practical aspects" is crucial to the success of any model-based effort to improve steam system performance and reduce sitewide utility costs.

Using the Right Information

Good information is essential to successful steam system modeling, monitoring and operational improvement. Ideally, this information should be representative of actual utility system operation and must be sufficiently detailed and accurate to support the model's construction of individual unit and overall plant steam balances. In addition, the model should track key influencing variables – such as plant and unit throughputs and other variables that are likely to correlate well with the day-to-day variations in energy use patterns throughout the plant. These correlations provide the basis for setting dynamic performance targets that can serve as baselines against which to assess overall plant and equipment performance.

Toward this end, *ESteam* is equipped with a plant data interface that allows available operating data to be imported into the model, where it is used to construct truly representative steam system mass and energy balances. [Alternatively, users are free to perform "what-if" studies based on user-specified values.] The importation of plant data can be performed automatically (e.g., on a daily or shift basis) to minimize the need for user intervention, but also can be done on user demand, even when the software is being utilized in an "off-line mode." This steam balance information is, in turn, captured by *ESteam*'s database function, thereby accumulating historical operating data on an ongoing basis. This historical data can be used to support the development and analysis of long-term energy-usage trends and/or selectively retrieved to permit system analyses for specific operating periods.

In practice, metered data on plant steam generation and use (and other energy use parameters) is invariably (and, not infrequently, woefully) incomplete. Consequently, *ESteam*'s plant data interface allows for the specification of default values for needed steam system parameters not currently being metered. These default values may be based on design specifications or operating experience and may either be fixed (but modifiable) values or, more desirably, scalable values that are automatically adjusted according to the values of key influencing variables such as plant or unit production rate.

The early stages of model development should include a careful review of available monitoring instrumentation and a focused effort to develop steam balances at the individual unit and overall plant levels that achieve 90-95% closure – i.e., for which no more than 10% (and, preferably, 5% or less) of steam production and use is unaccounted for. Veritech's experience is that no more than several man-weeks of system review and field checking are required to bring steam balance definition up to this standard. Importantly, this level of steam balance closure is almost always possible in a reasonably instrumented plant with judicious consideration given to meter accuracy and default values based on design or scaled estimations of unmetered users. In the process, additional instrumentation needs can be identified and evaluated. Moreover, such steam balance reviews frequently yield immediate benefits by identifying unnecessary or excessive steam usage due to forgotten vents or letdowns or uncontrolled and/or unmetered steam flows that can be reduced or eliminated.

Development of a plausible and well-defined base case steam balance is essential to verifying the accuracy of the model and establishing its credibility as a tool for projecting future steam system improvements and day to day operating changes that will affect profitability. Potential reductions in plant steam usage, though invariably substantial in monetary terms, may correspond to a numerically modest percentage of total steam usage. Model accuracy in duplicating current system performance will enable the model's quantification of potential and realized savings identified to be perceived and accepted as real, notwithstanding inevitable margins of error and uncertainty in overall steam balance closure.

Characterizing and Understanding Current Operation

Although it may seem obvious that steam system operation extends to virtually all corners of the plant, it is not uncommon for decisions affecting steam system operation and configuration to be made "locally", without proper consideration of the resulting net "global" impact. The steam system model can help to counter this tendency by presenting the user with a concise overall steam balance picture and then allowing the investigation of steam system use and generation to be tracked down to the individual process unit level and, ultimately, to specific steam users and generators.

Figure 1 shows a typical site steam balance diagram generated by *ESteam.* The diagram offers a "bird's-eye" view of the overall steam balance at each major header pressure level. Site-aggregated information on steam generation and on steam use is provided, along with data on letdown flows and venting, steam balance closure and losses, and overall steam system operating cost. As a whole, the diagram provides an instant visual orientation to the overall state of the steam system.

Similar information is available at the unit level by clicking the appropriate link in a drop-down menu (not shown in Figure 1) to produce the summary diagram for the specific process unit of interest.

Figure 1
Site Steam Balance Summary Diagram

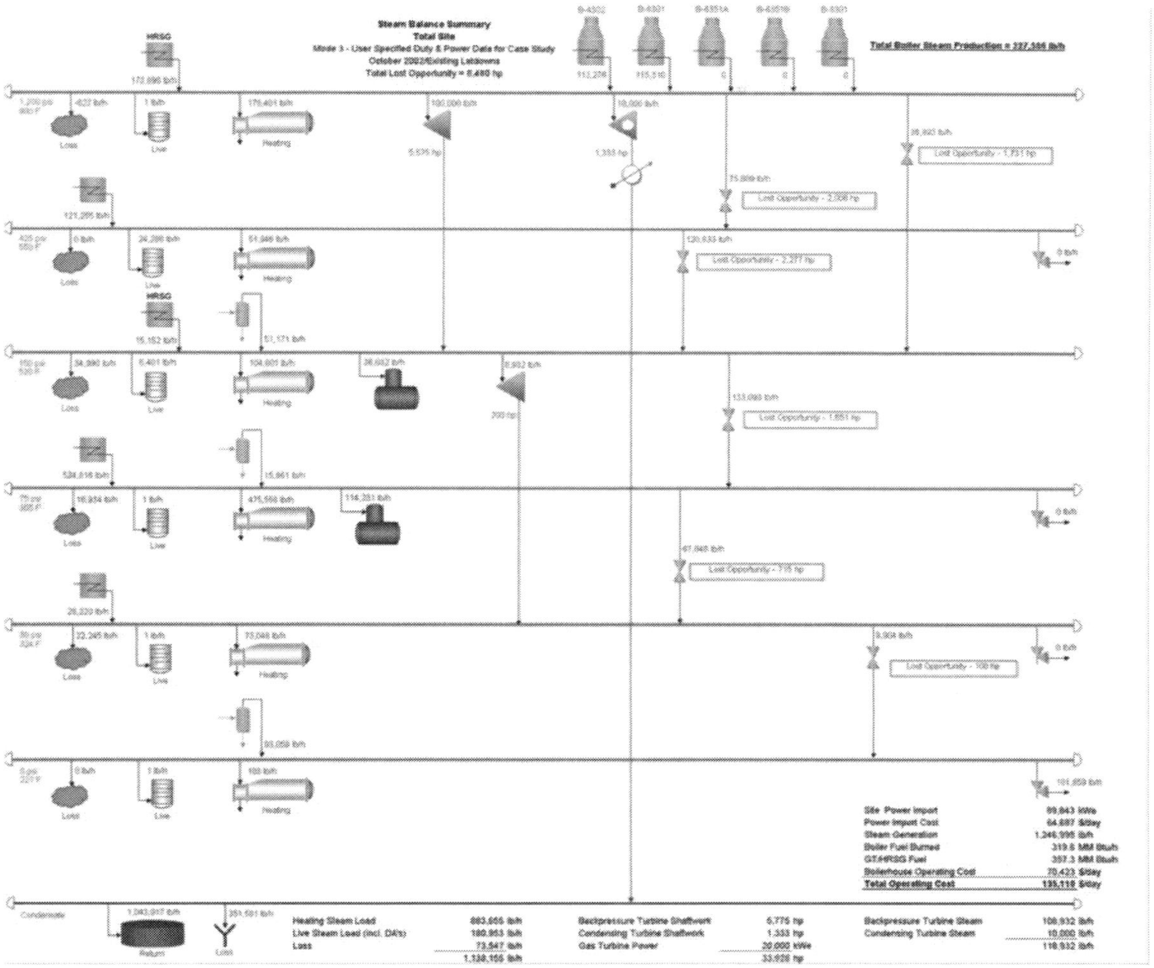

Clicking on one of the diagram icons, such as the icon representing aggregated high pressure heating steam use, will produce another level of detail (Figure 2a) for that portion of the unit steam balance. Still another click, on a specific heating steam user listed in the "Equipment Details" window, will access the model's equipment database containing full details on the item of interest. (Figure 2b) This "drill-down" capability allows models user to follow and gain an initial perception of the interactions among steam users and generators plantwide.

Figure 2
Detailed Steam Balance Information

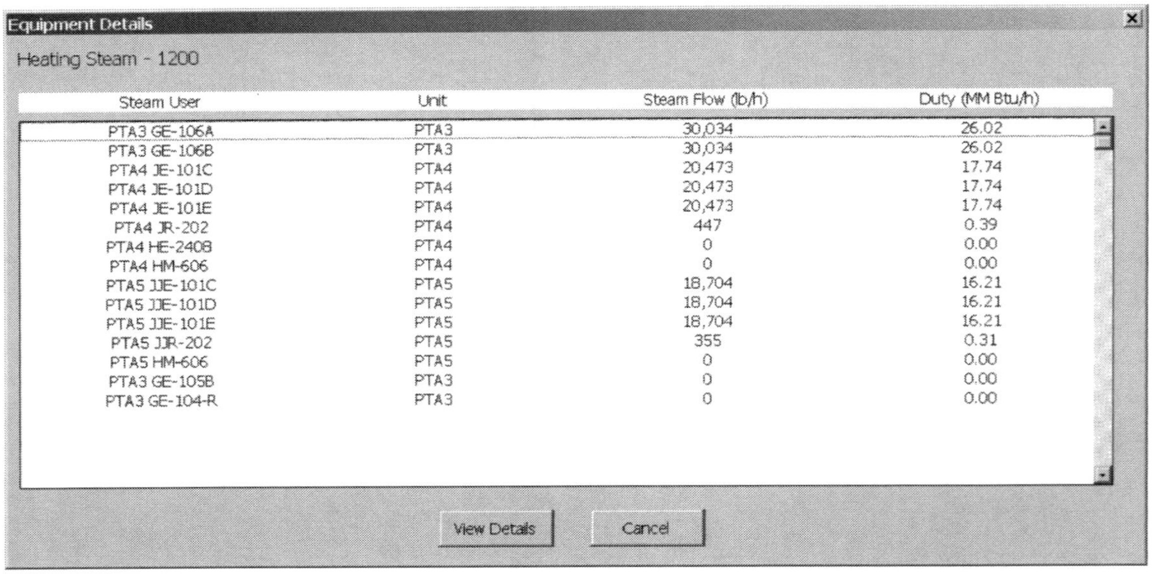

Equipment Details

Heating Steam - 1200

Steam User	Unit	Steam Flow (lb/h)	Duty (MM Btu/h)
PTA3 GE-106A	PTA3	30,034	26.02
PTA3 GE-106B	PTA3	30,034	26.02
PTA4 JE-101C	PTA4	20,473	17.74
PTA4 JE-101D	PTA4	20,473	17.74
PTA4 JE-101E	PTA4	20,473	17.74
PTA4 JR-202	PTA4	447	0.39
PTA4 HE-2408	PTA4	0	0.00
PTA4 HM-606	PTA4	0	0.00
PTA5 JJE-101C	PTA5	18,704	16.21
PTA5 JJE-101D	PTA5	18,704	16.21
PTA5 JJE-101E	PTA5	18,704	16.21
PTA5 JJR-202	PTA5	355	0.31
PTA5 HM-606	PTA5	0	0.00
PTA3 GE-105B	PTA3	0	0.00
PTA3 GE-104-R	PTA3	0	0.00

(2a)

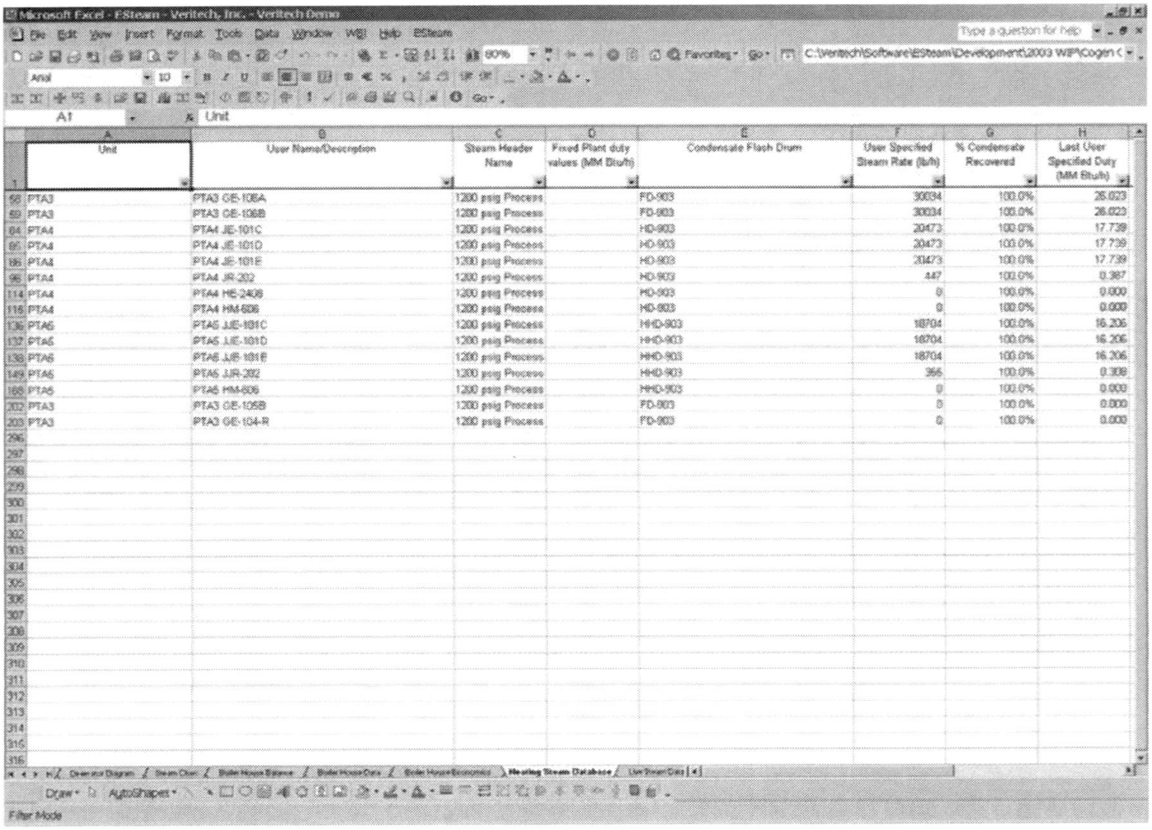

	Unit	User Name/Description	Steam Header Name	Fixed Plant duty values (MM Btu/h)	Condensate Flash Drum	User Specified Steam Rate (lb/h)	% Condensate Recovered	Last User Specified Duty (MM Btu/h)
58	PTA3	PTA3 GE-106A	1200 psig Process		FD-903	30034	100.0%	26.023
59	PTA3	PTA3 GE-106B	1200 psig Process		FD-903	30034	100.0%	26.023
84	PTA4	PTA4 JE-101C	1200 psig Process		HD-903	20473	100.0%	17.739
85	PTA4	PTA4 JE-101D	1200 psig Process		HD-903	20473	100.0%	17.739
86	PTA4	PTA4 JE-101E	1200 psig Process		HD-903	20473	100.0%	17.739
98	PTA4	PTA4 JR-202	1200 psig Process		HD-903	447	100.0%	0.387
114	PTA4	PTA4 HE-2408	1200 psig Process		HD-903	0	100.0%	0.000
116	PTA4	PTA4 HM-606	1200 psig Process		HD-903	0	100.0%	0.000
136	PTA5	PTA5 JJE-101C	1200 psig Process		HHD-903	18704	100.0%	16.206
137	PTA5	PTA5 JJE-101D	1200 psig Process		HHD-903	18704	100.0%	16.206
138	PTA5	PTA5 JJE-101E	1200 psig Process		HHD-903	18704	100.0%	16.206
149	PTA5	PTA5 JJR-202	1200 psig Process		HHD-903	366	100.0%	0.308
188	PTA5	PTA5 HM-606	1200 psig Process		HHD-903	0	100.0%	0.000
202	PTA3	PTA3 GE-105B	1200 psig Process		FD-903	0	100.0%	0.000
203	PTA3	PTA3 GE-104-R	1200 psig Process		FD-903	0	100.0%	0.000

(2b)

Also available to the steam user is a diagram (Figure 3) providing a concise summary of cogeneration operation at the site.

Figure 3
Cogeneration Summary Diagram

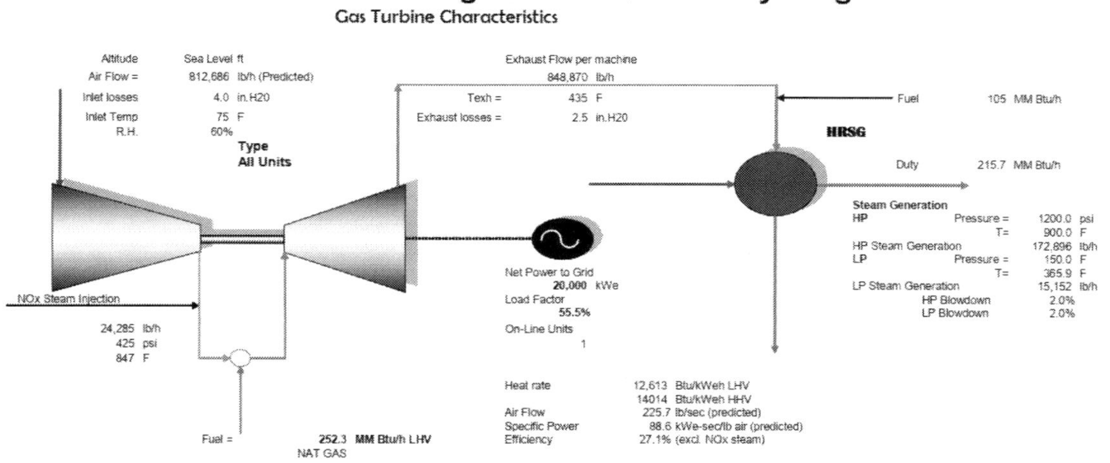

Summarizing System Operation and Economics

While the data gathering capabilities of steam system model are crucial, the resulting overabundance of raw information can be a barrier, rather than a bridge, to system understanding. Accordingly, it is essential to distill the vast amount of available information into concise reports that characterize what is going on. Figures 4 and 5 show two such reports available in *ESteam.*

The key aspect of these reports is the presentation of overall performance indices (See following section on "Key Performance Indicators".) For example:

- The roll-up of component operating costs into an overall site cost
- Costs per unit of production for boiler steam and cogenerated power
- Characterization of thermal efficiency as easily recognized key energy parameters such as Boiler Energy Ratio (Btu fuel per lb steam), Cogeneration Fuel Chargeable to Power, etc…
- Quantification of overall site improvement potential in $ terms as venting losses and lost opportunity from letdowns

Figure 4
Steam System Summary Report

STEAM SYSTEM SUMMARY REPORT

Current Period	04/01/03 08:27 to 04/02/03 08:27		
Basis	Fuel Cost ($/MM Btu LHV) - Average	$3.85	
	Electric Cost ($/kWeh) - Average	$0.030	
	Total Utilities Operating Cost	**$135,110**	/day
Electrical	Site Power Import	89,843	kWe
	Imported Power Cost	**$64,687**	/day
Boilerhouse Operation	Total Steam & Power Production Cost (incl.Cogen.)	**$70,423**	/day
	Boiler Steam Cost	**$37,412**	/day
	Total Cogeneration System Operating Cost	**$33,011**	/day
	Boilers		
	Boiler Fuel	319.6	MM Btu/h
	Boiler Steam Production	227,586	lb/h
	Boiler Energy Ratio	1.404	
	Average Boiler Steam Cost	**$6.85**	/ klb
	Gas Turbine/HRSG Cogeneration		
	Turbine Power Output	20,000	kWe
	Turbine Fuel	252.3	MM Btu/h
	HRSG Supplementary Fuel	105.0	MM Btu/h
	HRSG Steam Production		
	HP Steam	172,896	lb/h
	LP Steam	15,152	lb/h
	NOx Steam Injection	24,285	lb/h
	Cogen Fuel Chargeable to Power (FCP)	138.4	MM Btu/h
	Average Power Generation Cost	**$0.0266**	$/kWeh
Lost Opportunity Cost	**Letdowns**	**$2,092**	**$/day**
	Steam Venting	**$9,058**	**$/day**

Figure 5
Site Cogeneration Performance Report

Energy Performance - All Units

Turbine Fuel	252.3	MM Btu/h	
Power at Generator	20,000	kWe	
Turbine Heat Rate	12,613	Btu/kWeh LHV	
Turbine Efficiency	27.1%	(excl. NOx Steam)	
Turbine Exh. Temp.	435	F	
Unfired HP Steam Gen.	83,237	lb/h	Predicted from model
Unfired LP Steam Gen.	13,873	lb/h	Predicted from model
NOx Steam injection	24,285	lb/h	
	33.8	MM Btu/h	
Net Unfired Steam	72,825	lb/h	
	76.1	MM Btu/h	
Displaced boiler fuel	1,257.3	Btu/lb	
Offset Fuel (Unfired)	91.6	MM Btu/h	
Net Fuel Impact	160.7	MM Btu/h	
Fuel Chargeable to Power	160.7	MM Btu/h	(Excl. Supp. Firing)
	8035	Btu/kWeh LHV	
Fuel Chargeable to Power	138.4	MM Btu/h	(Incl. Supp. Firing)
	6918	Btu/kWeh LHV	
Energy Cost of Power	**2.663**	**c/kWeh**	

Supplemental Firing

Total Steam	163,763	lb/h
Unfired Steam	72,825	lb/h
Supp. Fired Stm	90,938	lb/h
Fuel Fired	105	MM Btu/h
Fuel Fired	1,154.6	Btu/lb
Suppl. Firing Eff	100.8%	
Blr Fuel Saved	245.644	Btu/lb
	22.3	MM Btu/h
	738,317	$/yr

Turbo-Generation

Unit		
Power Generated		
Condensing Power	1,333	hp
Extraction Power	5,775	hp
	5,300	kWe
Boiler Fuel Allocation	34.48	MM Btu/h
Costs		
Power Credit	$159.01	/hr
Boiler Fuel Costs	$132.73	/hr
Surface Condensate Costs	$3.58	/hr
Turbo-generated Power Cost	**2.57**	**c/kWeh**

Average Cost of Power Generation

Average Cost of Cogenerated Power	2.644	c/kWeh
Supplementary Firing Credit	(0.340)	c/kWeh
Sub-total (excl. O&M)	2.304	c/kWeh
O&M	0.000	c/kWeh
Total	2.304	c/kWeh

Cogeneration Economic Performance

Cost of turbine fuel	971	$/hr
Supplementary firing fuel	404	$/hr
Turbo-generation costs	136	$/hr
O&M Costs	0	$/hr
Total Costs	1,512	$/hr
Boiler fuel savings	843	$/hr
Power Savings	759	$/hr
Total Savings	1,602	$/hr
Net Savings	**90**	**$/hr**
	773	**k$/yr**

"Key Performance Indicators" (KPI)

The steam system model can also render raw steam balance data more meaningful by translating usage and costs into "key performance indicators (KPI)." Figure 6 shows an example of site level KPI's, which reduce overall steam balance information to aggregate energy consumption and cost information averaged on a week-to-date (WTD), month-to-date (MTD), and year-to-date (YTD) basis.

An important aspect of the KPI is the examination of *specific* steam and energy consumption and costs that place this information on a "per unit of plant production" basis, thereby providing a more coherent basis for comparing plant energy consumption and costs over time as plant throughput varies. Supplementing the actual performance figures are their comparison to targeted values, thereby providing a concise picture of how effectively energy is being used at the site over time.

Also highlighted in Figure 6 are the economic losses attributable to steam venting and letdown station flows, as well as "unaccounted/losses" and % condensate return, parameters which provide additional indication of how well the steam system is being operated and which may suggest areas requiring more extensive operational analysis.

Figure 6
Site Key Performance Indicators (KPI)

Main Menu — Unit Report

Update Report

Site Key Performance Indicators

Starting		05/05/03	05/12/03	05/01/03	01/01/03
Ending		05/11/03	05/13/03	05/13/03	05/13/03
YTD Days	132				
		Prev Week	WTD	MTD	YTD
Total Utilities Operating Cost	**$/d**	**$216,108**	**$233,112**	**$220,359**	**$220,359**
Specific Energy Cost	**$/Bbl**	**$2.401**	**$2.590**	**$2.448**	**$2.448**
Imported Power Cost	$/d	**$79,200**	**$79,200**	**$79,200**	**$79,200**
Boiler Steam Cost	$/d	**$79,539**	**$67,321**	**$76,485**	**$76,485**
Cogeneration Operating Costs	$/d	**$57,369**	**$86,590**	**$64,674**	**$64,674**
Boiler Steam Cost	$/klb	**$6.37**	**$6.02**	**$6.29**	**$6.29**
Specific Steam Energy Use	Btu/Bbl	145,646	145,646	145,646	145,646
Fuel Fired Cost	$/Bbl	$1.345	$1.542	$1.395	$1.395
Specific Fuel Use	Btu/Bbl	349,457	400,566	362,234	362,234
Boiler Steam, Actual	lb/h	520,301	465,622	506,631	506,631
Boiler Steam, Target	lb/h	487,500	487,500	487,500	487,500
Boiler Steam, Delta	lb/h	32,801	-21,878	19,131	19,131
Boiler Steam, Actual	lb/Bbl	138.75	124.17	135.10	135.10
Boiler Steam, Target	lb/Bbl	130	130	130	130
Boiler Steam, Delta	lb/Bbl	8.75	-5.83	5.10	5.10
Vent Steam Cost	**$/d**	**$5,584**	**$0**	**$4,188**	**$4,188**
Steam Vented, Actual	lb/h	62,870	0	47,153	47,153
Steam Vented, Target	lb/h	1,500	1,500	1,500	1,500
Steam Vented, Delta	lb/h	61,370	-1,500	45,653	45,653
Letdown Lost Opportunity	**$/d**	**$3,341**	**$2,098**	**$2,844**	**$2,844**
Recommended Savings	$/d				
Unaccounted/Losses	lb/h	26,015	23,281	25,332	25,332
Condensate Return	lb/h	60.7%	48.3%	56.5%	56.5%

Additional areas of opportunity can be identified using *ESteam's* optimization capabilities, which can simultaneously handle multiple boilers, multiple boiler fuels, and switching capability between turbines and motors for boiler house and process area drives.

As indicated above, steam system model optimization will generally identify incentives to shift boiler loadings, alter boiler house fuel mix, or swing turbine and motor drives. (Figure 7) Changes in cogeneration operating loads also may be indicated. While simple in concept, this is a difficult optimization due to the binary nature of operating status variables such as the selection of turbine drives and boiler or cogeneration operating status, and the various

Figure 7
Detailed Optimization Results

Report Date
Description ◉ All Drivers ○ Change Optimized ○ Change Modified

Unit	Driver Name	Driver Type	Status Imported	Status Optimized	Status Modified	Change Optimized	Change Modified
No.2 Cogen Plant	DUMMY BP TURBINE	Backpressure Turbine	OFF/OPT	ON/OPT		TURN ON	
No.2 Cogen Plant	DUMMY BP TURBINE 1	Backpressure Turbine	OFF/OPT	OFF/OPT			
No.1 Cogen Plant	FUTURE	Condensing Turbine	ON/OPT	ON/OPT			

Report Date
Description

Boiler	Output Imported	Output Optimized	Output Modified	Change Optimized	Change Modified
B-4302	184,506	0		DECREASE	
B-6301	189,493	25,577		DECREASE	
B-6351A	0	0		INCREASE	
B-6351B	0	0		INCREASE	
B-8301	0	363,537		INCREASE	

Description

Boiler	Fuel A	Fuel B	A/(A+B) Imported	A/(A+B) Optimized	A/(A+B) Modified	Change Optimized	Change Modified
B-4302	NAT GAS	FUEL OIL	1.00	1.00			
B-6301	NAT GAS	FUEL OIL	1.00	1.00			
B-6351A	NAT GAS	FUEL OIL	1.00	1.00			
B-6351B	NAT GAS	FUEL OIL	1.00	1.00			
B-8301	NAT GAS	FUEL OIL	1.00	1.00			

discontinuities introduced by the steam balance and equipment operating constraints. The optimizer chosen must be able to handle MINLP problems such as these.

To actually achieve the identified savings potential, however, optimization results must be translated into actions in the control room and in the field to effect the desired changes in boiler house load profiles and fuel selection, and/or to bring various drivers on line and shut others off. This reality places practical constraints on how system optimization should be investigated. These constraints include limitations on the frequency with which drivers can be swung and on the number of services that can reasonably be considered switchable at any given time. *ESteam* allows for alternative optimization calculations using modified constraints so that the impact of reducing the available degrees of freedom (for example, by making additional driver services unavailable for switching) can be rapidly assessed.

Operability Analysis to Identify Additional Opportunities

The steam model can contribute in other ways to the identification of opportunity areas deserving of closer operator attention or more detailed analysis by plant staff. As indicated above, paying close attention to KPI's can identify variability of operation that appears to be excessive relative to the variation in plant throughput. Excessive venting and letdown flows, poor steam balance closure (high losses or unaccounted steam usage), and low condensate recovery also represent situations in which operational resolution may yield high economic benefits, often with no or very modest investment.

To set meaningful operating targets for individual equipment items, reliable correlations against influencing variables can be developed. Using the model's database information, the historical variation in energy use/generation associated with specific equipment items can be examined over an extended time period and correlated with key production variables, such as throughput. Such analysis provides a clear basis for projecting the benefit that would accrue from more rigorous monitoring and targeting (M&T) to reduce the energy data scatter relative to current operation.

Some data will be found to be well correlated with throughput as the sole independent variable, as indicated by a high regression coefficient for the analysis. Data that is not well correlated suggest the need to determine the impact of one or more additional key influencing variables in order to improve the correlation. However, in the interim, weighting the savings potential in accordance with the current regression coefficient will provide an initial, conservative projection of the savings potential. Ultimately, more rigorous data analysis, using “data mining” methodologies, may establish additional relationships capable of better illuminating the interactions between an individual equipment item’s steam usage or generation and other plant variables.

The savings potential at this early stage of analysis may be determined from the degree of scatter relative to average observed performance. In many cases, however, there is a significant deterioration in energy efficiency as the plant throughput decreases. Rather than decreases in specific energy use, or constant specific energy, increases in specific energy use are observed. These trends present additional scope for savings as the reasons behind the observed trends are investigated and actions are taken, where appropriate, to improve energy performance.

Realization of the savings identified in this analysis requires additional effort and commitment from the site. M&T must be undertaken as a continuous activity that involves organizational effort and, potentially, investment in additional monitoring software to assist in the analysis of data and the dissemination of meaningful results and performance improvement actions. Ultimately, the savings will only come from taking actions based on performance deviations from target.

Ceramic Membrane Process for Upgrading Vacuum Residual Oil

Bruce Bishop and Robert Goldsmith
CeraMem Corporation
12 Clematis Avenue
Waltham, MA 02453

Robert Schucker and Keith Rawlins
Trans Ionics Corporation
2408 Timberloch Place, Suite D-5
The Woodlands, TX 77380

Key Words: vacuum residual oil, ultrafiltration, ceramic membrane, deasphalting

Prepared for presentation at the 2004 Spring Annual Meeting, New Orleans, LA., April 25-29

Seventh Topical Conference on Refinery Processing: Energy Conservation

10 March 2004

Unpublished

This research was supported, in whole or in part, by DOE Grant DE-FG02-00ER82970. Such support does not constitute an endorsement by DOE of the views expressed in the article.

Abstract

CeraMem Corporation has investigated the use of ceramic ultrafiltration membranes for upgrading vacuum residual oil for the purpose of feeding the upgraded oil to a catalytic cracking unit. The membrane process would replace supercritical solvent deasphalting and would reduce the need for solvent addition by a factor of 10 or more. This new process may reduce the costs for the upgrading vacuum residual oil while maintaining deasphalted oil quality and yield.

CeraMem's research was focused on two issues: vacuum residual oil modification to enhance membrane performance and evaluation of membrane performance at modest temperature and pressure to determine deasphalted oil quality and yield. Vulcanization with sulfur and addition of equilibrium catalyst fines were investigated for the purpose of increasing asphaltene retention by the ceramic ultrafiltration membranes. Several cutter stocks were blended with selected vacuum residual oil samples to determine if any could modify the solubility parameter of the asphaltenes in the oil solutions so that retention of the asphaltenes by the membranes could be increased. An additional benefit of these cutter stocks used as diluents was the ability to send them to the catalytic cracker and eliminate solvent recycling. CeraMem found that none of the modification processes attempted would significantly increase the effectiveness of the membrane process through changes in the residual oil characteristics.

Membrane performance was evaluated with a heptane to residual oil ratio of less than 1 to 1 at temperatures between 150°C and 200°C and a process pressure of up to 150 psig. High and stable membrane flux of up to 550 $kg/m^2/day$ was observed over several runs with no evidence of long term membrane fouling. Deasphalted oil yield in excess of 80% was achieved. Metal (Ni and V) and Conradson Carbon (microcarbon) analyses showed that the quality of the deasphalted oil permeated through the membrane was inferior to that projected for supercritical deasphalted oil but it is likely that permeate quality was specific to the type of residual oil processed. A comparison of the capital costs of a supercritical solvent deasphalter separation unit and a membrane upgrading system indicated that the costs were comparable. Additional development and optimization of the membranes and process will be required in order to create a large enough economic incentive for oil processors to switch to the membrane process.

Ceramic Membrane Process for Upgrading Vacuum Residual Oil

Bruce Bishop and Robert Goldsmith
CeraMem Corporation

Robert Schucker and Keith Rawlins
Trans Ionics Corporation

AIChE Spring Meeting, 2004 1

Acknowledgements

- DOE funding under SBIR grant number DE-FG02-00ER82970
- Garry Haacke and Jeremy Cardin at CeraMem
- Irv Wiehe under subcontract to Trans Ionics
- Crown Refinery for supplying the vacuum residual oil for the pilot plant tests

AIChE Spring Meeting, 2004 2

Rationale for Work

- Solvent deasphalting (SDA) used to remove asphaltenes and metals from vacuum pipestill residual oil prior to sending to catalytic cracker
- SDA uses a large quantity of solvent (up to 8:1 solvent to resid ratio) to precipitate asphaltenes
- SDA uses gravity settling to separate precipitated material from deasphalted oil (DAO)
- Membrane ultrafiltration (UF) may be used to effect separation with less solvent in smaller system

AIChE Spring Meeting, 2004 3

Project Objectives

1. Find UF conditions to achieve membrane flux of 350 $kg/m^2/d$ at 75% DAO recovery and with DAO quality equivalent to commercial solvent deasphalting
2. Demonstrate attractive economics relative to SDA for processing vacuum resid with solvent recycle or atmospheric resid without solvent recycle
3. Demonstrate stable membrane performance and no apparent deterioration over operation of at least 800 hours in total recycle mode.

AIChE Spring Meeting, 2004 4

Project Scope

- Evaluate the effects of feed pretreatment on UF membrane performance
- Assemble and test an UF membrane pilot plant
- Conduct membrane process life tests
- Perform process design and cost optimization

AIChE Spring Meeting, 2004 5

Structure of Asphaltenes

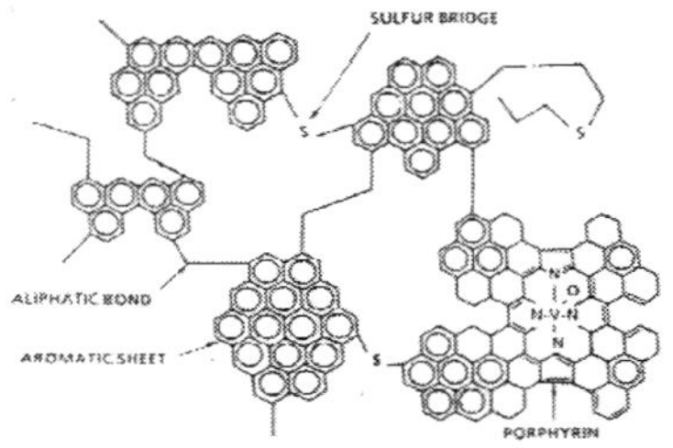

AIChE Spring Meeting, 2004 6

Membrane Separation of Asphaltenes

- Asphaltenes are associated into micellar structures
- Micelles can be separated with modest molecular weight cut-off UF membranes
- Treatments that increase the size of the micelles or cause phase separation will enhance ability to separate the asphaltenes from the oil

AIChE Spring Meeting, 2004 7

CeraMem UF Membranes: Lab Scale

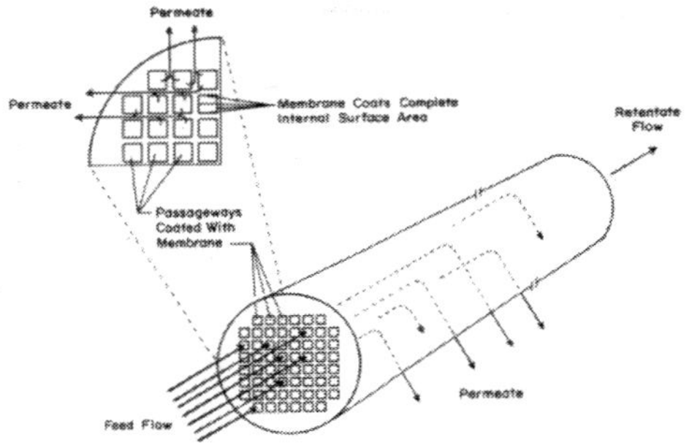

AIChE Spring Meeting, 2004 8

CeraMem UF Membranes: Lab Scale

1" dia. x 12" long element used in pilot testing

AIChE Spring Meeting, 2004 9

CeraMem UF Membranes: Full Size

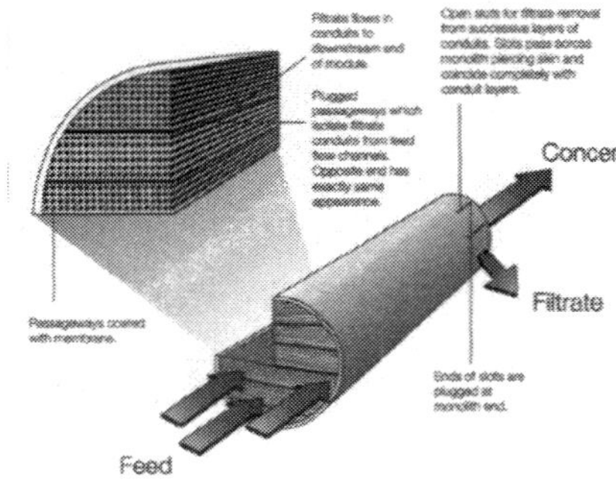

AIChE Spring Meeting, 2004 10

CeraMem UF Membranes: Full Size

5.6" dia. x 34" long; 115 ft^2 membrane area

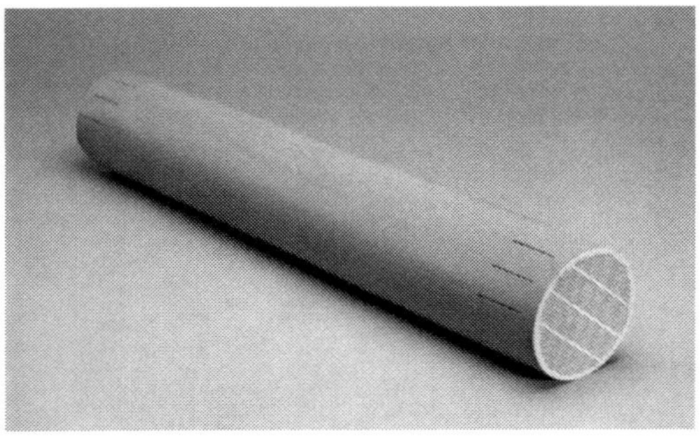

Pretreatment: Approaches

- Purpose is to increase size of asphaltene micelles in order to make them more filterable
- Vulcanization to link asphaltene structures with sulfur bonds
- Sorb asphaltenes onto fine catalyst particles
- Add diluents that would increase aggregation of asphaltenes in oil phase

Pretreatment: Resid Samples

	C_7 Asphaltenes (wt%)	Ni (ppm)	V (ppm)	Conradson Carbon (wt%)	Sulfur (wt%)	API
Vacuum Resid A	9.9	51	120	20.3	2.4	6.2
Vacuum Resid B	2.7	7	29	13.2	3.2	10.6
Vacuum Resid C	5.1	46	3	12.4	0.0	18.6
Vacuum Resid D	1.5	64	3	13.4	0.3	15.4
Atmospheric Resid E	3.0	85	31	8.3	1.3	12.5

Pretreatment: Diluent Samples

	C_7 Asphaltenes (wt%)	Ni (ppm)	V (ppm)	Conradson Carbon (wt%)	Average BP (°F)	API
Straight Run Gas Oil	0.0	0.4	0.2	0.0	270	25.7
Light Cat Gas Oil	0.0	0.4	0.1	0.0	500	35.0
Light Cat Cycle Oil	0.0	0.6	0.2	0.0	356	37.0
Vacuum Gas Oil	0.0	-	-	0.0	650	-

Pretreatment: Vulcanization

- Sulfur was added to mixture of resid and diluent, stirred and heated in autoclave, then solvent deasphalted.
- Purpose was to increase asphaltene aggregation through sulfur linkages.
- Results show vulcanization marginally increased asphaltene content at expense of DAO yield and quality.

Pretreatment: Vulcanization

Resid	Experiment #	Blanket	Time (min)	Temperature (°C)	Solvent	Wt % S	Wt % Asph (± 1 wt%)	Wt % DAO (± 1 wt%)
A	1	H_2	60	250	C_5	0.0	18	82
	2	H_2	60	250	C_5	0.5	19	83
	3	H_2	60	250	C_5	2	23	77
	4	H_2	60	250	C_7	0.0	8	92
	5	H_2	60	250	C_7	2	12	88
	6	N_2	60	250	C_7	2	10	91
	7	N_2	30	200	C_7	2	9	93
	8	N_2	30	150	C_7	5	12	88
	9	N_2	60	250	C_7	2	11	90

Pretreatment: Vulcanization

- Comparison of DAO with and without vulcanization shows an increase in DAO sulfur

Sample	Vulcanized	Diluent	Nickel (ppm)	Vanadium (ppm)	Micro Carbon (%)	Sulfur (%)
Resid C DAO	no	C_5	4	18	11.27	2.960
Resid C DAO	yes	C_7	4	18	13.62	3.218
Resid A DAO	no	C_5	17	40	13.91	2.083
Resid A DAO	yes	C_7	26	66	18.48	2.538

Pretreatment: Catalyst Fines

- Used equilibrium catalyst fines that had metals and carbon
- Catalyst fines were added to a mixture of pentane, SRGO, and resid immediately before UF testing

Pretreatment: Catalyst Fines

- Catalyst fines adsorbed some oil, but were not selective to asphaltenes
- No obvious improvement in UF flux was observed

Pretreatment: Solubility

- Developed thermodynamic solubility model to predict practical mixtures of resid and diluents that would increase the aggregation of asphaltenes
- None of the diluents tested was capable of developing asphaltene instability in any of the resid samples at low diluent to resid ratios

UF Pilot Plant: Design

- Single membrane stage
- One lab scale 50,000 MWCO silica UF membrane
- Batch and Total Recycle Mode
- 200°C and 150 psi operation

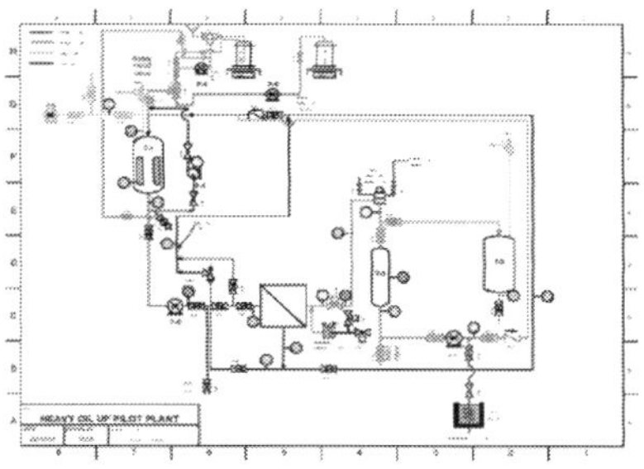

UF Pilot Plant: Final System

UF Pilot Plant: Experimental Materials

- Diluent: Heptane
- Resid: West Texas Vacuum Residual Oil
  - C_7 asphaltenes: 2.9 to 3.7 wt%
  - Nickel: 31 ppm by wgt
  - Vanadium: 49 ppm by wgt
  - Microcarbon residue: 13.7 wt%
  - Sulfur: 1.1 wt%
  - API: 20.0

UF Pilot Plant: Experimental Operation

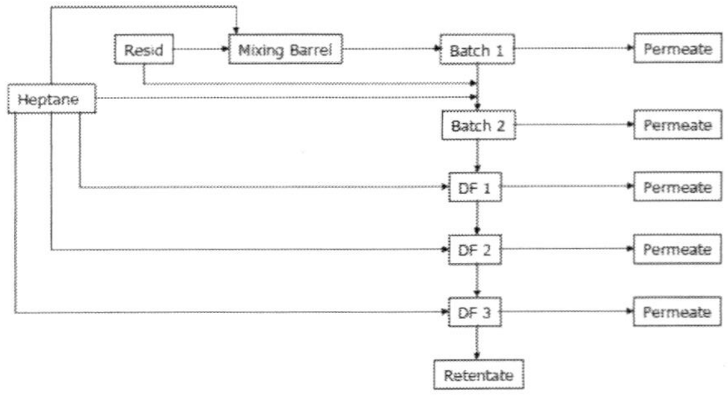

UF Pilot Plant: Results

Run A and Run B Test Parameters and Process Flux Results		
Parameter	Run A	Run B
Temperature (°C)	170	185
Transmembrane Pressure (psi)	50	50
Crossflow Velocity (m/s)	3.7	3.5
Heptane/Resid Ratio	Batch 1: 0.50:1 Batch 2: 0.37:1	Batch 1: 0.41:1 Batch 2: 0.33:1
Average DAO Flux (kg/m^2/day)	352	555

UF Pilot Plant: Results

Analytical Data for Permeate from Runs A and B		
	Cumulative wt% Removal	
Analyte	Run A	Run B
Asphaltenes	86	85
Nickel	53	53
Vanadium	60	61
Microcarbon	35	38
Purification of DAO not as good as that projected for SDA		

UF Pilot Plant: Flux Data for Run A

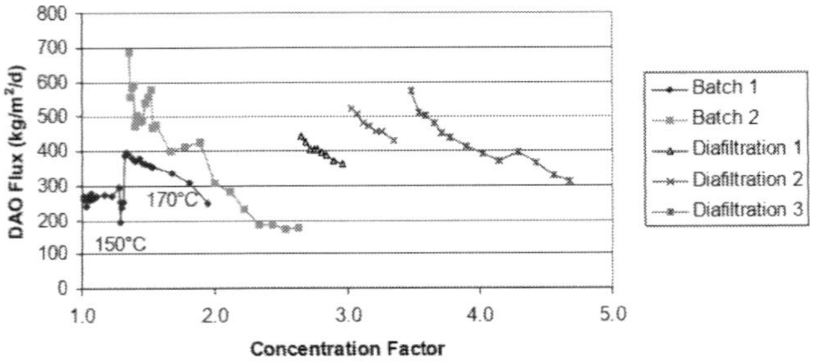

UF Pilot Plant: Flux Data for Run B

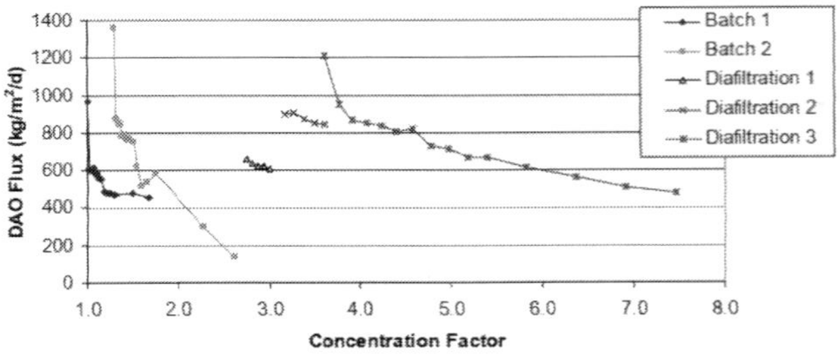

Economics: Target Costs

- Deasphalter unit capital cost: ~$550/bbl

(Total Deasphalter Process Equipment Capital Cost ~ $1100/bbl)

- Price Target for Membrane System: $300/bbl

AIChE Spring Meeting, 2004 29

Economics: 40,000 bbl/day Membrane System Design

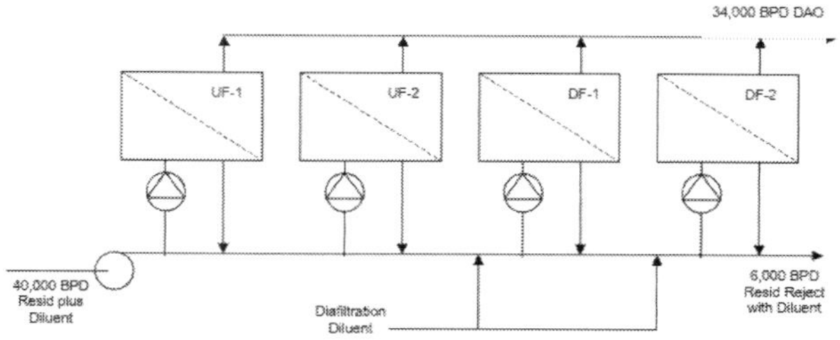

AIChE Spring Meeting, 2004 30

Economics: Membrane System CAPEX

Parameter	Value
Resid feed flow	40,000 BPD
DAO flow @ 85% recovery	34,000 BPD
DAO flux, including diafiltration	550 kg/m^2-day
Membrane Area, m^2	9,740 m^2
Membrane modules @ 11 m^2 per module	885
Membrane area cost	$6,500,000
System cost, including membranes	$17,000,000
Installation Cost, 50% of system (only) price	$5,250,000
System cost per BPD Resid Feed	$556

AIChE Spring Meeting, 2004 31

Economics: Membrane System OPEX

Pump Power (2 kW/module & $0.08/kW-hr)	$880,000
Direct Labor ($40/hr)	$640,000
Cleaning Solvent (8,400 gal/cleaning)	$50,000
Membrane Replacement (10 yrs, $430/m^2)	$420,000
Maintenance (5% of capital)	$850,000
Annual Operating Cost	$2,840,000
Operating Cost per Barrel of Feed	$0.19

AIChE Spring Meeting, 2004 32

Accomplishments and Key Results

- None of the pretreatment processes was successful in making the feed more filterable
- Pilot testing demonstrated process flux above 350 kg/m^2/day at very low diluent to resid ratio
- DAO yield of more than 75% was achieved
- In over 200 hours of testing, membranes did not appear to foul
- DAO quality less than desired for resid tested
- Capital cost for membrane system, based on limited process testing, appears comparable to that for solvent deasphalter

AIChE Spring Meeting, 2004 33

Projected Impacts on Industry

- Based on further development and optimization, the following may be achieved:
  - Lower cost for upgrading vacuum residual oil to quality necessary as catalytic cracker feed
  - Application of ceramic membranes to other non-aqueous applications such as lube oil recycle and separation of catalyst from Fischer-Tropsch wax

AIChE Spring Meeting, 2004 34

Paper Number 89e

THE FOULING LIMIT IN CRUDE OIL PRE-HEAT TRAIN DESIGN

G.T. Polley[1] and D.I. Wilson[2]
[1]www.pinchtechnology.com,
[2]Department of Chemical Engineering, Pembroke Street, Cambridge, CB2 3RA, UK

Paper presented at the Spring National AIChE Meeting, April 25-29, 2004

Session: Energy Conservation

7th International Conference on Refining Processes

UNPUBLISHED

The Fouling Limit in Crude Oil Pre-heat Train Design

G.T. Polley[1] and D.I. Wilson[2]
[1]www.pinchtechnology.com,
[2]Department of Chemical Engineering, Pembroke Street, Cambridge, CB2 3RA, UK

Recent years have seen an increase in our understanding of heat exchanger fouling, both in terms of mechanisms and quantitative modelling. In a quite wide range of cases the fouling mechanism has been found to be described reasonably well by a competition between processes that give rise to the formation of unwanted layers on heat exchanger surfaces and processes which act against it. The key result is that under some operating conditions a heat exchanger will foul and with others it will remain clean.

The challenge facing researchers in this area is twofold. The primary aim is the development of models which provide physically and pragmatically reasonable prediction of these conditions; the second aim is to develop means of characterising these models from plant data, because each system in practice will behave differently, and precision laboratory experiments are expensive.

Given these models, the design engineer is able to develop heat exchange systems in which fouling is minimised. For instance, in the case of fouling in pre-heat trains handling crude oil, the onset of fouling is now recognised to be strongly dependent on three factors: the composition of the crude, the velocity at which the crude passes through the exchangers, and the wall temperatures within the units. With most crude oils an increase in velocity results in an increase in the temperature of the wall at which fouling starts.

Both the velocity and wall temperature distribution are functions of the exchanger's design. The velocity at which the crude oil passes through the tubes of an exchanger is dependent on the tube pass arrangement. The wall temperature is dependant upon the temperatures of the fluids being processed (which is dictated by the structure of the pre-heat train – a 'global' design choice) and by the physical design of the exchanger (*i.e.* by 'local' design choices such as baffle arrangement *etc.*).

Experimental studies have shown that the temperatures encountered at the hot end of crude pre-heat trains give rise to fouling which, once established, can develop very rapidly. The rate of fouling increases as the extent to which the wall temperature exceeds that required for the initiation of fouling increases. Consequently, the designer should seek to develop a pre-heat train in which fouling is totally mitigated. However, there will be cases in which this is not physically possible for there is a limit on what can be achieved in design. In this paper we show how, given a quantitative model which represents the fouling process, this limit (which we will call the 'fouling limit') can be identified. We will discuss what can be done to extend this limit and how fouling can be mitigated in exchangers that operate beyond this limit.

Models for Predicting the Onset of Fouling in Pre-heat Trains

Ebert & Panchal [1997] coined the phrase 'fouling threshold' for crude oil systems, to describe combinations of operating variables (chiefly flow velocity and temperature) which gave rise to negligible rates of fouling. They also recognised its significance, in providing guidance for selecting exchanger design to mitigate (tube-side) crude oil fouling. They developed a quantitative model for the prediction of this threshold by postulating that fouling in crude oil is controlled by two competing mechanisms. The first, which is related to chemical reaction, promotes fouling while the second, which is related to the shear stress acting on the tube surface, acts to mitigate fouling. If the first mechanism dominates the second, deposition will occur. The threshold – a net rate of zero – occurs where these mechanisms are balanced; below the threshold, no significant fouling will occur. With threshold fouling models, as we move above and away from the threshold the dominance of the first mechanism can be expected to increase and that of the mitigating mechanism decrease, so that fouling becomes more severe.

It is important to note how this differs from the asymptotic fouling model described by Kern and Seaton [1959], which describes conditions where a deposit grows until changes in deposition results in a zero net rate of growth: asymptotic fouling yields a suggested extent of fouling which will not change much over time and leads to over-sizing of exchangers based on estimates of under-performance.

The equation later proposed by Panchal and co-workers [1999] to describe tube-side threshold fouling is:

$$\frac{dR_f}{dt} = \alpha \,\mathrm{Re}^{\beta}\, \mathrm{Pr}^{-0.33} \exp\left(\frac{-E}{RT_f}\right) - \gamma \tau_w \qquad (1)$$

where R_f is the fouling resistance, *Re* and *Pr* the Reynolds and Prandtl numbers of the fluid respectively, *R* is the gas constant, T_f is the film temperature and τ_w is the wall shear stress: α, β, γ and *E* (a chemical reaction activation energy) are dimensional model parameters.

Knudsen *et al.* [1997] subsequently reported a careful laboratory study on an Alaskan crude in which the existence of a fouling threshold was clearly demonstrated. Their experiments featured an annular test cell with a heated inner rod, with bulk velocities in the range of 3-10 ft/s (0.9-3 m/s), at two bulk temperatures (149 °C and 204 °C) and a range of heated surface temperatures. The table reproduced below indicates the conditions under which fouling effects were observed: no fouling was observed at the lower bulk temperature.

Table 1 Summary of threshold experiments by Knudsen *et al.*

Velocity m/s (ft/s)	Highest surface temperature with no fouling observed °C (°F)	Lowest surface temperature where fouling was observed °C (°F)
0.91 (3.0)	204 (400)	232 (450)
1.68 (5.5)	274 (525)	288 (550)
2.44 (8.0)	288 (550)	316 (600)
3.05 (10.0)	316 (600)	329 (625)

The experimental data reported at a velocity of 1.68 m/s and wall temperatures of 260 °C and 274 °C exhibited negative R_f values. This is most likely due to enhanced heat transfer from a thin layer of uneven deposit. Other workers (*e.g.* Wilson and Watkinson [1996]) have demonstrated that the initial stages of deposition can give rise to increased heat transfer in chemical reaction fouling via roughness effects.

Unfortunately, Ebert and Panchal's equation did not provide good prediction of these measured thresholds. Polley *et al.* [2002] postulated some simple modifications:

1. Wall temperature is used instead of film temperature in the reaction term
2. Retain the velocity dependency for the generation term, but use $Re^{-0.8}$
3. Assume that the removal term is proportional to $Re^{0.8}$

yielding:

$$\frac{dR_f}{dt} = \alpha \,\mathrm{Re}^{-0.8}\, \mathrm{Pr}^{-0.33} \exp\left(\frac{-E}{RT_w}\right) - \gamma\, \mathrm{Re}^{0.8} \qquad (2)$$

Most recently, Yeap *et al.* [2003] have analysed a database of measured crude oil fouling rates obtained from both pilot plant experiments and data reconciliation on monitoring of the performance of industrial exchangers. They have compared the predictions of both of the above equations and an equation based on a modification of one proposed by Epstein [1994] with this database. They found that the Modified Epstein Model yielded the best predictions. The Modified Epstein Model can also be used to predict fouling thresholds. Comparisons between thresholds predicted using this model, those predicted using the Modified Ebert-Panchal Model (equation 2) and the Knudsen data are made in Figure 1.

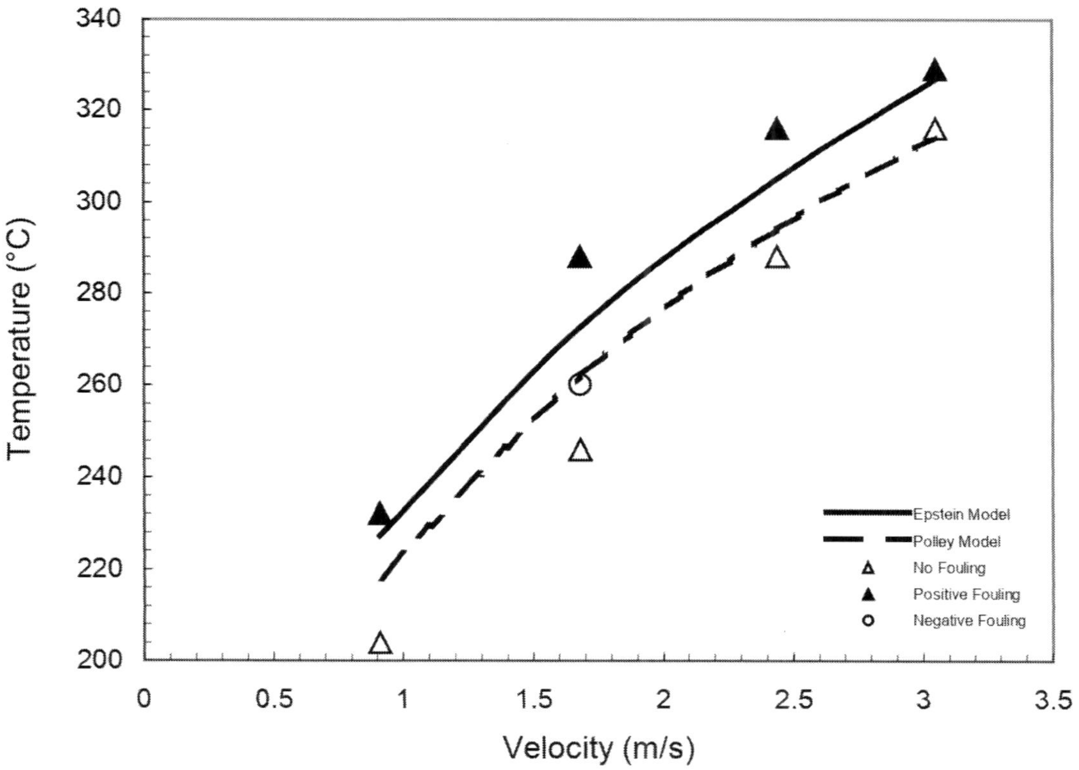

Figure 1 Comparison of fouling threshold models with the data of Knudsen *et al.*

Wall temperatures at which fouling first occurs are derived from the model equations by:

(i) Setting the tube-side velocity and estimating velocity dependant parameters;

(ii) Equating the two terms on the right hand side of the equation and solving to yield the film (Ebert-Panchal) or wall (Modified Ebert Panchal and Modified Epstein models) temperature at which fouling starts.

It must be noted that, to date, the threshold model concept has only been demonstrated for tube-side fouling. Studies of hot stream fouling – particularly for streams such as vacuum residues – and shell-side fouling have not been reported. The following analysis therefore assumes that fouling is dominated by crude oil passing through an exchanger's tubes.

Operation of exchangers above the fouling threshold

Each of the fouling models indicates (and this is supported by experimental data) a rapid increase in fouling rate as the extent by which the wall temperature exceeds the threshold increases. This will be accompanied by rapid increase in exchanger pressure drop and reduction in operating period between cleaning. All these factors indicate that the fouling threshold is an important design limitation.

The Temperature Field Plot

As already noted, heat recovery match selection will also influence the fouling that occurs in pre-heat trains as it determines the wall temperatures encountered by the crude. A simple means by which match selection can be compared with fouling propensity would therefore be of benefit. In Figure 2 we show the terminal temperatures of a heat exchanger on a plot of hot stream temperature (vertical axis) against cold stream temperature (horizontal axis). We can examine the likelihood of fouling occurring in the proposed exchanger by simply superimposing 'threshold temperature' lines upon this plot. These lines display the wall/film temperature at which the onset of fouling occurs for a given velocity. [In this figure the Ebert-Panchal model has been used: lines generated using the Modified Ebert-Panchal model, Equation 2, would be horizontal]. A range of lines, each corresponding to a different velocity, is used.

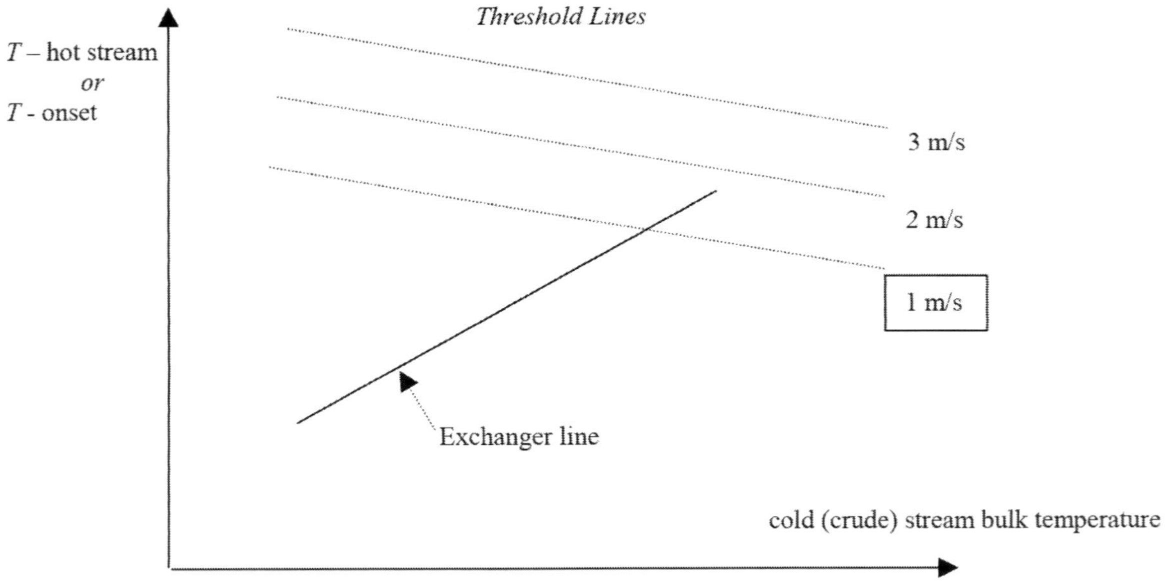

Figure 2 Temperature Field Plot construction

If the exchanger line is located completely below the threshold line, fouling will be suppressed in an exchanger operating at the velocity associated with that line. This allows us to determine, for a given match, the tube-side velocity necessary to suppress fouling within an exchanger handling that match.

If the exchanger line and the threshold line cross, fouling can be expected to occur within an exchanger which handles the proposed match if it is operating at or below the velocity associated with that threshold line. In this situation the designer has the options of:

(i) increasing tube side velocity to suppress fouling
(ii) reducing the heat load on the match by reducing the temperature at which the hot stream enters the exchanger
(iii) re-position the match such that the cold stream enters at a lower temperature
(iv) look for an alternative match.

For a conservative approach we can represent individual exchangers using the bulk fluid inlet and outlet temperatures. Given reasonable estimates of the overall heat transfer coefficient and the film heat transfer coefficient for the crude side we can derive lines based on exchanger wall temperature. The use of wall temperatures (as will be seen below) is a much less conservative approach.

Finally, in the construction used in Figure 2 we have assumed that the exchangers are in pure counter-flow (*i.e.* are single pass units). Multi-pass units exhibit different internal temperature profiles. As observed by Polley *et al.* [2002b], good practice is to ensure that the hot fluid enters the exchanger such that flows co-currently with the flow through the final pass of the unit. Given this arrangement, the procedure is again conservative.

Identification of Fouling Limit in new designs

One task that faces the engineer responsible for the design of a new plant is the determination of the heat recovery level that should be used. Since the temperature field plot has the same axes as the Driving Force Plot used in heat exchanger network design, the two plots can be superimposed on each other. The result may be that shown in Figure 3. Guidance on the derivation and uses of various process integration diagrams, including the Driving Force Plot, is given on the Pinchtechnology.com website.

In the case of a crude oil pre-heat train, the Driving Force Plot (the red line in Figure 3 – the green line, which is not relevant in this study, is parallel to the diagonal and off set by the minimum temperature difference for the identified utility consumption) shows the hot stream temperature (vertical axis) that is vertically aligned on the Composite Curves with a given crude temperature (horizontal axis). This tool is used to identify heat exchanger matches which make good use of available temperature driving force. Since the hot utility used to complete the heating of the crude is a furnace, the maximum temperature to which the crude can be brought through heat recovery is shown by the vertical line at A on the process curve.

The problem facing the designer is specification of the temperature target for the crude to attain through heat recovery. To date, this has usually been determined by examining the economic trade-off between energy consumption and capital cost of the heat exchangers used. Fouling considerations have largely been ignored. Unfortunately, the costs of fouling can far exceed those of energy: exchanger fouling is accompanied by increased flow resistances and this can result in reduced refinery throughput and lost production.

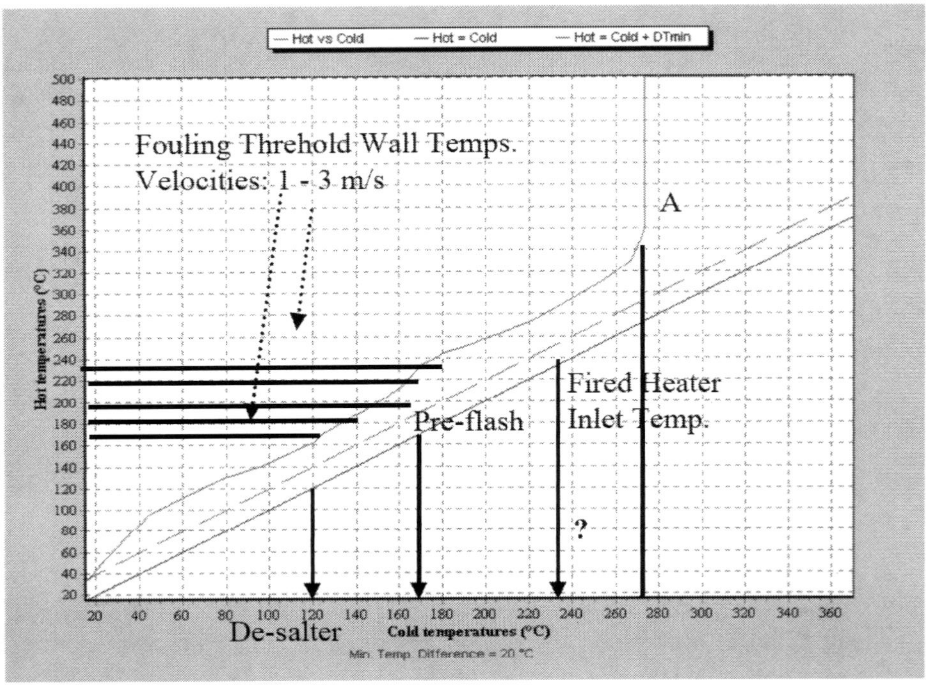

Figure 3 Combined Temperature Field and Driving Force Plots

The plot shown in Figure 3 provides a guide to the conditions under which fouling is likely to be a problem. The tube velocity at which an exchanger can be safely operated depends on the material used for the tubes. If that material is carbon steel, then the maximum velocity that can be used is around 3 m/s as higher values could give rise to erosion and/or corrosion.

From the plot we can determine the wall temperature at which fouling first occurs given the limiting velocity. We can also determine the film heat transfer coefficient for crude oil flowing through a tube of standard size (typically a nominal 25 mm at the hot end of a unit). Then, given a reasonable approximation for the overall heat transfer coefficient, we can construct a line showing the aligned wall temperature rather than bulk hot stream temperature. The crude oil temperature at which this wall temperature value equates with the fouling threshold yields the fouling limit. It is impossible to design a system that has a higher heat recovery and does not foul without recourse to special fouling mitigation measures.

This construction also provides us with guidance relating to the heat recovery matches that should be used at the hot end of the train.

Wall temperature is the controlling factor. The old rule of 'use the hottest stream at the hottest end of the train' is therefore not a safe rule. In the absence of other mitigation techniques (see below) the stream having bulk temperature given by point B in Figure 4 should be used at the hot end of the train. Streams having higher temperatures than this should be used lower down the train where the cooler crude results in lower (acceptable) wall temperatures.

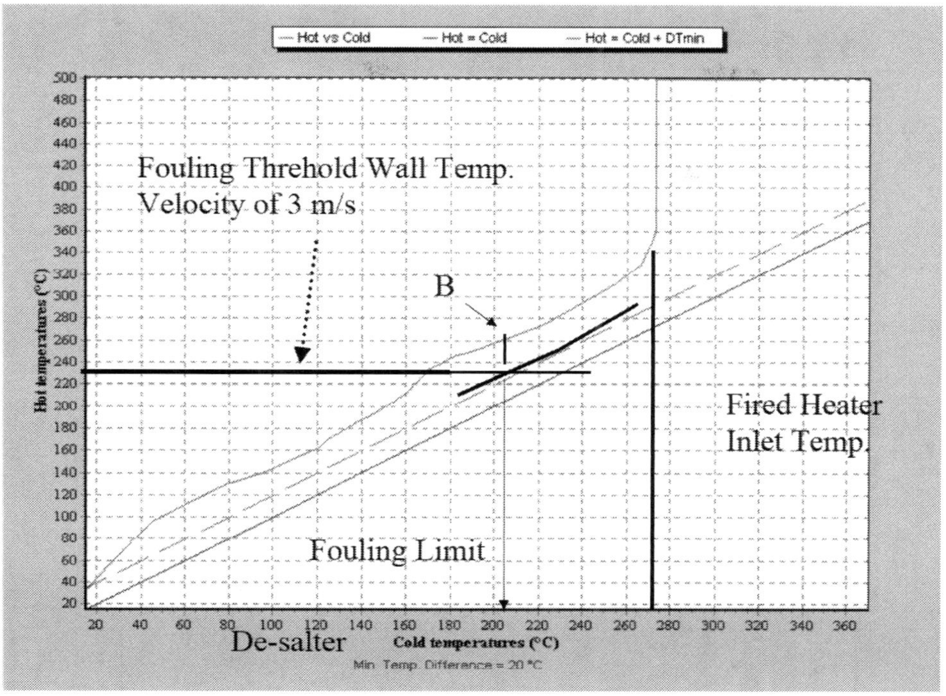

Figure 4 Determination of Fouling Limit

Identification and use of Fouling Limit in Retrofit

In retrofit situations the design of existing exchangers is known. Consequently, film heat transfer coefficients and exchanger wall temperatures can be calculated. Figure 5 shows a temperature field plot showing the threshold line (for 3 m/s), driving force plot and the temperature profiles (in terms of inlet/outlet bulk temperatures) of the units in an existing preheat train.

A separate purely thermodynamic analysis using traditional pinch technology techniques indicates that re-structuring of the train and the introduction of additional exchangers could result in an energy saving of over 15 MW. Furthermore, the economic analysis indicated that this saving could be achieved with a two year payback period.

Using the Temperature Field Plot we can examine the fouling propensity of each of the exchangers. For instance, the exchanger positioned at the hottest end of the train actually features a tube velocity of 0.88 m/s and as such it was operating above the fouling threshold (see Figure 6). This exchanger fouled badly. However, this is due to the design of the exchanger. It can be seen that replacement by a unit operating at a higher tube velocity would resolve the fouling problem.

If we assume that the change in design does affect the ratio of the two film coefficients (in reality there will be a beneficial effect for the tube-side coefficient increases and the wall temperature line moves downward) we can use the Temperature Field Plot to see how far we can move the modified design to the right before fouling is encountered (Figure 7).

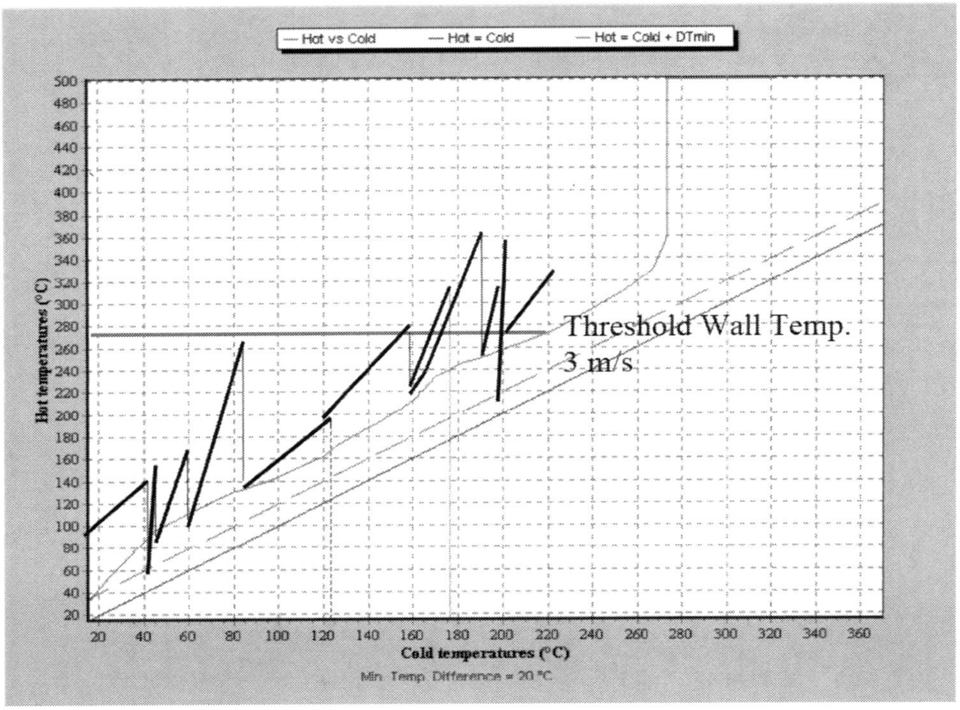

Figure 5 Temperature field plot for existing Pre-heat Train

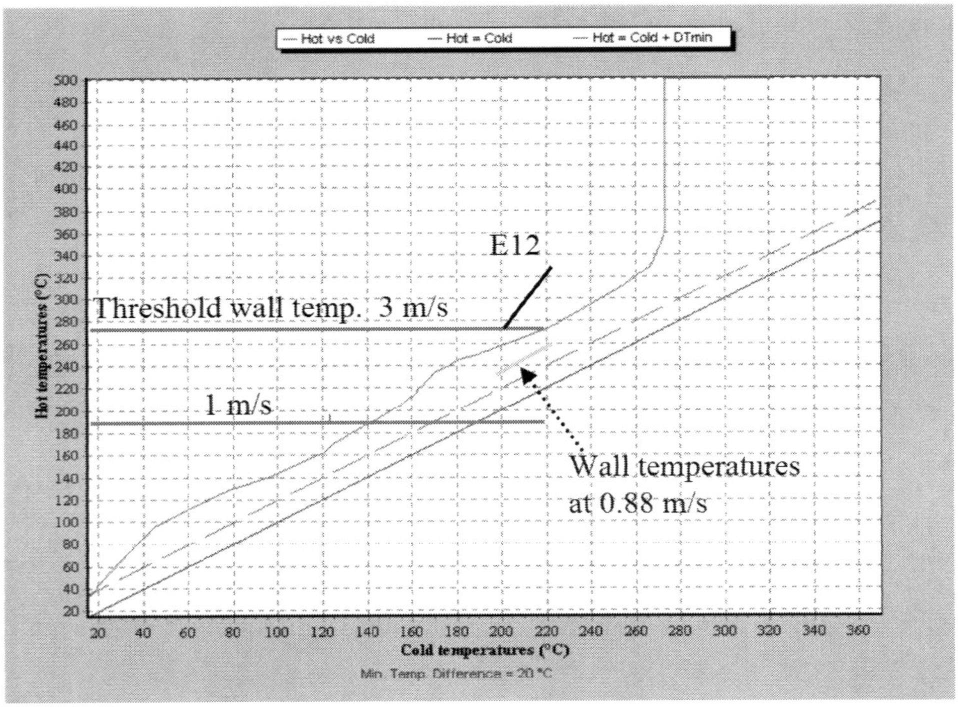

Figure 6 Operation of hottest exchanger, E12

We can do this for each exchanger in turn and thereby identify two things:

1. Which match should be positioned immediately prior to the fired heater
2. The temperature to which the crude can be raised before fouling is unavoidable.

In this case E12 is the 'critical' exchanger and it is found that the crude can be raised to 260°C before fouling is unavoidable. Once more we have identified the Fouling Limit. In order to safely operate above this limit, the designer needs to introduce one or more of the mitigation technologies described below.

Most significantly, an increase in crude oil temperature to 260°C involves an additional energy saving of 5.9 MW, not the 15 MW suggested using traditional pinch analysis. The application of a traditional pinch analysis would have resulted in a revamped pre-heat train in which fouling would be a significant problem.

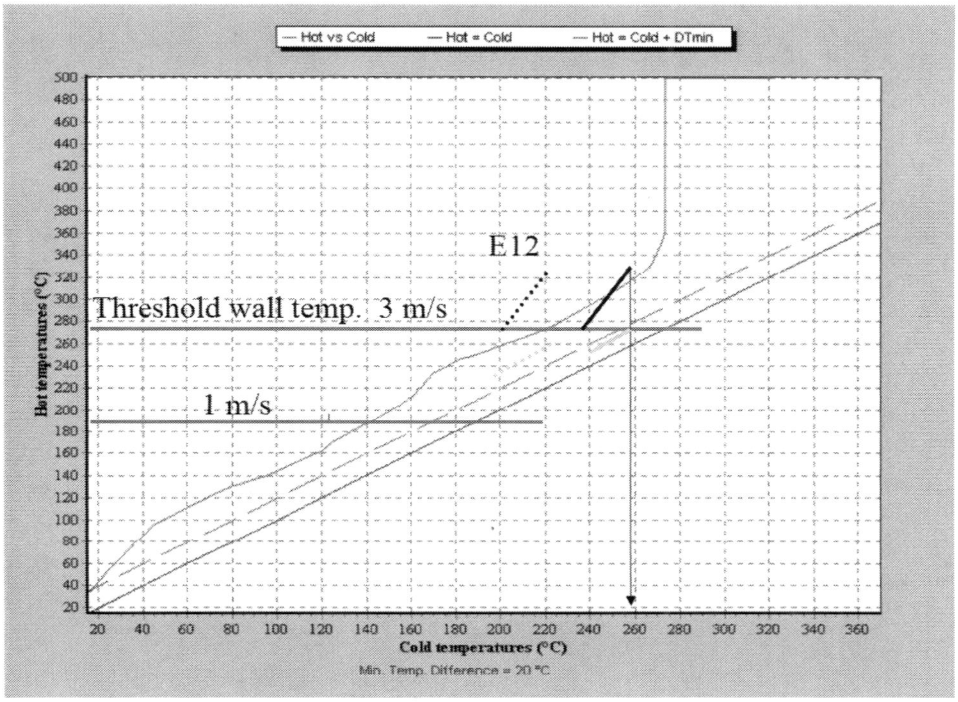

Figure 7 Maximum 'shift' for exchanger E12

Extending the Fouling Limit

The fouling limit can be extended in two basic ways:

1. Removal of tube velocity constraint by switching crude to shell-side of the unit
2. Adjustment of the fouling characteristics of the crude

Such action can be undertaken in a pragmatic ('try it and see') manner. However, the ability to identify better options and generate confidence in their selection requires further systematic R&D work.

At the moment, only the Modified Ebert Panchal model can be extended in order to provide a prediction of the fouling in geometry other than the plain tube and, even then, its predictions must be considered tentative. Clearly, research into fouling models must include alternative geometries. In the meantime, the Modified Ebert Panchal Model has been incorporated into ESDU's EXPRESS computer program. Using this program the engineer can determine how close a given design is to the fouling threshold or, in the case of new exchangers, geometry that lies below the fouling threshold.

There is strong evidence to suggest that the processing of crude on the shell-side of an exchanger has advantages in mitigating fouling. Some companies have used exchangers fitted with normal segmental baffles in this way for some time. Others have recently reported good experience with twisted tube bundles and with units fitted with helical baffles. The EXPRESS program handles helically baffled exchangers as well as units fitted with traditional segmental baffles.

Another mitigation strategy is to modify the fouling characteristics of a crude using chemical additives. However, in order to utilise additives effectively we need to know what effect they have on the fouling threshold. It is clear that more work is required in this area.

Operating Beyond the Fouling Limit

Finally, there is the old adage 'if you can't beat it do your best to live with it'. This can mean the adoption of other fouling mitigation technologies. A good example is the use of tube inserts. We note here that identification of the fouling limit allows us to identify exactly where and how tube inserts should be used.

Figure 8 reproduces the results of an industrial preheat train study reported by Bories and Patureaux (2003). In this study, existing exchangers were monitored before and after the installation of Turbotal™ tube inserts. The vertical line shows the time of installation of the inserts. Prior to the installation fouling was rapid and progressive. Following the installation of the inserts three of the four units exhibited asymptotic fouling behaviour. The fourth showed (indicated by the green points) featured a residue stream passing through the shell and may, therefore, have been subject to fouling on both sides of the unit. In this units the installation of the inserts resulted in a significant reduction in the overall fouling rate and level.

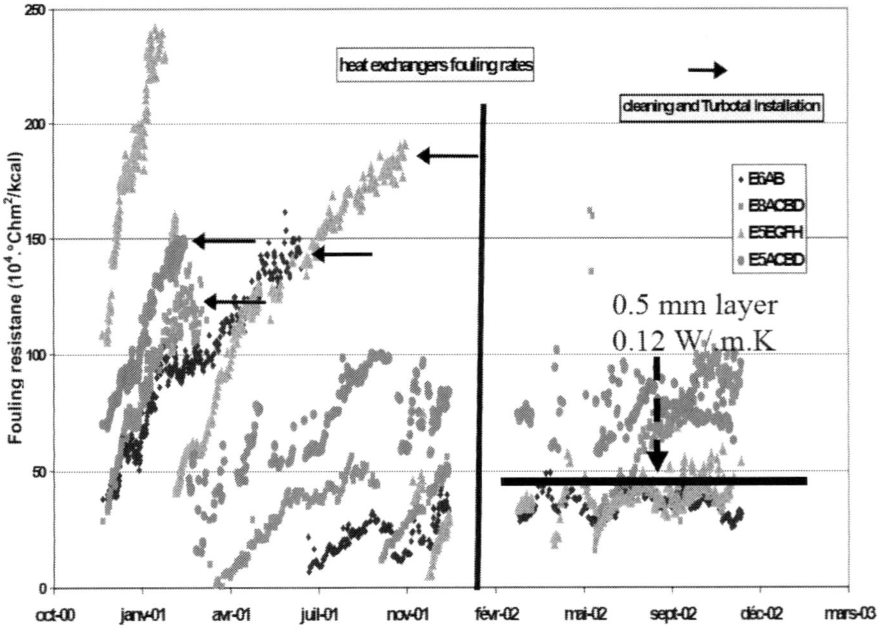

Figure 8 Plant data demonstrating fouling mitigation using Turbotal™ inserts

The asymptotes for two of the units, at 0.12 m K/W corresponds to a fouling layer thickness of ~0.5 mm. This must be close to the clearance between the rotating wire and the clean wall. So, we would conclude that the asymptote is due to cleaning action of the wire.

The authors of this study did not discuss the question of exchanger pressure drop. It can be expected that the installation of the inserts resulted in increased flow resistance and pressure drop at the clean condition. However, modelling conducted by the authors of this paper (unpublished work) suggests that the pressure drop encountered once the fouling asymptote has been reached is significantly less than that eventually reached after just two months operation of the original units.

Conclusions

Recent research has demonstrated the existence of a threshold below which fouling does not occur during the heating of crude oil. This threshold is dependent upon velocity and wall temperature and can be expected to be different for differing crude compositions. At the temperatures encountered at the hot end of crude pre-heat trains fouling can develop rapidly once the threshold is exceeded.

Since there are limitations on the velocity at which an exchanger can operate, there is a limit to the temperature that crude oil can be raised before it becomes impossible to mitigate fouling through normal exchanger design. We have endeavoured to show how that limit can be identified and the implications that the limit has on the identification of pre-heat train structure.

The limit can be extended by passing the crude oil through the shell of the exchanger (rather than the tube) or by modifying the crude reaction chemistry via additives. Operation of conventional units above the fouling limit requires the use of tube inserts.

References

Bories, M. and Patreaux, T., 2003, Preheat train crude distillation fouling propensity evaluation by Ebert and Panchal model, *Proc.* Heat Exchanger Fouling & Cleaning: Fundamentals & Applications, Santa Fe, July 2003.

Ebert, W.A. and Panchal, C.B., 1997, Analysis of Exxon crude-oil slip stream coking data in *Fouling Mitigation of Industrial Heat Exchange Equipment*, Panchal, C.B., Bott, T.R., Somerscales, E.F.C. and Toyama, S.. (eds.), Begell House, NY, pp 451-460.

Epstein, N., 1994, A model of the initial chemical reaction fouling rate for flow within a heated tube, and its verification, *Proc. 10th Int. Heat Trans. Conf.*, Brighton, IChemE, Vol. 4, pp 225-229.

Kern, D.Q. and Seaton, R.E., 1959, A theoretical analysis of thermal surface fouling, Brit. Chem. Eng., 4, pp 258-262.

Knudsen, J.G., Lin, D. and Ebert, W.A., 1997, The determination of the threshold fouling curve for a crude oil, in: Bott, T.R., Melo, L.F., Panchal, C.B. and Somerscales, E.F.C., (eds.) *Understanding Heat Exchanger Fouling and its Mitigation*, Begell House, NY, pp 265-272.

Panchal, C.B., Kuru, W.C., Liao, C.F., Ebert, W.A. and Palen, J.W., 1999, Threshold conditions for crude oil fouling, in: Bott, T.R., Melo, L.F., Panchal, C.B. and Somerscales, E.F.C., (eds.) *Understanding Heat Exchanger Fouling and its Mitigation*, Begell House, NY, pp 273-279.

Polley, G.T., Wilson, D.I., Yeap, B.L. and Pugh, S.J., 2002b, Use of crude oil threshold data in heat exchanger design, *Applied Thermal Engineering*, 22, pp 763-776.

Polley, G.T., Wilson, D.I., Yeap, B.L. and Pugh, S.J., 2002, Evaluation of laboratory crude oil fouling data for application to refinery pre-heat trains, *Applied Thermal Engineering*, 22, pp 777-788.

Wilson, D.I., Polley, G.T. and Pugh, S.J., 2002, Mitigation of crude oil preheat train fouling by design, *Heat Transfer Engineering*, 23, pp 24-37.

Wilson, D.I. and Watkinson, A.P., 1996, A study of autoxidation fouling in heat exchangers, *Can. J. Chem. Eng.*, 74, pp 236-246.

Yeap, B.L., Wilson D.I., Polley G.T. and Pugh S.J., 2003, Mitigation of crude oil refinery heat exchanger fouling through retrofits based on thermo-hydraulic fouling models, *Chem. Eng. Res. Des.* (*in press*).

Optimisation and Top-Level Analysis of Industrial Utility Systems

Varbanov,P., Smith,R., Perry,S.
Department of Process Integration - UMIST
UMIST PO Box 88, Manchester M60 1QD, UNITED KINGDOM
email: P.Varbanov@student.umist.ac.uk

Abstract

A new methodology for optimisation and analysis of industrial utility systems is presented. New methods have been developed for the modelling of steam turbines, gas turbines and boilers. In particular, the new models give a better description of part-load performance than previously published ones. These have been incorporated into an overall optimisation model for utility cogeneration systems. This has been successfully applied to optimisation and top-level analysis of existing systems. The optimisation is performed using a successive MILP procedure. In the case of top-level analysis, a stepwise optimisation procedure is applied to determine the true marginal value of steam and the limits on steam savings imposed by the utility system. This provides a strategy for steam saving on total sites and modifications to the utility system.

Keywords

Energy efficiency, utility system optimisation, top-level analysis, marginal steam price, SMILP

1 Introduction

Utility systems are an important part of most processing sites. On large processing sites the cost of fuel and power can be very significant offering a potential for significant savings. These can often be accomplished without capital expenditure by more effective day-to-day management of the system. Thus, a robust model of utility systems is an important tool in determining the strategy for energy reduction on the site and for modifications to the utility system.

Much work has been published on the design and optimisation of utility systems. Nishio et al (1980) have proposed a heuristic design method for utility systems based on estimation of thermodynamic losses and the irreversibility in the system. Petroulas and Reklaitis (1980) addressed the synthesis of plant utility systems by decomposing the design task into header selection and driver allocation. Another approach using MILP has been suggested by Papoulias and Grossman (1983) for constant process heat and power demand for a site. There is a number of other MILP based methods. These include the works by Iyer and Grossmann (1995), Singh (1997), Singh et al. (1998), which assign fixed efficiency to steam and gas turbines, boilers and heat recovery steam generators.

Mavromatis and Kokossis (1998a) introduced more sophisticated models for steam turbines, later extended by Shang (2000) who also added a gas turbine and boiler models reflecting part-load efficiency variation. He also established a synthesis framework involving selection of steam levels. One attempt to improve the modelling accuracy by Bruno et al (1998) resulted in a MINLP procedure.

A graphical method for targeting site-level heat recovery, based on the concept of the site heat source and heat sink profiles, was first introduced by Dhole and Linnhoff (1992). This has been further developed by Raissi (1994) and Klemes et al. (1997 to allow targeting for cogeneration of heat and power, and for fuel consumption.

The steam turbine model first developed by Mavromatis (1996) and Mavromatis and Kokossis (1998), and extended by Shang (2000), have a number of shortcomings. Firstly, the turbine models do not account for all factors influencing the performance. Secondly, the part-load modelling assumptions are too simplistic.

2 Improved steam turbine model

2.1 General full-load performance of steam turbines

Steam turbine performance is affected by a number of factors. The most significant among them are:

- Turbine size in terms of maximum power load

- Pressure drop across the turbine
- Current load

The overall turbine efficiency can be represented as consisting of two components: the isentropic efficiency and the machine efficiency. The latter is higher than the former (Siddhartha et al., 1999) and changes in a relatively narrow range, while the isentropic efficiency varies substantially with load.

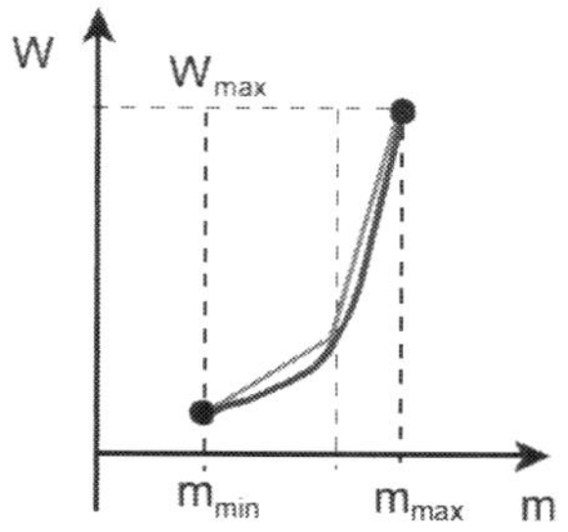

Figure 1. Willan's lines

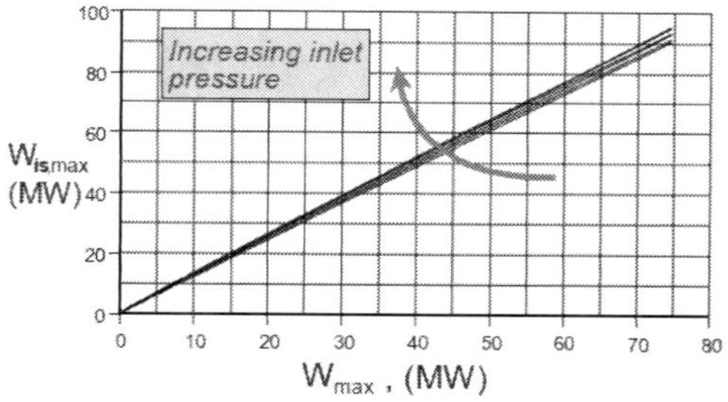

Figure 2. Isentropic vs. shaft power

The steam turbine efficiency varies with load, being higher at higher loads. This behaviour can be approximated by a series of straight lines, known as the Willan's lines (Figure 1). Thus, the actual steam turbine performance is replaced with one or more straight lines with different, but fixed coefficients (Figure 1). The performance equation in each modelling interval has the form given in Equation 1.

$$W = n \cdot m - W_{int} \tag{1}$$

Peterson and Mann (1985) present curves of the overall efficiency of industrial steam turbines at maximum load, each curve representing data for different steam turbines at full-load. The efficiency curves can be represented as plots of the isentropic power versus the actual shaft power. In this form, the shape of the plots appears to be linear (Figure 2). Thus, the isentropic power from a steam turbine can be related to the maximum shaft power through the following generic equation:

$$W_{is,max} = A + B \cdot W_{max} \tag{2}$$

Table 1. Regression coefficients for developed steam turbine model

	Back Pressure Turbines		Condensing Turbines	
	$W_{max} < 2$ MW	$W_{max} > 2$ MW	$W_{max} < 2$ MW	$W_{max} > 2$ MW
b_0 (MW)	0	0	0	-0.463
b_1 (MW/°C)	0.00108	0.00423	0.000662	0.00353
b_2 (–)	1.097	1.155	1.191	1.220
b_3 (1/°C)	0.00172	0.000538	0.000759	0.000148

The lines in Figure 2 have been subjected to regression analysis, based on the saturation temperature difference across the turbine, Equations 3 and 4.

$$A = b_0 + b_1 \cdot \Delta T_{sat} \tag{3}$$

$$B = b_2 + b_3 \cdot \Delta T_{sat} \tag{4}$$

The coefficients obtained by regression from the data of Peterson and Mann (1985) are given in Table 1. For existing utility systems, it is recommended that an independent regression for each particular steam turbine is performed using operating data.

2.2 *Part-load performance of steam turbines*

After assuming a proportional relationship of the Willan's line intercept on the maximum load for the current modelling interval, the following expressions can be derived for the slope and intercept of the Willan's line:

$$n = \frac{L+1}{B} \cdot \left(\Delta h_{is} - \frac{A}{m_{max}} \right) \tag{5}$$

$$W_{int} = \frac{L}{B} \cdot (\Delta h_{is} \cdot m_{max} - A) \quad (6)$$

Equations 5 and 6 represent the complete model, primarily intended to be used for design. For the case of modelling existing utility systems, however, the coefficient *A* can be set to zero and thus excluded from the analysis, since it has a meaning only when comparing many steam turbines for system design.

3 Improved gas turbine model

3.1 Turbine configuration

A gas turbine model has been developed in a similar way to that for steam turbines. Configurations allow steam injection, Figure 3.

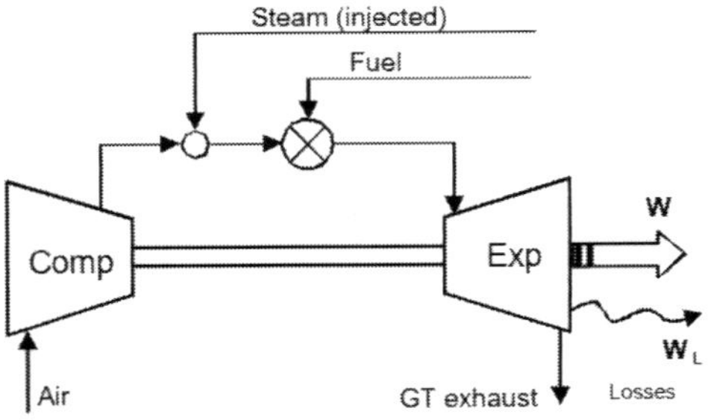

Figure 3. A gas turbine with steam injection

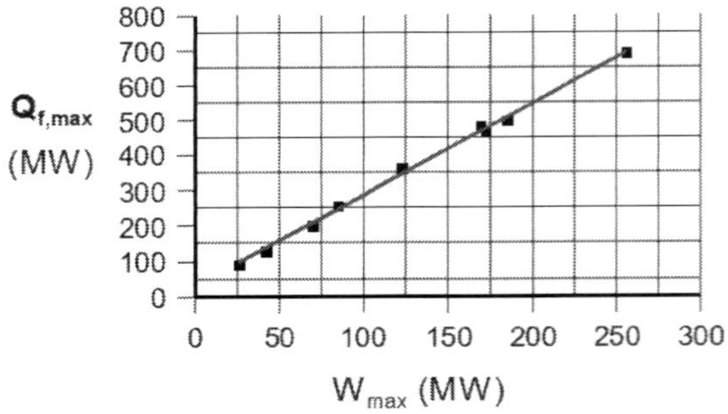

Figure 4. Fuel heat for industrial gas turbines at full load

3.2 Factors determining gas turbine performance

Similar to steam turbines, a combination of factors determines the performance of gas turbines. The factors considered are the turbine size, the current load and the ambient temperature. One of the major factors, affecting gas turbine performance is its size in terms of rated power production at ISO conditions (P = 1 atm, T = 15 °C and 60% relative humidity). A relationship can be established between the heat flow from fuel combustion and the turbine ISO rating. Data on gas turbine performance is published by manufacturers, for example, General Electric reference library documents (Brooks, 2001), see. Figure 4. The plot in Figure 4 is linear. Thus, the following equation can be formulated:

$$Q_{f,max} = A_{gt} + B_{gt} \cdot W_{gt,max} \quad (7)$$

Values for the coefficients of this relationship can be obtained by regression. The following values for the coefficients A_{gt} and B_{gt}, obtained from the data set (Brooks, 2001) plotted in Figure 4, have been regressed:

A_{gt} = 21.9917 (MW); B_{gt} = 2.6683 (–)

26.1 MW $\le W_{gt,max} \le$ 255.6 MW

3.3 Performance trends at part-load

The modelling equation for gas turbine part–load is also based on the Willan's line approximation (Equation 8). It is valid for each modelling interval, including the interval maximum load.

$$W_{gt} = n_{gt} \cdot m_f - W_{gt,int} \quad W_{gt,max} = n_{gt} \cdot m_{f,max} - W_{gt,int} \text{ (maximum load)} \quad (8)$$

The following expressions for the slope and the intercept of the gas turbine Willan's line are obtained

$$n_{gt} = \frac{L_{gt} + 1}{B_{gt}} \cdot \left(NHV - \frac{A_{gt}}{m_{f,max}} \right) \quad (9)$$

$$W_{gt,int} = \frac{L_{gt}}{B_{gt}} \cdot (NHV \cdot m_{f,max} - A_{gt}) \quad (10)$$

4 Boiler model

The steam boilers can be modelled in two alternative ways – with a constant efficiency model or with a variable efficiency model. The variable efficiency model can be formulated in a linear fashion, as follows:

$$Q_{BF} = \Delta h_{gen} \cdot \left[(b_{Blr} + 1) \cdot m_{stm} + a_{Blr} \cdot m_{stm,max} \right] + R_{BD} \cdot m_{stm} \cdot \Delta h_{pre} \tag{11}$$

The model allows the losses created by the boiler blowdown to be accounted for separately from the other losses. The performance coefficients in this model are $a = 0.0126$ and $b = 0.2156$ (Shang 2000).

5 Optimisation framework

5.1 Major degrees of freedom and trade-offs

In order to understand the system-level features, it is necessary to identify the degrees of freedom and trade-offs in utility plants. Some operational degrees of freedom are found in firing machines such as boilers and heat recovery steam generators, which can use different fuels and have different efficiencies. Inside the steam system, different flow patterns between steam turbines and letdowns affect the efficiency of power generation. The final options are condensing power generation and steam venting.

5.2 Operating scenarios

The utility system simulation and optimisation models assume fixed site. Usually operating and market conditions change with time. The most important factors that change are the electricity tariffs, the fuel feed-stocks and their prices and the ambient conditions. There are two types of electricity price variations:

- **Seasonal**. These variations are related to the changing ambient conditions. Usually there are two distinct price periods – summer and winter.
- **Current**. Supply companies usually divide every 24 hour period into a number of periods, typically peak, semi-peak and off-peak time intervals with different power prices.

Also, the cost of imported power might differ from the price of export power.

5.3 Implications of the hardware models on the optimisation task

The described utility system model leads to MINLP optimisation formulations. In order to overcome the arising computation problems, the values of the steam header enthalpies and gas turbine exhausts are fixed during the optimisation to produce a linear optimisation model. This is followed by a rigorous simulation after every optimisation step. The linear optimisation is repeated, followed again by simulation, and so on until convergence is achieved. Thus, a Successive Mixed Integer Linear Programming (SMILP) procedure is established. This is characterised by rapid convergence.

6 Case study

Consider the utility system in Figure 5. This includes a gas turbine with heat recovery steam generator, two steam boilers, four steam turbines for electricity generation and two direct drive steam turbines.

6.1 Scenario 1

The initial operation is shown in Figure 5. This has been subjected to optimisation and the result shown in Figure 6. The total operating cost of the site is reduced by 7.3508 MM$/y, a 14% reduction with respect to the initial operation.

6.2 Scenario 2

In the second scenario, a situation with significantly lower LP steam consumption is analysed. The initial operation is given in Figure 7. The optimised operation for Scenario 2 is shown in Figure 8. For this scenario, the on-site power production is decreased, leading to increased power import. The overall reduction in the utility system operating cost is 6.3593 MM$/y, again a 14% reduction with respect to the initial operation.

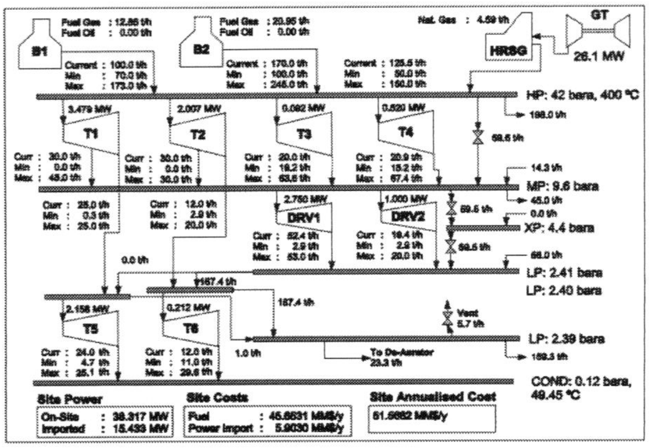

Figure 5. Initial operation for Scenario 1

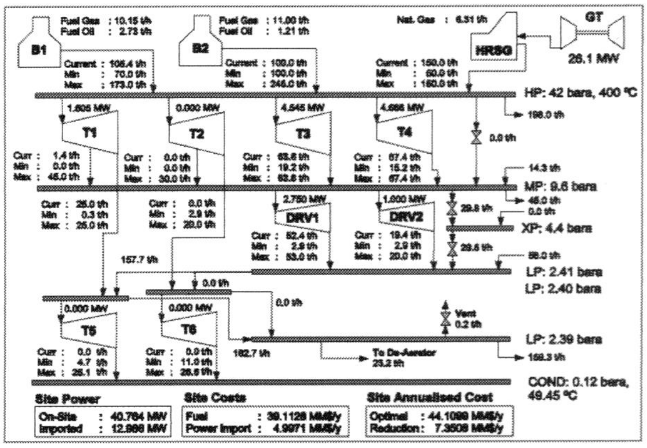

Figure 6. Optimal operation for Scenario 1

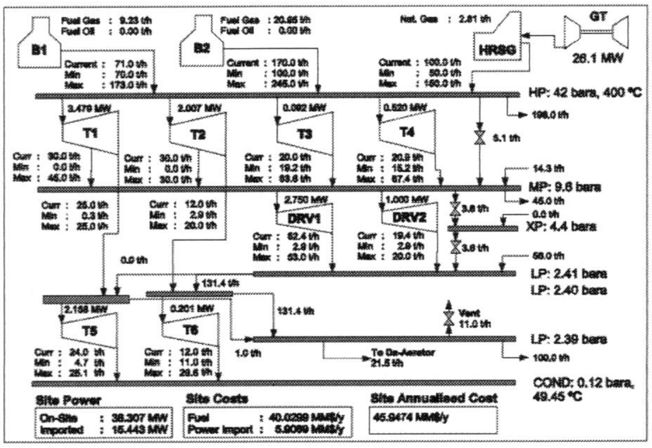

Figure 7. Initial operation for Scenario 2

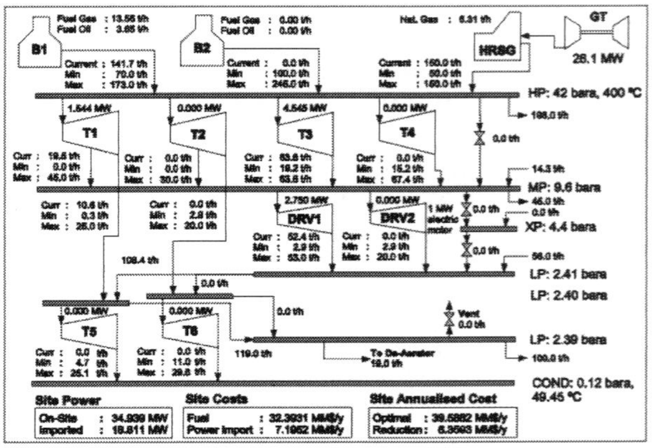

Figure 8. Optimal operation for Scenario 2

6.3 Discussion

The above two scenarios feature opposite trends for on-site power generation after optimisation. Scenario 1, for which the LP steam consumption is higher, tends to increase the power generation. For scenario 2, the on-site generation decreases. The reason for the different cost trends is that the power import is cheaper than the on-site power generation. However, the much higher efficiency of the combined power and heat generation makes part of the on-site generated power even cheaper. As a result, any steam turbine load above the one satisfying the steam requirement is not profitable.

7 Top-Level Analysis

7.1 Stepwise optimisation procedure

Energy efficiency measures applied to an industrial site usually can reduce the substantially the utility energy demand from site processes. In order to estimate their payback and profitability, however, it is necessary to evaluate if they are actually saving costs, and if yes – how much.

The complex interactions and trade-offs inside the utility system can be accounted for by exploiting the optimisation procedure described above. The concept of steam marginal price is used as an indicator in the analysis. It is defined as the change in utility system operating cost per unit change in steam demand of a given header (change in steam header balance):

$$PM_{header} = \frac{\Delta Cost}{\Delta m_{stm,header}} \tag{12}$$

Thus, the potential economic benefits of steam savings on the site are assessed using the same procedure regardless of the site scale or the modelling tools employed. The change in the operating cost is taken between the optimal operation before the steam demand change and the optimal operation after the steam demand change for the current step. Obviously, the result is context-specific and, for different operating conditions, each steam header will have a different marginal steam price. Even further, the marginal steam price can change for a given header depending on how much steam is saved.

Top-level analysis follows a stepwise optimisation procedure. To illustrate the approach, it is assumed that the objective of the analysis is to reduce steam demand through energy conservation in the site processes. The initial step of the procedure is to optimise the operation for the initial steam demands. This allows the true steam prices to be obtained in the subsequent steps. Next a loop is performed, comprising two major parts. The first part determines the header with highest steam marginal price for the current steam demands and selects it for analysis. The second part exploits the capacity for a potential decrease in demand at the selected header. The demand is lowered in a stepwise manner with operational optimisation performed at each step. If a constraint on usage (or generation) is reached, or the marginal price for the header changes, the steps of demand decrease (generation increase) are terminated. At this point, a test is performed to determine whether further decrease of process steam demands (increase in generation) at any mains is possible. If so, the loop moves to the next iteration. Otherwise, the procedure is terminated.

7.2 Top-Level Analysis of Generator-Only Systems

First consider a system with only steam turbine generators. The steam flows through turbines of this type can change smoothly between their minimum and maximum limits because the shaft power is not fixed but is allowed to vary within certain limits. This results in a continuous decrease in the site operating cost. In the case of analysis for steam saving potential, this means a monotonic profile of the steam marginal prices. Figure 9 shows a typical variation of marginal steam price with the extent of steam savings for systems with only generator steam turbines.

7.3 Systems with Direct Drive Turbines

Direct drive steam turbines impose the additional constraint of fixed power to be produced by the machine. As a result, the steam flows through driver turbines are virtually constant while they are in operation. However, if the process steam demands to the considered utility system features drop , then powering certain drivers with steam turbines may become less economic than to switch the drive to electric motors running on imported or site-generated power. As a result for such sites, with the extent of the steam savings, some turbines may be turned on and off, and the profiles of marginal steam prices take rectangular shapes. This type of behaviour is illustrated in Figure 10.

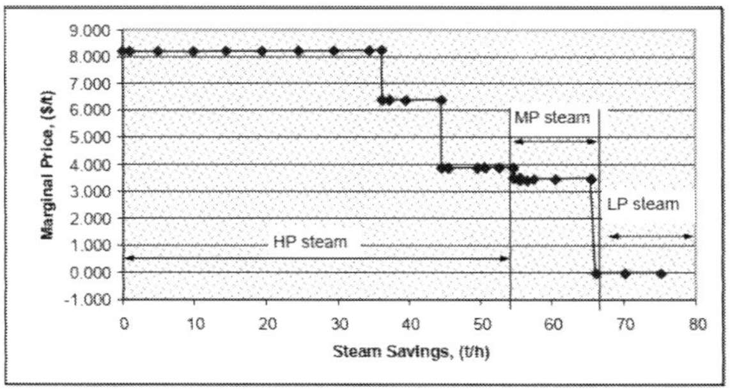

Figure 9. Steam marginal prices – generators only

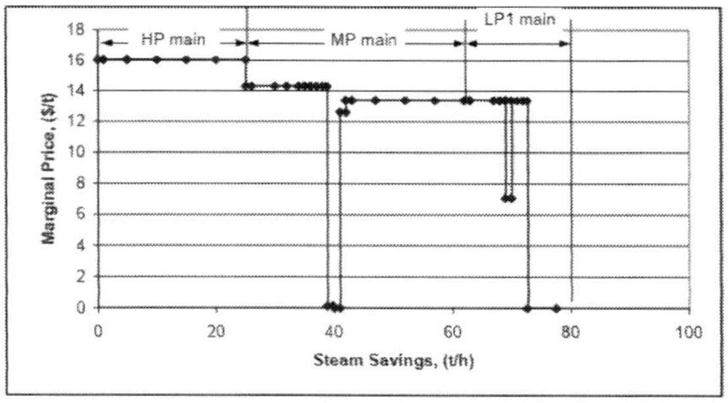

Figure 10. Steam marginal prices - drivers

8 Conclusions

Improved methods for utility system modelling and optimisation have been presented. The improved equipment models have tailored for use in optimising system operation and design. Further, a procedure for robust operational optimisation has been formulated. This is based on a successive MILP approach to solve an MINLP problem. The results from the optimisation case study indicate that, for different site operating scenarios, the optimal steam flows, power generation and fuel firing may show quite different trends. This behaviour is closely linked to the problem of steam marginal price. It is evident that the latter is different for different steam headers and even for the same header for different process steam demands.

The top-level analysis procedure provides steam marginal prices and allows retrofit and operational decisions to be made. The procedure allows automated simulation/optimisation tools to be exploited. This decouples the low-level system modelling tasks from the high-level system analysis and allows the procedure to remain the same as the models evolve. Thus, the procedure enables the analysis of systems with complex configurations, involving gas turbines, heat recovery steam generators, multiple boilers and multiple steam turbines. Even the most complex systems can be subjected to the analysis.

The procedure determines the true value of steam on a site and can be used to develop a strategy for energy saving on the site. The true value of steam depends on many factors including how much is being used. This contradicts the common practice of attributing a single value to steam at a given level for the purpose of energy management and the distribution of costs between different business areas.

References

Brooks, F.J., 2001, GE Gas Turbine Performance Characteristics., [Online], [Accessed April 2001], General Electric Reference library – GER-3576H, Available from World Wide Web : <http://www.gepower.com/publications/gers/GER3567H.pdf>.

Bruno, J.C., Fernandez, F., Castells, F. , Grossmann, I.E., 1998, A Rigorous MINLP Model for the Optimal Synthesis and Operation of Utility Plants. *Trans IChemE, Part A, Chem Eng Res Des*, 76(March): 246-258.

Dhole, V.R. and Linnhoff, B., 1992, Total site targets for fuel, co-generation, emissions and cooling. *ESCAPE –II Conference*, Toulose, 1992, France.

Iyer, R.R., Grossmann, I.E., 1998, Synthesis and Operational Planning of Utility Systems for Multiperiod Operation., *Computers Chem. Engng.*, 22(7/8): 979-993.

Klemes, J., Dhole,V.R., Raissi, K., Perry, S.J., Puigjaner, L., 1997, Targeting and Design Methodology for Reduction of Fuel, Power and CO_2 on Total Sites. *Applied Thermal Engineering*, 17(8/10): 993-1003.

Mavromatis, S.P., 1996, *Conceptual Design and Operation of Industrial Steam Turbine Networks*, Ph.D. thesis, UMIST, UK.

Mavromatis, S.P., Kokossis., A.C., 1998, Conseptual Optimisation of Utility Networks for Operational variations – I. Targets and Level Optimisation. *Chem.Eng.Sci.*, 53(8): 1585-1608.

Nishio, M., Itoh, J., Umeda, T., 1980, A thermodynamic approach to Steam-Power System Design, *Industrial & Engineering Chemistry Process Design and Development* 19(2): 306.

Papoulias, S.A., Grossmann, I.E., 1983, A structural Optimisation Approach in Process Synthesis – I. Utility Systems, *Computers and Chem. Engng.*, **7**(6): 695-706.

Peterson, J.F., Mann, W.L., 1985, Steam System Design: How it Evolves. *Chemical Engineering*, 92(21): 62-74.

Petroulas, T., Reklaitis, G.V., 1980, Computer-Aided Synthesis and Design of Plant Utility Systems. *AIChE Journal*, 30(1): 69-78.

Raissi, K., 1994, *Total Site Integration*, Ph.D. Thesis, UMIST, UK.

Shang., Z., 2000, *Analysis and Optimisation of Total Site Utility Systems*. Ph.D. Thesis, UMIST, UK.

Siddhartha, M., Rajkumar, N., 1999, Performance Enhancement in Coal Fired Thermal Power Plants. Part II: Steam Turbines. *Int. J. Energy Res.*, 23: 489-515

Singh, H., 1997, *Flue Gas Minimisation from Total Sites*. Ph.D. Thesis, UMIST, UK.

Singh,H., Smith, R., Zhu, X.X., 1998, Economic achievement of environmental regulation in chemical process industries. *Comput. Chem. Engng.*, 33(Supplement): S741-S744.

Paper No. [90a]

Distillate Hydrotreating to ULSD
The Impact of Aromatics

Laura Jones and Peter Kokayeff

UOP LLC, Des Plaines, Illinois, USA

Prepared for Presentation at the AIChE 2004 Spring National Meeting, April 27, 2004

[90] - Advances in Hydro/Desulfurization

Distillate Hydrotreating to ULSD
The Impact of Aromatics

Abstract

Due to ultra low sulfur diesel legislation, distillate hydrotreating (HDT) has become a key concern in the refining industry. Rate limiting parameters that had previously been overlooked, such as aromatic inhibition, may need to be considered for the production of ULSD (<10 wppm S). As hydro-desulfurization (HDS) and aromatics saturation (HDA) reactions occur in parallel in a shared environment they compete for the necessary resources, namely hydrogen and catalyst sites. A series of pilot plant and numerical simulation experiments were completed in order to explore how the competition for hydrogen and catalyst sites impacts the rate of HDS in the presence of polycyclic aromatics. A diagnostic pilot plant test showed that external mass transfer was not rate limiting. Intra-particle mass and energy transport were investigated by calculating the concentration and temperature gradients in a catalyst pellet. The effectiveness factors for HDS and HDA decreased in the presence of increased aromatics concentrations; however, the magnitude of the changes was insufficient to conclude that the competition for hydrogen was the sole mechanism needed to explain the suppression of HDS activity. Furthermore, the pilot plant data indicated that the presence of poly-aromatics has a more significant impact on the slower HDS reactions (sterically hindered DBTs). As a result, it was concluded that aromatics inhibition (competitive adsorption on catalyst sites) impacted the rate of HDS most significantly. Although the consequences are significant, the impact of aromatic inhibition was minor relative to the analogous nitrogen inhibition.

Introduction

The chemistry of ultra low sulfur diesel hydrotreating has become a key issue in the refining industry. Ultra low sulfur diesel (ULSD) regulations requiring <15 ppm sulfur will be in effect in North America by 2006 and regulations requiring <10 ppm sulfur will be in effect in Europe by 2009. In order to comply with these regulations essentially all of the organo-sulfur species must be removed including the substituted dibenzothiophenes (DBT) and other refractory sulfur species. The mechanisms and kinetics for deep desulfurization are more complicated than the simple first order reactions that hold for the less refractory sulfur species, e.g., thiols, sulfides and disulfides, due to the treatment of strerically hindered sulfur species. These sulfur species remain present at 500 and 50 ppm levels of sulfur and in some cases are left untouched. For this reason some sources have suggested that it may be as difficult to reduce the sulfur levels from 50 ppm to 10 ppm as it was to change from 500 ppm to 50 ppm.[1]

Multiple reactions occur in parallel on the hydrotreating (HDT) catalyst surface including hydrodesulfurization (HDS), hydrodenitrogenation (HDN), and aromatic saturation/hydrogenation (HDA). UOP has invested significant resources to develop a distillate hydrotreating kinetic model for ULSD production. In the development of the kinetic model the key parameters influencing the HDS rate were carefully investigated. Nitrogen was already known to inhibit HDS; therefore, efforts were focused on the interaction of HDS and HDA, and the availability of hydrogen.

Hydro-Desulfurization

In modeling HDS the sulfur species are generically characterized as "easy" or "difficult" sulfurs. The sulfur species are grouped based on the relative difficulty of sulfur extraction from the compound. In general, "easy" sulfur species (boiling range <610 °F) experience little steric hindrance and are converted quickly. Conversion of the "difficult" sulfur species (boiling range >610 °F) is much slower than the "easy" sulfurs due to steric hindrance and more complex reaction pathways. Several species have been identified as being particularly difficult to convert in

[1] Torrisi Jr., S., T. Remans, J. Swain. *"The Challenging Chemistry of Ultra-Low-Sulfur Diesel"*, *World Refining*, 24-29, Dec. 2002.

HDS, mostly substituted dibenzothiophenes (DBTs). In order to illustrate the difference in kinetics 4,6-dimethyldibenzothiophene (4,6-DMDBT), 4-ethyl,6-methyl dibenzothiophene, and C_3-DBT-A will be used as representatives of the "difficult" sulfurs. Although the specific structure of C_3-DBT-A has not been identified, it has been established that it is a C_3 substituted dibenzothiophene.

The HDS chemistry for easy sulfur is reasonably represented by an irreversible first order reaction with hydrogen inhibited by H_2S which competes for catalyst sites. It is widely accepted that there are two primary desulfurization paths for difficult sulfur species, direct desulfurization (also known as sulfur extraction) and hydrogenation (also known as saturation).[2] Direct desulfurization follows the same chemistry as the easy sulfur. Unfortunately the sulfur is not readily accessible to catalyst sites for substituted DBTs due to the rigidity of the fused aromatic rings and steric hindrance. As a result, the second prevalent reaction route includes a reversible hydrogenation reaction preceding the sulfur extraction. The saturation of one of the aromatic rings increases the flexibility of the molecule causing the sulfur to be more accessible for desulfurization. Desulfurization of difficult sulfur species is known to be inhibited by both H_2S and organo-nitrogen; however, the impact of aromatics on the overall rate of HDS has not been as fully understood.

Aromatic Saturation

Aromatic saturation is a significant reaction in distillate hydrotreating. The aromatic content of the feedstock varies widely depending on the source of the material; however, in all cases the aromatic compounds undergo saturation during hydrotreating. For modeling purposes the mechanism that was adopted was a series of reversible reactions

$$A + H_2 \xleftrightarrow{k1/k2} B + H_2 \xleftrightarrow{k3/k4} C + H_2 \xleftrightarrow{k5/k6} D \qquad (1)$$

where A, B, and C represent poly (tri+), di-, and mono- aromatic compounds respectively, and D represents cyclic compounds (saturated aromatics). Due in part to aromatics saturation, the hydrogen consumption increases far beyond the stoichiometric requirement for desulfurization as ULSD sulfur levels are targeted. Although hydrogenation is limited by the thermal equilibrium at high temperatures, the rate of aromatic saturation is ~100 times faster than HDS reactions for difficult sulfur and also varies with the number of fused rings.

Background

Significant controversy exists in the literature regarding inhibition of HDS by polycyclic aromatic compounds; there is further controversy over the significance and magnitude of aromatics inhibition among those who confirm that it exists. Inhibition of 4,6-DMDBT desulfurization by the addition of naphthalene was observed by Farag et al.[2] The conversion of 4,6-DMDBT over a CoMo catalyst decreased with increasing naphthalene concentrations present (0, 5, and 10 wt-%). Similar results are reported by Isoda et al in a study of a variety of catalysts, measuring 4,6-DMDBT conversion in the presence of 0, 5, and 10 wt-% naphthalene.[3] Based on the changes in the desulfurization product distribution in the presence of naphthalene it is thought that the inhibition is stronger for the hydrogenation route than for the direct extraction pathway.

A series of experiments to explore the competitive adsorption of organo-nitrogen and aromatics compounds and the implications for deep hydrodesulfurization was executed by Reinhoudt and colleagues.[4] The base feedstock for the work was 0.15 wt-% 4-E,6-MDBT (4-ethyl,6-methyl dibenzothiophene) in *n*-cyclohexane spiked with DMDS (dimthyl-disulfide) to increase the H_2S present. The conversion of 4-E,6-MDBT over two catalysts was

---

[2] Farag, H., K. Sakanishi, I. Mochida, and D. D. Whitehurst, Kinetic Analyses and Inhibition by Naphthalene and H_2S in Hydrodesulfurizaiont of 4,6-Dimethyldibenzothiophene (4,6-DMDBT) over CoMo-Based Carbon Catalyst, *Energy & Fuels*, 13, 449-453, 1999.

[3] Isoda, T., S. Nagao, X. Ma, Y. Korai, and I. Mochida. Hydrodesulfurization of Refractory Sulfur Species. 1. Selective Hydrodesulfurization of 4,6-Dimehtyldibenzothiophene in the Major Presence of Naphthalene over $CoMo/Al_2O_3$ and Ru/Al_2O_3 Blend Catalysts. Energy & Fuels, 10, 482-486, 1996.

[4] Reinhoudt, H.R., R. Troost, A.D. van Langeveld, S.T. Sie, J.A.R. van Veen, J.A. Moulijin, Catalysts for Second-Stage Deep Hydrodesulfurization of Gas Oils, *Fuel Processing Technology*, 61, 133-147, 1999.

measured for the base feedstock and two additional blends containing 275 ppm-wt carbazole and 3 wt-% tetralin respectively. Although on a mole basis the amount of tetralin present was 10 times greater than carbazole, the carbazole had a much more dramatic effect on the rate constant. Based on these results it was concluded that nitrogen inhibits HDS activity, but the influence of tetralin was insignificant. The adsorption constants reported by LaVopa and Satterfied, 0.011 and 0.51 kPa^{-1} for napthalene and carbazole respectively, are consistent with Reinhoudt's conclusions.[5]

Further complicating the issue of aromatic inhibition is the variation in the degree of inhibition depending on the number of fused rings. Kabe et al studied the desulfurization of dibenzothiphenes in the presence of di-cyclic and tri-cyclic aromatic compounds.[6] The base feedstock, a straight run light gas oil (21.9 wt-% mono, 16.6 wt-% Di, 1.2 wt% Tri+ aromatics), was spiked with 5 wt-% phenanthrene (3-ring aromatic) and 1-methylnaphthalene (2-ring aromatic). No inhibition was observed in the presence of 1-MN; however, a significant decrease in HDS activity was observed in the presence of phenanthrene. The adsorption constant increases as the number of fused rings increases, LaVopa reported a 6-fold increase in the adsorption constant when the number of fused rings is increased from 2 to 3 (0.064 kPa^{-1} for phenanthrene – compared to 0.011 kPa^{-1} for naphthalene).[5] In their literature review Grigis and Gates suggest that the adsorption constant may increase by as much as an order of magnitude when the number of fused rings is increased from 3 to 4.[7] The work of Koltai et al indicates that the molecular structure may be even more significant than the number of fused rings.[8] The rate of HDS for 4,6-DMDBT (n-dodecane solvent) was measured in the presence of anthracene, pheneanthrene, fluorene, acridine, and carbazole. Fluorene, a 2-ring aromatic with a structure similar to DBT, significantly impacted the rate of HDS while the inhibiting affect of anthracene, a 3-ring aromatic, was negligible.

The conclusions regarding the influence of aromatic compounds on the rate of HDS in the literature vary even among those reviewing HDS on a broad scale. Grigis et al concluded, "inhibition by aromatic hydrocarbons is not negligible."[7] On the other hand, Van Looij et al concluded from their study that the influence of polycyclic aromatics is negligible (at concentrations exceeding 10000 ppm).[9] The conflicting conclusions about the influence of aromatics, coupled with the contrast with organo-nitrogen inhibition cause the mechanism by which aromatics influence HDS to be questioned. Is there an explanation for the varied results reported in the literature?

Although there is controversy in the literature as to the nature, magnitude, and significance of the influence of aromatics on HDS kinetics, it is clear that the presence of aromatics impacts HDS. Highly aromatic feedstocks, such as LCOs, are known to be difficult to treat compared to straight run or coker material with lower aromatic concentrations. The relative difficulty was suspected to be due in part to the high aromatic concentrations. As HDS and HDA reactions occur in parallel in a shared environment there is competition for the resources needed to complete the reactions, namely hydrogen and catalyst sites (inhibition). A series of pilot plant and numerical simulation experiments were completed in order to explore the impact of the competition for hydrogen and possible inhibition of HDS by polycyclic aromatics. Polycyclic aromatic compounds were chosen to represent the aromatic species based on the stronger inhibition and adsorption reported in the literature as the number of fused rings increases.

---

[5] LaVopa, V., C.N. Satterfield, Poisoning of Thiophene Hydrodesulfurization by Nitrogen Compunds, *Journal of Catalysis*, 110, 375-387, 1988.

[6] Kabe, T., K. Akamatsu, A. Ishihara, S. Otsuki, M. Godo, Q. Zhang, and W. Qian, Deep Hydrodesulfurization of Light Gas Oil. 1. Kinetics and Mechanisms of Dibenzothiophene Hydrodesulfurization, Ind. Eng. Chem. Res., 36, 5146-5152, 1997.

[7] Grigis, M.J., B.C. Gates, Reactivities, Reaction Networks, and Kinetics in High-Pressure Catalytic Hydroprocessing, Ind. Eng. Chem. Res., 30, 2021-2058, 1991.

[8] Koltai, T., M. Macaud, A. Guevara, E. Schulz, M. Lemaire, R. Bacaud, M. Vrinat, Comparative inhibiting effect of polycondensed aromatics and nitrogen compounds on the hydrodesulfurization of alkyldibenzothiophenes, *Applied Catalysis A: General*, 231, 253-261, 2002.

[9] van Looij, F., P. van der Laan, W.H.J. Stork, D.J. DiCamillo, J. Swain, Key Parameters in Deep Hydrodesulfurization of Diesel Fuel, *Applied Catalysis A: General*, 170, 1-12, 1998.

Reactions Competing for Hydrogen

The overall rate of reaction in a fixed-bed reactor is controlled by a rate-limiting step that is slow relative to the rest of the process[10]. If the rate-limiting step is either the chemical reaction or the adsorption or desorption from the catalyst surface, the rate is kinetically controlled. Mass-transfer control can be either intra-particle or external. Intra-particle control occurs when the slowest step is the transport of reactants into the catalyst pores from the external surface. External mass-transfer control occurs when the rate-limiting step is the transfer of reactants from the bulk fluid to the catalyst surface. Petersen states in *Chemical Reaction Analysis* "if the diffusion to the external surface limits the over-all reaction rate, then internal diffusion must also limit the reaction rate"[11]. Furthermore, Petersen states "that the resistance to mass transfer within a porous catalyst is greater than the resistance between the external surface and the bulk fluid phases".

Historically, hydrotreating has been thought of as mass-transfer controlled. As a result, the hydrotreating catalysts are *small* shaped extrudates, maximizing surface area per catalyst volume. Reaction rates, however, are modeled on the basis of the bulk fluid properties and hydrogen partial pressure, ignoring any mass transfer effects - interphase or diffusion of reactants through the pore structure of the catalyst pellet. Hydrogen is the common reactant for all reactions occurring in distillate hydrotreating. If hydrogen were the limiting reagent, it is possible that the reaction environment in the catalyst pores could become hydrogen deficient thus reducing the HDS rate. In that case, the higher the aromatics concentration the more deficient the hydrogen supply and hence a greater impact on the rate.

Stoichiometric Considerations

In conventional trickle-bed hydrotreaters the hydrogen readily available for reaction is dissolved in the liquid phase. The solubility of H_2 in a diesel feed at 650 °F and 700 psi is ~50 SCF/BBLor 0.13 lbmoles/bbl. One barrel of a hypothetical distillate feedstock (using a density of 320 lbs/B) that contains 1.0 wt-% easy sulfur and 0.1 wt-% difficult sulfur contains 0.1 lbmoles easy and 0.01 lbmoles difficult sulfur species. The stoichiometric hydrogen requirement for 100% desulfurization is 0.24 lbmoles. If the feedstock also has aromatics present the stoichiometric hydrogen demand increases. For a feedstock that has 30 wt-% aromatics the the stoichiometric hydrogen requirement increases to 1.24 lbmol/BBL(saturating only 1 ring of the polycyclic aromatics). Similarly, for a feedstock containing 70 wt-% aromatics the stoichiometric hydrogen requirement would be 2.64 lbmol/bbl. Notice that the hydrogen solubility is only 5-10% of the stoichiometric requirement! This simple stoichiometric calculation raises an interesting issue. Although excess hydrogen is available in the gas phase, the reactions are occurring in a hydrogen deficient enviornment relative to the stoichimetric requirements for 100% conversion.

The Nature of Mass Transport Limitations in HDT Trickle-bed reactors

The first step in moving toward a rigorous evaluation of the hydrogen availability was to investigate the nature of the mass transport limitations. The diagnostic technique chosen required the measurement of the conversion as a function of the space velocity at two different catalyst bed length-to-diameter ratios corresponding to two, high and low, linear velocities. The observed conversions were plotted as a function of the residence time for both the high and low velocity data series. The curves are expected to coincide if the conversion is not limited by the external mass transfer to the catalyst surface.[12] If the external mass transfer influences the conversion, the high velocity series is expected to have higher conversion than the low velocity series. It is important to note that this diagnostic test applies to external mass transfer only. The lack of external mass transfer does not mean that there are not intra-particle diffusion limitations. The data from this simple diagnostic test certainly did not provide

---

[10] Rawlings, James B. and John G. Ekerdt, *Chemical Reactor Analysis and Design Fundamentals*, Madison, Nob Hill Publishing, 2000, 287-327.
[11] Petersen, Eugene E, *Chemical Reaction Analysis*, Englewood Cliffs, Prentice-Hall, Inc., 1965, 135.
[12] Hill, Jr., Charles G., *An Introduction to Chemical Engineering Kinetics & Reactor Design*, New York, John Wiley & Sons, 1977, 178-181, 457.

evidence of external mass transfer limiting the conversion of sulfur. Albiet, the possiblity of internal mass transfer limiting the conversion was not eliminated.

More extensive work on the nature of the mass transfer limitations in hydrotreating has been reported in the literature. As early as the 1950s the question of diffusional limitations in hydrodesulfurization was being addressed.[13] Based on experiments varying the catalyst pellet size (0.35 mm to 5 mm diameter) it was concluded that the diffusion effects inside the pellet were rate limiting, although it was unknown whether the diffusional limitation was wholly or partly limiting. Van Deemter also pointed out that diffusional limitations are greater for more reactive compounds than for those that react slowly. Consistent with the current study, no external mass transfer limitations were observed by Ancheyta et al in a similar test.[14] Two experiments were conducted to check for the presence of external mass transfer limitations. The 2.3 mm diameter catalyst was tested under the following conditions: LHSV 0.4-2.5 hr^{-1} with rates of 75-150 ml/hr and temperatures of 643-715 °F. In the first experiment, conversions of HDS, HDN, and HDA were measured as a function of space velocity for two different catalyst loadings. In the second, conversion was measured at LHSV of 1.0 and 2.5 hr^{-1} with rates ranging from 75-150 ml/hr. In both cases no external mass transfer limitiations were detected.

Satterfield and colleagues investigated both external and intra-particle mass transfer limitations.[15] The observed reaction rate was not signficantly effected by the liquid flow rate; any differences were within the magnitude of experimental error. The resulting conclusion was that diffusional limitations in the liquid phase (external mass transfer) of trickle bed reactors were not expected to be significant, again consistent with the current findings. On the other hand, the existence of substantial intra-particle diffusion limitations were confirmed by the calculation of the activation energy. Furthermore, the size of the catalyst pellet impacted the observed reaction rate; reducing the catalyst size increased the observed rate indicating intra-particle diffusion limitations. Increased conversion as a result of reducing the feed viscosity provided further evidence of the significance of pore diffusion.

Isothermal Single Catalyst Pellet Calculations

Fixed-bed reactors operate somewhere between the two extremes of kinetic control and intra-particle transport control.[10] As a result, intra-particle transport must be considered for heterogeneous catalyst systems – such as a fixed-bed reactor. In general, the transport properties may refer to either energy or mass transport. Initially only mass transport was considered in an isothermal catalyst pellet, the hydrogen profile within the catalyst pellet was the key interest. The general equation of continuity solved to find the concentration profiles within the catalyst pellet is

$$\nabla \cdot (D_{ej} \nabla c_j) = \sum_i R_{\upsilon i,j} \tag{2}$$

where D_{ej} is the effective diffusivity, c_j is the concentration, and $R_{\upsilon i,j}$ is the rate of consumption per catalyst volume of reactant j by reaction i. The subscript j refers to component j of a multi-component system. Similarly, the subscript i refers to reaction i for a system of coupled reactions. In cylindrical coordinates Equation 2 becomes

$$\frac{1}{r}\frac{\partial}{\partial r}\left(r D_{ej} \frac{\partial c_j}{\partial r} \right) = \sum_i R_{\upsilon i,j} \tag{3}$$

where r is the radial position in the catalyst pellet.

[13] Van Deemter, J. J., *Trickle hydrodesulphurization – a case history,* Koninklijke/Shell-Laboratorium, Amsterdam (Shell Research N.V.), 215-223, 1965.
[14] J. Ancheyta,, G. Marroquín, M. J. Angeles, M. J. Macías, I. Pitault, M. Forissier, and R. D. Morales, *Some Experimental Observations of Mass Transfer Limitations in a Trickle-Bed Hydrotreating Pilot Reactor,* Energy & Fuels, *16,* 1059-1067, 2002.
[15] Satterfield, C. N., A. A. Pelossof, T. K. Sherwood. *Mass Transfer Limitations in a Trickle-Bed Reactor.* AIChE Journal, 15, (2), 226-234, March 1969.

Even for a simplified set of three irreversible reactions Equation 3 becomes a set of three coupled second order differential equations. In order to solve the continuity equation (Equation 3) two boundary conditions are needed. The boundary conditions for a cylindrical catalyst pellet are as follows

$$\frac{\partial c_j}{\partial r} = 0 \text{ at } r = 0 \text{ and } c_j = c_{js} \text{ at } r = R \tag{4}$$

where subscript s refers to the outer surface ($r = R$).

Equations 3 and 4 constitute a class of differential equations known as two point boundary value problems due to the split boundary conditions, one at one end and the other at the other end of the limits of the independent variable. Only a small number of such problems can be solved analytically, most require numerical solutions. But even for the case of a numerical integration the split boundary conditions pose a problem since the integration can not be started at either end. While a number of methods may be applied to this type of problem, e.g. any of the methods of weighted residuals (orthogonal collocation being the most popular for reaction/diffusion problems), the "shooting" method was used. In the "shooting" method a guess is made of value of the concentration at the pellet center, e.g. $c_j = c_{jo}$ at $r = 0$, and the equation solved to determine the value of the concentration at the surface. The concentration at the pellet center is then adjusted and the process repeated until the calculated value of the concentration at the surface matches the surface boundary condition.

Di-aromatics saturation and desulfurization of difficult and easy sulfur species were included in the single pellet calculations. The appropriate kinetic expressions were substituted into Equation 3 for each species. For di-aromatics the mass balance is

$$D_A \frac{1}{r}\frac{d}{dr}\left(r\frac{dC_A}{dr}\right) = f(T, P, C_A, C_H) \tag{5}$$

where the subscripts A and H refer to aromatics and hydrogen respectively. The mass balance is a bit more complicated for the easy sulfur species due to the inhibition by hydrogen sulfide (H_2S),

$$D_S \frac{1}{r}\frac{d}{dr}\left(r\frac{dC_S}{dr}\right) = f(T, P, C_S, C_H, C_{H2S}) \tag{6}$$

where the subscript S refers to the easy sulfur species. The desulfurization kinetics for difficult sulfur species included two reversible steps, adsorption onto the catalyst sites and saturation, and two inhibition terms, nitrogen (N) and H_2S. Equations 3 becomes

$$D_{S2} \frac{1}{r}\frac{d}{dr}\left(r\frac{dC_{S2}}{dr}\right) = f\left(T, P, C_{S2}, C_H, C_{H_2S}, C_N\right) \tag{7}$$

for the difficult sulfur species (subscript $S2$). The rate constants (k_i) and equilibrium constants (K_{EQ}) were taken from previous modeling work. Sulfur species boiling from 550-610 °F and C_3-DBT-A were chosen to represent the easy and difficult sulfur species respectively. When considering three independent reactions, the system is fully defined once three components are specified as in Equations 5-7. As a result, the hydrogen concentration profile was defined in terms of the concentrations of the sulfur and aromatics concentrations as follows

$$C_{H2} - C_{H2B} = \sum_j \alpha_j * \left(\frac{D_j}{D_{H2}}\right) * \left(C_j - C_{jB}\right) \tag{8}$$

Assuming c_{js} is equal to the bulk fluid concentration (c_{jB}), the two-point boundary value problem can be solved numerically as an initial value problem using a non-linear optimizer to find c_{jo} (subscript o refers to the center of the catalyst pellet, r=0). The values of c_{jo} are chosen (optimized) so that the calculated c_{js} is equal to c_{jB}. The assumption that $c_{js} = c_{jB}$ is valid only when external mass transport is not rate limiting, which is the true for a trickle-bed hydrotreater based on the previous discussion.

Isothermal results

As stated previously the solubility of hydrogen in the liquid phase limits the amount of hydrogen readily available to less than the stoichiometric requirements even at the outside of the pill. Although the diffusion of hydrogen is considered fast relative to the other reactants in the liquid[13], it was hypothesized that if the hydrogen gradient within the pellet was significant it may influence the rate of reaction especially when high concentrations of aromatics are present. In order to explore this hypothesis the two-point boundary value problem was solved to determine the concentration profiles within the catalyst pellet and effectiveness factors for each of the reactions as the bulk concentrations of sulfur and aromatic species were varied.

The bulk diffusion constants (D_j) for both the sulfur and polycyclic aromatic compounds were assumed to be 5 x 10^{-5} cm^2/sec. Assuming that hydrogen diffuses faster and more easily than the bulkier compounds the diffusivity was taken to be 2 x 10^{-4} cm^2/sec. The effective diffusivity for diffusion in a porous solid is defined as

$$D_{ej} = \frac{D_j \theta \kappa_p}{\tau \kappa_r} \tag{9}$$

where τ is the tortuosity factor and θ is the void volume for the solid, κ_p is the equilibrium partition factor, and κ_r is the enhanced drag factor.[16] The drag and equilibrium partition factors correct for effects of the size of the molecule. As a result, both κ_p to κ_r are functions of the ratio of the molecular and pore diameters. Applying the drag and equilibrium partition factors leads to

$$D_{ej} = \left(\frac{\theta}{\tau}\right) 2 * 10^{-5} \text{ cm}^2\text{/sec} \tag{10}$$

for the sulfur and polycyclic aromatic compounds based on an average catalyst pore diameter of 80 Å and molecular diameter of 12 Å. The κ_p and κ_r are approximately unity for hydrogen due to its small size relative to the pore diameter. The tortuosity factor accounts for the slowing of the diffusion process due to the curvature, length, and shape of the pores.[17] The tortuosity and catalyst void fraction were assumed to be 4.0 and 0.5 respectively. These numbers are in the ranges typical for porous catalysts.

As noted previously, hydrotreating catalysts are prepared as small size tri-lobes, with a nominal dimension of 1/20" from the base to the top. A tri-lobe may be approximated by three cylinders fused together. An inspection of a 1/20" tri-lobe shaped catalyst reveals that the diameter of one of the cylinders is ~ 2/3 of the nominal dimension or 0.085 cm. The radius is then 0.0425 cm. In order to account for the path length into the center of the structure the radius was taken as 0.05 cm for purposes of the diffusion/reaction calculations within the catalyst pellet.

The variation of the hydrogen concentration profile within the catalyst pill as the sulfur and aromatics concentrations were increased was considered first. The hydrogen concentrations relative to the bulk concentration are shown as the percent of diffusion length (100% is at the catalyst surface and 0% is at the center of the pill). These profiles are shown for increasing aromatics concentrations in the presence of 1.4 wt-% sulfur in Figure 1. As expected the relative hydrogen concentration decreased as the radial position in the catalyst approached the center of the pellet. Additionally, relative hydrogen concentration at the center of the pellet decreased with increasing aromatics and sulfur concentrations in the bulk. The hydrogen consumption is highest in the extreme case of 1.4 wt-% sulfur and 70 wt-% aromatics; hence, the relative hydrogen concentration at the center of the pill is the lowest, 69%. At the other extreme, 1.4 wt-% sulfur and 5 wt-% aromatics, the relative hydrogen concentration at the center of the pill was 85%.

---

[16] Lapidus, Leon, Neal R. Amundson, *Chemical Reactor Theory – A Review*, Englewood Cliffs, Prentice-Hall, Inc., 1977, 199-200.
[17] Satterfield, C.N., *Mass Transfer in Heterogeneous Catalysis*, Cambridge, M.I.T. Press, 1970, 33-39.

There is clearly an impact on the hydrogen concentration within the catalyst pill as the aromatics concentration is increased; however, in order to evaluate the effects of this change in hydrogen concentration on the reaction rates the effectiveness factors were calculated for each reaction, HDA and HDS (easy and difficult). The effectiveness factors are shown as a function of the poly-aromatic concentration in Figure 2. The effectiveness factor was the lowest for the conversion of easy sulfur followed by the aromatics saturation reaction. This is consistent with van Deemter's statement that the fastest reactions will be most impacted by diffusion limitations.[13] The change in the effectiveness factor for the difficult (C_3-DBT-A) and easy sulfur was only 3-5% as the concentration of poly-aromatics was increased from 5 wt-% to 70 wt-%.

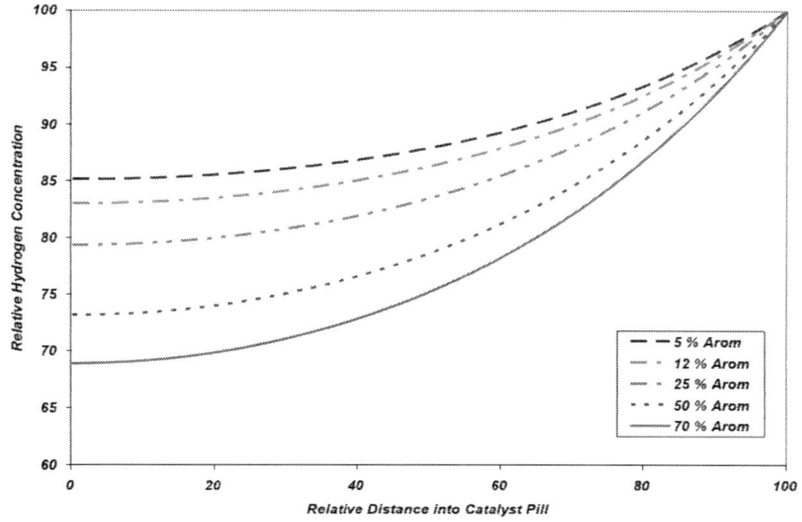

Figure 1: Intra-particle hydrogen concentration profile for an isothermal catalyst pellet in the presence of 1.4 wt% sulfur and varied poly-aromatic concentration.

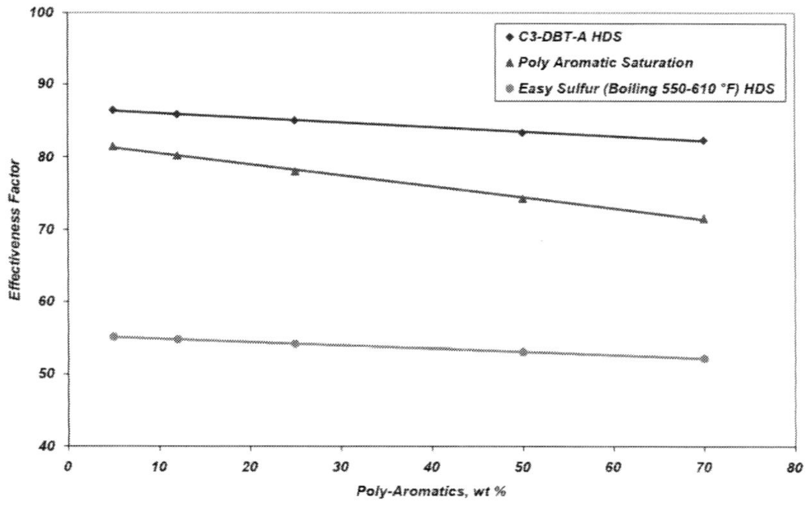

Figure 2: Isothermal catalyst pellet HDA effectiveness factors for poly-aromatics, C_3-DBT-A, and easy sulfur boiling 550-610 °F as a function of the bulk poly-aromatic concentration with 1.4 wt-% total sulfur.

Non-Isothermal Single Catalyst Pellet Calculations

In order to further explore the process occurring within the catalyst pores, the gradients were calculated for a non-isothermal case. In order to calculate the non-isothermal intra-particle temperature and concentration gradients the mass and energy balances must be solved simultaneously. The energy balance at steady-state conditions is

$$\nabla \cdot (k_e \nabla T) = \sum_{i=1}^{n_r} \Delta H_{Ri} r_{\upsilon i} \quad (11)$$

where ΔH_{Ri} and r_{vi} are the heat and rate of reaction i, n_r is the number of reactions, and k_e is the effective thermal conductivity of the pellet.[10] The non-isothermal case was solved by the same means as the isothermal pellet; however, the temperature was adjusted based on the energy balance at each integration step. By combining Equations 2 and 11 the temperature gradient may be defined in terms of the concentration gradients. The effective thermal conductivity is dependent on the properties of both the liquid and the porous solid. Typical values, according to Lee, are 10^{-2}-10^{-4} cal/cm.sec.°C for liquids and 10^{-3}-10^{-4} cal/cm.sec.°C for porous solids.[18] The liquid thermal conductivity estimated from Maxwell's correlation of the variation of thermal conductivity of petroleum liquids with temperature was 2.6×10^{-4} cal/cm.sec.°C at 670 °F.[19] The effective thermal conductivity can be approximated by taking a geometric average of the solid and liquid conductivities

$$k_e = k_s \left(\frac{k_f}{k_s} \right)^{\varepsilon_p} \quad (12)$$

where the subscripts s and f refer to the solid and fluid, and ε_p refers the to porosity of the catalyst.[12] The effective thermal conductivity, k_e, is the key parameter in determining how large the temperature rise within the catalyst pill is, a low thermal conductivity results in a larger temperature rise.

The temperature profile within the catalyst pellet is shown in Figure 3 for the non-isothermal case with k_e 10^{-4} cal/cm.sec.°C and the following bulk phase parameters: 670 °F, 700 psig, 1.4 wt-% easy sulfur, 0.1 wt-% C_3-DBT-A, and 5-70 wt-% poly-aromatics. The largest temperature rise was only ~1 °F, corresponding to 70 wt-% aromatics. This is consistent with the conclusions in the literature that the internal temperature gradient for heterogeneous catalyst systems involving liquid is low.[20] Based on this small temperature gradient the results of the calculations for the isothermal case apply in this case as well.

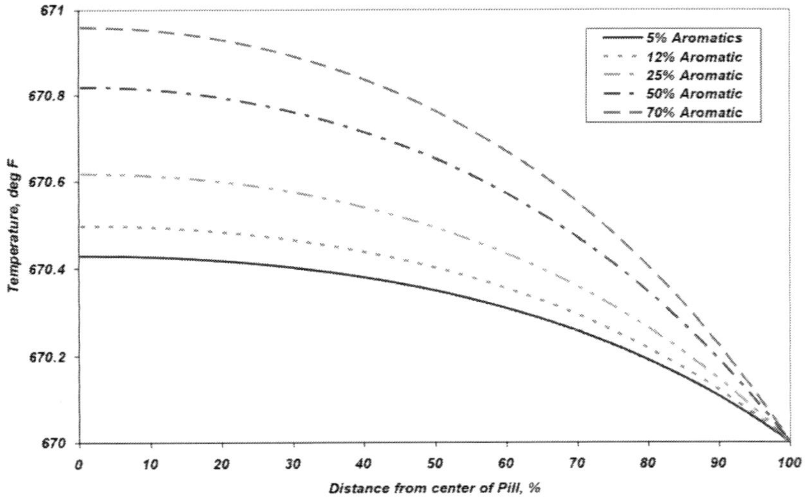

Figure 3: Intra-particle temperature profile for a non-isothermal catalyst pellet in the presence of 1.4 wt% sulfur and varied poly-aromatic concentration. (k_e = 10^{-4} cal/cm.sec.°C, T_b = 670 °F, P = 700 psig)

[18] Lee, Hong H., *Heterogeneous Reactor Design*, Boston, Butterworth Publishers, 1985, 112.
[19] Maxwell, J. B. *Data Book on Hydrocarbonsi,* New York, D. Van Nostrand Company, Inc., 1950, 207-213.
[20] Trambouze, P., H. Van Landenghem, J.P. Wauquier, *Chemical Reactors: Design/Engineering/Operation*, Houston, gulf Publishing Company, 1988, 363.

Aromatic Effects in Pilot Plant Data

A pilot plant study was conducted to quantify the effects of poly-aromatics on HDS. Two feedstocks were processed over NiMo catalyst at 700 psig, LHSV 1.4 hr^{-1}, and H_2/Oil 2000 SCF/bbl. The two feedstocks were 50/50 blends of a straight run distillate one with hydrotreated products from past pilot plant work (HDT Products) and the other with a low sulfur LCO (LS LCO). The SR/LS LCO blend had 31.4 wt-% poly-aromatics compared to 7.3 wt-% poly-aromatics in the SR/HDT product blend. The properties of the components used for blending and feedstocks are shown in Table 1.

Table 1: Blend component and feedstock properties.

	Straight Run	**HDT Products***	**LS LCO**	**Straight Run + HDT Product**	**Straight Run + LS LCO**
Gravity, API	33.73	39.28	21.39	36.32	27.44
Total Sulfur	1.94 wt-%	29 ppm-wt	612 ppm-wt	0.85 wt-%	0.89 wt-%
4,6-DMDBT, ppm-wt	-	-	-	63	74
C_3-DBT-A, ppm-wt	-	-	-	72	76
Nitrogen, ppm-wt	332	2	97	153	201
1-Ring Aromatics, wt-%	16.8	19.9	19.5	17.4	18.3
2-Ring Aromatics, wt-%	13.2	1.6	37.9	6.7	27.4
Poly Aromatics, wt-%	2.7	0.2	6.1	0.6	4.0
Distillation, °F					
IBP / 5wt%	294 / 417	258 / 421	103 / 330	222 / 353	183 / 368
50 / 70	573 / 611	544 / 575	500 / 548	535 / 586	484 / 537
95 / EBP	683 / 735	642 / 652	643 / 710	670 / 729	671 / 720

*The properties shown for the HDT products represent the average of individual test analyses.

The impact of the poly-aromatics was seen most clearly when the product effluent sulfurs of the SR/HDT Product blend (low poly-aromatic) and the SR/LS LCO blend (high poly-aromatic) were compared, Figure 4. With the feed sulfur and nitrogen essentially the same, the products had significantly different sulfur concentrations. The SR/LS LCO product sulfur was 2-3 times higher than the HDT product blend over the entire temperature range. Sulfur simulated distillations were completed in order to determine the species of sulfur that were present. The reactor effluent concentrations of easy sulfur (boiling <610 °F), 4,6-DMDBT, and C_3–DBT-A are shown in Figure 5 for both blends processed at the same conditions. Both of the strerically hindered sulfur species (4,6-DMDBT and C_3–DBT-A) were significantly impacted by the increased aromatics presence. The concentrations of the DBTs approximately double! The conversion of 4,6-DMDBT was ~11% lower for the SR/LS LCO blend than the SR/HDT product blend. Similarly, the C_3-DBT-A conversion dropped ~16% in presence of the higher aromatics concentration. The concentration of easy sulfur nearly tripled as a result of the increased aromatics concentration; however, this change in concentration corresponded to a <1% change in conversion.

Although the increased hydrogen consumption in the presence of aromatics may influence the HDS rates via competition for hydrogen the diffusion/reaction pill calculations and pilot plant data indicate that it is unlikely to be the sole mechanism of the aromatics influence. The effectiveness factors did not decrease sufficiently as the aromatic concentration increased to account for the differences in the reactor effluent sulfur concentrations. Furthermore, the presence of the poly-aromatics had a more dramatic impact on the slow reacting compounds (substituted DBTs) implying that inhibition (competition for catalyst sites) was the primary mechanism of the aromatic interference with HDS. If pore diffusion resulting in competition for hydrogen was the primary mechanism of HDS-HDA interactions the faster reaction would be expected to be impacted more significantly than the slower reactions. Nevertheless, pore diffusion should not be completely dismissed; it may have significant consequences for ULSD production. The impact of the aromatics on non-sterically hindered (fast reacting) sulfur species may have been small in absolute terms; however, at 10 ppm-wt sulfur even minor influences have monumental consequences.

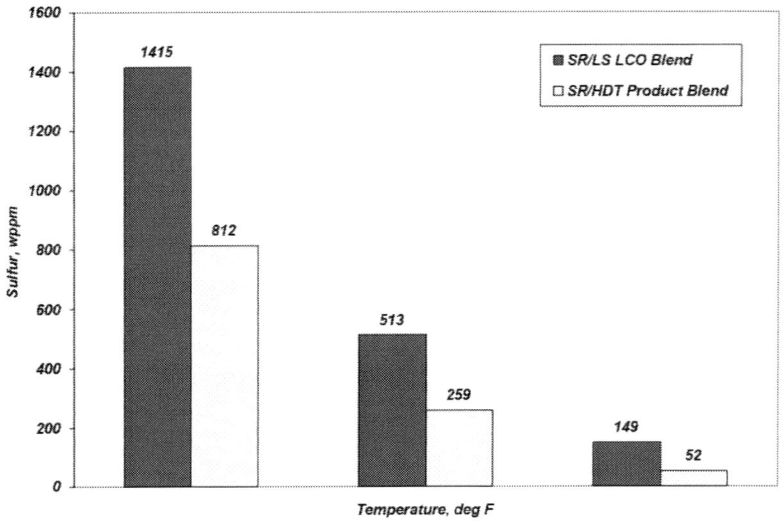

Figure 4: Comparison of reactor effluent sulfur concentration for LS LCO/SR and SR/HDT product blends, LHSV=1.4 hr^{-1}.

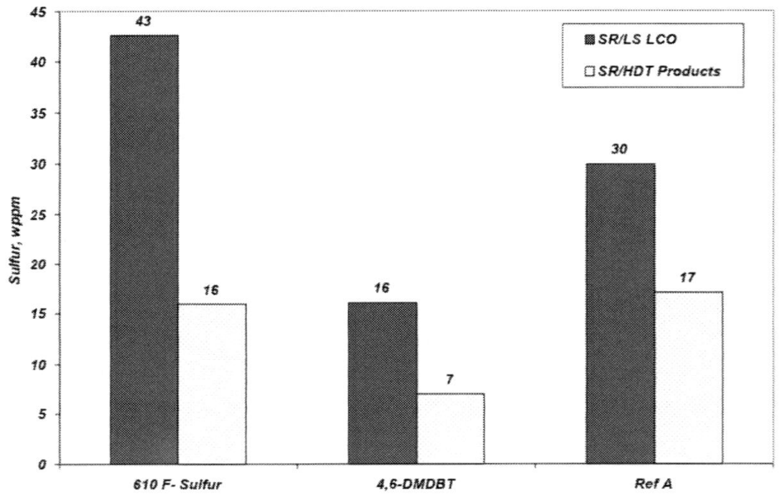

Figure 5: Reactor effluent sulfur speciation as measured by sulfur simulated distillation for SR/LS LCO and SR/HDT Product blends.

Inhibition by Aromatics

Inhibition terms (competitive adsorption constants) were introduced into the ULSD hydrotreating model in order to investigate the consequences of aromatics inhibition. The predicted concentration profile of 4,6-DMDBT was plotted as a function of percent of the bed length at various poly-aromatic concentrations, Figure 6. At the front end of the reactor the nitrogen inhibition dominates and as a result the effects of the aromatics are not observed. However, the curves diverge as the nitrogen drops below 500 ppm-wt. With aromatic inhibition included in the model the SOR temperature required for ULSD production can be predicted at various aromatics concentrations, shown in Figure 7 for a feedstock with 1.227 wt-% total sulfur at 900 psig. The temperature required with 0 wt-% aromatics is analogous to having no aromatic inhibition. By increasing the poly-aromatic concentration to 70 wt-% the start of run temperature required increased by 42 °F. The strength of aromatics inhibition is thought to be species dependent. As the number of fused rings increases the adsorption affinity dramatically increases. As a result, application of aromatic inhibition in modeling must be done with caution; effects may be exaggerated or understated depending on the distribution of the aromatic species present in the feedstock.

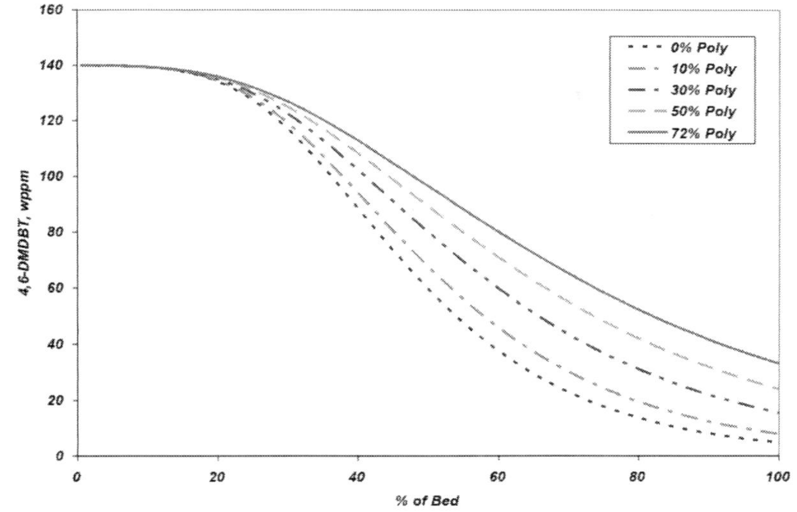

Figure 6: 4,6-DMDBT reactor profile predicted by UOP ULSD kinetic model modified to include poly-aromatic inhibition. Model feedstock contained 9249 wppm easy sulfur and varied poly-aromatic concentrations. Operating conditions were 700 psig and LHSV 1.5 hr^{-1}.

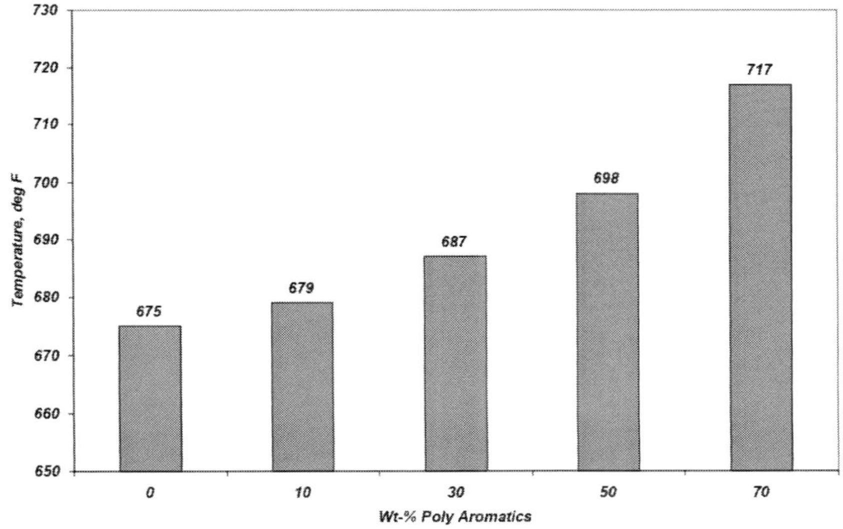

Figure 7: Temperatures required to achieve 10 wppm sulfur predicted by UOP ULSD kinetic model including aromatic inhibition at various poly aromatic concentrations, 900 psig.

Aromatic Inhibition vs. Nitrogen Inhibition

If aromatic inhibition exists the question still remains, why is the concentration of polycyclic aromatics needed to noticeably slow HDS so large relative to that of nitrogen? The experimental data appear to support the idea that HDS is inhibited by polycyclic aromatic compounds; however, the inhibition is weak compared to the inhibition by nitrogen. For 4,6-DMDBT the aromatic inhibition term was nearly two orders of magnitude lower than the analogous term for organo-nitrogen. In a study comparing inhibition by naphthalene and 3-ethylcarbazole

(3ECBZ) Ho found that 3ECBZ had a dramatic impact on the conversion of 4.6-DMDBT while naphthalene had a minor yet not negligible effect.[21]

Adsorption of Sulfur, Nitrogen, and Aromatics

A factor that influences the degree of inhibition is the relative affinity of the molecules for adsorption onto the catalyst sites. The equilibrium adsorption constants of the molecules competing for the available catalyst sites are directly proportional to the percent of the surface that is covered by a particular species at equilibrium. The adsorption constants shown in Figure 8 for naphthalene, phenanthrene, and carbazole were calculated by LaVopa et al.[5] The DBT adsorption constant was estimated from the napthlene:DBT ratio reported by Girgis.[7] The adsorption constant for carbazole is an order of magnitude larger than the other compounds. As a result, organo-nitrogen is highly favored in the competition for catalyst sites. The difference in the magnitude of nitrogen and aromatics inhibition follows the same trend as the adsorption equilibrium. Ho also points to the differences in the adsorption as an explanation for the different inhibiting effects of organonitrogen and aromatics.[21,22]

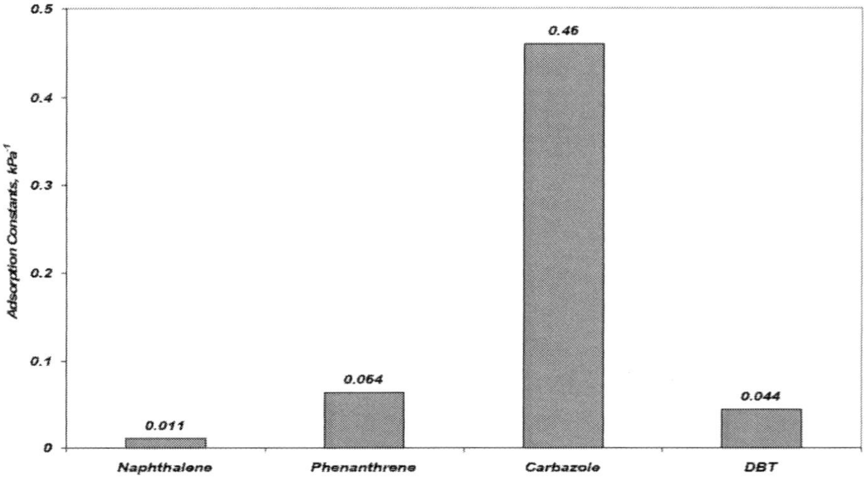

Figure 8: Adsorption constants for naphthalene, phenanthrene, carbozole, and dibenzothiophene (DBT).[5,7]

Catalyst Sites – Two Types

The adsorption constants are one parameter that influences the degree of inhibition; however, the existence of two types of catalyst sites (as discussed in the literature[2,7]) suggests a complementary hypothesis. Farag and colleagues treated 4,6-DMDBT over Co-Mo catalyst in the presence of 0, 5, and 10 wt-% naphthalene.[2] The HDS selectivity for the hydrogenation and direct desulfurization routes was measured by the product concentrations of 3,3'-dimethylphenylcyclohexane (HYD product) and 3,3'-dimethylbyphenyl (DDS product). Both the HYD and DDS desulfurization routes were retarded by the presence of naphthalene; however, the selectivity for DDS increased as the naphthalene concentration increased meaning that the HYD route was more significantly inhibited. Furthermore, in the presence of H_2S the DDS activity decreased sharply while HYD increased. The combination of the H_2S and naphthalene data are strong evidence that there are two types of catalyst sites, a hydrogentation type at which both DDS and HYD can occur and a second type where only DDS proceed. Assuming that HDN requires both types of catalyst sites, nitrogen species compete with the sulfur for ALL of the catalyst sites. Aromatic saturation, on the other hand, will only occur at hydrogenation type sites. As a result, unlike nitrogen, aromatics only compete with sulfur for the hydrogenation type sites; therefore, aromatic inhibition is expected to be weaker than nitrogen inhibition.

---

[21] Ho, The C., Inhibiting effects in hydrodesulfurzation of 4,6-diethyldibenzothiophene, *Journal of Catalysis*, 219, 442-451, 2003.

[22] Ho, The C., Property-reactivity correlation for HDS of middle distillates, *Applied Catalysis A: General*, 244, 115-128, 2003.

Conclusions

Based on the work that has been completed to date it is clear that the presence of poly-cyclic aromatics influences the rate of hydrodesulfurization. Although the influences of the competition for hydrogen can not be dismissed completely, there is stronger evidence for inhibition by poly-aromatics (competition for catalyst sites). The diffusion of hydrogen into the catalyst pellet is fast enough to prevent the concentration from decreasing below 65% of the bulk concentration. The stoichiometric hydrogen requirement increases with increasing aromatics concentrations. However, the supply of hydrogen, i.e. the bulk concentration in the liquid phase, is constant assuming that it is continually replenished to the saturation limit for hydrogen as a result of the contact with the gas phase. The slight decreases in the HDS and HDA effectiveness factors did not appear to be significant enough to account for the results obtained from experiments with varied feed concentrations of aromatics. The magnitude of aromatics inhibition was small relative to inhibition by organo-nitrogen; however, an inspection of the adsorption constants and the types of catalysts sites available provides possible explanations for the differences. The introduction of poly-aromatic inhibition terms to the HDS kinetic model demonstrated that aromatics impact the operating conditions required for ULSD production. The model results demonstrated that the impact of aromatics can not be ignored.

In the end, it is clear that HDS is inhibited by poly-aromatics; however, the significance of the influence of the competition for hydrogen on the observed HDS rate remains unresolved. It may not be possible to distinguish the difference between the competition for catalyst sites and hydrogen. If the primary concern is whether or not HDS activity in the presence of aromatics can be modeled accurately, is it necessary to fully understand the interaction of HDS and HDA to model that interaction? No. If a fixed bed reactor is simulated with a series of stirred tanks, data may be generated showing decreased desulfurization activity as the aromatic concentration was increased as a result of the competition for hydrogen, Figure 9 (▲). The data generated by reactions competing for a co-reactant (H_2) were fit with an inhibition term modeling competitive adsorption onto a catalyst surface, shown by the lines in Figure 9. The close agreement between the data (generated by competition for hydrogen) and the fit to this data using an inhibition term in the model leads to the conclusion that the precise cause of the depression of the desulfurization rate by the presence of poly-aromatics is not critical for modeling purposes.

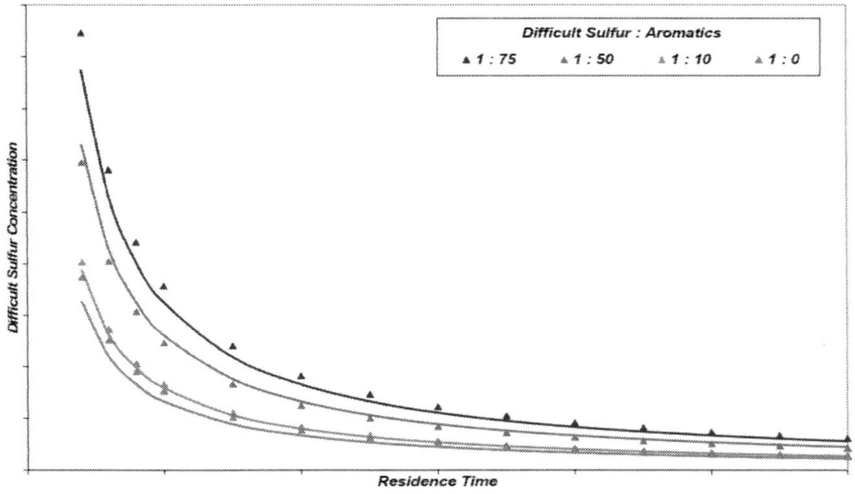

Figure 9: Data generated by modeling competition for hydrogen in a series of stirred tanks (▲), modeled by aromatic inhibition neglecting competition for hydrogen (–).

CATALYST EFFECTIVENESS FACTOR DURING DEEP HDS OF LCO

Jinwen Chen, Hong Yang, and Zbigniew Ring

National Centre for Upgrading Technology (NCUT), One Oil Patch Drive
Devon, Alberta, Canada, T9G 1A8

Prepared for Presentation at the 2004 AIChE National Spring Meeting, Session 90: Advances in Hydro/Desulphurization, April 25-29, 2004, New Orleans, LA, USA

April 2004

ABSTRACT

A comparative HDS study was conducted in a pilot plant hydrotreater with a full-size commercial NiMo/Al_2O_3 hydrotreating catalyst (extrudates of 1.6 mm in diameter and 5 mm in length) and its crushed catalyst particles (35-60 mesh) under similar operating conditions. The feed used was light cycle oil from a commercial fluid catalytic cracking unit. Identification and quantification of the hardest-to-remove sulphur compounds (i.e. dibenzothiophene and its alkyl substituted derivatives) in the feed and hydrotreated products was conducted using gas-chromatograph atomic-emission detector. It was found that in the full-size commercial catalyst, significant intra-particle diffusion resistance limited the overall rate of sulphur removal for all the dibenzothiophenic compounds investigated. The effectiveness factors ranged from only 0.300 to 0.822 in the temperature range of commercial importance studied here. In the crushed catalyst, with the effectiveness factors between 0.813 and 0.985, the intra-particle diffusion effect was almost eliminated. As expected, the more reactive sulphur compounds with intrinsic reactivities that were little affected by the steric hindrance of methyl and/or ethyl substituents had lover effectiveness factors because of stronger intra-particle diffusion resistance. The less reactive sulphur compounds, with reactivities hindered to a significant degree, had higher effective factors.

1. INTRODUCTION

According to the latest environmental regulations, sulphur content in diesel and other transportation fuels worldwide has to be significantly reduced from current levels in order to further reduce SO_2 pollution caused by engine exhaust emissions (Song, 2003; Plantenga, et al., 2002). Although, hydrodesulphurization (HDS) is still the dominant refining technology used to remove sulphur from petroleum fractions (Song, 2003), alternative technologies are being studied and developed, particularly to reach new stringent sulphur specifications in finishing fuel blending stocks (Babich and Moulijn, 2003; Levy, et al., 2002).

Catalyst particles with diameters of 0.8 to 1.6 mm and average lengths of 5 – 6 mm are commonly used in commercial trickle-bed hydrotreating/HDS reactors. This range of catalyst sizes reflects the tradeoff between the desires to provide high catalyst utilization (high catalyst effectiveness factors) on one hand, and to maintain a manageable pressure drop in the reactor on the other. In this range of catalyst particle sizes, the intra-particle diffusion resistance significantly reduces the overall HDS reaction rate. Furthermore, the intra-particle diffusion effect varies with operating conditions (particularly with temperature) and different sulphur compounds. This issue has not been adequately addressed in the vast HDS literature, although many HDS studies have been conducted separately with either full-size catalysts (Chen and Ring, 2004a; Chen et al., 2003a; Kumar et al., 2001; Lecrenay et al., 1997) or crushed catalysts (Kabe et al., 1997; Vanrysselberghe and Froment, 1998).

This study focused on the experimental determination of the effectiveness factors for different dibenzothiophenes (DBTs) commonly found in cracked light distillates. It was conducted in support of the National Centre for Upgrading Technology's (NCUT's) hydroprocessing model development activities, but serves here as a reminder of the important effect of physical phenomena on desulphurization rates. These phenomena, including intra-particle diffusion (dealt with in this paper) and phase equilibrium in the reactor (Chen and Ring, 2004; Ring et al., 2004) are usually neglected in the very large number of papers dealing with the kinetics analysis of HDS reactions and even some commercial hydroprocessing models.

2. EXPERIMENTAL

This experimental program was performed in a pilot plant hydrotreating unit. A detailed description of the experimental setup, procedures, and analytical methods used have been provided elsewhere (Chen et al., 2003a; Chen and Ring, 2004a). Only a brief summary of the experimental program is given below.

The hydrotreating reactor, a stainless steel tube 1 m long and 2.54 cm in diameter (I.D.) was equipped with a 0.64 cm (O.D.) centrally located thermowell that housed 12 thermocouples for detailed measurement of the axial temperature profile in the reactor. Two hydrotreating experiments were conducted. In the first, the reactor was packed with 140 ml of full-size commercial NiMo/Al_2O_3 hydrotreating catalyst (extrudates 1.6 mm in diameter and, on average, 5 mm in length). This catalyst was diluted with 0.2 mm glass beads in the 1:1 volume ratio. The total catalyst bed height was 57 cm. In the second experiment, the reactor was packed with 118 ml of 35 – 60 mesh (0.25-0.5 mm) catalyst obtained by crushing the full-size catalyst. No glass beads were used to dilute the crushed catalyst. The bed height in the second experiment was 24 cm.

Both experiments were conducted with co-current down-flow of gas and liquid in the trickle-flow regime. Prior to the hydrotreating experiments, the fresh catalysts was sulphided and stabilized by processing the cracked feedstock for about 500 hours under operating conditions similar to those used in the experiments. A flat axial temperature profile was maintained throughout the program. The feedstock and products were analyzed for C and H, simulated distillation (SimDis), and density using standard ASTM methods. These analytical data, together with the process data acquired (i.e. the feed, product, hydrogen and vent gas rates, vent gas composition, etc.), were subsequently used to perform detailed carbon, hydrogen, and overall mass balance calculations. At each set of operating conditions at steady state, the liquid product was gathered during two 3-hour back-to-back mass balance periods. Since it was shown that this kind of experiments have routinely generated repeatable results (Chen et al., 2003a), to reduce the related analytical work at most of the conditions, a sample of only one liquid product was analysed in detail. The other sample was analyzed only when unexpected results were observed.

The key physical properties of the fluid catalytic cracking light cycle oil (LCO) feeds are shown in Table 1.

Total nitrogen and sulphur in the feed and hydrotreated effluents were analyzed with a DOHRMAN-chemiluminescence instrument (ASYM D4629) and a gas chromatography with a sulphur chemiluminescence detector (GC-SCD, ASTM D5623), respectively. The quantification of individual sulphur compounds in the feed and hydrotreated effluents was achieved with a gas chromatograph with an atomic emission detector (GC-AED, sulphur channel 181 nm) using the external-standard method. The identification of individual compounds was achieved by matching the retention times of their corresponding peaks with those of standard compounds collected in NCUT's database. More details on sulphur speciation and quantification can be found elsewhere (Chen et al., 2003a, Chen and Ring, 2004a). About 70 sulphur compounds were positively identified in the LCO feed. This paper focuses on 29 compounds including dibenzothiophene (DBT) and its one-, two-, or three-alkyl (methyl, ethyl) substituted DBTs.

Table 1. Composition and physical properties of the LCO feeds

	Feed 1 (used for full-size catalyst)	Feed 2 (used for crushed catalyst)
Density (15°C), g/ml	0.960	0.936
Carbon, wt%	89.13	88.20
Hydrogen, wt%	9.84	10.84
Total sulphur, wppm	8686.2	7072.8
Total nitrogen, wppm	730.9	660.7
Aromatics, wt%	83.76	71.27
Paraffins, wt%	6.80	8.91
Cycloparaffins, wt%	8.87	19.82
SimDis, °C		
IBP	113.8	128.4
10wt%	230.9	225
30wt%	258.8	255.3
50wt%	289.1	288.5
70wt%	328.7	326.4
90wt%	373.5	372.4
FBP	431.9	430.8

This experimental study was conducted in the reaction temperature range of 350-400°C and liquid space velocities in the range of 1.58-4.5/h, while other conditions were fixed at the total pressure of 69.1 atm, gas-to-oil ratio of 1000 NL/kg, and H_2S concentration in the treat gas of 0.5vol%.

3. RESULTS AND DISCUSSIONS

3.1. Sulphur speciation

The sulphur chromatogram of one of the LCO feedstocks used in this study is shown in Figure 1. The individual peaks in the chromatogram mostly correspond to thiophenic, benzothiophenic, dibenzothiophenic, and benzo-naphthothiophenic (NTHs) compounds with various degrees of substitution and relatively short paraffinic chains at different locations. In general, the peaks with shorter retention times represent thiophenic and benzothiophenic compounds while those with longer retention times represent dibenzothiophenic and naphthothiophenic compounds.

Figures 2 and 3 show the sulphur chromatograms of the hydotreated products obtained at reactor temperature of 350°C with the full-size and the crushed catalyst beds, respectively. A comparison of the feed and effluent chromatograms indicates that the thiophenic, bezothiophenic compounds, and even DBT were completely removed in both reactor effluents but alkyl-substituted DBTs and NTHs persisted. Under similar operating conditions the crushed catalyst removed more sulphur (Figure 3) than the full-size catalyst (Figure 2) pointing to improved effectiveness of the crushed catalyst discussed in detail below.

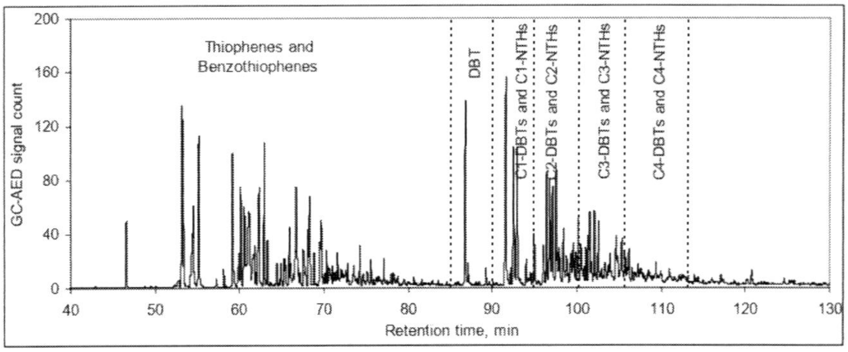

Figure 1. Sulphur chromatogram of one of the LCO feeds

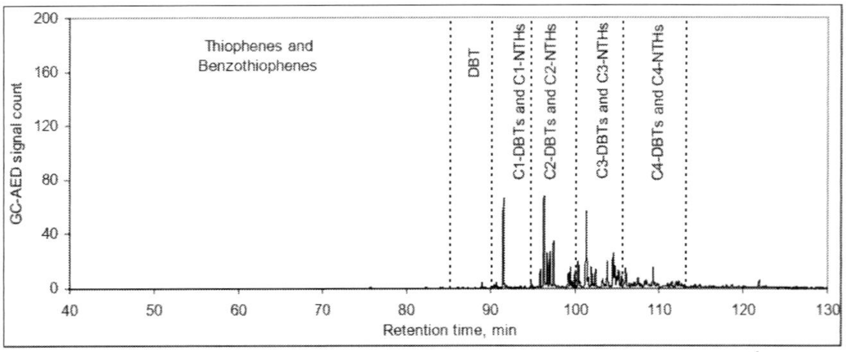

Figure 2. Sulphur chromatogram of effluent at 350°C, SV=1.58 h^{-1} (full-size catalyst)

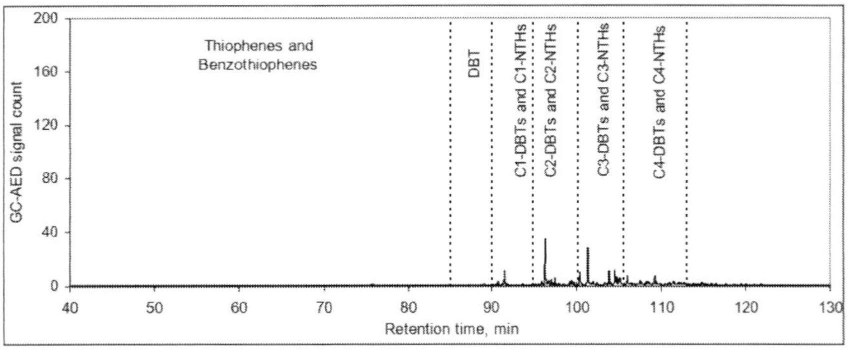

Figure 3. Sulphur chromatogram of effluent at 350°C, SV=1.70 h^{-1}(crushed catalyst)

3.2. Pseudo-First Order Kinetics

In this study, HDS experiments were performed at several temperatures ranging from 350 to 400°C for both full-size catalyst bed and crushed catalyst bed. As reported in our previous studies (Chen et al., 2003a; Chen and Ring 2004a), the total sulphur and nitrogen removal, as well as the removal of individual sulphur compounds, were well described by the pseudo-first order reaction kinetics up to 385°C. At higher temperatures, the effect of hydrogenation/dehydrogenation equilibrium modified by the effect of vapour-liquid phase equilibrium became significant (Ring et al., 2004; Chen and Ring 2004b) and the pseudo-first order kinetics were not applicable any more. In this case, more sophisticated kinetics models will have to be used in conjunction with detailed phase equilibrium treatment. We reported that the two-path HDS reaction kinetics model accounting for hydrogenolysis and hydrogenation was capable of partially describing the reduced temperature sensitivity of the reaction rate at higher temperatures (Chen et al., 2003b) even without modeling phase equilibrium. The results of modeling these data using the actual liquid-phase concentrations of individual sulphur compounds is currently underway and the results will be reported in the future.

Figures 4, 5 and 6 show the pseudo-first order reaction rate constants for HDS of total sulphur, 4-MDBT, and 2,4,8-TMDBT in the temperature range of 350 to 385°C with both full-size and crushed catalysts. The straight lines fitting the experimental points in the *Lnk* vs. *1/T* coordinates indicate that the pseudo-first order reaction assumption in this temperature range is reasonably good. At the same temperatures, the reaction rate constants obtained with the crushed catalyst is much higher than that corresponding to the full-size catalyst. This is an indication that substantial intra-particle diffusion limitation was present for the full-size catalyst.

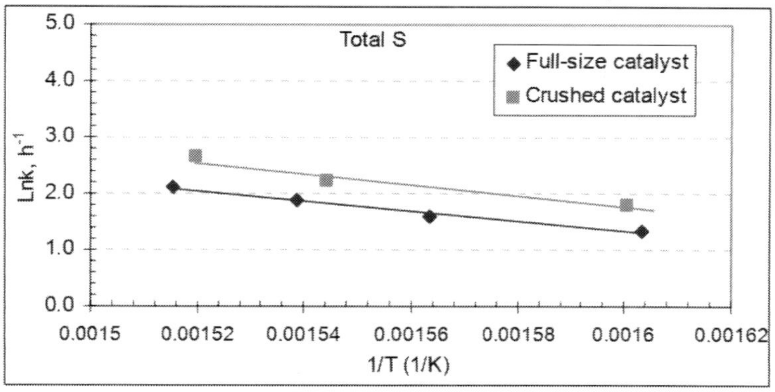

Figure 4. Pseudo-first order rate constants for total sulphur at different temperatures

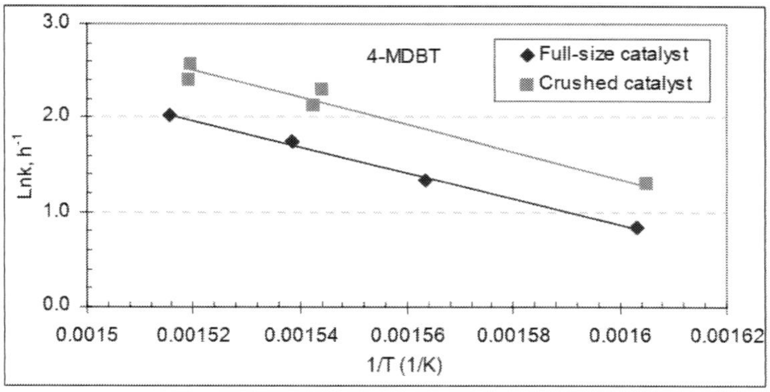

Figure 5. Pseudo-first order rate constants for 4-MDBT at different temperatures

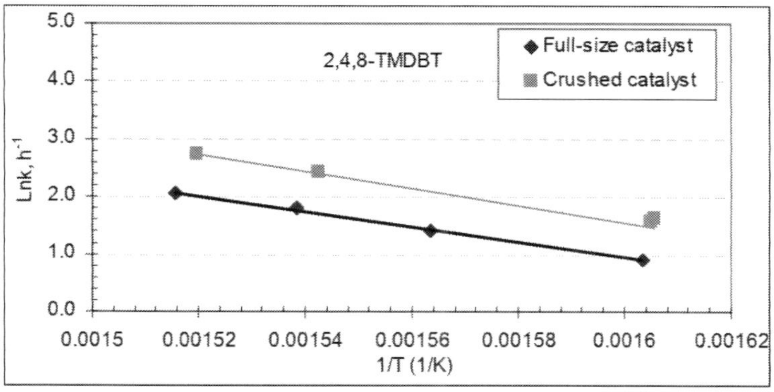

Figure 6. Pseudo-first order rate constants for 2,4,8-TMDBT at different temperatures

3.3. Catalyst Effectiveness Factors

Since it was not certain that the crushed catalyst particles used in this study were small enough to eliminate intra-particle diffusion effects in the 350 to 385°C temperature range, it was assumed that this limitation was present in both cases. Therefore, the catalyst effectiveness factors for catalysts of both sizes were calculated according to the following equations:

$$k = \eta k_{\text{int}\,rinsic} \quad (1)$$

where

$$\eta = \frac{I_1(2\phi)}{\phi I_0(2\phi)} \quad \text{(for cylinders of infinite length, full-size catalyst)} \quad (2)$$

$$\eta = \frac{1}{\phi}\left(\frac{1}{\tan k(3\phi)} - \frac{1}{3\phi}\right) \quad \text{(for spheres, crushed catalyst)} \quad (3)$$

$$\phi = \frac{V_p}{a_p}\sqrt{\frac{k}{D_e}} \quad (4)$$

In the above equations, k and $k_{intrinsic}$ are the apparent and intrinsic pseudo-first order reaction rate constants, η is the catalyst effectiveness factor, ϕ is the Thiele modulus, D_e is the effective diffusivity in the catalyst particle, and V_p and S_p are the volume and surface area of the catalyst particle, respectively. By applying the above equations to both the full-size and crushed catalyst data at the same temperature, $k_{intrinsic}$ and D_e were calculated for each individual sulphur compound and then the catalyst effectiveness factor was calculated for each.

The pseudo-first order reaction rate constants at different temperatures were calculated for each individual sulphur compound according to its concentrations in the feed and hydrotreated effluent, and the space velocity used. In order to filter out experimental errors and reliably estimate the individual effectiveness factors from the limited number of experimental points, the reaction rate constants at temperatures of 352, 362, 372, and 382°C were calculated by fitting the Arrhenius equation to the experimental reaction rate constant data calculated, as described above. In the process, the activation energies and the pre-exponential factors were also obtained.

Figure 7 shows the catalyst effectiveness factors at different temperatures for DBT, 4-MDBT, 4,6-DMDBT and 2,4,8-TMDBT. As expected, the effectiveness factors decrease with increasing temperature. This is because temperature has a larger effect on the intrinsic reaction rate constant than the intra-particle diffusion coefficient. The calculated effectiveness factors for each individual sulphur compound at different temperatures are listed in Table 2. For the 25 DBTs, the effectiveness factors ranged from 0.300 for 2,7-DMDBT to 0.822 for 4,6-DMDBT in the full-size catalyst, indicating significant intra-particle resistance is present for all the sulphur compounds. As expected, the intra-particle diffusion effect was much less pronounced in the crushed catalyst but it was not eliminated. The effectiveness factors exceeded 0.9 for most of the sulphur compounds, except for 2-MDBT, 3-MDBT, 2,7-DMDBT, 2,8-DMDBT and 3,7-DMDBT. DBTs with relatively higher HDS reactivities, such as DBT, 2- and 3-MDBT, and 2,7-DMDBT, had lower effectiveness factors than DBTs with relatively lower HDS reactivities such as 4,6-DMDT, 2,4,6-TMDBT, 3,4,6-TMDBT, and 4-E-6-M-DBT. This observation can be explained with Equation (4). The differences in the molecular size among all the DBTs examined in this study were relatively small, and therefore the differences in their

Table 2. Catalyst effectiveness factors for total sulphur, total nitrogen and individual sulphur compounds

Sulphur compound	Effectiveness factor, full size catalyst				Effectiveness factor, crushed catalyst			
	T=352°C	T=362°C	T=372°C	T=382°C	T=352°C	T=362°C	T=372°C	T=382°C
Total S	0.633	0.615	0.598	0.582	0.959	0.956	0.953	0.949
Total N	0.750	0.731	0.713	0.696	0.977	0.975	0.972	0.970
DBT	0.484	0.472	0.460	0.449	0.922	0.917	0.913	0.908
1-MDBT	0.524	0.494	0.466	0.440	0.934	0.925	0.915	0.904
2-MDBT	0.387				0.878			
3-MDBT	0.374				0.870			
4-MDBT	0.595	0.582	0.568	0.556	0.952	0.949	0.946	0.943
1,4-DMDBT	0.644	0.615	0.589	0.563	0.961	0.956	0.950	0.945
1,6-DMDBT	0.631	0.619	0.606	0.595	0.959	0.957	0.954	0.952
2,4-DMDBT	0.564	0.528	0.494	0.462	0.945	0.935	0.925	0.914
2,6-DMDBT	0.644	0.598	0.557	0.518	0.961	0.953	0.943	0.933
2,7-DMDBT	0.300				0.813			
2,8-DMDBT	0.390	0.515	0.489	0.464	0.880			
3,4-DMDBT	0.544	0.532	0.507	0.483	0.940	0.932	0.923	0.914
3,6-EDBT	0.559				0.943	0.936	0.929	0.921
3,7-DMDBT	0.364				0.865			
4,6-DMDBT	0.822	0.796	0.771	0.748	0.985	0.983	0.980	0.977
4-EDBT	0.696	0.629	0.570	0.516	0.970	0.959	0.946	0.932
1,3,7-TMDBT	0.588	0.572	0.556	0.541	0.950	0.947	0.943	0.939
1,4,6-TMDBT	0.689	0.675	0.663	0.651	0.969	0.967	0.965	0.963
1,4,7-TMDBT	0.731	0.699	0.670	0.642	0.975	0.970	0.966	0.961
2,4,6-TMDBT	0.734	0.717	0.700	0.684	0.975	0.973	0.970	0.968
2,4,7-TMDBT	0.539	0.487	0.439	0.397	0.938	0.922	0.904	0.884
2,4,8-TMDBT	0.502	0.481	0.460	0.441	0.928	0.920	0.913	0.905
3,4,6-TMDBT	0.780	0.766	0.752	0.739	0.981	0.979	0.978	0.976
3,4,7-TMDBT	0.710	0.678	0.647	0.619	0.972	0.967	0.962	0.957
4-E-6-M-DBT	0.758	0.716	0.676	0.640	0.978	0.973	0.967	0.961

intra-particle diffusivities (D_es) were also small. However, these DBTs had very different intrinsic kinetics rate constants ($k_{intrinsic}$) as observed in this study as well as in our previous studies (Chen et al., 2003a; Chen and Ring, 2004a). DBTs with no or little steric hindrance such as DBT, 2- and 3-MDBT, and 2,7-DMDBT, had much higher $k_{intrinsic}$ than those with strong steric hindrance such as 4,6-DMDT, 2,4,6-TMDBT, and 3,4,6-TMDBT, and 4-E-6-M-DBT. Higher $k_{intrinsic}$ resulted in higher Thiele modulus and, thus, lower catalyst effectiveness factor.

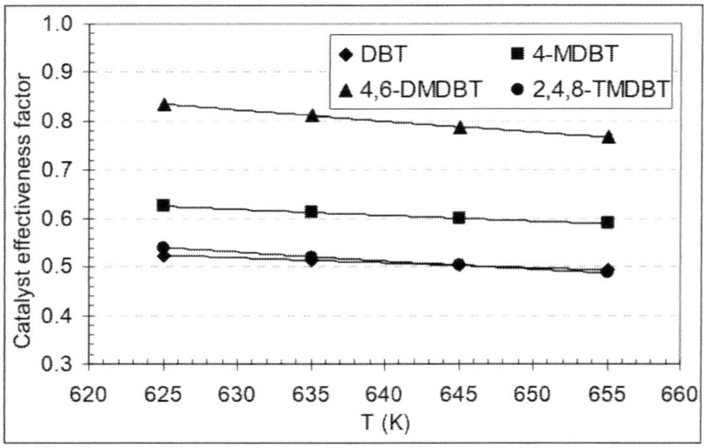

Figure 7. Catalyst effectiveness factors for DBT, 4-MDBT, 4,6-MDBT, and 2,4,8-TMDBT at different temperatures

4. CONCLUSIONS

A comparative HDS study was conducted in a pilot plant hydrotreater with a commercial hydrotreating catalyst in the forms of commercial-size extrudates and crushed particles under similar operating conditions. Significant intra-particle diffusion limitations were present in the full-size commercial catalyst for all the DBTs investigated in this study. The corresponding catalyst effectiveness factors for these DBTs ranged between 0.300 and 0.822 in the temperature range studied. The intra-particle diffusion resistance was much less pronounced but present in the crushed catalyst. The effectiveness factors for the crushed catalyst ranged from 0.813 to 0.985 in the same temperature range. As expected, the catalyst effectiveness factors decreased with temperature. DBTs with little or no steric hindrance had lower effectiveness factors (stronger intra-particle diffusion effects), whereas DBTs with strong steric hindrance had higher effective factors (weaker intra-particle diffusion effects).

ACKNOWLEDGEMENT

The authors are grateful to the following individuals for their help and support: Renata Szynkarczuk and Dennis Carson for conducting the pilot plant experiments, Yevgenia Briker and Cecile Lay for sulphur speciation analysis, and Norm Sacuta for paper editing. Partial funding for NCUT has been provided by the Canadian Program for Energy Research and Development (PERD), the Alberta Research Council (ARC) and the Alberta Energy Research Institute (AERI).

REFERENCES

Babich, I. V. and Moulijn, J. A., "Science and Technology of Novel Processes for Deep Desulfurization of Oil Refinery Streams: A Review", Fuel, 82(6), 607-632, 2003.

Chen, J., Te, M., Yang, H., and Ring, R. "Hydrodesulphurization of Dibenzothiophenic Compounds in a Light Cycle Oil", Petrol. Sci. Tech., 21(5&6), 911-935, 2003a.

Chen, J., Yang, H., and Ring, Z. "HDS Kinetics of Refractory Sulphur Species in Bitumen Derived Feedstocks", paper collected in the proceedings of 2003 AIChE Spring Meeting "6th International Conference of Refinery and Second International Aromatics Processing", Page 490-499, March 30 – April 3, 2003b, New Orleans, LA, USA.

Chen, J. and Ring, Z., "HDS Reactivities of Dibenzothiophenic Compounds in a LC-Finer LGO and H_2S/NH_3 Inhibition Effects", Fuel, 83, 305-313, 2004a.

Chen, J. and Ring, Z, "Vapor-Liquid Phase Equilibrium During Hydrotreating of Light Cycle Oil", presentation at 2004 AIChE Spring Meeting, New Orleans, LA, USA, April 25-29, 2004b.

Kabe, T., Akamatsu, K., Ishihara, A., Otsuki, S., Godo, M., Zhang, Q., and Qian, W., "Deep Hydrodesulphurization of Light Gas Oil. 1. Kinetics and Mechanisms of Dibenzothiophene Hydrodesulphurization", Ind. Eng. Chem. Res., 36, 5146-5152, 1997.

Kumar, V. R., Balaraman, K. S., Rao, V. S. R., and Ananth, M. S., "Performance Study of Centain Commercial Catalysts in Hydrodesulphurization of Diesel Oils", Petrol. Sci. Tech., 19, 1029-1038, 2001.

Lecrenay, E., Sakanishi, K., and Mochida, I., "Catalytic Hydrodesulphurization of Gas Oil and Model Sulphur Compounds over Commercial and Laboratory-made CoMo and NiMo Catalysts: Activity and Reaction Scheme", Catal. Today, 39, 13-20, 1997.

Levy, R. E., Rappas, A. S., Lee, F-M, Nero, V. P., DeCanio, S. J., "Non Conventional Desulfurization", Hydro. Eng., 7(7), 25-28, 2002.

Plantenga, F. L., Brevoord, E., Mayo, S., Inoue, Y., and Tokumoto, H., "Production of Ultra Low Sulfur Diesel with STARS™ & NEBULA™ Catalysts" Oil Gas European Magazine, 3, 37-38, 2002.

Ring, Z., Chen, J., Yang H., and Du, H., "HDS – An Old Topic with a New Approach", presentation and paper in proceedings of 2004 AIChE Spring Meeting (7th International Conference on Refinery Processing), New Orleans, LA, USA, April 25-29, 2004.

Song, C., "An Overview of New Approaches to Deep Desulfurization for Ultra-Clean Gasoline, Diesel Fuel and Jet Fuel", Catal. Today, 86, 211-263, 2003.

Vanrysselberghe, V. and Froment, G. F., "Kinetics Modeling of Hydrodesulfurization of Oil fractions: Light Cycle Oil", Ind. Eng. Chem. Res., 37, 4231-4240, 1998.

TransFining™ – A NOVEL PROCESS FOR LOW SULFUR FUELS

Robert C. Schucker

Trans Ionics Corporation
2408 Timberloch Place
Suite D-5
The Woodlands, TX 77380-1020
rcschucker@transionics.com

Prepared for Presentation at the 2004 Spring Meeting of the
American Institute of Chemical Engineers
April 25-29, 2004
New Orleans, LA

Seventh Topical Conference on Refinery Processing
Advances in Hydrodesulfurization

Abstract

The production of ultra low sulfur (ULS) gasoline and diesel, at a cost that is sustainable long term and in quantities that will continue to meet the demands of U. S. drivers, is potentially the single most significant and costly challenge that will be faced by the refining industry during this decade. On one hand, sulfur in fuels is a known problem. It is not only directly responsible for increased levels of NOx in exhaust gases; but it is also a poison for many of the newer catalytic converters that are being introduced by automakers. Paradoxically, supplies of crude oils are becoming increasingly heavier and more sour, thus requiring more processing. As a result, refiners are already having to hydrotreat feeds more severely just to stay at the current sulfur levels.

Considerable effort is being expended by refiners and other organizations to develop technology that will allow them to meet the mandated sulfur specifications of 30 ppm S for gasoline and 15 ppm S for diesel on time. New lower pressure naphtha (gasoline) hydrotreating technology is being developed and capital funds are already being allocated to build additional higher pressure distillate hydrotreaters. In addition, reactive distillation, adsorption and pervaporation-based membrane technologies have all been announced. There is general agreement in the industry that the major hurdle is not the initial reduction of sulfur in gasoline to 120 ppm and diesel to 100 ppm but rather, the reduction from those levels to 30 and 15 ppm, respectively. Based on observations in the European refining arena, the ultimate reality is probably zero sulfur in all fuels – a target that cannot be achieved economically using existing catalytic hydrotreating technology.

Trans Ionics Corporation, under SBIR grants from the Department of Energy, is developing a novel, non-catalytic process, called TransFining™, to effectively desulfurize liquids fuels to ultra low levels (<10 ppm S) at intermediate temperatures (~260°C) and low pressures (<300 psig H_2). Unlike conventional catalytic hydrotreating, TransFining™ also works well on fuel oil, vacuum residual oils and heavy crudes, making it an extremely flexible process. This paper will detail the results of recent experimental studies and preliminary economic analyses that show TransFining™ to be a more cost effective option than catalytic hydrotreating for low sulfur fuel oil and ultra low sulfur fuels.

1. Introduction

The production of ultra low sulfur (ULS) gasoline and diesel, at a cost that is sustainable long term and in quantities that will continue to meet the demands of U. S. drivers, is potentially the single most significant and costly challenge that will be faced by the refining industry during this decade. On one hand, sulfur in fuels is a known problem. It is not only directly responsible for increased levels of NO_x in exhaust gases; but it is also a poison for many of the newer catalytic converters that are being introduced by automakers. Paradoxically, supplies of crude oils are becoming increasingly heavier and more sour (higher sulfur levels) thus requiring more processing. As a result, refiners are already having to hydrotreat feeds more severely just to stay at the current sulfur levels.

Sulfur reduction in transportation fuels is primarily being driven by changes in governmental regulations worldwide. The U.S. and foreign governments have mandated scheduled reductions in sulfur for both gasoline and diesel, which many refiners will find difficult to meet. On December 21, 1999 the EPA issued its Tier 2 regulations requiring a reduction of sulfur in gasoline to an average of 30 ppm effective January 1, 2006. Diesel has been impacted to an even greater extent. On May 17, 2000, the EPA issued a Notice of Proposed Rulemaking calling for a 97% reduction in the maximum sulfur level in diesel (from 500 ppm to 15 ppm) effective June 2006. Since this is a maximum, realistically, this translates into an average of only 7-10 ppm S in diesel.

Considerable effort is being expended by refiners and other organizations to develop technology that will allow them to meet the mandated sulfur specifications on time. New lower pressure naphtha (gasoline) hydrotreating technology is being developed and capital funds are already being allocated to build additional higher pressure distillate hydrotreaters. In addition, new reactive distillation, adsorption and pervaporation-based membrane processes have also been announced. There is general agreement in the industry that the major hurdle is not the initial reduction of sulfur in gasoline to 120 ppm and diesel to 100 ppm but rather, the reduction from those levels to 30 and 15 ppm, respectively. Refiners today have basically two choices of technology for lowering the sulfur content of transportation fuels: (1) those based on hydrotreating and (2) those that are not. The applicability of these processes depends to a large extent on whether the feed stream is gasoline or diesel.

Trans Ionics Corporation of The Woodlands, TX has developed two processes for sulfur removal which fall into the non-hydrotreating category. TranSep™ is a patented membrane pervaporation process for removal of sulfur from gasoline boiling range feeds; and TransFining™ is a unique process that uses elemental sodium to remove sulfur from feeds ranging in boiling point from distillate to heavy oil and bitumen. This paper will focus on the development of TransFining™ for desulfurization of a range of feeds.

2. Background

The use of sodium metal to treat feeds was first suggested and patented seventy-five years ago [Hofass 1929]. The procedure employed metallic sodium at medium pressure (>700 psig) hydrogen and elevated temperatures (> 575 °F). This was followed by work at Standard Oil of Indiana [Ruthruff 1933], DuPont [Vose 1934] and a major effort by Esso Research and Engineering Co. [Kimberlin 1963, Welty 1974, Bearden and Fink 1974, Baird and Bearden

1977 and Bearden 1978] in the mid 1970's investigating the effects of sodium treating on residual oil.

In addition to the extensive work done on heavy oil desulfurization, sufficient work has been done on naphtha and diesel streams to show that desulfurization using sodium is also possible. Earlier work also done at Esso [Vanderbilt 1957] demonstrated that sulfur in light cat naphtha (LCN) could be reduced from 210 ppm to 2 ppm using 1 wt. % Na at 410 °F. Later work at Exxon Research and Development Laboratories [Howard et al. 1977] also showed that 90% desulfurization of LCN could be achieved with **no loss in RON** (a feat which naphtha hydrotreating technology still cannot accomplish).

Similar results are achievable with distillate fuels. Sternberg and coworkers [Sternberg et al. 1974] showed that 99% of sulfur could be removed from dibenzothiophene (a molecule typically found in diesel) at 200 psig hydrogen pressure and 350 °C.

The chemistry of sodium treating to remove sulfur is straightforward as shown in the reaction below:

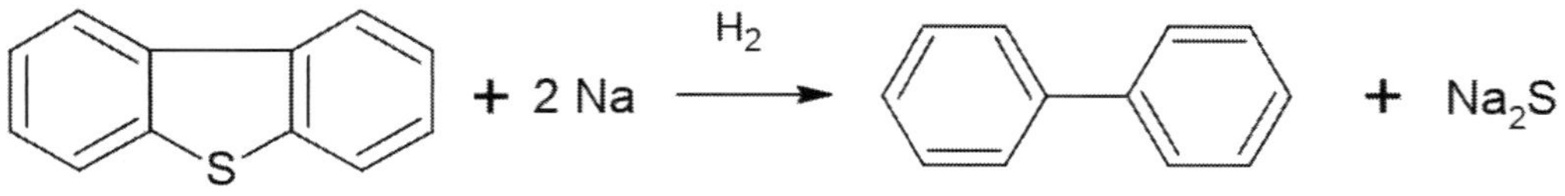

Because the sodium is a stoichiometric reagent and not a catalyst, it is consumed in the reaction producing one mole of Na_2S for every two moles of sodium consumed. More importantly, only one mole of hydrogen (H_2) is required for every mole of sulfur removed, leading to extremely low hydrogen consumption. Required pressure is low (< 300 psig) because it is simply necessary to have hydrogen available to stabilize the intermediates produced to keep them from undergoing recombination (which would produce a higher molecular weight product).

These data all show conclusively that heavy oils, naphtha and diesel can all be desulfurized under mild conditions by the use of alkali metals like sodium. So why then was this technology never commercialized? We believe the clear answer is economics. Even for trim desulfurization, the cost of sodium is too high to allow it to be used on a once-through basis (there is also a safety issue with the storage of large quantities of molten sodium and a further issue of recovering and disposing of the sodium sulfide product). In order to be economically viable, sodium must be regenerated and recycled to the feed. One way of achieving this was to use the sodium-sulfur battery developed by Ford Motor Company in 1969 just prior to the time this desulfurization work was being done in the mid 1970's. It was anticipated that Ford's battery research would result in an improved electrochemical cell that could economically and efficiently recover the sodium. Unfortunately this was not realized during that time; and the technology was never commercialized. Trans Ionics, however, has recently developed a novel and improved sodium sulfur battery that is the heart of its TransFining™ desulfurization process; and this paper is the first to describe both the process and technological innovations behind it.

3. The TransFining™ Process

TransFining™ has been in active development since early 2001 under the sponsorship of the U. S. Department of Energy. Since work done over the past thirty years has shown conclusively that sodium treating was capable of removing sulfur from a wide range of feeds, the primary focus of this work has been on the development of a new and improved sodium-sulfur battery to regenerate the active sodium reagent. A schematic of the overall process is shown in Figure 1.

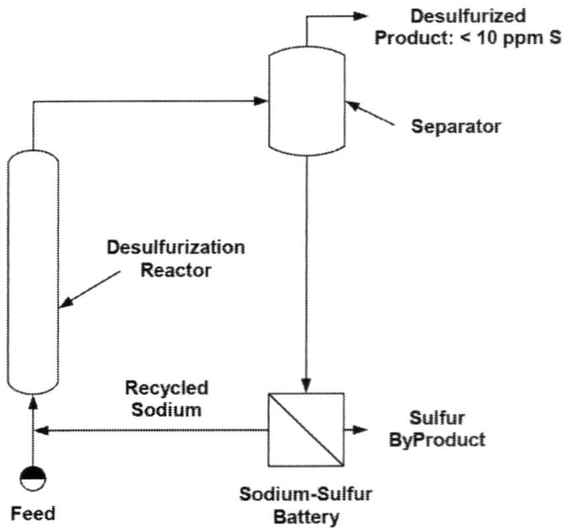

Figure 1. Schematic of TransFining™

In this process, feed is mixed with elemental sodium using a proprietary procedure and then passed through the desulfurization reactor. Experimental work has shown that the reaction takes place at temperatures above 260 °C (500 °F) and furthermore that the reaction is extremely rapid. This results in a finely dispersed Na_2S product which is separated from the desulfurized product in a proprietary separator and sent to the sodium-sulfur battery for processing. Sodium regenerated in this way is recycled to the feed stream and the other product (elemental sulfur) is recovered for disposal.

4. Sodium Sulfur Battery

The basic construction of a sodium-sulfur battery is shown below in Figure 2. In this design the two electrodes are liquid (liquid sulfur and liquid sodium) rather than solid as is the case in lead-acid batteries. The battery must thus operate at temperatures above the melting points of these two materials and traditionally has been run around 350 °C (661 °F). Liquid sodium is electrically conductive and acts as its own electrode. Liquid sulfur, unfortunately, is not; and that compartment requires a conductive material such as graphite gauze (or the like) to carry current.

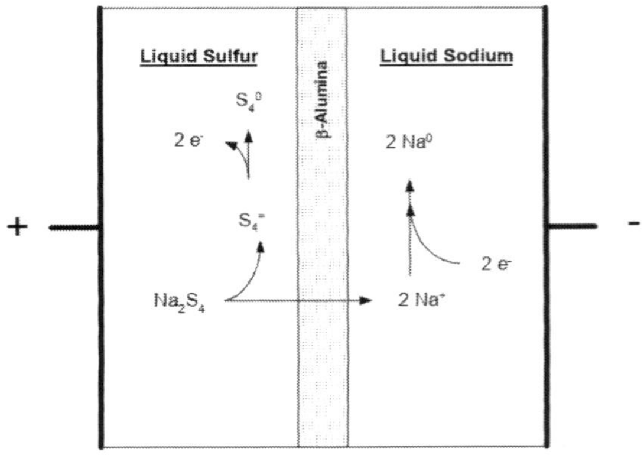

Figure 2. Sodium Sulfur Battery

The electrolyte dividing the two electrodes is solid β" - Al_2O_3, which conducts only sodium ions. Because, in the typical battery applications developed in past years, the electrolyte was usually self supporting, it was constrained to be thick enough to resist breakage (which could produce the dangerous mixing of sodium and sulfur). This means that electrolytes were normally 1-2 mm thick.

Under normal discharging conditions, sodium ions diffuse across the electrolyte and react with sulfur producing a complex mixture of sodium polysulfides and electricity at ~1.7 V. If a potential is applied instead, (charging mode at 2.1 V as shown in the drawing), the reaction is reversed, producing Na and S from sodium sulfide (which is the desired reaction for use in fuels desulfurization). Thus, this membrane, like all other membranes in which diffusion through the bulk is the rate-limiting step, would show marked improvement in sodium flux if the membrane electrolyte were significantly reduced in thickness.

The two major problems with sodium-sulfur batteries to date have been (1) their low power density (current per unit area expressed as A/cm^2) which is, in part, caused by the thickness of the electrolyte, and (2) their cost. Attempts to run at higher power densities in the past have resulted in a fractured β" - Al_2O_3 membrane.

The first commercial application of these batteries for electric utility loading leveling was in Japan at the Kawasaki Substation of its Kawasaki Electric Energy Storage Facility; and this system has been in operation since the end of 1998. Unfortunately, it cost ¥1.5 billion for only 500 kW (about $25,000/kW); and cost remains a major stumbling block to successful and widespread application of this technology.

5. Improved β"-Al_2O_3 Electrolyte

The key to Trans Ionics' sodium sulfur battery is the electrolyte. As mentioned previously, whereas most β"-Al_2O_3 electrolyte films made to date were relatively thick (800 – 2000 μ), the TransFining™ battery electrolyte is only 20 μ, a reduction of 40 to 100 over previous electrolytes. Since transport rate of sodium is inversely proportional to the film thickness, these films exhibit a higher transport rate than thicker films. An SEM photomicrograph of one of these films is shown in Figure 3. As can be seen, the electrolyte is fully dense (which is important since leakage of reactant through pinholes lowers the efficiency of the battery).

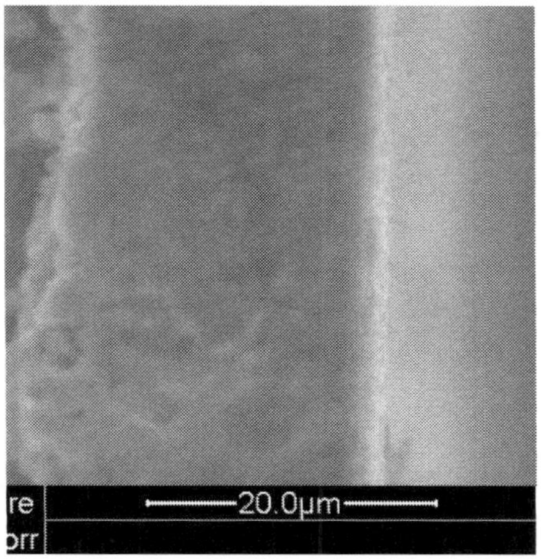

Figure 3. SEM Cross Section of Electrolyte

These films have been characterized by AC impedance spectroscopy and found to have conductivities equal to or better than those obtained by Ford Motor Company researchers who invented the sodium sulfur battery [Kummer and Weber 1971].

6. TransFining™ Economics

When viewed as an entire process, TransFining™ offers a number of significant advantages over conventional hydrotreating. The primary advantage is in capital and operating cost reduction. Since the process operates at < 300 psig and < 575 °F, there is no need for more expensive high temperature, higher pressure vessels. Secondly, since the

desulfurization reaction is extremely fast, the size of the primary reactor is reduced significantly. Residence times of 5 – 10 minutes are anticipated for complete conversion. Third, because the battery has a high efficiency, it is anticipated to be manufactured for a fraction of the cost of the Japanese batteries.
Operating cost reduction is also realized. Hydrogen consumption is only 1% of that used in conventional hydrotreating. With operation at lower temperatures and pressures, energy consumption is also significantly reduced.

A confidential study comparing TransFining™ to conventional hydrotreating for reduction of sulfur in an atmospheric residuum from 3.8 wt% to 0.1 wt% showed that the capital costs and annual operating costs were 38% and 35%, respectively, of conventional hydrotreating - a substantial reduction in both. Because of these attractive economic figures, TransFining™ is expected to find application in both refining operations for distillate, VGO and fuel oil desulfurization as well as in upstream heavy oil/bitumen upgrading in Canada and Venezuela. Trans Ionics Corporation anticipates having TransFining™ available for commercial sale in mid 2007.

Acknowledgements

We are grateful to the U. S. Department of Energy under SBIR grant DE-FG03-01ER83317 for funding to carry out a substantial amount of the development work to date.

References

Baird, W. C. and R. Bearden, "Desulfurization and Hydroconversion of Residua with Sodium Hydride and Hydrogen," *U. S. 4,003,824*, Jan. 18, 1977.

Bearden, R. and T. E. Fink; "Alkali Metal Desulfurization Process for Petroleum Oil Stocks Using Low Pressure Hydrogen," *U. S. 3,787,315*, Jan. 22, 1974.

Bearden, R.; "Combined Desulfurization and Conversion with Alkali Metals," *U. S. 4,076,613*, Feb. 28, 1978.

Hofass, M.; "Treatment by Pressure and Heat of Heavy Mineral Oils and Carbon Material," *U. S. 1,729,943*, Oct. 1, 1929.

Howard, K.A.; W. E. Winter; K. H. Moritz and J. D. Paynter; "Process for Desulfurization and Blending Naphtha," *U. S. 4,062,762*, Dec. 13, 1977.

Kimberlin, C. N,; W. J. Mattox and C. E. Adams; "Treatment of Petroleum residua with Sodium Alloys," *U. S. 3,093,575*, June 11, 1963.

Kummer, J.T. and N. Weber; "A Sodium-Sulfur Battery," Paper presented to MAP-PENNTAP Seminar, Philadelphia, PA, March 24, 1971.

Ruthruff, R. F.; "Desulphurizing Hydrocarbon Oils," *U. S. 1,938,672*, Dec. 12, 1933.

Schucker, Robert C., "Novel Electrochemical Membrane Reactor for the Desulfurization of Transportation Fuels," Small Business Innovation Research Grant DE-FG03-01ER83317, August 2001.

Sternberg, H. W.; C. L. Delle Donne; R. E. Markby and S. Friedman; "Reaction of Sodium with Dibenzothiophene. A Method for Desulfurization of Residua," *Ind. Eng. Chem., Proc. Des. Dev., 1974, 13*(4), 433-436.

Vanderbilt, B. M. "Desulfurization of and Refining of Naphthas by Metallic Sodium," *Ind. Eng. Chem., 1957, 49*(4), 696-703.

Vose, R. S.; "Process for Refining Hydrocarbon Oils," *U. S. 1,952,616*, Mar. 27, 1934.

Welty, A. B.: Process for the Desulfurization of Petroleum Oil Fractions," *U. S. 3,785,965*, Jan. 15, 1974.

WNiPd /$TiO_2AL_2O_3$ catalyst: hydrocracking selectivity in presence of aromatics

*Roberto Galiasso Tailleur and Juan A Ravigli Nascar**

PDVSA Intevep, Caracas 1010, Venezuela-galiassa@cantv.net

ABSTRACT

The production of a high quality diesel using one trough Hydrocracking process and a new WNiPd/TiO_2-Alumina (HCKI) catalyst has been analyzed to understand the impact of aromatics on selectivity. It was shown that HCKI behaves differently than WNi/Y-Alumina catalyst due to the presence of a different type of acid site. It was demonstrated that the former has higher selectivity into diesel fraction and lower susceptibility toward aromatics concentration than the later, and its products distribution are in between "amorphous" and zeolite type catalyst. In presence of aromatics the isomerization rates is increased as well as the survival of Di and Tri -branched-paraffin. It is proposed that most of the benefit in selectivity for HCKI catalyst is due to the Ti-Al-W metallic acid sites. A simplified model for the path reactions was outlined.

1. INTRODUCTION

The stringent regulation for Diesel fuels had increased the interest in One through Hydrocracking (OTHCK) process. The diesel high cetane number and a low cloud point are the key properties that these processes can control better. The forecast of diesel demand indicated that diesel production must be increased at rate higher than 3% per year during the next decade. From them the renewal interest of integrating an OTHCK process in the existing refinery where highly aromatics Vacuum Gas Oil are available. The main goal is to selectively crack the heavy gas oil paraffin in presence of large amount of aromatics [1]. In HCK the branching and cracking reactions occur on bifunctional partially wetted catalysts, in presence of a high hydrogen partial pressure. The most popular mechanism or reactions was proposed by Weisz [2] that pass through a carbenium ion that undergo skeletal isomerization and scission. Most of the previous isomerization Hydrocracking studies [3, 4, 5] have being performed by using gas or "liquid" phases conditions and pure paraffin, but without the presence of aromatics, sulfur, and nitrogen molecules likes occurred in the OTHCK of hydrodesulfurized vacuum gas oil (DVGO). Our research using commercial, and our own developed catalysts, with different DVGO feeds [1,6] and operating conditions indicated a low isomerization /cracking ratio in comparison with published results with pure normal and isoparaffin [3, 4, 5]. This paper is focused on studying the effect of aromatics on cracking and isomerization of paraffin using particular WNiPd on TiO_2-Alumina support treated with steam - ammonia.

2. EXPERIMENTAL PART

2.1 Pilot plant

The effects of temperature and space velocity on a HCK catalyst have been studied in a small-scale pilot plant (see detail about it in reference [1]). This unit has one 0.05liter down-flow fixed bed reactor that operates isothermally. The hydrogen and the hydrocarbon feed were preheated before entering the reactor. The DVGO with 350-530°C boiling range, as well as the blend of DVGO plus 15 or 30% aromatics added, were react at three space velocity (0.8, 1, 1,2 h^{-1}) and three temperatures (350, 360, and 370°C), but keeping the same pressure (12.4MPa) and H_2/HC ratio of 20 (molar). After reaction, the liquid product was fractionated in Naphtha, Diesel and non converted Vacuum Gas Oil (UVGO). The fractions were analyzed using GC (PONA) for Naphtha and a specially developed GC-MS program for Diesel and DVGO (temperature programmed using silica gel Gas Chromatography coupled with a low intensity mass spectrometry). This special analysis was able to discriminated families of aromatic, naphthen, and paraffin [7].

The other feed used in this study was prepared by blending n-C_{16} pure paraffin (Merck) with 30% of extract, plus 50wtppm of sulfur-as Dibenzothiophene -and 30wtppm of nitrogen -as Quinoline-. Again the same three temperatures and three space velocities mentioned above were explored. The products were fractionated in C_{16}^- and C_{16}^+ cuts. The hydrocarbons C_{16}^- were analyzed using a Varian Gas Chromatograph with two capillary columns (CP-Sil-5 QB and HI-PONA), a programmed temperature and a back flux system. In this way the conversion of n-C_{16} was first measured on the silica, and then the product distribution and the C_6-C_{10} isomers were determined using the PONA column. In addition

* Actual Address: Atanor SA, 327 Cordova, Buenos Aires, Argentine

NMR (nuclear magnetic resonance) analysis was used to determine an average degree of isomerization for heavy paraffin (C_{10}-C_{16}). The C_{16}^{+} cut was analyzed using the GC-MS method mentioned above.

The catalytic fixed bed was dried in air at 350°C, and then sulphided in situ using straight run gas oil where 1% of CS_2 was added at 300°C during 6 hours. The catalyst was used at particle diameter of 0.001m (cylinder), diluted with 50% inert material. This special precaution was taken to insure proper fluid dynamics according to our previous studies and those reported by De Brujin [8], between others. Based on previous experimental result [6], the trickle bed behaves as a plug flow isothermal reactor, with no inter or intra-particle mass and heat diffusion control of the reactions and with 0.75 wetting of the catalyst [9].

2.2 Aromatics extract preparation

The DVGO was used as feed to prepare the aromatics extract. A 100 kg was contacted with 5 kg of 98% purity sulfuric acid in a stirred tank reactor for 3 hours at 100°C at 0.1MPa. Then the reactor was cooled and 10 kg of sodium hydroxide had been added at room temperature, and then decanted for three days. Oiled phase had been removed from the reactor, and the remaining phase (sulphonated aromatics) was treated with 20 kg of water at 150°C and 0.1MPa for six hours. The reactor was cooled and then oiled phase had been removed, washed with three portions of fresh water till being sodium free, and analyzed by GC-MS method described above. Them they had being blended with original DVGO to increase aromatic content in the feed (see properties in Table 1)

In addition the extracted aromatics were diluted in n-C_{16} paraffin to obtain aromatics concentration of 30% by weight to perform fundamental studies about hydrocracking and hydrogenation kinetics.

2.3 Catalyst Preparation (HCKI and REF)

HCKI catalyst was prepared by co-precipitation of Aluminum gel and titanium sulfate (to have TiO_2 on Al_2O_3 ratio of 0.1 by weight in the final catalyst [10]). Then the powder was dried at 120°C and heated at 450°C in air. Pellet of 0.001-m diameter 0.005-m length was prepared using a controlled high-pressure extruder machine. The pellets were heated in air at 500°C, and then treated with ammonia and steam for three hours at 300°C. After that it was dried at 120°C, and active metals incorporated using two steps of incipient wetness impregnation: first with an aqueous solution of tungsten salt and second with other aqueous solution containing nickel and palladium salts. Then it was dried at 120°C and heated in air at 500°C for 5 hours. The HCKI catalyst was sulfide with diesel spiked with 2% of CS2 at 350°C for six hours. Final catalyst contains 6.2% of Sulfur, less than 0.1% of carbon and less than 50ppm of Na.

The Reference catalyst (REF: WNi/Y-Alumina) is a sample provided by a catalyst supplier and the characterization done under permission. It is only used for comparison purpose due to fundamental HCK studies were made with this type of zeolite. The Y zeolite has a Si to Al ratio of 10, and the pellets were presulphided in the same way that HCKI catalyst. It is important to notice that this catalyst doesn't contain Pd nor it was treated with steam-ammonia. Final sulphided REF catalyst has 6.5 wt% of sulfur.

2.3 Catalyst characterization

To understand the performances in the pilot plant unit the Hydrocracking and Reference catalysts were submitted to a series of analysis and test.

Chemical characterization: total metal analysis has being performed using atomic adsorption spectroscopy (Varian Techtron analyzer). Metals (bulk) were reported in % by weight of total metal oxides in the support (W, Ni, Al and Ti). See Table 1.

Physical method: Surface, pore volume, and average pores diameter were measured using standard nitrogen adsorption and mercury porosimetry method (Table 1)

XPS: Spectra were obtained in a Leybolh-Hereaus LHS-10 apparatus using AlKα with 50eV of power. XPS method has been used to assess the metal dispersion on supports using the peak area intensity (corrected) to measure atomic concentration, according to Deffosse et al [11] method. Binding energies in sulfide catalyst were $Ni2p_{1/2}$ at 854.2 and $Ni2p_{(3/2)}$ at 854.9eV, $W4f_{5/2}$ at 34.4 and $W_{7/2}$ at 32.5eV, Ti2p at 64.5eV and Al2p at 74.2eV. In this way W, Ni, Ti and Al were measured using the picks deconvolution to integrate the area and reported here as a ratio of metal to total metal in surface (Table2). Palladium was only detected as a large shoulder and don't measured.

ATPD. Ammonia Thermal Desorption Analysis was used to characterize the acidity strength of the catalysts. A Mc Bain microbalance was employed using 1 mg of sample. Ammonia was adsorbed at 250°C and the total mmole of NH_3/m_2 adsorbed was measured. Then the sample was heated using a ramp of 6°C/min and the amount of NH_3/m_2 remained on surface calculated at three temperatures (200/300/ 500°C). See Figure 1 for information.

FTIR. A Perkin Elmer 2865 Spectrometer with Furrier transformer capabilities was used to determine the acidity of the HCK catalysts. The typical plot of absorbency as a function of increasing wavelength were obtained for each catalyst and reported in Figure 2. The acids adsorption bands where recorded in the range of 3500 –3750 cm^{-1} because this region are associated to the stretching bands of the OH groups. Strong acid bands appeared in the region of 3600-3650 and weak ones in the region of 3550 cm^{-1} [12].

Pyridine adsorption. Pyridine was adsorbed at room temperature on the catalyst. FTIR had been performed using the same equipment than above mentioned and the resulting bands at 1558, 1676, and 1999 cm^{-1} recorded as a function of WS_2 content (Figure 3).

3. Results

3.1 Catalysts properties

Table 1 Catalyst properties

catalysts	HCKI	REF
NiO wt%	5.3	5.1
WO_3 wt%	13.6	12.8
PdO_2 wt%	0.2	-
Y zeolite wt%	-	15
TiO_2	15	-
Surface m^2/g	180	190
Pore Volume cm^3/g	0.44	0.48

Table 1 show that (HCKI) contains W and Ni sulfide species and Pd, as "metal" active phase [10]. It provides a stronger hydrogenating function than standard REF catalyst, as it will be shown later. The support (10wt% of TiO_2 in γAlumina) contributed with the "acidic" function; it was treated with steam-ammonia that modified its selectivity to diesel [13]. The acid support is prepared in the way that some Ti is exposed on the surface and it is accessibly. HCKI physical properties analysis shows a medium average pore diameter catalyst, with an intermediate surface area. Reference catalyst doesn't contain Pd nor was treated with steam ammonia, but it contain around the same amount of W, Ni and 15% Y (dealuminated) zeolite in γ-Alumina, and it has similar physical properties than HCKI .

3.2 Metal dispersion

Table 2 Metals dispersion IMe/(I Total)

Metals ratios	Signal	MHCK	REF
NiS_y I / I total	Ni2p(1/2)	0.55	0.51
WSx I / I total	W4f (5/2)	10.33	8.99
SiO_2 I / I total	Al 2p	12.8	11.4
TiO2 I / I total	Ti 2p	4.7	5.5

The metals dispersion was measured by XPS. The emission lines W4f (5/2-7/2) indicated the presence of W^{+4} centered at 34.3 and 32.5eV respectively; in addition it shows small shoulders at 38 and 35.8eV assigned to W^{+6} (non reduced species). The XPS spectra have a Ni $2p_{(3/2)}$ emission line centered at 854eV that is assigned to NiS, but there is also a shoulder that indicated other non-sulfide species (NiO?). Pd is assumed to be associated to Ni-W cluster since it changes the rate of sulphidation-reduction of them [14]. It can be seen in table 2 that surface atomic ratio is different than bulk atomic ratio. The W dispersion in sulfide REF catalyst is lower than in HCKI catalyst. It also has lower sulfur content. Those are associated to its lack of Pd effect on sulphidation and reduction reactions. There is a small shift in the W and Ni signal between the two catalysts.

Previous XPS studies demonstrated that Ti signal increased when Pd is present. That might be associated to the higher dispersion of W and to the interaction between W, Ti and Al. Sulphided catalyst present a Ti broad band at 65eV, but this information (and other gathered by FTIR and ESR) isn't useful to discriminate the Ti surrounding interaction.

3.3 Acidity

Figure 1 that shows the Ammonia Thermo-Desorption indicated a higher HKCI acidity than REF catalyst. In the ATPD, the higher the temperature, the lower is the amount of ammonia that remains adsorbed in both samples. Most of the ammonia is adsorbed in a week type acid center and it desorbs when is heated. In this way HCKI has the strongest acid center that retains ammonia adsorbed at high temperature (OTHCK reaction temperature), probably in Lewis type of sites. Metals on HCKI sulfide catalyst contribute to Lewis and Brönsted acid centers [13] as well as the Ti-W-Al metallic acid sites out of alumina framework. Ammonia treatment increases that sites (see NMR results on [13]), most of them might appear on surface by migration of –Ti- from the framework. Comparing I_{Al}/I_{Total} XPS signal against

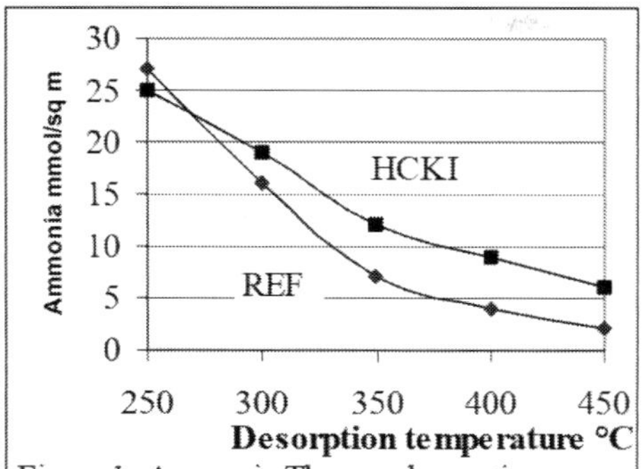

Figure 1 Ammonia Thermo desorption

the REF catalyst it could be verified a larger Aluminum dispersion on HCKI in comparison with REF catalyst.

To complement the acid sites characterization, a FTIR spectroscopy (Figure 2) was performed on sulphided catalyst. Both catalysts present a not well defined picks in the region from 3550 to 3770 cm^{-1}. The picks at 3559 are associated to Lewis acid centers, others at 3590 to a non identified acid species, and those of 3645 cm^{-1} assigned to Brönsted sites [15]. It can be seen that the HCKI catalyst shows a low amount of acid centers in the region of 3555 and 3650 cm^{-1}. In comparison, REF catalyst (dotted line in Figure2) presents no one. This band might be associated to –Ti-Al-. When WS2 content was increased in the HCKI catalyst the 3590 cm^{-1} signal strengthen as it's shown in figure 3 right axis in relative basis.

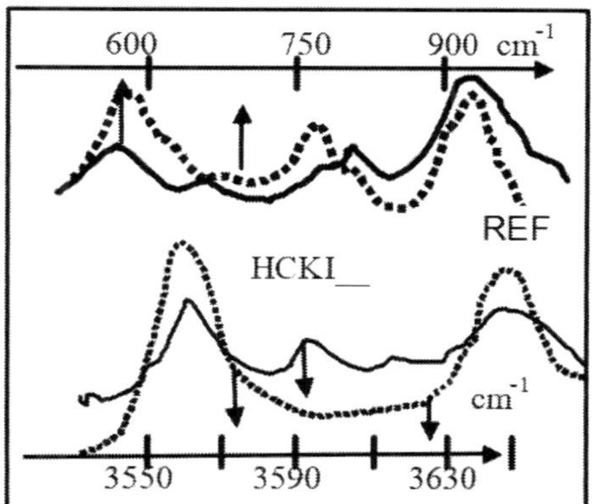

Figure 2 FTIR absorbance analysis in the framework (upper lines) and in the stretching region (lower lines)

In addition Pyridine adsorption studies done in both HCK1 and REF catalysts the presence of a new adsorption bands around 1547 cm^{-1} only in the first one. This particular pyridine adsorption band also increased with WS_2 content on HCK1 catalyst. It is associated to –SH Brönsted acid sites linked to metal clusters (Me). It is well know [16] the interactions between Ti-W-Al and they might be critical for the formation of "metal" acid centers. Topsoe [15] had report a metallic Brönsted sites on NiMo/Al2O3 catalyst that might support our hypothesis here.

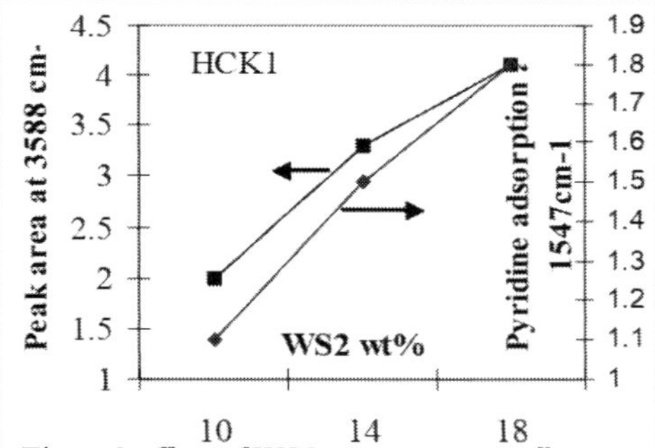

Figure 3 effect of WS2 content on metalic acid center adsorption.

3.4 Hydrocracking reactions

The Hydrocracking paths of reaction proceed through a first stage where the paraffin is dehydro genated in "metallic type sites" (reversible reaction), and then a second stage where the isomerization and cracking reaction take place on "acidic" type center. The isomerization is a reversible reaction and the cracking is an irreversible one. It was reported that the isomerization would require a Brönsted type center to produce a secondary carbenium ion, as intermediate to the alkenes formation, or crack throughout a β-scission to produce lower alkenes and a carbenium ion [17]. Higher is the temperature higher the cracking reaction. The previous catalyst characterization [13] have shown that the metallic type of sites may be very complex, being W^{+4}, Ni^{2+} and Pd sulfide cluster's component able to produced the hydrogenation-dehydrogenation stages quite fast. By other side there are acid sites in the external layer of TiO_2-Al_2O_3 support (out-of-framework) and in W-Ni-Al oxy-sulfide metallic clusters that may change cracking-isomerization balance in comparison with REF catalyst.

Aromatics hydrogenation and hydrogenolisis follow an in series reactions path with adsorbed species. The poly-ring aromatics might be bounded to the surface using a π interaction with metal, and the hydrogen by σ one, nearby in the same cluster. There were many discussions about the "metal active phase" used for hydrogenation and further hydrogenolisis reactions of the naphthen ring to produce an alkyl aromatics and alkyl-naphthen molecules. No attempt is made here to clarify the point but we are convinced, based in our fundamental studies with n-C_{16} hydrocracking, that Pd plays an important role in the ring opening activity.

3.4.1 Feed Characterization.

Table 3 Feeds used in the pilot plant

Components	DVGO	Arom.	II	III
Aromatics Added wt%	0.00	0	15.0	30.0
density kg/l	0.9123	0.9542	0.9595	0.9220
sulfur wt ppm	54	111	61	67
nitrogeno wtppm	13	25	15	16
Aromatics wt%	32.7	83.3	39.3	44.4
Mono-ring-Aromatic wt%	11.4	31.7	14.0	16.1
Di-rings-aromatics wt%	14.5	36.7	17.4	19.6
Tri-ring-aromatics wt%	6.8	14.9	7.9	8.7
Naphthenes wt %	38.6	16.2	35.7	33.4
Mono-ring-Naphthene wt%	12.6	13.2	12.7	12.7
Di-rings Napthene wt%	15.8	3	14.1	12.8
Tri-rings-Naphthene wt%	10.2	0	8.9	7.8
Paraffin wt%	16.2	0	14.1	12.5
Isoparaffin wt%	12.5	0	10.9	9.6

The composition of the extract of aromatics is presented in Table 3 that indicated the presence of a highly proportions of mono and Di-rings-aromatics. The extraction method selected those aromatics with α and or γ open position for sulphoning reaction, and for that isn't totally representative of all aromatics present in DVGO. Mainly they behave differently in ring opening reactions than those with α/γ alkyl occupied position. Nevertheless the extract represent 76% of aromatics present in DVGO and it covers almost all the distillation range (350-530°C). When the extract is added to DVGO, the Mono, Di, and Tri-ring-aromatics distribution had changed, as it can be seen in fourth and fifth columns in comparison with second ones.

3.4.2 Reaction of DVGO on HCK and REF catalyst

Table 4 Conversion (wt%) and selectivity with DVGO

Catalyst		HCK1		REF	
Temperature	LHSV h^{-1}	Conv.	Selec.*	Conv.	Selec.*
	0.8	47	5.61	54.1	3.6
	1	39	5.77	43.8	3.63
360°C	1.2	30	5.85	38.4	3.69
	0.8	55.0	5.22	61.2	3.0
	1	43.3	5.33	32.3	2.8
370°C	1.2	32.1	5.42	25.1	2.5
	0.8	60	5.1	66.3	2.4
	1	56	5.15	59.8	2.0
380°C	1.2	49	5.22	53.4	1.6
Selec*: wt Diesel/ wt Naphtha					

The first set of experiment was done with DVGO, and three temperatures and three space velocities (other conditions being constant). Table 4 presents in the third and fifth columns the conversions (defined by the weight of DVGO converted divided the initial one) for HCK and REF catalysts respectively. The fourth and sixth columns show the selectivity (ratio in weight of Diesel to Naphtha produced). The other columns present the operating conditions. As it was expected conversion increase with severity on both catalysts. In addition the higher the temperature, the lighter is the cracked product, in agreement with previous results with DVGO [12]. By comparing the selectivity of the two catalysts it can be seen that REF (zeolite) produce more Naphtha and it is more sensible to the temperature and space velocity than HCKI one. High temperature and low space velocity favor the HCKI catalyst to produce Diesel. The quality of the diesel produced will be discussed later.

The hydrogenation reactions happen at the same time than paraffin and isoparaffin conversion, and the "family" of aromatics originally present in the DVGO is transformed into other "family" of naphthen and paraffin. In addition Naphthen are also transformed into iso-paraffin, following a well establish path of reaction. The analysis of the families of component to track the reaction is presented below, as an example in Figure 4, 5 and 6 for naphtha, diesel and UVGO.

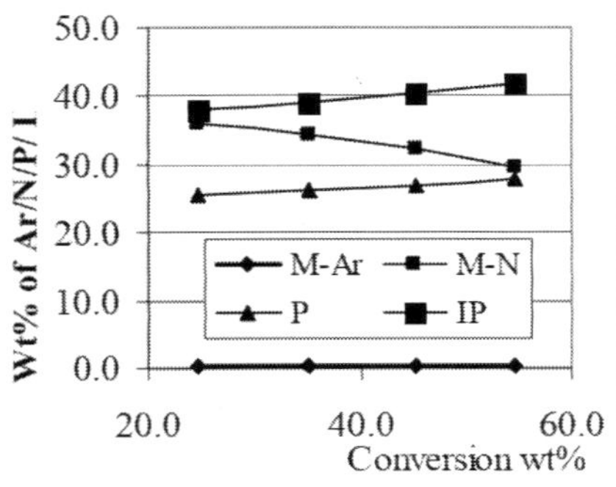

Figure 4 Ar / N / P and IP in naphtha

Figure 4 shows the Ar, N, P, IP as a function of conversion obtained at 370°C, 1 h^{-1} and 12.4MPa (HCKI). In can be seen a slightly increases of isoparaffin and naphthen, and a slightly decreases (10%) in the Diesel /Naphtha ratio when DVGO conversion goes up. In Naphtha the isoparaffin (IP) where discriminated in mono-branched, Di-branched, and Tri-branched. Higher the conversion higher is the content of less branched paraffin in Naphtha.

Figure 5 shows the composition of Diesel as a function of conversion, obtained at the operating conditions mentioned above. Aromatic and naphthen in Diesel comes via dealkylation of heavier alkyl-aromatics and alkyl-naphthen-molecules, while n-paraffin and isoparaffin are formed via controlled hydrogenolisis of the same compounds present in DVGO. These reactions took place in series an in parallel to reactions of aromatic hydrogenation and naphthen ring opening.

Figure 6 present the composition of UVGO. At this level of conversion it seems that naphthen are nearly constant, while aromatics decreases and isoparaffin increases as a function of conversion. The progress of the Hydrogenation reaction seems to be limited by the hydrogenolisis of the aromatics and naphthen.

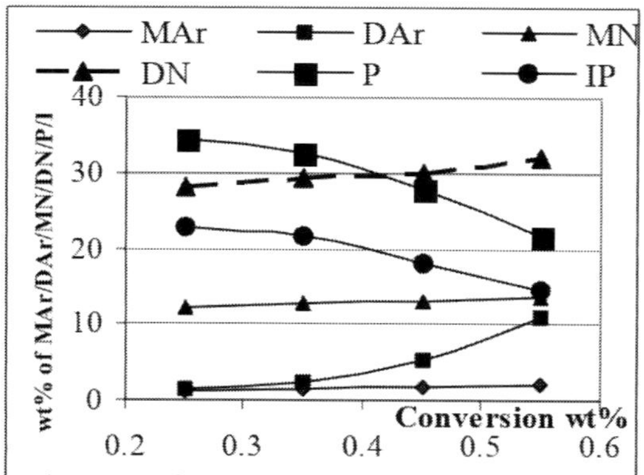

Figure 5 Diesel composition

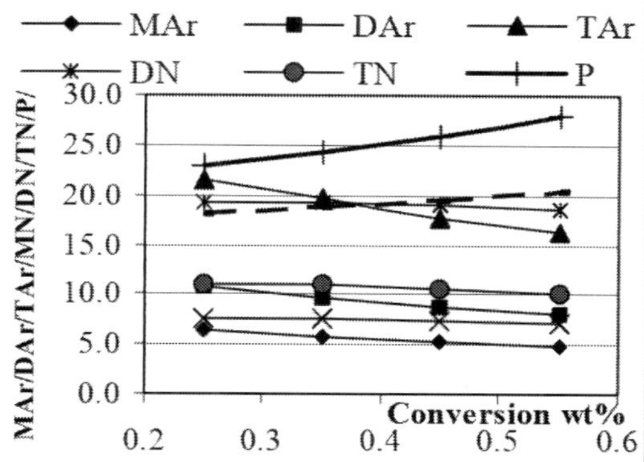

Figure 6 DVGO composition

3.43 Reaction of DVGO plus extract (HCKI and REF catalysts)

To study the effect of aromatics, 15% and 30% of extract were added to the original DVGO. Figure 7 shows theses effect on activity (left axis) and selectivity right axis). It can be notice the depressing effect on activity and the improvement in selectivity toward producing higher amount of diesel, due to higher amount of aromatics in the DVGO. To recover the activity it was necessary to increase 10°C for 15wt% and 16°C of temperature for 30 wt % of aromatics added. The HCKI activity was less affected than REF catalyst.

The conversion into diesel seems to depend more in the hydrogenolisis of paraffin and naphthen, and dealkylation of aromatic and naphthen than on conventional paraffin cracking reaction, almost at this hydrogen partial pressure and for HCKI catalyst. Detailed analysis of the products, similar than described in figure 4, 5, 6, supported the above preliminary interpretation. The UVGO properties are presented in Table 5 as an example

.

Table 5 Hydrogenation and hydrogenolisis of Aromatics

Wt % of armatic extract	15		30	
Catalysts	HCKI	REF	HCKI	REF
Mono-ring-Aromatics wt%	6.20	10.3	9.12	13.4
Di-rings-aromatics wt%	12.60	15.0	11.06	17.4
Tri-ring-aromatics wt%	12.90	14.0	13.00	15.8
Mono-ring-Naphthenes wt%	7.30	6.3	6.50	7.2
Di-rings Napthenes wt%	16.73	15.4	17.45	13.4
Tri-rings-Naphthenes wt%	10.40	14	12.40	12.6
Paraffins wt%	15.00	13	13.00	10.2
Isoparaffins wt%	18.85	12	17.58	10

The analysis indicated a large difference composition after reaction with HCK or REF catalyst, for 15% and for 30% of aromatics added. Always HCKI´s products have more paraffin and naphthen and lower aromatics than REF´s one, showing high hydrogenation and hydrogenolisis capability. By looking into the naphtha Isoparaffin composition, it can be seen that even for REF catalyst that always has more secondary and tertiary-cracking reactions, the amount of Di-branched paraffin goes up with aromatics content. But for both catalysts the addition of Aromatic reduces total I/N paraffin ratio in the naphtha produced. This phenomenon is more noticeable at high temperature (>370°C) and a high conversion (>50%) where I/N paraffin ratio go up with the carbon number. All of that suggest that the β-scission mechanism increase with temperature limiting the shift (type A Isomerization) of the alkyl-carbenium ion into a secondary carbenium ion [18, 19, 20]. What is new with HCKI catalyst is the presence of Tri-branched isomers in C_{10}-C_{15} carbon numbers (HNMR analysis). Metallic acid sites- probably in large pores or on external surface of the particle - allow the formation of stable tertiary ions, and prior cracking a rapid desorption of these isomers. They may be formed using active Brönsted "metallic acid sites" that are not present in REF catalyst (zeolite). It was reported in the literature that the ion involved in β-scission might be or ter-ter, or sec-ter and ter-sec for heavier hydrocarbons than C7. The relative Isomerization reaction rate to produce poly branched heavy paraffin seems to increase with molecular weight, but we are in the preliminary stages of quantifying the degree of isomerization in the diesel fraction using a new shape selective gas chromatography that is under development.

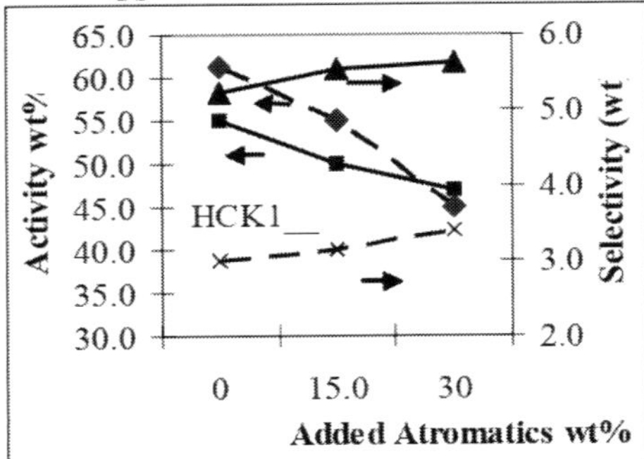

Figure 7 Effect of aromatics, 370°C, LHSV:1

3.4.2 Effect of Aromatics in hexadecane cracking (HCK catalyst)

The effect of the aromatic adsorption was also confirmed based on the results with hexadecane cracking (n-C_{16} paraffin) where the analysis is simpler than DVGO.

Table 6. Effect of Aromatics on I/N-paraffins ratio			
Carbon number/% of Ar	0	15%	30
Temperature °C	350	361	370
C_7	4.2	4.5	4.3
C_8	4.0	4.8	4.4
C_9	3.9	5.3	4.6
C_{10}	3.8	5.5	4.9
C_{16} Conversion % molar	50	49	52

Table 6 present an example of the information gather by converting n-C_{16} alone (second column) or with 15% or 30% wt% (third and fourth column). The experiments were done at the same level of conversion (50%) by adjusting the reaction temperature, other operating condition was held constant (LHSV: $0.8h^{-1}$, P: 18.2MPa, H_2/HC: 20, HCKI). It can be seen that 11°C or 20°C are required to recover the same paraffin conversion, when 15 or 30wt% of aromatic are respectively added to the feed. The presence of Aromatics gave a more flat carbon distribution curves, and lower amount of light material formed. Table 6 shows I/N paraffin ratio for C_7-C_{10} hydrocarbons products. Here, there are two causes for the change in the ratio: the aromatics and the temperature. When aromatics are present, the isomerization improves due to preferential adsorption of them on the strongest sites that limited the cracking, while the "metallic" acid sites remain available for isomerization. But the temperature has being also risen to obtain similar conversion that give, in consequence, a higher cracking rate that reduced I/N ratio. The final result is a similar I/N ratio for low molecular weight paraffin but it goes up with the number of carbons in molecules.

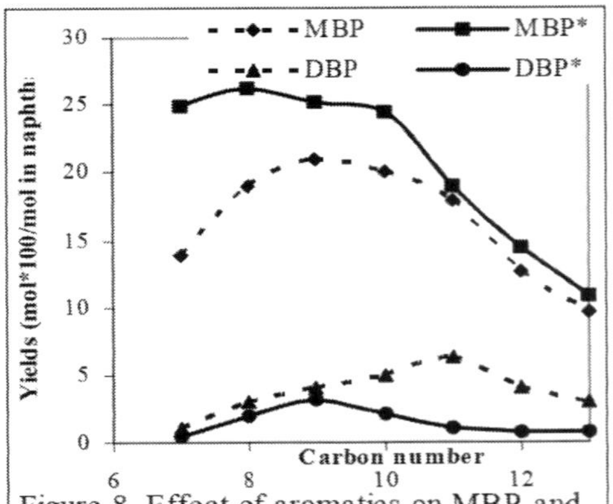

Figure 8 Effect of aromatics on MBP and DBP distributions in naphtha fraction.

The aromatic adsorption, is more important for acid sites in zeolite type catalyst (REF catalyst which results is not shown) than for HCKI's "metals acid sites" that haven't molecular shapes diffusion control.

3.5 Bases for a simplified Kinetics model

Aromatic molecules (Alkyl-Di and Tri-rings) are deeply hydrogenated on this catalyst following the well know in series reactions. (ca: alkyl-Naphthalene was hydrogenated to Alkyl-Tetraline and then to Alkyl-Decaline). In parallel the Alkyl-Tri or Di or Mono-ring aromatics dealkylated to lower molecular weight alkylated aromatics molecules. The Naphthen (Mono and Di-rings) associated to aromatics rings suffers a hydrogenolisis reaction on metal sites and produce alkyl- Mono or Di-rings-aromatic or alkyl-naphthen-aromatics molecules (ca: alkyl-tetraline forms di-alkyl-benzene). The lower parts of figure 9 summarize the mechanism for hydrogenation, hydrogenolisis and hydrodealkylation reactions.

The aromatic that undergo hydrogenation along the trickle bed reactor and desorbs, produces two effects: increases the metal sites and "metallic" acid sites available for isomerization and hydrogenolisis, and reduce the consumption of adsorbed hydrogen. Aromatic adsorption not also competes with i-alkenes saturation, but might help the "survival" of secondary and tertiary alkenes that are produced on them.

Our degree of understanding of reactions path mechanism is limited due to large number of molecules present in the feed, for that a simplified apparent kinetics model is developed instead of more fundamental one suggested by (Raemdonck and Froment [23]) for Hydrocracking case. A matrix of thirty fifth apparent liquid phase reactions for each DVGO pseudo-component number (10) might be used. The path of in series-parallel reactions considered that all the olefins isomerization occurred in series and the cracking in parallel on acid sites. Desorbed isomers and cracked molecules will no react again (it means that no secondary reaction take place at low conversion). A Lagmüir type rate of reaction expressions was used that include in the denominator only the alkyl-aromatics $-Km_{Ar}$ / Ka_{Ar}- and alkyl-naphthen-aromatic $-Km_{NAr}$ / Ka_{Ar}- adsorption coeffi cients. Metal site has differently coefficients than acid site in the heterogeneous adsorption terms. The model neglects H_2S, NH_3, alkenes, cycle-naphthen and paraffin adsorption terms. Bimolecular reactions and the poly and Mono-cycle-alkanes ring opening hydrogenolisis are disregarded.

Based on fundamental studies performed with n-C_{16} it is proposed that the reactions follow an apparent order of reaction of one for all hydrocarbons. Apparent order of reaction respect to hydrogen is 0.6 for Hydrocracking, 0.5 for hydrogenolisis of cycle-paraffin (ring opening) and 0.8 for hydrodealkylation of alkyl-naphthen or alkyl-aromatic reactions.

The paths of reactions is presented in Figure 9.

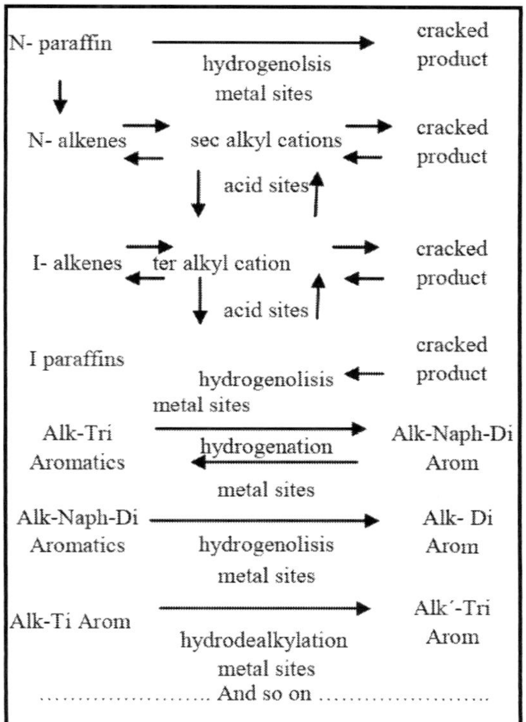

Figure 9 Path of reaction for Hydrocracking

3.5.1 Kinetics equation for the model

The main paths for the N-paraffin ($_{np}$) reaction go through n-alkenes ($_{no}$) formation on metals sites and the rate of reactions is represented by:

$$-R_{m_{np-no}} = \frac{(k_m 1^i_{np-no} C_{np} - k_m 2^i_{no} C_{no} C_{H2})}{[1 + K_{m_{Ar}} C_{Ar} + K_{m_{NAr}} C_{Nar}]} \quad (1)$$

Where m is for metal sites and i super-script for the carbon number, *km1* and *km2* are the rate constant (Subscript indicated the direction of the reaction). *Km* is the adsorption constant on metal sites (subscript indicate the compound. C̲ is the concentration (mol/l- in liquid phase), that takes in account the vaporization of th species via Henry law. The N-alkenes desorbs and are adsorbed on nearby acid sites. The adsorbed n-alkenes follow the isomerization way that produced the sec and ter-cations. Here two sites have to be considered: sites (a) that use the strong Brönsted acid function on surface (–OH), being constrained by pore sizes that induced mainly sec-cations. Sites b) that use the "metallic"-acid function (-SH) it produces mainly ter-cations. Both sec and ter-cations may be cracked to lighter products or deprotonized to produce i-alkenes. The rates of reaction of mono- branched olefins ($_{mbo}$), di-branched ($_{dbo}$), and tri-branched olefins ($_{tbo}$) are represented by the following equation 2, 3 and 4 for site (a):

$$Ra_{no-mbo} = \frac{(ka3^{i}_{no-mbo} C_{no} - ka4^{i}_{mbo} C_{mbo} - ka9^{i}_{mbo} C_{mbo})}{[1 + Ka_{Ar} C_{Ar} + Ka_{NAr} C_{Nar}]} \quad (2) \qquad Ra_{mbo-dbo} = \frac{(ka5^{i}_{mbo-dbo} C_{dbo} - ka6^{i}_{dbo} C_{dbo} - ka10^{i}_{dbo} C_{dbo})}{[1 + Ka_{Ar} C_{Ar} + Ka_{NAr} C_{Nar}]} \quad (3)$$

$$Ra_{dbo-tbo} = \frac{(ka7^{i}_{dbo-tbo} C_{dbo} - ka8^{i}_{tbo} C_{tbo} - ka11^{i}_{tbo} C_{tbo})}{[1 + Ka_{Ar} C_{Ar} + Ka_{NAr} C_{Nar}]} \quad (4)$$

Where *ka3* to *ka8* are rate constants for isomerization and *ka9* to *ka10* the cracking rate constants. *Ka* is the adsorption constant on Brönsted acid sites. The same equation applied for (b) but the constants *kb and Kb* are different. The α is the fraction of ((a) / (a)+(b)) acid sites to total acid sites, and C'_{no} equal αC_{no}.

Finally i-alkenes desorb from both acid sites and adsorb on metal sites where they are protonized to i-paraffin. The equation for mono- branched olefins is:

$$Rm_{mbo-mbp} = \frac{(km15^{i}_{mbo-mbp} C_{mbo} C_{H2} - km16^{i}_{mbp} C_{mbp})}{[1 + Km_{Ar} C_{Ar} + Km_{NAr} C_{Nar}]} \quad (5)$$

The i-paraffin can also be hydrogenolized to lighter products (irreversibly reaction).

$$Rm_{mbp-light} = \frac{(km17^{i}_{mbp-mbp'} C_{mbp} C^{0.5}_{H2})}{[1 + Km_{Ar} C_{Ar} + Km_{NAr} C_{Nar}]} \quad (6) \qquad Rm_{dbp-dbp'} = \frac{(km18^{i}_{dbp-dbp''} C_{dbp} C^{0.5}_{H2})}{[1 + Km_{Ar} C_{Ar} + Km_{NAr} C_{Nar}]} \quad (7)$$

$$Rm_{tbp-tbp'} = \frac{(km19^{i}_{tbp-tbp'} C_{tbp} C^{0.5}_{H2})}{[1 + Km_{Ar} C_{Ar} + Km_{NAr} C_{Nar}]} \quad (8)$$

The isomerization and cracking constant rates are a function of carbon number and temperature.

Additional equations are required for aromatics and naphthen dealkylation. ATAr represents the alkyl-tri rings-aromatics, ADAr the alkyl-di-rings-aromatics ANDAr represent the alkyl-Napthen-Di-ring-aromatic and so on:

$$Rm_{ATAr} = \frac{(km20^{i}_{ATAr} C_{ATAr} C^{2}_{H2} - km21^{i}_{ANDAr} C_{ANDAr})}{[1 + Km_{Ar} C_{Ar} + Km_{NAr} C_{Nar}]} \quad (9) \qquad Rm_{ANDAr} = \frac{(km22^{i}_{ANDAr} C_{ANDAr} C^{2}_{H2} - km23^{i}_{ADNAr} C_{ADNAr})}{[1 + Km_{Ar} C_{Ar} + Km_{NAr} C_{Nar}]} \quad (10)$$

$$Rm_{ADNAr} = \frac{(km24^{i}_{ADNAr} C_{ADNAr} C^{3}_{H2} - km25^{i}_{ADNAr} C_{ADNAr})}{[1 + Km_{Ar} C_{Ar} + Km_{NAr} C_{Nar}]} \quad (11)$$

Where *km20* to *km*25 are the reaction rate constants for hydrogenation. The same denominator is used in equations 1, 5, 6 and 7.

$$Rm_{ATAr} = \frac{(km26^{i}_{ATAr} C_{ATAr} C^{0.8}_{H2})}{[1 + Km_{Ar} C_{Ar} + Km_{NAr} C_{Nar}]} \quad (13) \qquad Rm_{ANDAr} = \frac{(km27^{i}_{ANDAr} C_{ANDAr} C^{0.8}_{H2})}{[1 + Km_{Ar} C_{Ar} + Km_{NAr} C_{Nar}]} \quad (14)$$

$$Rm_{ADNAr} = \frac{(km28^{i}_{ADNAr} C_{ADNAr} C^{0.8}_{H2})}{[1 + Km_{Ar} C_{Ar} + Km_{NAr} C_{Nar}]} \quad (15) \qquad Rm_{ATAr} = \frac{(km29^{i}_{ATAr} C_{ATAr} C^{0.8}_{H2})}{[1 + Km_{Ar} C_{Ar} + Km_{NAr} C_{Nar}]} \quad (16)$$

$$Rm_{ATN} = \frac{(km30^{i}_{ATN} C_{ATN} C^{0.8}_{H2})}{[1 + Km_{Ar} C_{Ar} + Km_{NAr} C_{Nar}]} \quad (17) \qquad Rm_{ADN} = \frac{(km31^{i}_{ADN} C_{ADN} C^{0.8}_{H2})}{[1 + Km_{Ar} C_{Ar} + Km_{NAr} C_{Nar}]} \quad (18)$$

$$Rm_{AN} = \frac{(km32^{i}_{AN} C_{AN} C^{0.8}_{H2})}{[1 + Km_{Ar} C_{Ar} + Km_{NAr} C_{Nar}]} \quad (19)$$

Where k_m26^{i} to k_m32^{i} *are* the constants for hydrodealkylation. The ^{i} is expressed as function of molecular weight to account for the reactivity.

The last set equation need is for hydrogenolisis of naphthen rings which are also irreversible reactions.

$$Rm_{ANDAr-DASDAr} = \frac{(k_m 33^i_{ANDAr-DADAr} C_{ANDAr} C_{H2}^{0.8})}{[1 + Km_{Ar} C_{Ar} + Km_{NAr} C_{Nar}]} \quad (20) \qquad Rm_{ADNAr-DAAr} = \frac{(k_m 34^i_{ADNAr-DAAr} C_{ADNAr} C_{H2}^{0.8})}{[1 + Km_{Ar} C_{Ar} + Km_{NAr} C_{Nar}]} \quad (21)$$

Where k_m33^i to k_m34^i are the hydrogenolisis reactions constants

The mass balance differential equation for a plug flow reactor that considerer the matrix of lumped reactions 1 to 34 can be solved using a numerical method as Runge Kutta Felberg. Initially the rate constants for C_{20} (dehydrogenation-isomerization, hydrogenation and hydrogenolisis reactions were provided), as well as a polynomial function that predict rates constant as a function of carbon number (from $C_{20\ to}$ to C_{30}) and the α value. All constants in the polynomial were adjusted using the experimental value obtained at different temperature and space velocity. The DVGO conversion range used in the model go between 25 to 60%. The search for the constant value that reproduces the experimental product distribution data was done using a genetic algorithm [23]. Other details and results will be published later [24].

Even when the model oversimplified the path of reaction three main aspects, other than aromatic impact, had being additionally recognized.

- Dehydrogenation stage seems to be far to the equilibrium in contradiction that others author had being used, almost at this hydrogen partial pressure and for this catalyst.
- Vaporization of light hydrocarbon partially reduced the secondary and tertiary cracking
- Particle wetting along the reactor may modified the selectivity
- The presence of H_2S that favor the built up of metallic Brönsted acid sites.

The simplified model is able to distinguish the aromatic effects using two isomerization-cracking centers that open a way for catalyst formulation and optimization of the HCK toward diesel production.

4. Conclusion

It was demonstrated that our HCK catalyst behaves differently than conventional WNi/Y-Alumina catalyst and that aromatic affect differently the behavior of both catalysts.

- It produces at the same conversion, different product distribution and degree of isomerization.
- The HCKI catalyst has a higher metal dispersion, and the presence of a new type of "metallic" acid sites, in comparison with REF catalyst. That gave lieu to higher conversion, heavier product distribution and lower effect of aromatics in isomerization.
- It also has the advantage to produce high naphthen ring opening and tertiary branched paraffin production.
- To model HCK reactions a simplified path of reaction with two acids sites has been proposed. The aromatic adsorption that effect activity and selectivity has been considered here by using a heterogeneous kinetics.

Acknowledgment

The authors want to thank the Intevep permission to publish these results as well as the professionals that help us to characterize the catalyst and the conversion products.

References

1. Galiasso T R., Di Marco M., and Salazar A., 13th World Petroleum Congress, Proceedings, 233-245 (1991).
2. Weisz P.B., *Adv. in Catal.* **13**, 137-141, (1962).
3. Vansina H., Baltanas M.A., Froment G.F., *Ind. Eng. Chem. Prod. Res. Dev.*, **22**, 526-533, (1983).
4. Martens J.A., Souverijns W., Verrelst W., Parton R., Froment G.F., Jacobs J.A.; *Angew*, Ed Engl., **22**, 34-45, (1995).
5. Martens J.A, Parton R., Uytterhoven L., Jacobs P. A., Froment G.F. ;*Appl.Catal.*,**76**, 95-101, (1991)
6. Galiasso T R; Preprint 101, Chemical Eng. World Congress, Melbourne, Australia, 223-234, (2001).

7. Mendez A. "Vacuum gas Oil analysis by GC-MS technique". Personal communication February, (2001).
8. De Brujin A., *The 6th. Int. Congress on Catalysis*, London, paper B34, 127-133, (1976).
9. Galiasso T. R, and Pieretti J., "Wetting of solid in a "liquid phase" reactor using gas-liquid distribu tor plates". Sent to *Chem. Engng. Com.* (2004)
10. Prada S. R., Galiasso T. R., Romero G., Reyes E., US patent 4465792 (1989)
11. Deffose C., Canesson P., Rouxhhet P.G., and Delmon B.J, *J. Cat.* **25**, 407.414, (1972).
12. Uytterhoeven J.B., Schoonheydt R., Liengme V., and Hall W.K., **J. Cat**. 13, 322-344, (1969).
13. Galiasso T.R., Proceeding of *Scientific bases for the preparation of Heterogeneous Catalysts*, Else vier, Ed. Gaigneaux et al., **143**, 321-333 (2002)
14. Galiasso T. R., and Prada S R., "Effect of sulphiding WNiPd/Y-Alumina on milhydrocracking catalyst". Sent to *Catalyst Letter* (2004).
15. Topsoe H., Clausen H., Topsoe N., Zeuthen P., *Stud. of Surf. Sci Cat.*, **53**, 77-100, (1989).
16. Economides N., Master Thesis, University of Cincinnatti (USA), (1998).
17. Scherrer J., and Bass J., *J. Cat.*, **28**, 101-112 (1973)
18. Coonradt H.L., and Garwod W.E, *Ind. Eng. Chem. Process Des. Dev.* 3, 38-43(1964).
19. Weitkamp J., and Ernst S., *Guidelines for mastering the properties of Molecular Sieves*, Plenum Press (Barthomeuf et al) New York, 343, (1990).
20. Martens J.A., Tielen M., Jacobs P.A., *Cat. Today* **1**, 435-442, (1987).
21. Martens G.G, Marin B.G., Martens J.A., Jacobs P.A., Baron G.; **J. of Cat**., *195*, 353-343, (2000).
22. Van Raembdonck, K. K., and Froment, G. F.; Hydrocracking, (Ed:Raasanti), chapB12, (1992).
23. T. Park and G. Froment; Computer Chem. Engng., 22, (1994), 103.
24. Paretti T., and R. Galiasso T.; "Kinetics model for DVGO Hydrocracking in presence of aromatics", in preparation (2004).

Molecular Modelling and Hydrogen Integration of Diesel Hydrotreating Reactor

Jianjun Sun, Nan Zhang

Department of Process Integration
UMIST, PO Box 88
Manchester, M60, 1QD
U. K.

Paper to be submitted to TA007 Advances in Hydro/Desulfurization

Key words: Molecular model, Diesel hydrotreating, Hydrogen network, Desulphurisation, Hydrogen pinch

Abstract

A molecular kinetic model for diesel hydrotreating process is based on the methodology developed in UMIST. This method uses the molecular type homologous series (MTHS) matrix to characterise the streams of feed and products of hydrotreating processes [1]. Within the matrix, hydrodesulphurisation (HDS), hydrodenitrogenation (HDN) and aromatics hydrogenation are included for hydrotreating reactions of elements. The linear free energy relationships and molecular weight relationships are used to calculate fundamental rate constants for the reaction network. The fundamentally derived reaction model can then be calibrated to observed data by tuning a limited set of parameters such as effectiveness factor. The deactivation model is also incorporated. Good model predictions for the properties of products are achieved.

In this work the diesel hydrotreater is used as a basic case to investigate the effects of the interactions with the hydrogen network. The molecular model of the diesel hydrotreater is combined with the hydrogen pinch analysis to investigate possible modifications that will improve the performance of the diesel hydrotreater. One case study has been carried out. By the deactivation model, different scenarios of the existing diesel hydrotreating processes are investigated with another important factor, run-length. The scenarios include using extra hydrogen and producing ultra-low sulphur diesel. The pattern of best operating conditions is obtained.

1. Introduction

Current trends show that there is an increasing demand for diesel production. Along with more demands, the concern about environment issues has led to stricter legislation on the quality of diesel. Therefore, refiners must improve the performance of the diesel hydrotreating processes by means of optimizing the operation, revamp or building new processes. Hydrogen is the key stream that refinery can use to overcome the problems. As the marginal value of products decreases, refiners also want to obtain more profit through the analysis of interactions between the hydrotreating processes and the hydrogen network. Both actions lead to an aspiration for a detailed kinetic model of diesel hydrotreating process. The kinetic model of the diesel hydrotreating must involve two aspects: characterisation of oil streams based on molecular information and reactions between molecules.

2. Molecular characterisation of streams of the diesel hydrotreater

2.1 Molecular type homologous series matrix

This work uses a matrix representation to describe the types and sizes of molecules in streams of the diesel hydrotreater, which is consistent with the characterisation methods. One example of such a matrix is given in Figure 1. The columns of the matrix are molecular types existing in a petroleum fraction:

nP, iP: normal and iso-paraffins;

nO, iO: normal and iso-olefins;

1N, 2N, 3N: five-carbon or six-carbon one-ring to three-ring naphthenic compounds: e.g. decalin 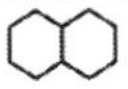 belongs to the 2N column;

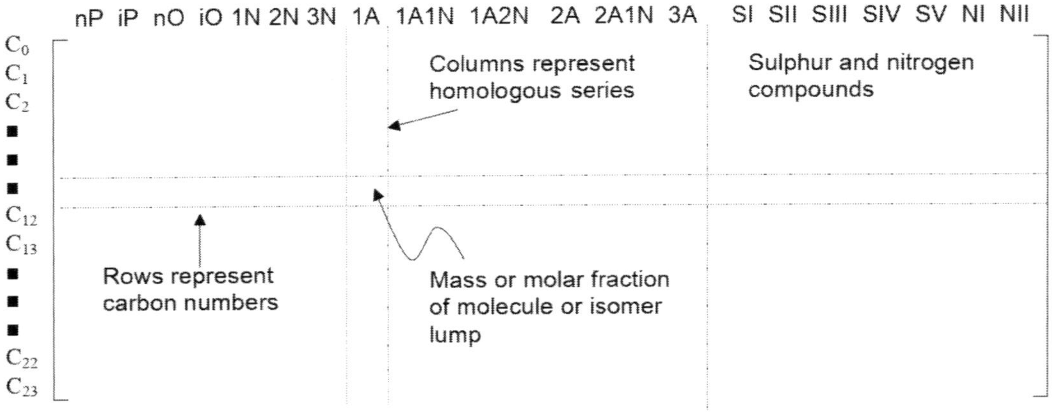

Figure 1: Matrix representation of a middle distillate

A, 2A, 3A: one-ring to four-ring aromatic compounds: e.g. 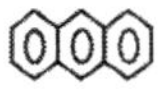 and 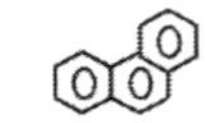 are isomers that are both included in the same lump of the 3A column;

1A1N6, 1A2N5, 2A1N: compounds containing both naphthenic and aromatic rings, e.g. the 1A1N column contains 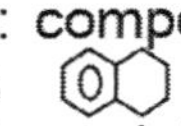 and 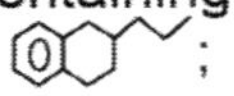;

SI, SII, SIII, SIV, SV: five sulphur lumps:

SI: mercaptans, e.g. CH_3SH, C_2H_5SH;

SII: sulphides, disulphides, thiophenes and benzothiophenes, e.g. CS_2, CH_3-S-C_2H_5 all belong to this column;

SIII: dibenzothiophenes not substituted at the 4 or 6 position, e.g. 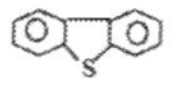 ;

SIV: dibenzothiophenes substituted at the 4 or 6 position, e.g. 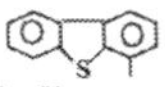 ;

SV: dibenzothiophenes substituted at both 4 and 6 positions, e.g. 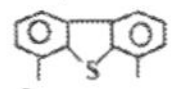 .

NI: basic nitrogen compounds, e.g. pyridine  and quinoline 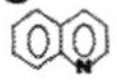 ;

NII: non-basic nitrogen compounds, e.g. indole 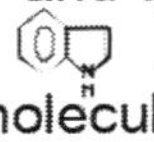 .

The rows of the matrix account for the size of the molecules, which can be conveniently captured by carbon numbers. C_0 row can be used to capture concentrations of H_2, H_2S, NH_3 and H_2O.

2.2. Predicting bulk properties from molecular compositions

Figure 2 shows the procedures of calculating bulk properties based on the molecular matrix of streams. Firstly, by using the molecular structure-property correlations [2], molecular fundamental properties (e.g. molecular weight, density and boiling point) can be calculated. Then all molecules in a stream are ranked by boiling temperature from low to high, and volume percent of each molecule is calculated from the given molecular compositions. According to the boiling points, these molecules are accumulated by volume percent to generate a distillation curve of the stream. By molecular composition based correlations and distillation curves, bulk properties of a stream can be calculated.

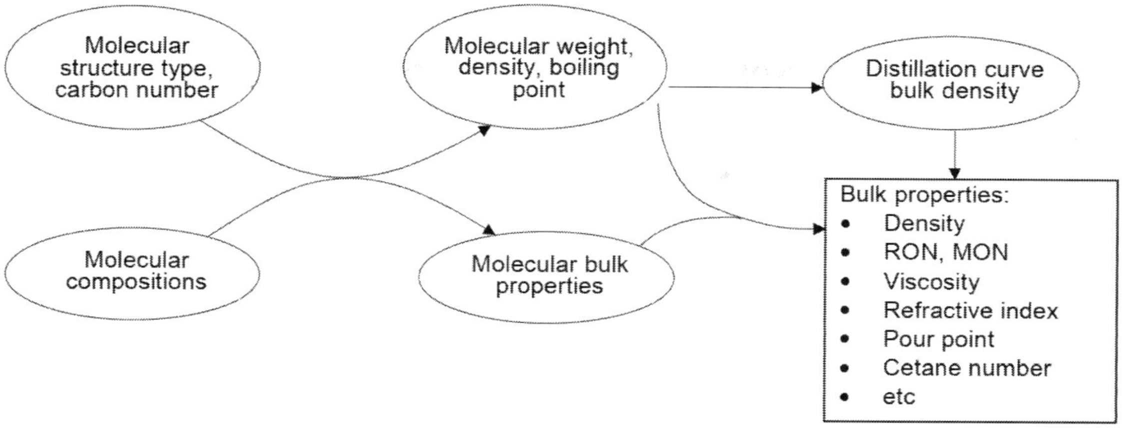

Figure 2. Calculation procedures translating molecular composition to bulk properties

2.3 Transformation methodology

Although it is possible to obtain detailed compositional analysis of the streams in a refinery, these data are rare and cannot be found for the specified range of the streams of units. Hence, there is a need to generate the molecular information of these streams from the known data. Figure 3 summarises the new methodology developed in characterisation of refining streams [3]. It consists of two major sections: generation of database and transformation of bulk properties into molecular information. In the database generation part, Molecular Type Homologous Series (MTHS) matrix is chosen to represent the molecular information of the feed streams. According to bulk properties of streams and published analytical data, sample matrices are generated. In the automatic selection part, an optimisation model is developed to calculate the ratios of sample matrices.

Analysis of various middle distillates from different petroleum fractions shows that the distribution of the sulphur-containing compounds over the boiling range is similar, regardless of fraction origin or total sulphur content. This is explained from the same classes of sulphur-containing compounds being involved [4]. In this work, the normalised sulphur distributions,

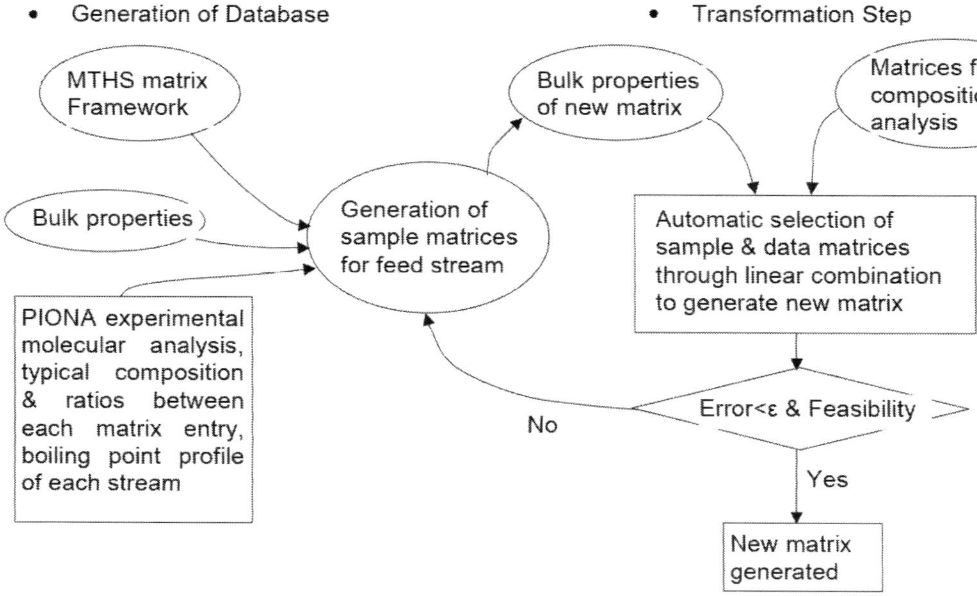

Figure 3. Schematic representation of the transformation methodology

sulphur content and experimental composition analysis data are used to derive sulphur compounds distribution in feed streams. The analyses of nitrogen compounds in middle distillates are quite challenging. In atmospheric gas oil (AGO), a 1/0.75/2.5 ratio of quinoline, indole, carbazole and their derivatives is identified. In light cycle oil (LCO), a 1/2.3/12.2 ratio of aniline, indole, carbazole and their methylated compounds is found [5].

3. Reaction model of the diesel hydrotreater

3.1 Reaction networks and models

A reaction family is defined as a specific rearrangement of molecular structure by the molecules in the petroleum fractions. In diesel hydrotreating processes, the large number of reactions can be classified mainly into 3 reaction families: hydrodesulphurisation (HDS), aromatics hydrogenation, hydrodenitrogenation (HDN).

Hydrodesulphurisation: The sulphur compounds in the diesel oil are classified into five groups according to their different HDS activities. In this work, an HDS model is included, which describes the disappearance kinetics of the sulphur compounds with two contributions, one through hydrogenolysis and the other through hydrogenation [6]. Each reaction takes place on a different site of the catalyst surface, denoted σ for hydrogenolysis and τ for hydrogenation. The model is derived based on the LHHW formalism assuming that the surface reaction is irreversible and constitutes the desulphurisation for both hydrogenolysis and hydrogenation of the sulphur groups. The possibilities of competitive or non-competitive adsorption of the sulphur-containing compounds, hydrogen sulphide and hydrogen are considered.

$$r_i = k_{i,\sigma} \frac{K_{i,\sigma} P_i K_{H_2,\sigma} P_{H_2}}{\left(1 + \sum_{j=}^{4} K_{j,\sigma} + K_{H_2S} P_{H_2S}\right)^2 \left(1 + K_{H_2} P_{H_2}\right)} + k_{i,\tau} \frac{K_{i,\tau} P_i K_{H_2,\tau} P_{H_2}}{\left(1 + \sum_{j=1}^{4} K_{j,\tau} P_j + K_I P_I\right)} \quad (1)$$

Where k_i is reaction constant, $E_{a,j}$ is the activation energy, K_i is adsorption constant, $H_{a,j}$ is the heat of adsorption, r_i is rate constant, P_i is partial pressure or concentration of species i.

Aromatic hydrogenation: The feedstocks of diesel hydrotreating reactors generally contain one to three rings aromatic compounds. The reaction networks of these aromatics are found in from the literature [7] and presented in Figure 4.

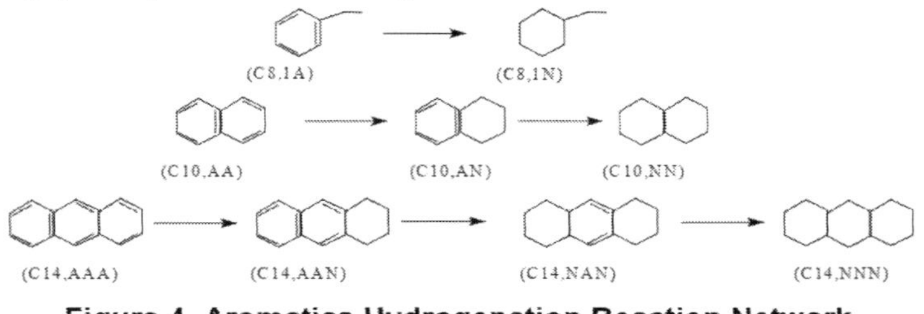

Figure 4. Aromatics Hydrogenation Reaction Network

Hydrodenitrogenation:

The hydrodenitrogenation (HDN) of heterocyclic nitrogen compounds generally involves the following reaction pathways: (1) hydrogenation of nitrogen heterorings, (2) hydrogenation of aromatic rings, and (3) hydrogenolysis of C-N bonds. The reaction speed of the first step generally is fast and equilibrium can be reached. Therefore the simplified hydrodenitrogenation network focuses on hydrogenolysis of C-N bonds. For example, the reaction pathway of indole is:

$$\text{Indole} \rightarrow \text{Aniline} + \text{Paraffin} \qquad \text{Indole} \rightarrow \text{Aaromatic} + NH_3$$

L-H equations are used to describe the reaction constants of aromatic hydrogenation and HDN. For nitrogen compounds, model parameters were estimated using experimental data for a range of single compounds. In a reaction family, a homologous series of molecules in MTHS matrix is subject to the same reaction. Within each column of the matrix, components differ only by the substituted groups. Differences in reactivity are attributable to these substituted groups. Kinetics correlations from structure-reactivity relationships [8] can be applied to each of the matrix columns. For sulphur and nitrogen compounds, correlations are developed between the reaction rate constant with molecular properties such as molecular weight or boiling point. Finally, the model is calibrated by the process data using effectiveness factors of the model.

3.2 Modelling diesel hydrotreating reactor

Trickle flow fixed bed reactors are widely used in the diesel hydrotreating processes. Reaction rate is limited by chemical kinetics. Mass transfer between gas, liquid and solid phases can be ignored. The reactor is co-current plug flow and in adiabatic conditions. The liquid volume in the reactor remains constant. Both mass balance and energy balance are considered.

$$\frac{dC_j}{dz} = r_j(C) \qquad (2) \qquad\qquad \frac{dT}{dz} = \frac{\Sigma(-\Delta H_r)_j * r_j(C)}{\Sigma C_{p_j} * C_j} \qquad (3)$$

Where $r_j(C)$ is reaction constant of molecule j, known from the kinetic model; H_r is reaction enthalpy; C_p is heat capacity. The values of these parameters can be obtained from published data, the functional group method or other correlations. Final solution is obtained by solving a set of differential equations.

3.4 Deactivation Model

Deactivation has a significant effect on the profitability of hydrotreating processes. Hydrotreating catalysts are primarily deactivated by the suppression of active sites by carbon containing deposits. In the deactivation model the amount of coke deposited is defined as a function of time and process conditions. Then the activity of catalysts is related to the coke content of the catalyst by a power-law relationship. In this deactivation model, the deposition of coke is formed via two parallel routes [9]: (i) thermal condensation reactions of aromatic moieties and (ii) catalytic dehydrogenation reactions. The aromatic compounds of oil streams are assumed that represent the coke precursor existed in the streams. For the catalytic coke generation a simple Langmuir-type equation (4) is used.

$$r_c = \frac{k_c K_{ads} C_q}{1 + K_{ads} C_q} \qquad (4) \qquad\qquad k_c = k_{c,0}(1 - D / D_{c,max}) \qquad (5)$$

Where D is the amount of coke deposited on the catalyst, $D_{c,max}$ is the maximum amount of catalytic coke, $K_{c,o}$ is initial activity for catalytic coke formation, K_{ads} is adsorption constant of coke precursor, C_q is the concentration of coke precursor, can be set as the aromatic content in the oil stream.

Coke accumulated through thermal reactions is simulated with the equation 6 of a first order reaction.

$r_t = k_t C_q / P_{H_2}$ (6)

Where k_t is rate constant for thermal coke formation, P_{H2} is hydrogen partial pressure.

The total rate of coke deposition r follows $r = r_c + r_t$ (7)

Then the total content of coke on the catalyst can be integrated the total rate with time. The deactivation coefficient is used to relate the activity to the coke content on the catalyst. It is defined as: $\alpha = r_c / r_0$ (8)

The power-law relationship is used to calculate the deactivation coefficient with the total coke content on the catalyst. $\alpha = 1 - a(C)^b$ (9)

Where α is deactivation coefficient, r_c is reaction constant at time period c, r_0 is reaction constant without deactivation, a and b coefficients, C is the weight ratio of coke on the catalyst.

In order to simulate the operation of the diesel hydrotreater, the best temperature increasing principle is set. In reality, operators need to increase the temperatures of the hydrotreater to compensate for the activity loss of the catalyst caused by coke deposition. Following procedures are adopted. In one operating round (from operation start to catalyst replacement), the whole time range is divided into multiple periods. Each period is set as two weeks. In each period, minimum increase of reactor temperature that can meet the sulphur specification of product would be raised.

4. Case studies

Two case studies are carried out to show the effects of molecular model of diesel hydrotreater. One reactor of diesel hydrotreating process is producing diesel of 2000ppm sulphur. Table 1 compares the model-predicted data with process data.

Table 1. Process Data and Molecular Model

	Process data	Model predicted data
Sulphur conversion	73.31	73.34
Nitrogen conversion	33.29	33.8
Product density (g/cm^3)	0.869	0.876
Distillation point, °C		
10	192	185
30	244	232
50	271	271
70	302	300
90	337	338
95	352	358

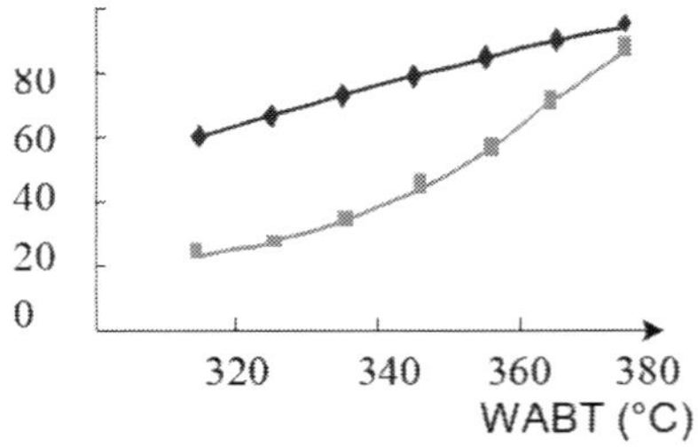

Figure 5. Conversion of Sulphur and Nitrogen with respect to WABT

From Fig. 5, it can be found that raising temperature to 375°C can meet the target of 500ppm sulphur. However, it will reduce 80 percent of the run-length. According to Fig. 6., reducing LHSV from 2.7 to 1.0 can produce 500ppm diesel. And it is also found that it is not possible to reach the 50ppm target only by changing operating conditions.

In the second case, light cycle oil can be mixed into the feedstock of the diesel hydrotreater. The blending ratio can be increased up to 35vol.%. Table 2 lists the sulphur contents change with respect to different LCO blending ratios. When 35vol.% LCO is blended, the sulphur content in the feedstock will be doubled. From Fig.7, the life is shortened to almost half comparing with the no blending operation. Blending LCO has strong effect on the catalyst life.

Table 2. Sulphur content distribution vs. different LCO blending ratio

Blending ratio (vol.%)	0	10	20	30	35
Sulphur in feed (wt%)	0.3	0.41	0.52	0.62	0.687

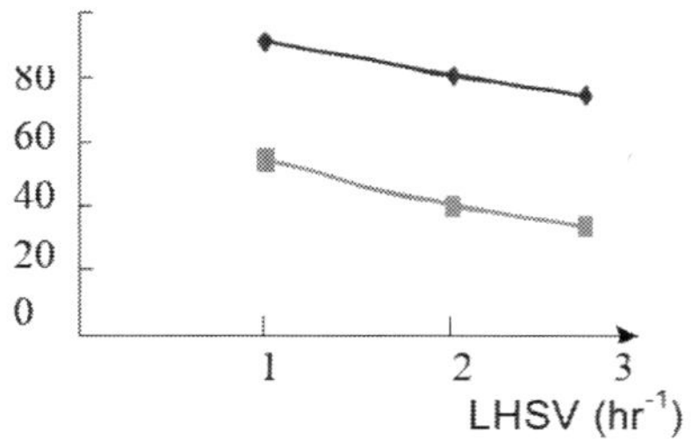

Figure 6. Conversion of Sulphur and Nitrogen with respect to LHSV

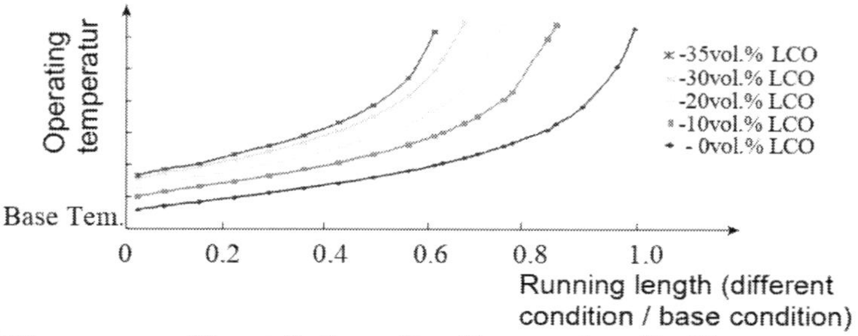

Figure 7. Operating Temperature Profiles with Respect to Run Length

5. Integration analysis of the diesel hydrotreating process and hydrogen network

Hydrogen pinch analysis helps to analyse hydrogen distribution systems of refineries and find the minimum hydrogen needed by the net [10]. Targets could be conservative as the process operating conditions are fixed. But in the hydrotreating processes, the change of operating conditions plays an important role in product quality as well as hydrogen consumption. A case study is used to introduce the procedures of investigation and the result of integration analysis. The process data were obtained from one BP refinery. The target process of the investigation is the diesel hydrotreating process. Currently the product of the diesel hydrotreating process is 50ppm sulphur diesel. As part of the new clean fuels project, the refinery must produce ultra low sulphur diesel (ULSD) in the future. And refiners hope to increase the profit margin of the diesel hydrotreater by blending into LCO. The effects on run length of LCO blending are shown in Fig. 7.

The investigation is focused on the hydrogen network and the interaction effect between the hydrogen network and diesel hydrotreating process. Based on the background, the target of the investigation includes possible modifications to improve the performance of diesel hydrotreater, the best way to meet the future specifications of diesel product and the effect of light cycle oil (LCO) in the diesel hydrotreater.

According to the targets, the procedures of analysis include the following steps:

1. Extract processes' data of the hydrogen network
2. Analyse the distribution of the hydrogen network by the hydrogen pinch technology
3. Use a molecular model to evaluate the operating conditions.
4. Use a deactivating model to simulate the operating periods under different conditions.
5. Analyse marginal benefits of different operating conditions and compare with the current operating condition (base case).
6. Check the feasibility of hydrogen network with different operating conditions and identify any possible improvement.

The process data were obtained from one refinery. The existing hydrogen network system is shown in figure 8. The hydrogen surplus diagram (Figure 9) shows that the system is feasible as the supply meets the demand. It means that there is positive surplus for all purities within the surplus diagram. From fig. 9, the 3.42 t/h surplus of hydrogen represents the best possible saving scenario and set a target for further improvement without significant capital investment cost. The result proves that currently the hydrogen network can supply more hydrogen to the hydrogen consumers. That means more hydrogen can either be saved or used to improve the economics of the refinery. Obviously, the refinery would like to use the surplus hydrogen to increase the profit. There are two different ways to use hydrogen to improve performance of the diesel hydrotreater:

1. Increase the throughput of diesel hydrotreater.
2. Blend cheaper feedstock, such as light cycle oil.

From these two aspects, the procedures introduced above were followed to study the possible improvement of the diesel process.

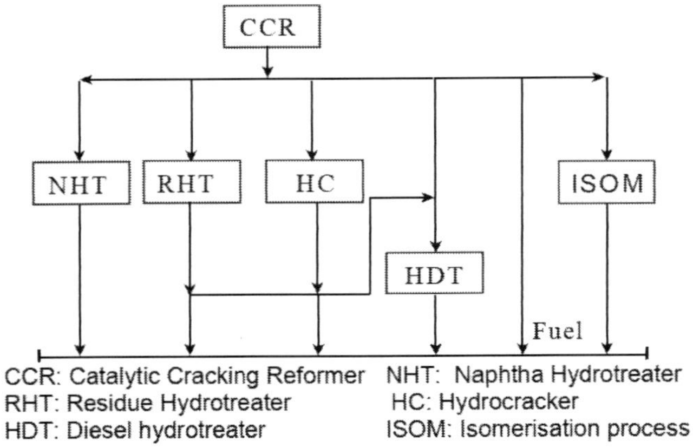

Figure 8. The Hydrogen Network of the Refinery

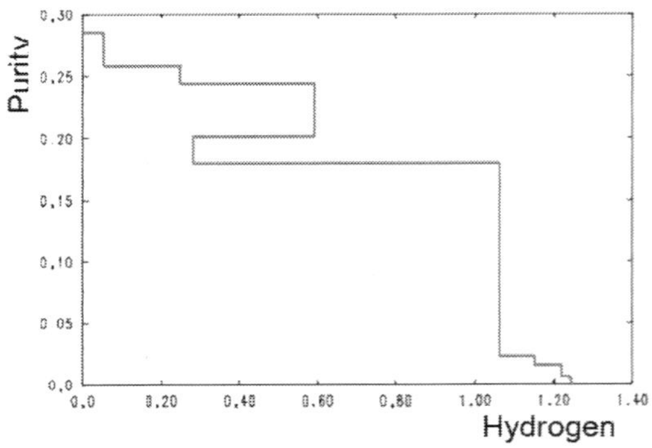

Figure 9. Hydrogen Surplus Diagram

Effects of increasing throughput on hydrogen network were analysed by following the procedures introduced above. Fig. 10 shows the economic data of increasing throughput of the diesel hydrotreater. The hydrogen cost and catalyst cost of diesel hydrotreater increases with more throughput. But increasing cost is smaller than increasing profit of changing throughput, so the profit of diesel hydrotreater increases within the whole range of increasing throughput. Then, the feasibility of the hydrogen network under the operation of different throughputs was tested. According to the hydrogen surplus diagram (figure 11), the hydrogen network cannot support more than 15 vol.% increase in throughput.

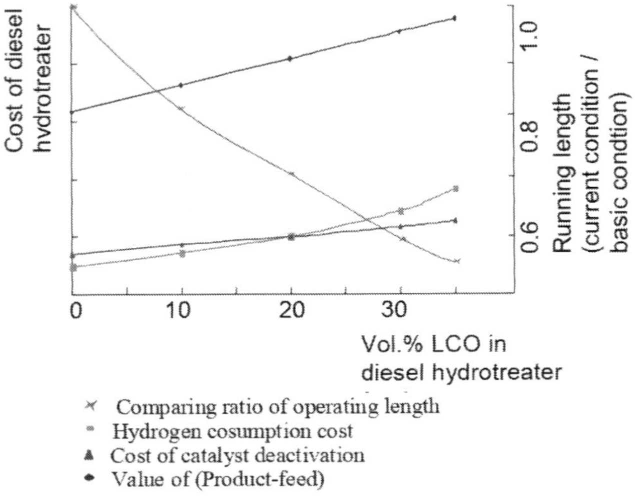

Figure 10. Cost of Diesel Hydrotreater with respect to LCO blending ratio

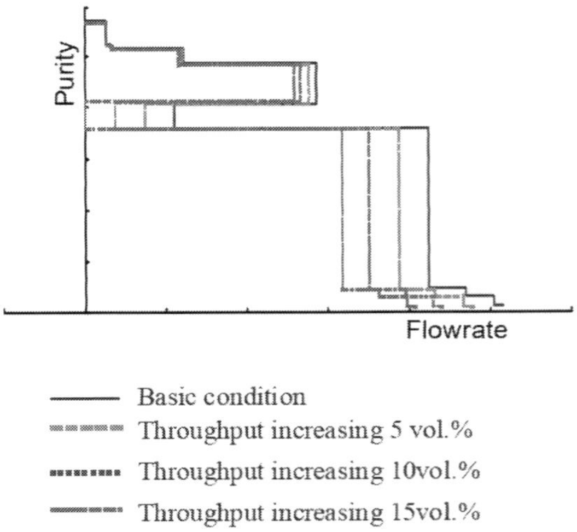

Figure 11. Hydrogen surplus diagrams with respect to different throughput

Another option of improving profit of HDT is blending LCO into the feedstock. The blending ratio could vary between 0 - 35%. The operating length of HDT decreases greatly with more LCO blending that is shown in figure 7. Figure 11 shows the profit of HDT with different LCO blending ratios. The margin value of the product increases linearly with the blending ratio of LCO. The cost of hydrogen consumption has a great impact on the profit. According to figure 11, the best blending ratio of LCO is 20 vol.%. After calculating the profits of different LCO blending ratios, the feasibility of hydrogen network was analysed. From figure 12, the surplus hydrogen only can support 5 vol.% LCO blending. Therefore, with the existing hydrogen network, 20 vol.% blending is not an available target. The best blending ratio of LCO is 5 vol.%. If hydrogen can be imported, the blending ratio can be increased up to 20 vol.%.

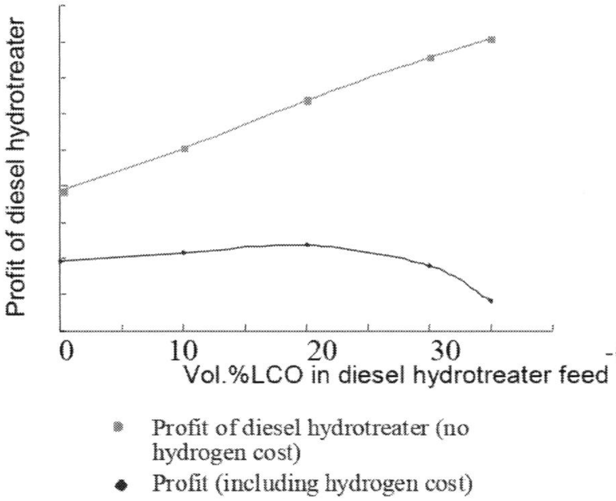

Figure 11. Profit of Diesel Hydrotreater with respect to LCO blending ratio

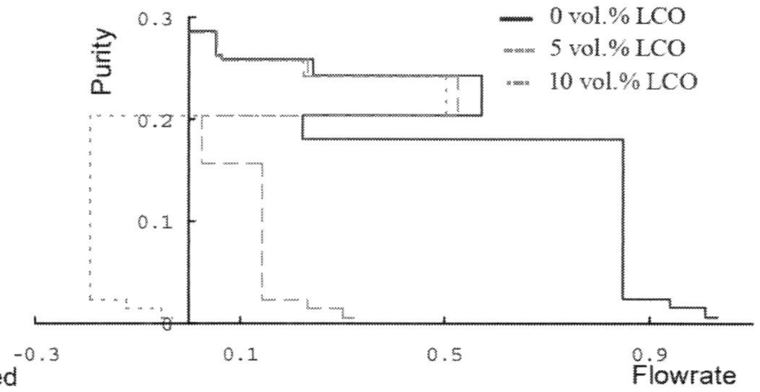

Figure 12. Hydrogen Surplus Diagram for Different LCO Blending Ratio

After the scenarios based on the current conditions were investigated, the questions of ULSD production come to the stage. A molecular model of HDT was used to simulate the operating conditions of future scenarios. Table 3 lists different conditions of reducing throughput (compared with the current condition). In the table, throughput reduction ratio 0.62 is set as the basic condition to compare with other throughputs. For the data in table 3, the lower throughput, the lower starting temperature of HDT and the longer run length is set out. Figure 13 shows the margin profit of combined operating conditions. The results include the marginal profit of the product, the cost of deactivation, the cost of hydrogen consumption and the hydrogen network analysis. The feasibility of hydrogen network under different operating conditions was checked. The available operating area is marked out by the feasible line in Figure 13. In the left area of the feasible line, the supply of the hydrogen network can meet the demands of hydrogen consumers. The best operating condition is also located in that area. The best operating condition is obtained by reducing throughput to 70%. LCO blending ratio can be increased to 10 vol.% while the throughput is reduced to 0.62.

Table 3 Conditions of Reducing Throughput

(Basic condition: the throughput reduction ratio of 0.62)

Throughput reduction ratio	0.765	0.708	0.663	0.64	0.62
Starting tem. (°C)	+6.3	+3.0	+1.4	+1.0	Base T.
Catalyst life ratio	0.77	0.84	0.92	0.95	1.0

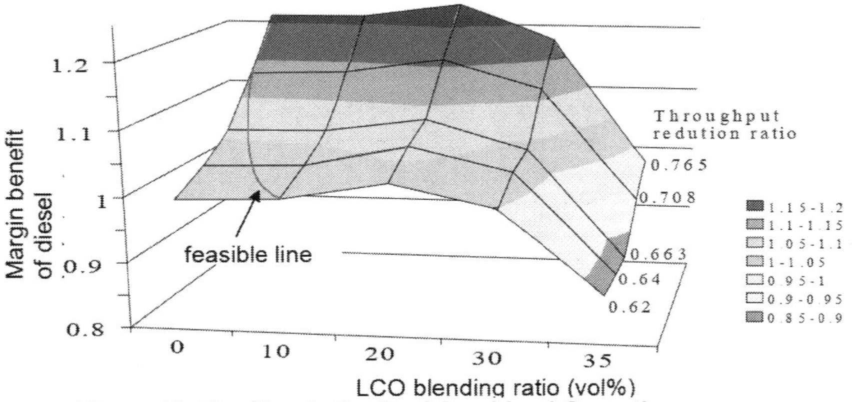

Figure 13. The Margin Profit of Combined Operating Conditions

6. Conclusions

Process models today are required to predict not just the yields of major products, but also the composition and quality of products, and detailed boiling point distributions. Molecular-based models are developed since molecular detail provides more chemistry information of products of diesel. Feed and product streams are described as molecular type homologous series matrices. Reaction networks are based on the chemistry knowledge. The model can accurately predict the sulphur and nitrogen conversion, as well as other bulk properties such distillation curve of streams. The deactivation model is developed to describe the coke deposition on the catalyst. The model can be used to analyse the future scenarios.

Hydrogen pinch analysis is a powerful tool for analysing the hydrogen network of a refinery. With the hydrogen surplus diagram, the minimum hydrogen demand of the hydrogen network can be identified. The analysis of hydrogen surplus can also help to find the possible improvement of the existing hydrogen network. Integration of the diesel hydrotreating process and hydrogen network can improve the economics of the diesel hydrotreater. Analysis of hydrotreating processes and hydrogen network can find better solutions to meet future specifications. The result of optimisation of hydrotreaters is more reliable based on the integration of the hydrogen network.

Acknowledgement

The authors express their gratitude to Sinopec. Co. and BP Company for their permission to publish the results presented.

References:

1. Peng, B., 1999, Molecular Modelling of Petroleum Processes, PhD thesis, UMIST
2. Zhang, Y., 1999, A Molecular Approach for Refinery Modelling, PhD thesis, UMIST
3. Mi Mi, S., 2003, PhD thesis, UMIST
4. Amorelli, A. et al., 1992, Estimate Feedstock Processability, Hydrocarbon Pro., 71(6), 93
5. Mushrush G. W., Beal E. J., Hardy D. R., Hughes J. M., 1999, Fuel Pro. Technol., 61:197
6. Angelica Hidalgo-Vivas,1999, PhD thesis, UMIST
7. Girgis M. J. and Gates B. C., 1991, Ind. Eng. Chem. Res., 30, 2021-2058.
8. Neurock, M., Klein, M. T. 1993. Polyc. Arom. Compounds, 3, 231-246.
9. D. Dong, et al., 1997,Catalysis Today 37, p267-275
10. Joao Alves,1999, Design of Refinery Hydrogen Distribution Systems, PhD thesis, UMIST

Effect of Catalyst Type and Operating Severity on the Products Quality in Hydrotreating Kuwait Atmospheric Residue

A. Marafi[1] A. Hauser[2] and A. Stanislaus[1]

[1] *Petroleum Refining Department. Kuwait Institute for Scientific Research. P.O. Box: 24885. 13109, Safat. Kuwait.*
[2] *Central Analytical laboratories. Kuwait Institute for Scientific Research. P.O. Box: 24885. 13109, Safat. Kuwait.*

A paper to be presented at AIChE Annual Spring Meeting, the 7th Topical Conference on Refining Processing. Advances in Hydro/Desulfurization, April 25-29, 2004 New Orleans, LA, USA

ABSTRACT

Upgrading of heavy petroleum oils and residues by hydrotreating, in particular, by the atmospheric residue desulfurization process (ARDS) to produce low sulfur fuel oil (LSFO) and purified feedstocks for down stream conversion units for more valuable clean environmentally friendly transportation fuels have gained considerable importance worldwide in recent years. Graded catalyst systems in multiple reactors are used in the ARDS process in order to achieve HDS, HDM, HDN and conversion of residues to distillates at desired levels. Three types of catalysts, namely, a HDM catalyst for metal removal in the front reactor, a HDM/HDS catalyst with balanced HDM and HDS activities in the middle section and a very active HDS/HDN catalyst in the last reactor, are commonly used in the ARDS process. In such process, the characteristics of the feedstocks processed in different reactors are significantly different. Thus, for example, while the feed entering the front reactor is original atmospheric residue, the feed to the second reactor is a demetallized and partially hydrotreated product from the first reactor. The quality of the feed entering the second reactor is strongly dependent on the operating severity in the first reactor and can have a strong influence on the performance of the catalysts in the following reactor with regard to various conversions and deactivation rate. In the present work, a comparative study was made on the effect of operating temperature on the quality of products obtained during hydrotreating of Kuwaiti atmospheric residue (KU-AR) over two industrial hydrotreating catalysts, namely, Mo/Al_2O_3 HDM catalyst and Ni-Mo/ Al_2O_3 HDS catalyst that are loaded in the front and middle reactors, respectively, of a commercial ARDS unit. For each catalyst, experiments were conducted in a fixed bed pilot plant unit at H_2 Pressure: 120 bar; LHSV=1.0 h^{-1}; H_2/Oil: 570 ml/ml; Temperature: 360-420°C. The liquid oil products from different runs conducted at various reaction temperatures over the two types of catalysts were collected and characterized for its quality. Furthermore, the product oil was fractionated to asphaltenes, and maltenes and each fraction was characterized by various techniques including ^{13}C NMR and elemental analysis. Special attention was paid to the changes in the characteristics of asphaltenes as a function of operating temperature. The results obtained from the comparison study are presented and discussed.

INTRODUCTION

Atmospheric residue desulfurization (ARDS) process is used extensively in Kuwaiti refineries for the upgrading of petroleum residues (1,2). The process involves hydrotreatment of the residual oil feed in a series of fixed bed reactors loaded with different types of catalysts. Catalysts for such hydrotreating processes are chosen on the basis of activity, selectivity and life. Catalyst life is dependent on the rate of deactivation by coke and metal deposits and sintering of the active phase. The performance of the overall ARDS process with regard to various reactions such as HDM, HDS, HDN, asphaltenes cracking and conversion to distillate as well as the life-on-stream is clearly linked to the performance of the catalyst in different reactors (3,4). In such processes, the characteristics of the feedstocks processed in different reactors are significantly different. Thus, for example, while the feed entering the front reactor is original atmospheric residue, the feed to the second reactor is a demetallized and partially hydrotreated product from the first reactor. The quality of the feed entering the second reactor is strongly dependent on the operating severity in the first reactor and can have a strong influence on the performance of the catalysts in the following reactor with regard to various conversions and deactivation rate. Information regarding the activity, selectivity, product quality and deactivation of the individual catalysts are therefore, highly desirable for optimizing reactor loading in the multiple catalyst systems. In the present study, performance comparison tests were carried out on two types of industrial catalysts, namely HDM (MoO_3/Al_2O_3) and HDS ($Ni/Mo/Al_2O_3$) catalysts that are loaded in the front and middle reactors of the commercial ARDS process, using Kuwaiti atmospheric residue (KU-AR) as a feed in a fixed-bed pilot plant. The primary objective of the work was to carry out systematic studies to assess the quality of the liquid oil products obtained at various reaction temperatures over both types of catalysts. The product oil was fractionated to its major fractions, nnamely, asphaltenes, and maltenes and each fraction was characterized by various techniques including ^{13}C NMR. Special attention was paid to the changes in the characteristics of asphaltenes as a function of operating temperature.

EXPERIMENTAL

Materials. The characteristics of the feed KU-AR are shown in Table 1. Two industrial hydroprocessing catalysts one MoO_3/Al_2O_3 (HDM catalyst with surface area =200 m^2/ gm; pore volume =0.64 cc/gm) and the other, $NiMo/Al_2O_3$ (HDS catalyst with surface area = 230 m^2/ gm; pore volume = 0.53 cc/gm) in the form of extrudates were used for conducting the comparison studies.

Experimental Techniques. The hydrotreating experiments were conducted in a fixed-bed reactor unit (manufactured by Vinci Technologies) using Kuwaiti atmospheric residue as feed. Experiments were carried out for each catalyst individually. A volume of catalyst charge diluted with an equal amount of carborundum was used for each run. The catalyst was presulfided by a standard procedure using straight run gas oil containing 3wt% dimethyl disulfide (DMDS) by a standard procedure before injecting the feed. After the presulfiding, conditions for each run were adjusted to desired operating temperature, pressure, hydrogen flow rate and LSHV. Table 2 shows the various operating conditions for the studies. During the course of each run, product samples were collected every 12 hours and analyzed for sulfur, metals (V and Ni), nitrogen, asphaltenes, CCR, viscosity, density and distillate yield. The oil product samples were further subjected to fractionation to the components, namely, asphaltenes and maltenes using standard seperation procedure (5).

Table 1 KU-AR Feed Characteristics

Feed property	Values
API	12.27
Total sulfur wt%	4.30
Asphaltene wt%	3.60
Total nitrogen ppm	2670

Kinematic viscosity @ 50°C cSt	871.2
CCR	12.20
Ni wt ppm	21
V wt ppm	69

Table 2. Run Conditions for Temperature Effect Studies

Process parameter	Temperature Effect Study
Temperature (° C)	360-420
Pressure (bar)	120
LHSV (hr)	1
H_2/Oil (ml/ml)	570
Time on stream (hr)	120

RESULTS AND DISCUSSIONS

Initial studies focussed on the effect of operating temperature on the overall oil product qualites over both HDM and HDS catalysts. The characteristics of the oil products obtained from various runs for both catalysts are presented on Table 3. The results reveal that an increase on the reaction temperature in the range 360-420°C enhances the removal of various hetroatoms present in the KU-AR to different extent. In general, the results indicate that the operating temperature has a favourable effect on increasing various reaction conversions purifing the oil product more from S, CCR, N, V, and Ni contaminants. Moreover, the results show that the density of the oil products decreases with increasing tempertusre for both catalysts, indicating increasing conversion of the heavy feed to lighter products at higher temperture. However, the oil products obtained from hydrotreating of KU-AR over the HDM catalyst possessed lower metal, asphaltenes, and higher S, N,CCR and density compared with the oil product obtained from HDS catalyst. These results are expected since the HDM catalyst posseses large pores, which facilitites the diffusion of metal containg species and asphaltene molecules into the catalyst pellet (6,7). On the other hand, the HDS catalyst has a higher hydrogenation activity and is dedicated for sulfur, CCR and moderate range of nitrogen removals.

The results clearly indicate that, as the reactor temperature increased, the products are more purified with lower S, V, Ni, CCR, asphaltenes, and N levels and become lighter suggesting that they could be processed more easily in the following reactor of an ARDS unit with multiple reactors. However, besides the contaminant levels, the structural characteristics of the various fractions of the product oil, particularly that of the asphaltene fraction, could play a dominant role in the ease of its processibility as well as on catalyst deactivation in the subsequent reactors in series. Therefore, the nature of changes in the characteristics of different fractions of the products obtained from hydrotreating of KU-AR over both types of catalysts were examined by ^{13}C NMR spectroscopy. Two fractions, namely maltenes and asphaltenes were separated from the liquid products collected from various runs. The fractions were suibjected to detalied characterization using elemental and ^{13}C NMR analysis.

Table 3. Reaction Temperature Effect on the liquid product quality in hydrotreating of KU-AR over HDM and HDS caatalysts.

	360°C	380°C	400°C	420°C
HDM Catalyst				
S,wt%	3.23	2.79	2.62	1.55
CCR, wt%	10.50	9.52	8.54	6.83
Asph, wt %	2.22	1.68	1.21	-
N, wt %	2600	2570	2486	2255
V, wtppm	35	23	11	6
Ni, wtppm	12	10	8	4
Density @ 15°C, gm/ml	0.9837	0.9655	0.9609	0.9225
HDS Catalyst				
S,wt%	1.38	1.08	0.56	0.22
CCR, wt%	7.56	6.95	4.76	3.66
Asph, wt %	2.93	2.40	1.75	1.32
N, wt %	2510	2430	1960	1780
V, wtppm	26	22	12	5
Ni, wtppm	10	8	6	4
Density @ 15°C, gm/ml	0.9737	0.9510	0.9444	0.9100

Table 4 compares the yield (wt%) of maltenes and asphaltenes obtained by hydrotreatment of KU-AR over two types of catalysts at various reaction temperatures (360–420^0 C). The results clearly indicate that the HDM catalyst has a higher activity for asphaltenes removal than the HDS catalyst. This is not unexpected since the HDM catalyst is designed with larger pores for metal and asphaltenes removals.

Table 4. Yield Comparison between HDM and HDS Catalysts

	360°C	380°C	400°C	420°C
HDM Catalyst				
Maltenes, wt %	97.80	98.32	98.79	-
Asphaltenes, wt%	2.2	1.68	1.21	-
HDS Catalyst				
Maltenes, wt %	97.07	97.61	98.25	98.68
Asphaltenes, wt%	2.93	2.41	1.75	1.32

Elemental Analysis: The asphaltenes fractions from both HDS and HDM catalysts were further subjected to elemental analysis. The results of H/C, N/C, and S/C for the asphaltenes fraction as a function of temperature are given in Table 5. H/C in the asphaltenes fraction for both catalysts decreases as the temperature increases with higher values for the HDS catalyst (Table 5). The S/C and N/C molar ratios in the asphaltenes fraction are higher for the HDM catalyst than the HDS catalyst at all temperatures. The results also indicate that the HDS catalyst possess the higher hydrogenation activity than the HDM catalyst. These results are consistant with the quality of the oil products obtained from hydrotreating KU-AR bver both catalysts as discussed eariler.

Table 5. H/C,N/C and S/C molar ratios in asphaltenes fraction in the product oil obtained from hydrotreating of KU-AR over HDM and HDS Catalysts as a function of reaction temperature

	360°C	380°C	400°C	420°C
HDM Catalyst				
H/C molar ratio	1.11	1.09	0.85	0.82
N/C molar ratio	0.010	0.011	0.012	0.013
S/C molar ratio	0.04	0.035	0.025	0.02
HDS Catalyst				
H/C molar ratio	1.22	1.15	1.20	1.19
N/C molar ratio	0.006	0.007	0..008	0.009
S/C molar ratio	0.03	0.033	0.015	0.017

^{13}C NMR Analysis: The aromaticity and structural characteristics of asphaltenes remaining in the product oil resulting from hydrotreatment of KU-AR over the two types of catalysts were further examined by ^{13}C NMR analysis. The results show that the HDS catalyst possesses a higher hydrogenation activity than the the HDM catalyst. Thus, over the whole temperature range, the asphaltenes in the products hydrotreated over the HDS catalyst possess about 10% less aromatic carbon than the asphaltenes in the HDM catalyst products (Fig.1a). Regarding the temperature dependance, both catalysts show the same trends for reducing aromatic carbon in the unconverted asphaltenes remaining in the products. As for the types of aromatic carbon in the unconverted asphaltenes, the following observation can be made from the results presented in Fig. 1 (b,c) :

- The HDM asphaltenes contain higher quaternary and tertiary aromatic carbons than the HDS asphaltenes.
- The tertiary aromatic carbon increases faster with increasing temperature in the HDM asphaltenes than in the HDS asphaltenes.

Substantial differences are also noticed in the degree of condensation and alkyl substitution in the product aspgaltenes for the two catalysts (Figs 2 and 3). At low temperatures, the HDS asphaltenes are more condensed than the HDM asphatenes. At higher operating temperature, the degree of condensation increases at a faster rate in the HDM asphatenes than in the HDS asphaltenes (Fig. 3). Alkyl substitution is lower in the HDM asphaltenes are less than in HDS asphaltenes (Fig. 4). Although for both catalysts, the asphaltenes remaining in the products become more condensed with less alkyl substitution with increasing operating severity, higher aromaticity and larger degree of condensation are noticed for the HDM asphaltenes compared to the HDS asphaltenes. This may be attributed to the differences in the hydrogenation function of the two catalysts. The unpromoted Mo/Al_2O_3 HDM catalyst has a lower hydrogenation activity than the Ni-Mo/Al_2O_3 HDS catalyst. Because of the lower hydrogenation function of the unpromoted Mo/Al_2O_3 HDM catalyst, thermal cracking and condensation may occur predominantly at higher temperature, and lead to highly condensed asphaltenes that are more refractory and less reactive and have a high propensity for coke formation. The results obtained in this study are considered to be very valuable for the optimization of the operating conditions and its impact on the quality of the product oil as well as its fractions, namely, asphaltenes and maltenes during hydroprocessing of KU-AR in an ARDS process.

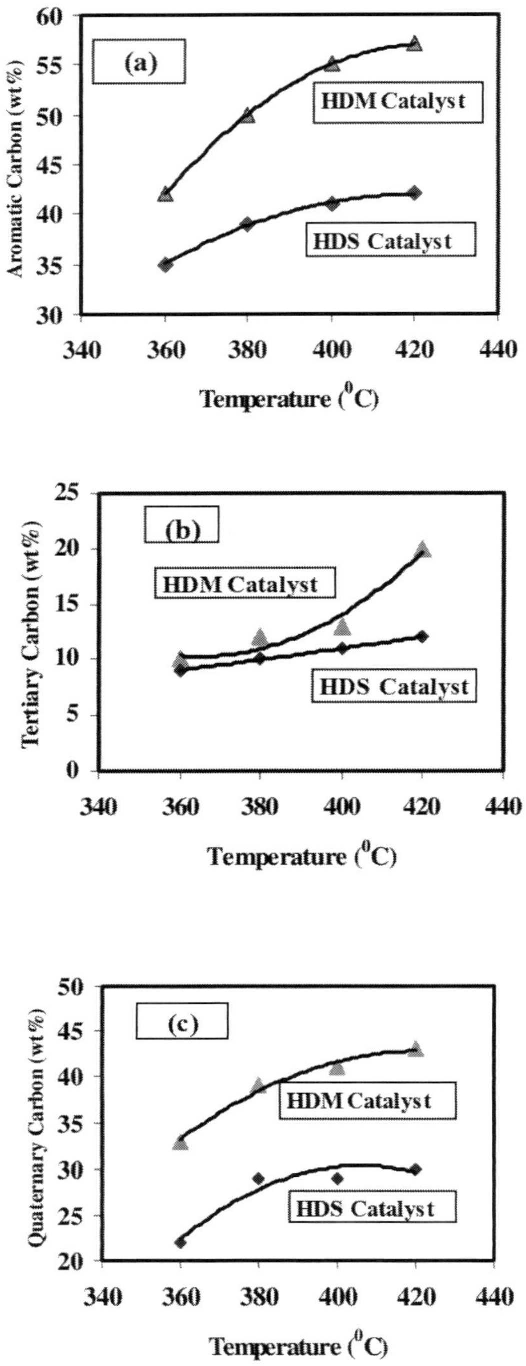

Fig. 1 Distribution of (a) total aromatic carbon (b) tertiary aromatic carbon and (c) quaternary aromatic carbon in the asphaltenes fraction remining in the oil product: Comparison between the HDM and HDS catalysts as a function of reaction temperature

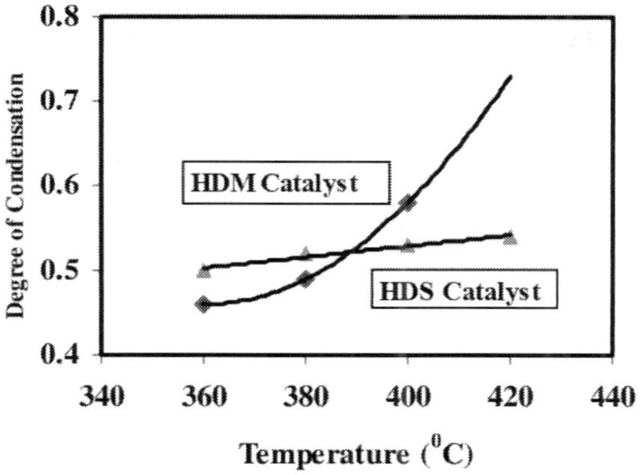

Fig. 2 Condensation of the asphaltenes fraction remining in the oil product: Comparison between the HDM and HDS catalysts as a function of reaction temperature

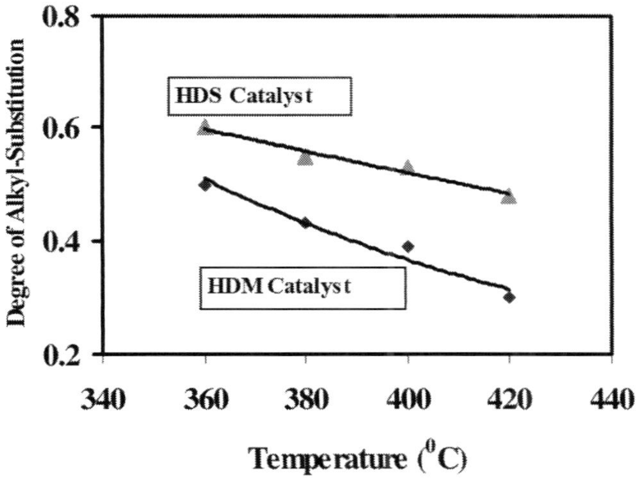

Fig. 3 Alkyl-substitution of the asphaltenes fraction remining in the oil product: Comparison between the HDM and HDS catalysts as a function of reaction temperature

CONCLUSIONS

In the present study, we investigated the effect of operating severity on the quality of liquid products obtained during catalytic hydrotreating of Kuwait atmospheric residue (KU-AR). Two types of industrial hydrotreating catalysts, namely, Mo/Al_2O_3 (HDM) and Ni-Mo/Al_2O_3 (HDS) catalysts, were used in the study in order to understand the role of catalysts in influencing the characteristics of the liquid products. The liquid products from different runs were fractionated into asphaltenes and maltenes and the asphaltene fraction was characterized by various techniques, including ^{13}C NMR. Special attention was paid to the changes in the nature of asphaltenes as a function of operating temperature during KU-AR hydrotreating. The results revealed that, besides the usual heteroatoms removal reactions, such as HDS, HDM, and HDN, asphaltenes and resins in the residual oil feed were converted to saturates and aromatics during the hydrotreating process. The asphaltenes remaining in the product oil become more condensed and less alkyl-substituted with increasing operating severity over both types of catalysts. The aromaticity and degree of condensation in asphaltenes was higher for Mo/Al_2O_3 (HDM) than $Ni\text{-}Mo/Al_2O_3$ (HDS) catalysts. The catalyst's hydrogenation activity appears to have a dominant role in the nature of the asphaltenes in the product oil. Because of the lower hydrogenation function of the unpromoted Mo/Al_2O_3 HDM type catalyst, thermal cracking and condensation may occur, predominantly at higher temperatures, and lead to highly condensed asphaltenes and more refractory and less reactive and have a high propensity for coke formation. The results are considered to be very valuable for the optimization of the operating conditions and its impact on the quality of the product oil as well in catalyst selection for hydroprocessing of KU-AR in an ARDS process.

ACKNOWLEDGEMENTS

This work is a joint project between Japan Cooperation Petroleum Center, JCCP (funded by the Ministry of Economic, Trade and Industry,METI), Japan and Kuwait Institute for Scientific Research, KISR, Kuwait. It bears the KISR Project No. PF010C.

REFERENCES

(1) Abbas, A.K., S.R. Chaudhuri and S. Bhattacharya. 1998. Performance optimization of atmospheric residue desulphurization units. Proceedings, 8th Annual Saudi-Japanese Symposium, Dhahran, Saudi Arabia, November 29-30, pp. 175-189.

(2) Al-Nasser, A. S., R. Chaudhari and S. Bhattacharya. 1996. Mina Abdulla refinery experience with atmospheric residue desullfurization (ARDS). In Catalysis in Petroleum Refining and Petrochemical Industries. Amsterdam, The Netherlands, Elsevier: pp. 171-180.

(3) Adams, C. T; A. A. Del Peggio; H. Schaper; W. H. J. Stock and W. K. Shiflett. Hydroprocessing Catalyst Selection. Hydrocarbon Processing (September): 57-61 (1989)

(4) Furimsky, E. Selection of Catalysts and Reactors for Hydroprocessing. Applied Catalysis A: General 171: 177-206 (1998)

(5) K. Matsushita, A. Marafi A. Hauser, and A. Stanislaus "Relation between Relative Solubility of Asphaltenes in the Product Oil'. Fuel 2004 (Accepted for Publication)

(6) Marafi. A., S. Fukase, M. Al-Marri and A. Stanislaus. A Comparative Study of the Effect of Catalyst Type on Hydrotreating Kinetics of Kuwaiti Atmospheric Residue. Energy and Fuel. Vol.17, Issue No. 3, May , 661-668 (2003)

(7) Quan, R. J; A. Ware; C. W. Hang and J. Wei. Hydrodemetallation of Petroleum. Advances in Chemical Engineering14: 95-257 (1988)

Integrated View to Understanding the FCC NO_x Puzzle

Frank S. Rosser, Mark W. Schnaith and Patrick D. Walker
UOP LLC
Des Plaines, Illinois, USA

ABSTRACT

Traditional regenerator optimization has focused on control of two key process variables: carbon on regenerated catalyst and afterburn. Now a third process variable: flue gas NO_x emissions has been added. However, the most widely used methods of controlling regenerator afterburn – platinum based CO promoter and high excess oxygen – also strongly increase NO_x. It will be demonstrated that regenerators with superior air-catalyst mixing control afterburn while emitting substantially less NO_x than typical bubbling bed regenerators. Proven technology that can be used to optimize all three process variables will be discussed.

INTRODUCTION

The flue gas stack from a fluid catalytic cracking (FCC) unit emits a number of pollutants. These include particulates, carbon monoxide (CO), sulfur oxides (SO_x), specifically SO_2 and SO_3 and nitric oxides (NO_x), specifically NO and NO_2.

NO_x is implicated as a cause of smog. This occurs when NO_x reacts with volatile organic chemicals and sunlight to form ozone. For this reason, regulatory agencies have increased attention on reducing NO_x emissions from stationary sources. These include petroleum refineries and coal burning power plants as well as others. For FCC units in the United States, national emission standards for particulates and CO are listed in the EPA's MACT II ruling and recorded in the federal register.[1] National emission standards for SO_x are listed in the EPA's New Source Performance Standard and also recorded in the federal register.[2] National emission standards for NO_x have not been established by the EPA. For many refineries, the NO_x emission limit from the FCC unit is established by a consent decree between the EPA and the refinery. The NO_x emission limit is different for each location, but most are around 20 ppmv.

There are several ways to control NO_x produced by an FCC regenerator. In situ methods include equipment design, operating conditions and additives, which include non platinum based CO promoters and NO_x reducing additives. Feed hydrotreating can reduce total feed nitrogen 20 to 90% depending on severity and consequently reduce the NO_x potential of the feed. Flue gas treating methods include selective catalytic reduction systems, selective non catalytic reduction systems and ozone injection systems. The focus of this paper is equipment design and operating conditions.

ORIGIN OF NO_x IN FCC REGENERATOR FLUE GAS

By definition, NO_x includes NO plus NO_2. Other nitrogen bearing oxides such as N_2O are also present in FCC unit flue gas, but are not considered to be NO_x and thus not controlled. It is generally understood that the predominant NO_x species inside an FCC regenerator is NO which is further oxidized to NO_2 upon release to the atmosphere.[3]

There are two ways NO_x can be produced: i) oxidation of N_2 which enters with the combustion air, also known as thermal NO_x and ii) oxidation of nitrogen compounds in the fuel, also known as fuel NO_x. In the case of FCC units, the fuel is coke, which is deposited on the catalyst during feedstock conversion inside the reactor. Thermal NO_x occurs at very low rates below 1600°F.[4] Consequently thermal NO_x is not a concern for FCC unit regenerators, most of which operate below 1400°F.

One may speculate that during regeneration, the peak intraparticle temperature will far exceed the bulk temperature resulting in thermal NO_x. Combustion of pulverized coal in power generators can indeed produce intraparticle temperatures which exceed the bulk temperature by as much as 700°F.[5] However, high intraparticle temperatures are unlikely within FCC catalyst due to the small particle size and low coke concentration. In order to demonstrate that thermal NO_x is not significant in an FCC regenerator, Dishman et al. performed an experiment where nitrogen free coke was combusted at typical FCC regenerator temperatures. The resulting flue gas contained essentially no NO_x.[6]

TYPICAL FCC UNIT NITROGEN BALANCE

Figure 1 shows the nitrogen balance obtained by UOP for FCC Unit "A." About 40 wt-% of the feed nitrogen exits the reactor as nitrogen heterocompounds in the liquid products and about 10 wt-% of the feed nitrogen exits the reactor as ammonia collecting in the sour water. The remaining 50 wt-% of the feed nitrogen exits the reactor as coke on spent catalyst delivered to the regenerator.

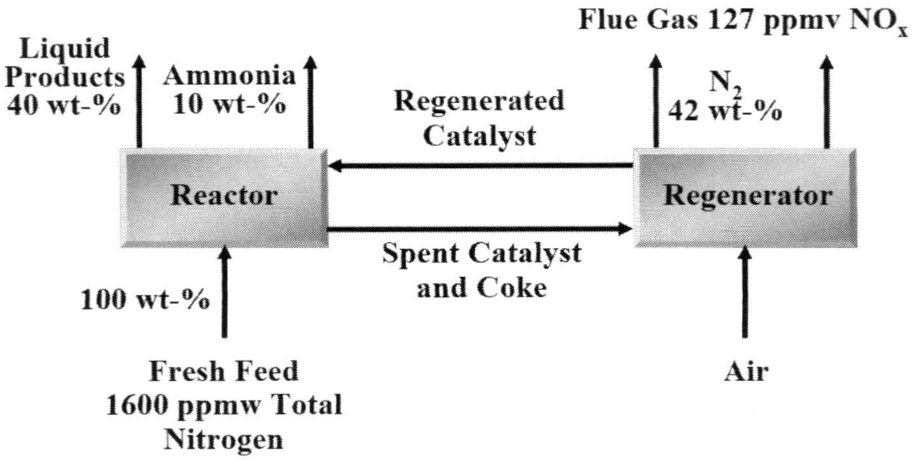

Figure 1

Unit "A" – Nitrogen Distribution among Reactor Products and Flue Gas

The flue gas in this example contains 127 ppmv NO_x while the feed contains 1600 ppmw total nitrogen. The nitrogen portion of NO_x in the flue gas amounts to about 3 wt-% of the total feed nitrogen which is equivalent to 6 wt-% of the coke nitrogen. The remaining coke nitrogen starts out as NO_x or some intermediate nitrogen species, but is quickly reduced to N_2. The overall result is that 94 wt-% of the coke nitrogen is reduced to N_2. For this FCC unit, reduction of coke nitrogen would have to increase from 94 wt-% to 99 wt-% to attain 20 ppmv NO_x in the flue gas. The large fraction of N_2 in combustion air makes it impractical to directly track feed nitrogen reduction to N_2 in a commercial FCC operation. However, riser pilot plant testing has proven that this does indeed occur.[3,4] This testing was conducted using pure argon and oxygen to regenerate the catalyst. The flue gas contained NO_x plus substantial amounts of N_2.

REGENERATOR MODE OF OPERATION

There are two general modes of operating an FCC unit regenerator: partial combustion mode and full combustion mode. In partial combustion mode, the flue gas contains several percent carbon monoxide. NO_x leaving a regenerator operating in partial combustion mode tend to be lower than NO_x leaving a regenerator operating in complete combustion mode. This is strong evidence that CO reacts with NO to form N_2. For units operating in partial combustion mode, CO in the flue gas must later be burned in an incinerator to meet CO emission limits. CO incinerators fire supplemental fuel and operate at 1600 to 1800°F. Consequently CO incinerators produce thermal NO_x in addition to the fuel NO_x already created by burning coke.

For FCC units operating in full combustion mode, CO incinerators are usually not in place. Therefore the regenerator itself must oxidize the CO to control afterburn and keep CO emissions below the regulatory limit. The MACT II emission limit for CO emissions from an FCC regenerator is 500 ppmv, but state or local limits may be lower. Platinum based CO promoter is frequently used to control afterburn and CO emissions. Reference 7 provides a very good description of the role of platinum based CO promoter in FCC regenerators.[7]

Oxidation of CO to CO_2 proceeds rapidly in the presence of excess oxygen. Hence oxidation of CO directly competes with reduction of NO_x via CO in the reaction $2NO + 2CO \rightarrow N_2 + 2CO_2$. Also, reduction of NO_x via carbon in the reaction $2NO + C \rightarrow N_2 + CO_2$ is significantly slower when the CO concentration is low.[8] CO promoter will increase the rate of CO oxidation and therefore work against NO_x reduction. Therefore the simultaneous minimization of both CO and NO in an FCC regenerator operating in complete combustion mode is difficult to achieve. Some researchers have looked at reaction equilibrium to better understand the limits of NO_x reduction. Calculations by Zhao et al. for combustion of coke on FCC catalyst indicate an equilibrium NO concentration of 8 to 10 ppmv.[3] Wojtowicz et al. conducted equilibrium calculations for coal combustion and showed similar results at typical FCC regenerator operating temperatures.[9] Commercial data shows that some advanced FCC regenerator designs can indeed approach these limiting values but most do not.

Types of Coke

Coke is characterized in three different types: i) catalytic coke; ii) feed residue coke and iii) catalyst circulation coke.[10]

Catalytic coke is hydrogen deficient coke located primarily within the pore structure and is a product of cracking mechanisms. This coke can be sub-divided into two categories: i) coke produced from the catalytic activity of the catalyst itself and ii) coke produced from the catalytic action of the metals deposited upon the catalyst. Contaminant metals accumulated on the catalyst surface will act as a dehydrogenation catalyst to further condense coke molecules.

Feed residue coke is primarily from large multi-ring aromatics (asphaltenes) which will not vaporize and rather deposits directly on the catalyst. The contribution from feed residue coke changes only with feedstock quality.

Catalyst circulation coke is hydrogen rich coke adsorbed on the catalyst surface or in the emulsion phase. This coke can be viewed as supplemental fuel which is burned in the regenerator to meet the heat demand of the reactor. The reactor temperature controller automatically adjusts this fuel rate by adjusting the catalyst circulation rate.

The laboratory method typically used to separate coke for quantitative analysis is solvent extraction resulting in only two types of coke. The portion of the total coke which can not be extracted is called hard coke. Everything else is called soft coke.

The amount of soft coke adsorbed on the spent catalyst can vary widely depending on the efficiency of the spent catalyst stripper. Based on a review of eight commercial FCC units, soft coke was found to range from 2.5 wt-% to 33.5 wt-% of the total coke. Soft coke nitrogen accounts for 0.27 wt-% to 5.24 wt-% of the total coke nitrogen. Refer to Table 1 for details.[11]

Table 1

Nitrogen Distribution between Soft Coke and Hard Coke

Sample	*1*	*2*	*3*	*4*	*5*	*6*	*7*	*8*
wt-% Soft Coke in Total Coke	**33.5**	**25.9**	**14.2**	**12.6**	**12.0**	**9.8**	**6.8**	**2.5**
wt-% Soft Coke N in Total Coke N	**99.73**	**0.27**	**5.24**	**1.06**	**2.29**	**0.47**	**NA**	**NA**

Basic and Non Basic Nitrogen Compounds

Heterocyclic nitrogen compounds are the dominant nitrogen compounds in FCC feedstock. These can be classified as basic and non basic. Basic nitrogen compounds are predominantly pyridine type structures, which include pyridines, quinolines and acridines. Non basic nitrogen compounds are predominantly pyrroles, typically indoles and carbazoles.[12] The chemical structure for these compounds are described in Figure 2.

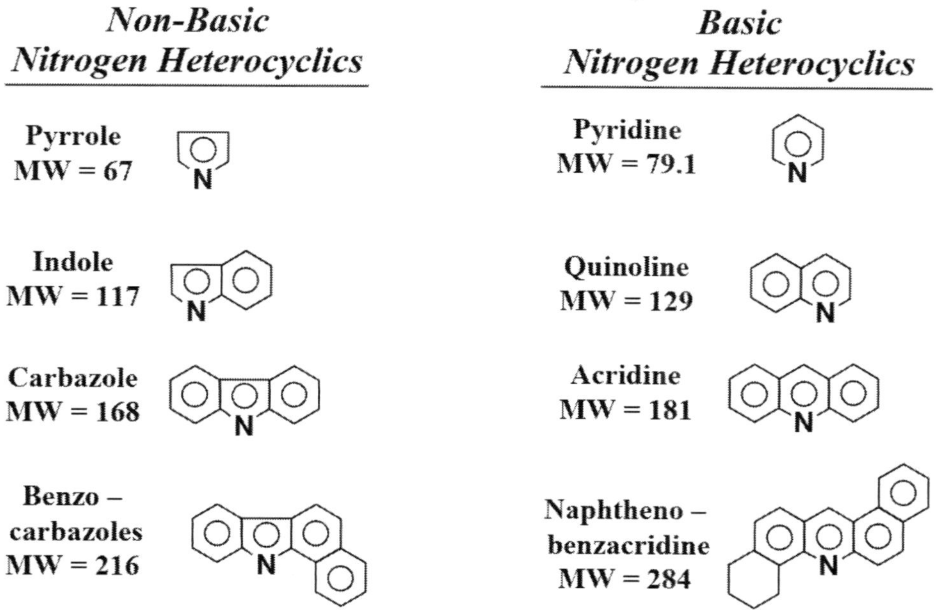

Figure 2

Chemical Structure of Feed Nitrogen

Basic nitrogen compounds have long been recognized as a catalyst poison and therefore an important component of coke.[13] It has been suggested that the majority of nitrogen compounds associated with coke are due to these adsorbed basic nitrogen compounds.[3] The concentration of acid sites in the catalyst exceeds the concentration of basic nitrogen molecules by several orders of magnitude. It is thus tempting to speculate that all basic nitrogen will become coke.

Table 2 is a detailed analysis of the nitrogen compounds in the feed and in the liquid products for FCC Unit "B." The nitrogen compounds in the liquid product are indeed predominantly non-basic pyrroles. However, a significant presence of pyridines indicates that the attraction of these basic nitrogen compounds to the catalyst acid sites is being hindered in some way.

Basic nitrogen compounds in the liquid products are typically 4 to 6 ring aromatics. These large molecules may be diffusion limited due to the critical dimension of the molecule exceeding the catalyst's pore mouth opening. Venuto and Habib indicate that any molecule in excess of three rings will likely experience diffusion limitations.[14] Such limitations will be enhanced as coke lay down reduces the catalyst's pore mouth opening.[15] Figure 3 is an illustration of the hindrance caused by large heterocyclic molecules.

Table 2

Unit "B" – Nitrogen Compounds in Feed and 450+ °F Liquid Products

Component	*Feed wt-% N*	*MW*	*Liquid Product wt-% N on FF*	*MW*
Napthenopyridines	0.0006	225.3	—	—
Indoles	0.0006	232.38	—	—
Quinolines	0.0009	314.02	—	—
Naphthenoquinolines	0.0005	298.49	—	—
Carbazoles	0.0162	354.93	0.00867	224.74
Acridines	0.0162	353.3	0.00094	257.59
Naphthenobenzoquinolines	0.0131	364.51	0.00020	294.44
Benzocarbazoles	0.0268	324.35	0.00596	275.72
Benzacridines	0.0113	359.43	0.00139	306.12
Naphthenobenzacridine	0.0083	372.5	0.00112	324.14
Dibenzocarbazoles	0.0060	349.01	0.00153	318.04
Dibenzacridines	0.0030	374.9	0.00054	336.86
Acenaphobenzindole	0.0017	335.41	0.00035	341.61
Tribenzocarbazoles	0.0012	350.52	trace	367.18
Benzonaphthacridines	0.0020	347.73	0.00037	328.24
Diindenoacridines	0.0008	366.42	—	345.79
Dinaphthocarbazoles	trace	367.14	—	—
Total N wt-% Feed	0.1090		0.0211	
Total Basic N wt-% Feed	0.0567		.00456	

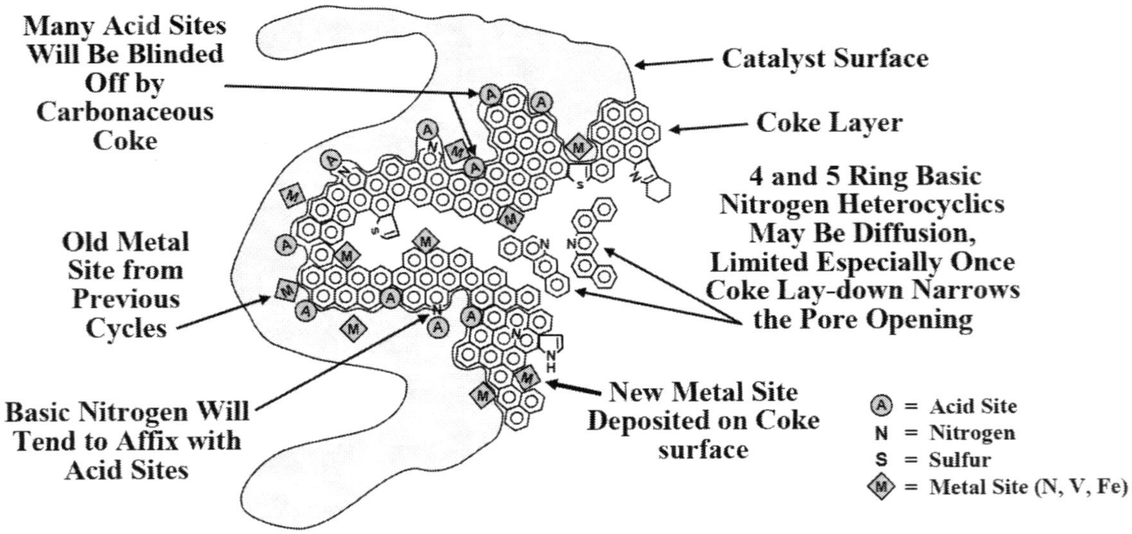

Figure 3

Hindrance of Large Heterocyclics

Other researchers have studied coked catalyst from several FCC units using MALDI-TOF Mass Spectrometry and were able to identify non-basic carbazoles and indoles in the coke.[16] Thus basic feed nitrogen is probably not substantially superior to total feed nitrogen as a predictor of NO_x potential.

COKE COMBUSTION

Figure 4 describes the principal mechanisms for combustion of coke and reduction of NO in an FCC regenerator. Many of these NO reduction mechanisms are surface reactions which require reactants to adsorb onto a free active carbon site.

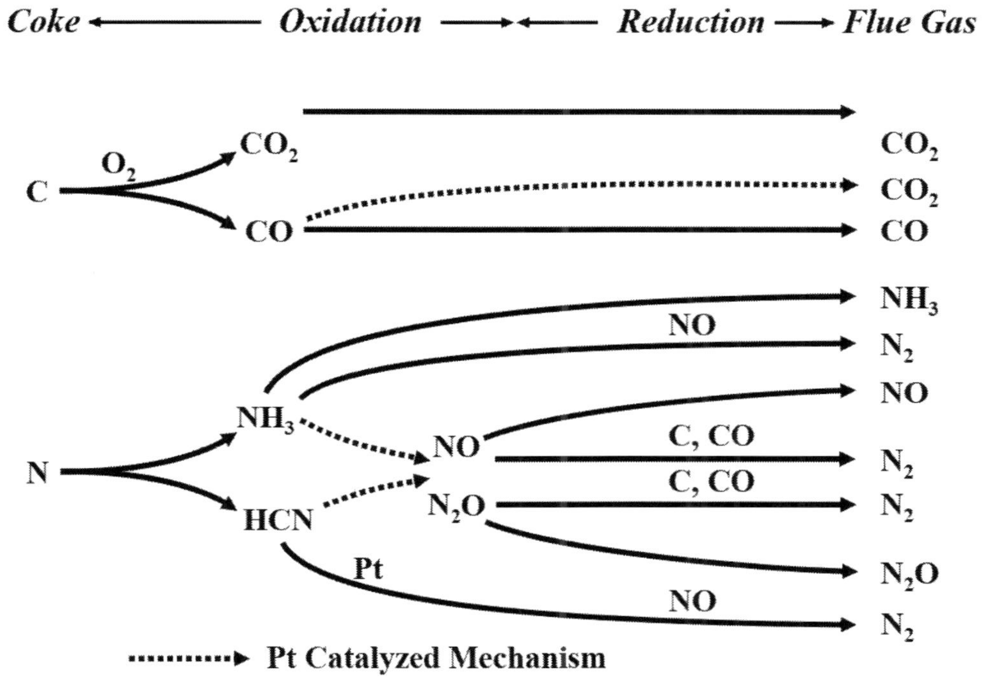

Figure 4

Primary Reactions and Products of Coke Combustion

SOFT COKE COMBUSTION

Combustion of soft coke will occur first. Soft coke nitrogen is a small percentage of the total coke nitrogen, typically less than 5 wt-% per Table 1. However soft coke nitrogen is quite substantial when compared with the nitrogen leaving the regenerator as NO_X. Some of the soft coke will contain nitrogen compounds, which convert to HCN and NH_3 intermediates and then further convert to NO, N_2O or N_2 through a series of homogenous reactions.[17] If platinum based CO promoter is present, it will oxidize the HCN and NH_3 to NO.[16] Since HCN and NH_3 can react with NO to form N_2, platinum based CO promoter can also inhibit NO reduction.

NO might adsorb onto a coke surface where it can be reduced via reaction with CO or carbon. However, this is unlikely because NO must compete with O_2 and CO for free active carbon sites. Furthermore the NO faces mass transfer limitations in moving from the bubble phase to the emulsion phase. It is therefore likely that NO formed from the soft coke fails to be effectively reduced in the regenerator.

HARD COKE COMBUSTION

Deposition of coke, and specifically deposition of coke nitrogen, will vary with pore size. Various researchers have grouped pore sizes into three categories, typically micropore, mesopore and macropore.[18] For discussion purposes, we will regroup these into two categories: large pores (greater than 200 Å) and small pores (less than 200 Å), and explore the reduction of NO in each.

Large Pores: After soft coke combustion is well underway, hard coke combustion will commence from the outside in. This will involve the adsorption of oxygen onto the coke surface, forming both NO and CO. For reduction of NO to proceed, a second NO molecule must be present. Nitrogen is randomly dispersed throughout the hard coke, so the likelihood of an adjacent NO being formed is small.[6] Therefore a second NO molecule must adsorb onto the coke surface from the gaseous phase. However, O_2 and CO are in abundance and competing strongly for the free active carbon sites. The NO formed in large pores therefore has a poor chance of reducing. NO desorbing from these large pores will then face the same competitive environment for reduction as does soft coke. Note that the presence of metals on catalyst (typically V, Ni, and Fe) can influence NO_x emissions. [19, 6]

Small pores: Inside the small pores, the impediments to NO reduction discussed thus far are not significant. These pores account for more than 80% of the total surface area and are the location of most of the acid sites. Recall that hard coke typically contains more than 95 wt-% of the coke nitrogen per Table 1. Hence the great majority of coke nitrogen is found in the small pores. These pores present a near perfect environment for reduction of NO. The reactants CO and NO will be at their highest relative concentrations and, due to the small pore volumes, the likelihood of collisions is enhanced. Researchers in char combustion have made similar observations regarding NO_x reduction in micropores formed in char.[17] Figure 5 is an illustration of coke distribution in the pore structure.

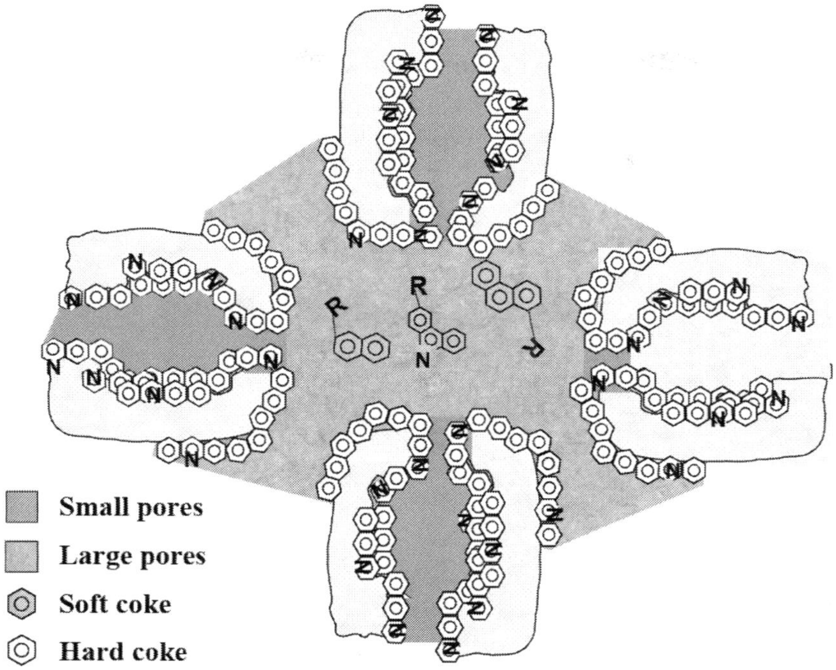

Figure 5

Coke Distribution in Catalyst Pore Structure

Even though soft coke nitrogen typically contains less than 5 wt-% of the total coke nitrogen, it probably contributes disproportionately to NO_x. Improving efficiency of the stripper will drive off soft coke and require that the catalyst circulation rate increase in order to restore the heat balance. This should increase hard coke and reduce soft coke and perhaps reduce NO_x emissions.

PROCESS VARIABLES

The process variables which have a large impact on the NO_x concentration in the flue gas are: i) feed nitrogen, ii) platinum level on the circulating catalyst and iii) flue gas excess oxygen. Figures 6 through 8 show the response of NO_x to each of these variables for three different FCC units.

Feed Nitrogen: A substantial change in feedstock such as adding resid will impact NO_x. Figure 6 shows this response for FCC Unit "C" which operates part of the time with VGO feedstock and part of the time with VGO plus resid feedstock. The VGO plus resid operation produces more than twice the NO_x of the VGO operation.

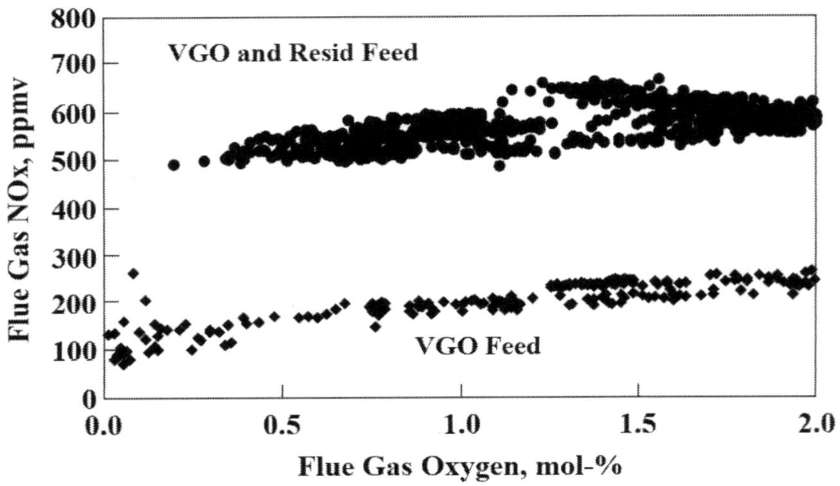

Figure 6

Unit "C" – Impact of Feed Resid

Hydrotreating will also result in a downward shift in NO_x. Hydrotreating reduces the total amount of nitrogen in the feed and therefore reduces the total amount of nitrogen that reaches the regenerator. The result will be lower overall NO_x compared with non-hydrotreated feedstock. Results will vary with feedstock properties and severity of hydrotreating.

Note that many FCC operators have reported no correlation between feedstock nitrogen and NO_x emissions as long as the upstream feedstock processing is unchanged.

Platinum Level on the Circulating Catalyst: Figure 7 is a plot of platinum on circulating catalyst versus NO_x emissions for FCC Unit "D." As platinum is increased, NO_x emissions increase. This same relationship has been widely reported by pilot plant studies[16] and by many FCC operators.

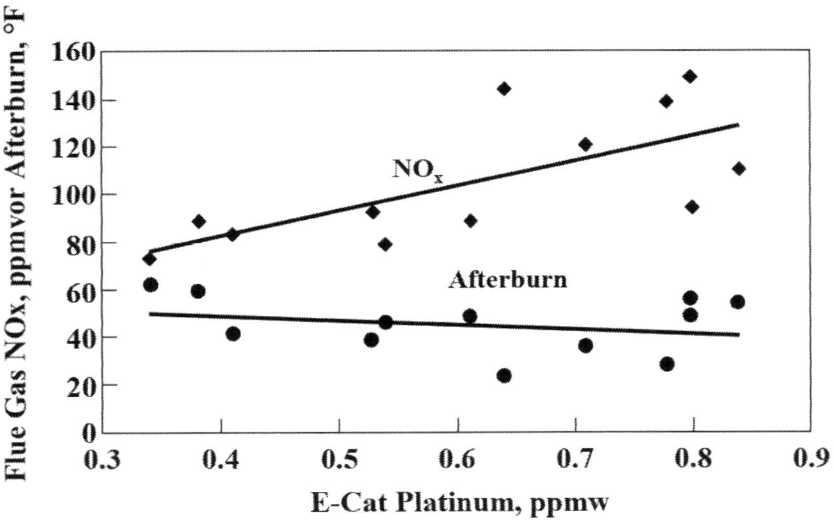

Figure 7

Unit "D" – Impact of CO Promoter

Flue Gas Excess Oxygen: Wojtowicz et al. conducted equilibrium calculations for NO_x produced from coal combustion at excess oxygen concentrations of 0.5 mol-% and 5.6 mol-%.[9] This work indicates that the predicted NO_x was negligible for the 0.5 mol-% excess O_2 case, and 50 ppmv for the 5.6 mol-% excess O_2 case. Dishman et al. found similar results in FCC pilot plant experiments in which they measured 16 ppmv NO_x at 0.5 mol-% excess oxygen and 33 ppmv NO_x at 4.0 mol-% excess oxygen.[6]

Figure 8 is a plot of flue gas NO_x response to excess oxygen for FCC Unit "E." This plot shows increasing NO_x emissions with increasing excess oxygen. Every unit we have studied shows a similar pattern, but usually with higher absolute values of NO_x. This particular data was from a UOP Combustor-style regenerator.

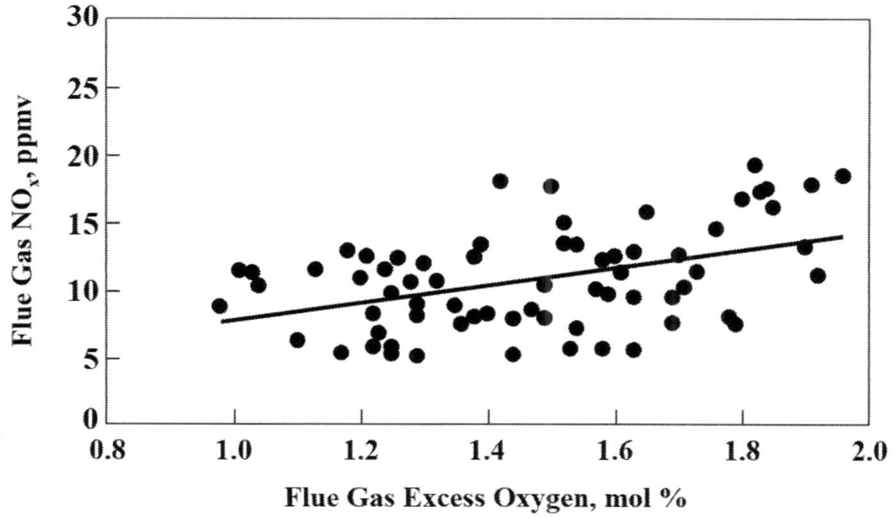

Figure 8

Unit "E" – Impact of Excess Oxygen

REGENERATOR DESIGN

There are two general types of FCC regenerator designs in service today: bubbling bed regenerators and combustor style regenerators.

BUBBLING BED REGENERATORS

Bubbling bed regenerators have been used to regenerate spent catalyst since the first FCC unit started up in 1942 and they are still widely used today. Following the introduction of CO promoter in 1974, many bubbling bed regenerators switched from partial combustion mode to complete combustion mode to capture yield advantages.

An important aspect of catalytic cracking is the requirement to move large amounts of fluidized catalyst back and forth between the reactor and the regenerator. The historical way to do this is to arrange the regenerator so the catalyst enters the top of the bed and leaves at the bottom per Figure 9. Most bubbling bed regenerators, but not all, are arranged this way. The combustion air flows up, counter-current to the catalyst which flows down.

Figure 9 also shows the internal process flow of a bubbling bed regenerator. A typical superficial velocity is around 3.0 ft/s. At this velocity, catalyst is entrained into the cyclones and flows back down into the dense bed at about 1.3 times the catalyst circulation rate.

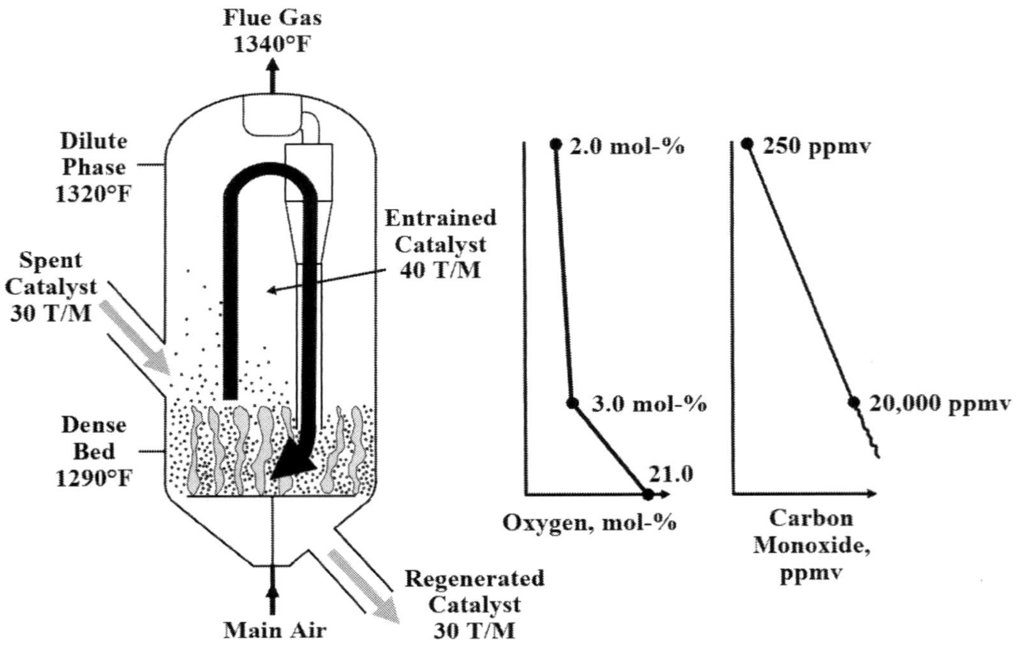

Figure 9

Bubbling Bed Regenerator

Internal Process Flow

For a typical bubbling bed regenerator, the temperature will increase about 30°F from the dense bed to the cyclone outlet. Knowing this temperature rise and the catalyst entrainment rate, one can calculate that the flue gas leaving the surface of the bed must have around 2 mol-% CO (or the heating value equivalent of carbon plus CO). This CO will require about 1 mol-% oxygen for combustion. Assuming the flue gas has 2 mol-% excess oxygen, there will likely be 3 mol-% oxygen at the surface of the bed and higher amounts below the surface.

COMBUSTOR-STYLE REGENERATORS

Figure 10 shows the internal process flow of a combustor-style regenerator. The combustor (lower vessel) is designed to completely burn the coke to CO_2 without CO promoter and with low excess oxygen. This is accomplished by excellent mixing and symmetrical design. Some combustor style regenerators do use small amounts of CO promoter because they operate at low temperature or way beyond original design capacity.

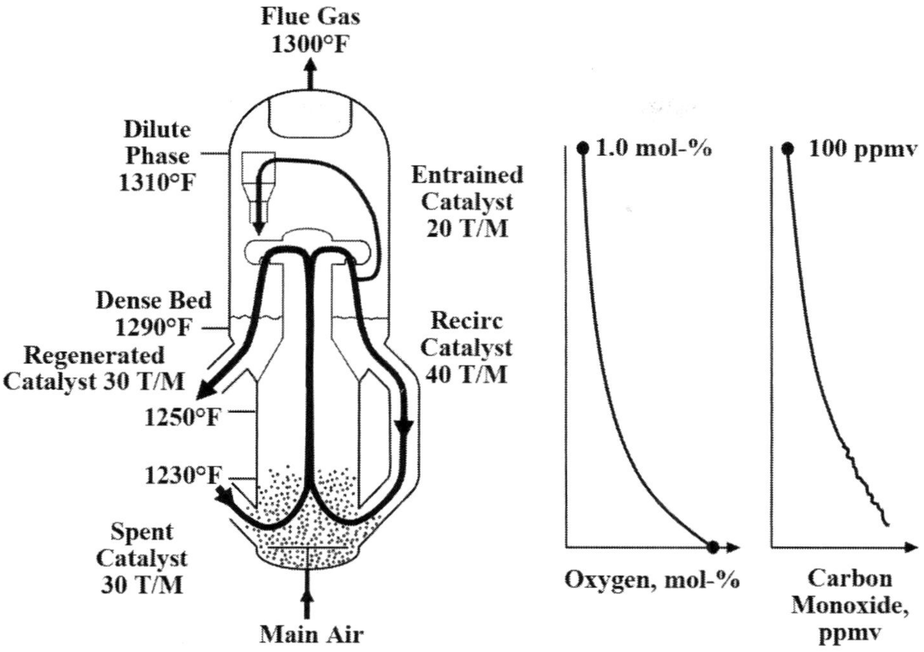

Figure 10

Combustor Style Regenerator

Internal Process Flow

The combustor is designed to operate at about twice the velocity of a typical bubbling bed regenerator. The higher velocity results in a lower density than in a bubbling bed so resistance to horizontal mixing is reduced. The lower part of the combustor is a highly backmixed fast fluidized bed. A portion of the hot regenerated catalyst is recirculated to the lower combustor to heat the incoming spent catalyst and to control the combustor density to the desired level. As the catalyst enters the combustor riser, the velocity is further increased and the two-phase mixture exits through symmetrical downturned arms. All of these features result in excellent afterburn performance. A typical high efficiency regenerator in service today uses less than 1 mol-% excess oxygen and results in less than 20°F afterburn.

UOP conducted a comprehensive survey of several FCC unit operators with high quality NO_x emission data, specifically units with continuous emission monitoring systems. Figure 11 shows the average reported NO_x as a function of the average excess oxygen for each unit in the survey. Each data point is a different unit, and the regenerator design is indicated for each data point - either combustor style or bubbling bed. Combustor style regenerators clearly produce 70 to 80 % lower NO_x than bubbling bed regenerators.

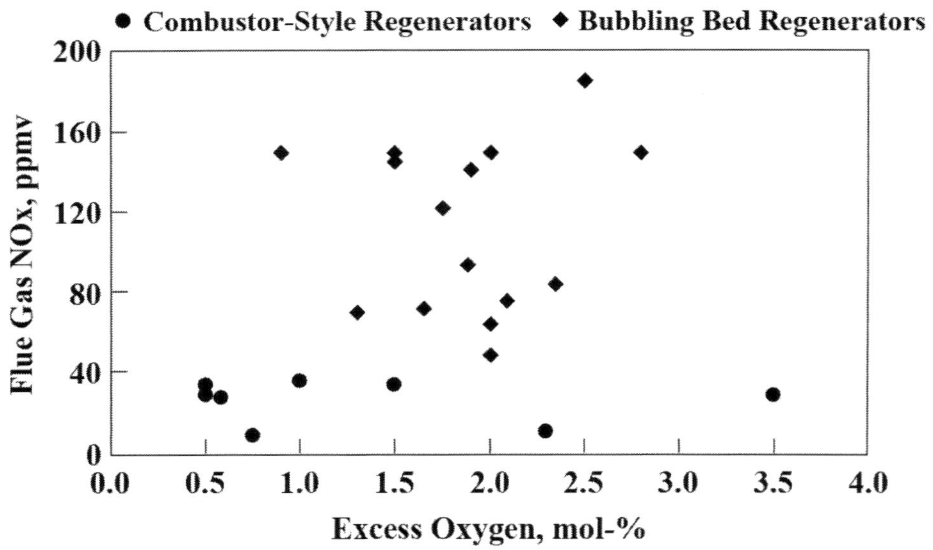

Figure 11

Impact of Regenerator Design

SUMMARY

The most important factor which influences NO_x emissions from an FCC regenerator is the regenerator design. Combustor-style regenerators provide excellent air-catalyst mixing and therefore can achieve low NO_x emissions without flue gas treating or catalyst additives. Combustor-style regenerators consistently achieve superior NO_x performance.

Other important factors which influence NO_x emissions from an FCC regenerator are excess oxygen and platinum based CO promoter use. Superior regenerator designs allow good control of afterburn while using minimal excess oxygen and minimal CO promoter. The amount of nitrogen in the feedstock is also an important factor which influences NO_x emissions. However, the impact of nitrogen in the feedstock is usually only observed when upstream processing of the feedstock is changed. Basic nitrogen is probably not substantially superior to total nitrogen as a predictor of NO_x performance.

Nitrogen in soft coke probably contributes disproportionately high to NO_x. Improving the efficiency of the spent catalyst stripper can reduce soft coke and may reduce NO_x emissions.

References

1. Federal Register, Title 40, Volume 67, Part 63, No. 70, Dated April 11, 2002: National Emission Standards for Hazardous Air Pollutants for Petroleum Refineries: Catalytic Cracking Units, Catalytic Reforming Units, and Sulfur Recovery Units; Final Rule.

2. Federal Register, Title 40, Volume 6, Part 60, Subpart J, No. 104, Revised July 1, 1999: Standards of Performance for New Stationary Sources, Subpart J – Standards of Performance for Petroleum Refineries, 60.104 - Standards for Sulfur Oxides

3. Zhao, X.; Peters, A.W.; Weatherbee, G.W; "Nitrogen Chemistry and NOx Control in a Fluid Catalytic Cracking Regenerator," Ind. Eng. Chem. Res. 1997, 36, 4535-4542

4. Peters, Alan W.; Zhao, Xinjin; Weatherbee, Gordon W. "The Origin of NO_x in the FCC Regenerator," AM 95-59, National Petroleum Refiners Association 1995 Annual Meeting

5. Chen, Z.; Lin M.; Ignowski, J.; Kelly, B.; Lionjewile, T.M; Agarwal, P.K.; "Mathematical Modeling of Fluidized Bed Combustion: N_2O and NO_x Emissions from the Combustion of Char," Fuel 2001 Volume 80 Pages 1259-1272

6. Dishman, K.L.; Doolin, P.K.; Tullock, L.D.; "NO_x Emissions in Fluid Catalytic Cracking Catalyst Regeneration," Ind. Eng. Chem. Res. 1998, 37, 4631-4636

7. Engelhard Corporation Catalyst Reports, Section 5: FCCU Mechanical and Operational Issues: "Understand CO Oxidation Operating Variables to Prevent Afterburn." Web site www.refiningonline.com/EngelhardKB, retrieved 9 Feb 2004

8. Aarna, Indrek; Suuberg, Eric M., "A Review of the Kinetics of the Nitric Oxide – Carbon Reaction," Fuel, Vol. 76, No. 6, page 475-491.

9. Wojtowicz, M.A.; Pels, J.R.; Moulijn, J.A.; "Combusiton of Coal as a Source of N_2O Emission," Fuel Process Technol., 1993, 34, 1-71

10. Upson, Lawrence; Daline, Ivan; Wichers, Wicher R.; "Heat Balance the Key to Cat Cracking," 3rd Katalistiks Fluid Cat Cracking Symposium, May 26-27, 1982.

11. Data courtesy of Akzo-Nobel (unpublished).

12. Silverman, L.D.; Winkler, W.S.; Tiethof, J.A.; Witoshkin, A.; "Matrix Effects in Catalytic Cracking," 1986 NPRA Annual Meeting

13. Fu, C.M.; Schaffeer, A.M.; "Effect of Nitrogen Compounds on Cracking Catalysts," Ind. Eng. Chem. Res., 1985, 29, page 68

14. Fluid Catalytic Cracking with Zeolite Catalysts, Paul Venuto and E.T. Habib; 1979 Marcel Dekker Inc, N.Y.

15. Occelli, M.L.; Olivier, J.P.; Auroux, A. "The Location and Effects of Coke Deposition in Fluid Cracking Catalysts during Gas Oil Cracking in Microactivity Test Conditions," Journal of Catalysis, (2002) Volume 209

16. Barth, J.O.; Jentys, A.; Lercher, J.A.; "Towards Understanding the Genesis and Removal of NO_x in FCC Regenerators," Proceedings 17th World Petroleum Congress, Rio de Janerio, Sept 2002

17. Thomas, K.M.; "The Release of Nitrogen Oxides During Char Combustion," Fuel, 1997, Volume 76 Number 6.

18. Handbook of Porous Solids Volume 1, Section 1.2 Definitions, Terminology, and Classifications of Pore Structures, Kenneth S.W. Sing and Ferdi Schuth, Published by Wiley-VCH 2002

19. Jones, Jenny M.; Zhu, Qian, Thomas, K. Mark, "Metalloporphyrin-Derived Carbons: Models for Investigating NOx Release from Coal Char Combustion," Carbon, 37, 1999 page 1123-1131

uop

UOP LLC
25 East Algonquin Road
Des Plaines, IL 60017-5017

UOP 4287B

Paper No. 92a

HDS: A New Approach to the Old Topic

Zbigniew Ring, Jinwen Chen, Hong Yang, Hongbin Du, Yevgenia Briker

National Centre for Upgrading Technology (Canada)

Prepared for Presentation at the 2004 AIChE National Spring Meeting,

Session 92: Desulphurization Tutorial, New Orleans, LA, USA

April 2004

Unpublished

Abstract

Because of increasingly stringent environmental regulations around the world, in the last several years, "ultra-low sulphur diesel (ULSD)" or even "zero-sulphur diesel" have been subjects of frequent discussions in the petroleum refining community. Refineries, catalyst vendors, technology developers and R&D institutions are all facing this significant fuel specification challenge. Because of the increasing pressure to reduce aromatics levels in diesel, it is expected that hydrodesulphurization (HDS) will remain the dominant commercial technology for achieving ultra-low sulphur levels in the future. This is in spite of alternative desulphurization technologies being developed and commercialized. Therefore, it is important to continue working on improvements to HDS-related technologies and help refineries achieve both their economic and environmental goals.

Unlike many other publications in this area, this tutorial will emphasize practical aspects of HDS research and mathematical modeling of HDS reactions in the context of other types of hydrotreating reactions. Particular attention will be given to the most HDS-resistant alkylated dibenzothiophenes. The theoretical background is illustrated with HDS experimental results generated at NCUT in the following areas: 1) methods of sulphur analysis; 2) new opportunities in sulphur speciation including the application of the quantitative structure–property relationships; 3) effect of phase equilibrium on kinetics analysis; 3) HDS catalyst effectiveness; 4) mathematical modeling of the reactivity of individual sulphur compounds in real feedstocks; and 5) modeling molecular interactions in real feedstocks.

Introduction

Hydrodesulphurization (HDS) of fuel blending stocks is indeed an old topic. Decades of research have been published covering every possible aspect of this group of processes, including:

- detailed consideration of advanced analytical methodologies;
- the role of the various types of catalytic sites;
- reaction mechanisms including steric effects and the commonly accepted two-route mechanism involving direct abstraction and hydrogenation followed by the removal of sulphur atom;
- HDS behaviour of various feedstock-catalyst systems;
- various attempts to model HDS kinetics, extract key model parameters and, based on those, predict HDS performance.

This huge body of experimental work covers both model and real systems. However, as we point out in this paper, not all important aspects of HDS have received sufficient attention over the years, and still new research opportunities turn up.

Our motivation in looking at HDS is the desire to build a predictive hydrotreater model that will account not only for the complex chemistry but also for the equally complex physics of the hydrotreater. In modeling, it is essential to be able to adequately describe and, hence, separate these two components, in order for the parameters of the predictive model to have physical meaning and, consequently, stay independent of operating conditions. In this sense, our paper is not so much about introducing a new approach to HDS as it is about looking at it from an angle that is rarely used. This paper examines HDS from the point of view of a model developer interested in the broad picture. It briefly reviews the work conducted in HDS at the National Centre for Upgrading Technology (NCUT) in Canada, especially in the areas we feel additional research is needed.

Analytical Methods

The knowledge of a feedstock's molecular composition plays a critical role in modeling. In particular, sulphur and nitrogen compounds, and various aromatics types are of key importance. In this section we focus on identification and quantification (speciation) of sulphur compounds, and leave the problem of finding an adequate molecular representation of the entire feedstock for another occasion.

Speciation of sulphur in petroleum fractions, potentially containing tens or even hundreds of thousands of sulphur species, is a formidable task for the analytical chemist. In fact, it is too much to ask of a chemist to completely speciate sulphur in such a complex mixture. To focus our discussion, we limit ourselves to the relatively simple case of cracked middle distillates, which are well represented by commercially important fluid catalytic cracking light cycle oil (LCO). The variety of sulphur compounds in LCO is greatly reduced compared to virgin material in a similar boiling range. In LCO, the content of easily hydrodesulphurized aliphatic sulphur species (e.g. mercaptans, sulphides, disulphides, etc) is low, and the longer aliphatic substitutents on the more difficult-to-remove benzothiophenes (BTs) and the most difficult-to-remove dibenzothiophenes (DBTs) are cracked off, predominantly leaving methyls and ethyls. But even in this relatively simple LCO case, the number of sulphur species is still in the thousands.

Traditionally, multiple chemical separations – including extraction, adsorption, precipitation, distillation, liquid chromatography, etc. – have been applied to fractionate the sulphur compounds contained in a sample, and allow for the quantification of the various sulphur types. Recently, elaborate methods involving selective oxidation, reduction and complexation combined with liquid chromatography were used to separate sulphur compounds from the hydrocarbon matrix, aliphatic sulphur from thiophenic sulphur types, as well as to isolate sulphur compounds according to aromatic ring number[1]. However, the detectors that are of practical importance in providing support for HDS model development and maintenance are probably limited to atomic emission (AED), pulsed Flame Photometric Detector (FPD), and chemiluminescence (SCD). Combined with gas chromatography, using high-resolution capillary columns, these detectors can quantify individual sulphur compounds without interference from the hydrocarbon matrix and using a single response factor.[2] Sometimes, solid phase extraction (SPE) LC pre-treatment of the sample is helpful. For example, if the removal of aliphatic sulphur overlapping with thiophenic sulphur is important, silver nitrate in an SPE cartridge can be used to separate the two sulphur types before injecting the sample into the GC. In some cases high resolution mass spectromentry (MS) or GC-MS techniques can also be helpful.

At NCUT, we use GC-SCD and GC-AED with SPE LC separation as the key tools in sulphur speciation. Conveniently, in the context of LCO, the BTs and DBTs with shorter substitutents are mostly distributed over two distinct ranges of GC retention times eliminating the overlap and simplifying speciation (see Figures 1a, b, and c). However, the identification of the various sulphur compounds in the sample is done by matching the retention time of a peek with a database of the available retention times. In this regard, the main concern for the GC-based methods is that relatively few sulphur compounds are available as analytical standards necessary to determine their corresponding retention times and, thus, build such a database.

Depauw and Froment[3] identified certain regularities in the chromatographic elution order and developed a linear regression model for the retention times of substituted BTs based on the positions of the various substituents. The model predicted elution time very well for di-, tri, and tetra- methyl substituted BTs, providing a useful way for identifying sulphur compounds in LCO. However, more work in this area is necessary to further develop this or a similar approach and extend it to DBTs and beyond. A relatively new methodology that holds promise in this area is the quantitative structure-property relationship (QSPR) approach, as described in the next section.

New Opportunities in Sulphur Speciation

The quantitative structure-property relationship (QSPR) and quantitative structure-property relationship (QSAR) approaches are derived using a statistical methodology developed to correlate various properties of pure compounds with a number of constitutional, topological, geometrical, and electronic descriptors of those compounds calculated from their geometrically optimized molecular structures. Widely used in new drug discovery, QSPR and QSAR have also been powerful tools for predicting the chromatographic behaviour of various classes of compounds[4-6]. But in principle, any chemical or physical property of a compound can be estimated by QSPR, with sometimes surprisingly good accuracy.

Motivated by the limited availability of analytical standards, we initiated a QSPR study to develop correlations between GC retention times and indices for various sulphur compounds to facilitate sulphur speciation. A set of 114 sulphur-containing compounds was selected for this study. This included 19 thiols, 13 sulfides and disulfides, 8 thiophenes, 29 BTs, 27 DBTs, and 13 other multi-ring sulphur compounds found in LCO. A large set of molecular descriptors was generated, ranked, and regressed with the retention times and indices using the commercial QSPR software CODESSA. The resulting multilinear regression models predicted, with good accuracy, both the retention times and retention indices for various sulphur compounds regardless of the type. Figure 2a show the good fit of the models. The descriptors included in the model reflected the geometric, topological, and electronic properties of the molecules that were related to the interactions between the solute and the stationary phase. In spite of the relatively good fit, the models had errors that were too large to identify a peek in the LCO sulphur chromatogram. The models could be further improved by limiting the data set to the 29 BTs and 27 DBTs of most concern in petroleum processing. These new multi-linear regression models predicted the retention times and indices of structurally similar compounds with considerably higher confidence (Figure 2b).

When the chemical standards are unavailable, it becomes very difficult to distinguish between the isomers that have similar elution times. However, identification of sulphur species can be achieved by simultaneous application of GC-MS to determine molecular masses, QSPR correlation to match retention times, and a QSAR correlation for the reactivity (as described in more detail below) of various DBT molecules used to verify the assignments. For example, the retention times for 2-EDBT and 3,6-DMDBT are very similar. Without the standards, it is Impossible to positively identify these species. However, their corresponding HDS pseudo-first order rate constants are quite different, which can be used to differentiate between the two peaks if the reaction data are available. We are currently using this methodology to expand our speciation database.

On the Importance of Phase Equilibrium in Kinetics Analysis

Probably because of the added difficulty involved in setting up phase equilibrium calculations, most of the studies published in the open literature completely neglect phase equilibrium as a factor affecting the kinetics analysis[7-9]. This kind of approach, apparently useful for other reasons, cannot be adopted in the process of building an HDS model. For example, neglecting phase equilibrium effect on the kinetics analysis may lead to the temperature dependency of such a typical parameter as the activation energy, which is expected to remain a constant for an elementary reaction.

Let's consider the physical picture of the flow of the treated liquid in a trickle-bed hydrotreater at the actual operating conditions. The compositions of the gas and liquid phases substantially differ from the compositions of the treat gas and hydrocarbon streams fed to the reactor. Under the high-temperature and high-pressure conditions, large amounts of hydrogen are dissolved in the liquid phase and large amounts of the lighter components of the hydrocarbon stream evaporate and join the gas phase. This not only affects the residence times (Figure 3) of the two phases but it also modifies the reaction driving force. Figure 4 was prepared to illustrate this point by showing phase equilibrium data obtained in the NCUT flow cell with an LCO and hydrotreated LCO (a reactor effluent)[10]. The figure shows the fraction of both the LCO feed and effluent in the liquid phase at the reactor conditions, together with the concentration of 4,6-DMDBT in the liquid phase, both as functions of temperature. In this figure, as reactor temperature increases, the amount of the liquid phase decreases and the concentration of 4,6-DMDBT increases. Hence, higher reactor temperature would be expected to aide sulphur removal in three major ways: through increasing the reaction rate (kinetics), residence time (evaporation), and the concentration of the key reactant in the liquid phase – the only phase that remains in direct contact with the catalyst. Figure 5 plots the corresponding conversion of 4,6-DMDBT. Clearly, despite all that help, the sensitivity of the conversion to temperature declines with temperature at around 400°C because of hydrogenation/dehydrogenation equilibrium limitations.

It is important to remember that, particularly at lower pressures and higher temperatures, the availability of the liquid phase in the experimental reactor may become so low that the catalyst may not be fully wetted by liquid. And full catalyst wetting is critically important to scale-up since, at the same space velocity, the superficial liquid velocity (liquid availability) in the commercial reactor may be as much as two orders of magnitude higher than in the experimental reactor. When, during temperature scouting in a pilot plant reactor, the catalyst is no longer fully wetted, the experimental results may become unreliable.

Catalyst Effectiveness

Classical reactor engineering teaches the importance of the limitation on the overall reaction rate due to the intra-particle diffusion resistance. In order to asses the importance of this limitation in a typical hydrotreating catalyst, we conducted a comparative pilot plant study at similar operating conditions, by hydrotreating an LCO using both a full-size commercial Ni-Mo/Al_2O_3 hydrotreating catalyst (extrudates of 1.6 mm in diameter and 5 mm in length) and the same catalyst crushed down to 35-60 mesh[11]. This study covered the commercially important 350 to 385°C temperature range. The analytical methods mentioned above enabled the determination of the catalyst effectiveness factors individually for 27 DBTs substituted with short paraffinic chains in various configurations and to varying degree (see Table 1).

This study showed that with the full-size commercial catalyst the intra-particle diffusion resistance significantly limited the overall rate of sulphur removal for all the DBTs investigated. The effectiveness factors ranged from only 0.300 to 0.822. In the crushed catalyst, with the effectiveness factors between 0.813 and 0.985, the intra-particle diffusion effect was almost eliminated. As expected, the catalyst effectiveness factors calculated for the relatively more reactive DBTs, with intrinsic reactivities little affected by the steric hindrance of methyl and/or ethyl substitutents, were lower (stronger intra-particle diffusion resistance). The less reactive DBTs with reactivities hindered to a significant degree had higher effectiveness factors.

In situations where the overall HDS rate is limited by the rate of intra-particle diffusion transport, operationally, the overall rate depends not only on the reactor temperature but also on the effective size of the catalyst particles used. Since in any commercial situation, the temperature is pushed to the metallurgy limits of the reactor to maximize the overall HDS rate, the only parameter that could potentially be used to increase catalyst effectiveness is the catalyst size. With the particle size also affecting the pressure drop in the commercial hydrotreater, the catalyst size is usually a result of a balancing act between catalyst effectiveness and pressure drop. Therefore, the intra-particle diffusional resistance should play a significant role in most commercial situations and, hence, catalyst effectiveness is an important factor that needs to be modeled in any attempt to build a predictive model of the hydrotreater.

Modeling the Reactivity of Individual Compounds

In an effort to better understand the relationship between the structure of the sulphur-containing molecule and its HDS, we have studied – using the density functional theory – the interactions of a series of substituted DBTs with the coordinatively unsaturated molybdenum atoms at the Mo-edge of a $Mo_{10}S_{18}$ nanocluster model. We also examined different adsorption configurations of substituted DBT on the Mo-edge of the $Mo_{10}S_{18}$ cluster[12]. The DBT-$Mo_{10}S_{18}$-cluster complex was subjected to the geometry optimization to find the most stable configuration. Adsorption energy, total electron density distribution and Mayer bond order were calculated and used to interpret the reactivity difference between various substituted DBTs (see Table 2).

The flat adsorption configurations had generally higher adsorption energies (76-82 kcal/mol) than those corresponding to the perpendicular configurations (27-34 kcal/ml). This indicated that the flat adsorption was thermodynamically more favourable than the perpendicular adsorption, due to the interactions of not only the sulphur atoms but also the thiophenic and aromatic rings with the Mo atoms on the catalyst surface. In the flat adsorption, DBT, 4-MDBT, 2,8-, 3,7- and 4,6-DMDBTs interacted similarly with the catalyst cluster, indicating that the methyl groups on the 4 and 6 positions did not hinder the binding of those molecules laying flat on the catalyst surface. However, in the perpendicular adsorption, the methyl group on the 4 and 6 positions prevented the bonding of the sulphur atom with the surface Mo atom, as evident from the total electron density distribution of the sulphur-compound-MoS_2-cluster complex shown in Figure 6. The results in Figure 6, calculated for the perpendicular adsorption and the optimized sulphur-molecule-$Mo_{10}S_{18}$ complex, show that for DBTs with methyl groups in the 2, 3, 7 and 8 positions there is complete overlap between the electron densities of the sulphur atom and the middle molybdenum atom of the cluster. A small gap was observed in the S-Mo bond for 4-MDBT, which indicated weaker interaction than for DBT due to the presence of the methyl group in the 4 position, and a complete lack of overlap was found for 4,6-DMDBT. These results imply that the interaction of the Mo atom with 4,6-

DMDBT through η_1S coordination is much less likely and, hence, hydrogenation followed by C-S cleavage is the dominant if not the only HDS pathway available for this molecule.

In an attempt to correlate the reactivity of various alkyl-substituted DBT molecules with their electronic properties, Isoda et al.[13] calculated the electron density of the sulphur atom on the HOMO and HOMO-5 orbitals (sigma orbital 5 levels below Highest Occupied Molecular Orbital) for a set of DBTs using WinMOPAC-PM3 software. They found good correlation between the electron density of sulphur atom on the HOMO-5 orbital and the HDS rate constants from an experiment with a CoMo/Al_2O_3 catalyst. No correlation was found between HDS rates of alkyl-substituted DBTs and the sulphur electron density on the HOMO orbital. Ma et al.[14] on the other hand, reported a good correlation between hydrogenolysis reactivity and the electron density of sulphur atom for DBTs without significant steric hindrance. They also related the hydrogenation reactivity of sulphur compounds with the bond order of the most unsaturated bond in the ring. We found that the pseudo-first-order HDS rate constants in Table 2 did not correlate with electron densities of the sulpur atom on the HOMO or HOMO-5 orbitals.

The Mayer bond order is a quantitative expression of the electron density overlap in optimized molecule-cluster complexes that accounts for both electronic and steric effects. The S-Mo bond orders for various substituted DBTs collected in Table 2 correlate with the reactivities in a nonlinear way. The Mayer bond orders increase in the following sequence: 4 and 6 substituted DBTs < 4 or 6 substituted DBTs < DBT < other substituted DBTs such as 1,3,7-TMDBT. Directionally, this is in good agreement with the HDS rate constants but more work would be required before the resulting correlation, shown in Figure 7, could be used to predict the rate constants with satisfactory accuracy. However, although this correlation is contributing to the understanding of molecular interactions of various DBT molecules with catalytic sites, it has a weakness from the modeling point of view. It involves assumptions about the models of both: the molecule and the catalyst cluster. It would be much preferred to predict the relative reactivity of the molecule (a ratio of the pre-exponential factors of the Arrhenius equation of the molecule in question and the reference molecule) from its original structure, unaffected by the interaction with the catalyst site. It is conceivable that, at least as the first approximation, the ability of the molecule to change its geometry due to the interaction with the catalyst site is a function of the properties of this molecule alone. This gives rise to the idea of modeling the reactivities of various DBT molecules on the specific catalyst relative to a single reference molecule for each homologous series (in this case DBTs).

In another study, we determined the reactivities of a large number of various DBTs (in terms of their pseudo-first-order reaction rate constants) by following the kinetics of their disappearance during hydrodesulphurization of a real feedstock on a peak-by-peak basis. Extending the approach taken by Isoda et al.[13], we applied the QSAR methodology to the resulting set of data consisting of pseudo-first-order HDS constants of 22 DBTs and, within the constraints of the method, found the charge and valency of the carbon next to the S atom to be the descriptors most highly correlated to the rate constants. The resulting multi-linear model correlated well with the experimental HDS rates (see Figure 8). This is just an illustration of opportunities in correlating HDS reactivities with the molecular structure.

Methyl substitutents, as electron donor groups, increase the electron density of DBTs, improving their adsorption on the Mo cluster. This is consistent with the expectation that the intrinsic or inherent reactivity of substituted DBTs increases with the degree of substitution as long as the steric hindrance effect is not increased by the introduction of new substitutents. This expectation resulted from another NCUT study on oxidative desulphurization[15]. This study looked at reactivities of DBT, 4-MDBT, and 4,6-DMDBT in oxidative desulphurization (ODS) using hydrogen peroxide, and supported ($Na_3PW_{12}O_{40}$) and homogeneous (HCOOH) catalysts. The ODS reactivities are compared in Table 3 with HDS reactivities of these three compounds. For supported catalysts in both HDS and ODS the reactivity ranking was the same (DBT > 4-MDBT > 4,6-DMDBT) but for the homogeneous catalyst the ranking completely reversed (4,6-DMDBT > 4-MDBT > DBT) without further investigation we assumed that the application of the homogeneous catalyst removed the steric effect. The rate constants in Table 2 are not always consistent with the expectation of improved reactivity of substituted DBTs, even in the expected absence of the steric hindrance effect. This is most likely because these reaction rate constants were calculated neglecting phase equilibrium effects. 4,6-DMDBT is the most sterically hindered DBT molecule and, on balance, despite two methyl substitutents, its reactivity is much lower than that of unsubstituted DBT. All this suggests that the reactivity of various DBT molecules should probably be correlated with at least two molecular descriptors: one capturing the degree of substitution and the other capturing the steric hindrance effect.

Our experience suggests that the reactivities of individual sulphur compounds should be correlated using data related to an elementary reaction (e.g. direct sulphur removal) and after factoring out the phase equilibrium effect. Furthermore, this should not be done in the automatic and inherently linear way offered by CODESSA. A scientific insight may lead to nonlinear correlations containing much fewer parameters but modeling individual reactivities much more accurately. However, at the exploratory stage, QSAR analysis performed using the commercial CODESSA software is a powerful tool that could be used to further examine correlating reactivities. QSAR could provide an extension to the linear free energy relationship (LFER) methodology in correlating kinetics parameters of chemical reactions. In the end, such correlations would be used to reduce the number of adjustable parameters in the predictive model of the hydrotreater.

Competition for Catalyst Sites in Real Feedstocks

One of the most challenging tasks is the modeling of interactions between various groups of compounds as they compete for the catalyst sites. HDS is inhibited by the presence of other hydrocarbons in the feed[16-22]. In general, the inhibition ranking is as follows: nitrogen compounds > organic sulphur compounds > polyaromatics ≈ oxygen compounds ≈ H_2S > monoaromatics. Among those hydrocarbons, organic nitrogen compounds such as quinoline, indole and carbazole are believed to be the strongest inhibitors. Most of the published nitrogen inhibition studies investigated either the suppression of HDS rate in real feedstocks (total sulphur) or model systems (individual sulphur compounds) in pure solvents. The former usually do not provide information sufficient for modeling purposes the latter neglect the catalyst interactions with the hydrocarbon matrix of the real feedstock.

We conducted a pilot-plant study that looked at nitrogen inhibition using five real feedstocks with different nitrogen types and concentration distributions. The feedstocks for this study were prepared by collecting different fractions when an LCO was passed through a silica gel column. The nitrogen concentration of the original LCO was 728.5 mg/L, those of the collected fractions ranged from 16.5 to 498.4 mg/L. Table 4 gives the properties of the five feedstocks. A crushed commercial $NiMo/Al_2O_3$ hydrotreating catalyst was used.

The effect of nitrogen on the removal of total sulphur, hard-to-remove sulphur (DBTs, naphthothiophenes, with one to four carbons in the substituent chains and other heavier unidentified compounds) and easy-to-remove sulphur compounds (thiophenes, BTs and DBT) is shown in Figure 9. When feed was changed from high nitrogen LCO to low nitrogen LCO, at the same operating conditions, the sulphur content in the final product gradually decreased. The total sulphur was reduced from 1.18 wt% to 498 mg/L for the original LCO (N=728.5 mg/L) and from 1.11 wt% to 121 mg/L for the ultra-low nitrogen feed F1 (N=16.5 mg/L). The easy sulphur was almost completely converted in all cases, which indicated that the presence of nitrogen compounds did not influence its removal rate significantly. However, the feed nitrogen significantly affected the rate of hard sulphur removal. Figures 10a and 10b present the concentration-temperature curves for DBT and 4,6-DMDBT using the original LCO and the ultra-low nitrogen LCO as feeds. The results show that the nitrogen content in the feed has a significant effect on the removal of 4,6-DMDBT. Much lower temperature is required to achieve the same conversion when using ultra-low nitrogen feed. On the other hand, the removal of DBT seems much less dependent of the feed type. The fact that nitrogen compounds inhibit the removal of 4,6-DMDBT stronger than DBT suggests that the HDS route involving hydrogenation, followed by hydrogenolysis, is much more strongly suppressed than the direct hydrogenolysis route. In the open literature, experimental information necessary to further explore and model these kinds of interactions are very rare and, in general, insufficient for modelling purposes.

References

1 Robbins, W. K. "Challenges in the analytical characterization of non-thiophenic and thiophenic sulphur compounds in petroleum". *Preprints by Division of Petroleum Chemistry,* **2000**, *45*, 68.

2 Te, M.; Fairbridge, C. and Ring, Z. "Various approach in kinetics modeling of real feedstock hydrodesulphurization". *Petroleum Science and Technology,* **2003**, *21*, 157.

3 Depauw, G. A. and Froment, G. F. "Molecular analysis of the sulphur components in a light cycle oil of a catalytic cracking unit by gas chromatography with mass spectrometric and atomic emission detection". *Journal of Chromatography A,* **1997**, *761*, 231.

4 Katritzky, A. R.; Maran, U.; Lobanov, V. S. and Karelson, M. "Structurally Diverse Quantitative Structure-Property Relationship Correlations of Technologically Relevant Physical Properties". *J. Chem. Inf. Comput. Sci.,* **2000**, *40*, 1.

5 Kaliszan, R. *Quantitative Structure-Chromatographic Retention Relationships*; J. Wiley & Sons: New York, **1987**.

6 Katritzky, A. R.; Petrukhin, R.; Tatham, D.; Basak, S.; Benfenati, E.; Karelson, M. and Maran, U. "Interpretation of Quantitative Structure-Property and -Activity Relationships". *J. Chem. Inf. Comput. Sci.,* **2001**, *41*, 679.

7 Bellos, G. D. and Papayannakos, N. G. "The Use of Three Phase Microreactor to Investigate HDS Kinetics", *Catal. Today,* **2003,** *70-80*, 349-355

8 de Jong, K. P. "Effects of Vapor-Liquid Equilibrium on Coke Deposition in Trickle-Bed Reactors during Heavy Oil Processing. 2. Modeling", *Ind. Eng. Chem. Res.,* **1994**, *33*, 3141-3145.

9 Nalitham, R. V.; Guin, J. A.; Tarrer, A. R. and Curtis, C. W. "Effect of Phase Behavior on Hydrotreater Performance: Simulation and Experimental Verification", *Ind. Eng. Chem. Process Des. Develop.*, **1985**, *24*, 598-607.

10 Chen, J. and Ring, Z. "Vapor-Liquid Phase Equilibrium During Hydrotreating of Light Cycle Oil". Presentation at 2004 AIChE Spring Meeting, New Orleans, LA, USA, April 25-29, **2004**.

11 Chen, J.; Yang, H. and Ring, Z. "Catalyst Effectiveness Factor During Deep HDS of LCO ". Presentation and paper in the proceedings of 2004 AIChE Spring Meeting, New Orleans, LA, USA, April 25-29, **2004**.

12 Yang, H.; Fairbridge, C. and Ring, Z. "Adsorption of Dibenzothiophene Derivatives over a MoS_2 Nanocluster-A Density Functional Theory Study of Structure-Reactivity Relations". *Energy & Fuels*, **2003**, *17*, 387-398.

13 Isoda, T.; Takase, Y.; Kusakabe, K. and Morooka, S. "Understanding HDS reactions based on quantum chemical calculation". *Preprints-American Chemical Society, Division of Petroleum Chemistry,* **1998**, *43*, 575.

14 Ma, X.; Sakanishi, K.; Isoda, T. and Mochida, I. "Quantum Chemical Calculation on the Desulphurization Reactivities of Heterocyclic Sulphur Compounds". *Energy & Fuels,* **1995**, *9*, 33.

15 Te, M.; Fairbridge, C. and Ring, Z. "Oxidation reactivities of dibenzothiophenes in polyoxometalate/H_2O_2 and formic acid/H_2O_2 systems". *Appl. Catal. A,* **2001**, *219*, 267-280.

16 Nagai, M. and Kabe T. "Selectivity of molybdenum catalyst in hydrodesulfurization, hydrodenitrogenation, and hydrodeoxygenation: effect of additives on dibenzothiophene hydrodesulfurization". *J. Catal.*, **1983**, *81,* 440-449.

17 LaVopa, V. and Satterfield, C.N. "Poisoning of thiophene hydrodesulfurization by nitrogen compounds". *J. Catal.,* **1988**, *110*, 375-387.

18 Van Looij, F.; van der Laan, P.; Stork, W.H.J.; DiCamillo, D.J. and Swain, J. "Key parameters in deep hydrodesulfurization of diesel fuel". *Appl. Catal. A,* **1998**, *170,* 1-12.

19 Knudsen, K.G.; Whitehurst, D.D. and Zeuthen, P., *Proceeding*, AIChE Spring Meeting, Atlanta, GA, March 5-9, **2000**. 329-334.

20 Laredo S., G.C.; De Los Reyes H., J.A.; Cano D., J. L. and Castillo M., J.J. "Inhibition effects of nitrogen compounds on the hydrodesulfurization of dibenzothiophene". *Appl. Catal. A,* **2001**, *207*, 103-112.

21 Zeuthen, P.; Knudsen, K.G. and Whitehurst, D.D. "Organic nitrogen compounds in gas oil blends, their hydrotreated products and the importance to hydrotreatment". *Catal. Today,* **2001**, *65*, 307-314.

22 Koltai, T.; Macaud, M.; Guevara, A.; Schulz, E.; Lemaire, M.; Bacaud, R. and Vrinat, M. "Comparative inhibiting effect of polycondensed aromatics and nitrogen compounds on the hydrodesulfurization of alkyldibenzothiophenes". *Appl. Catal. A,* **2002**, *231,* 253-261.

Table 1. Catalyst effectiveness factor for total sulphur, total nitrogen and individual sulphur compounds

Sulphur compound	Effectiveness factors, full-size commercial catalyst			
	T=352°C	T=362°C	T=372°C	T=382°C
Total S	0.633	0.615	0.598	0.582
DBT	0.484	0.472	0.460	0.449
1-MDBT	0.524	0.494	0.466	0.440
2-MDBT	0.387			
3-MDBT	0.374			
4-MDBT	0.595	0.582	0.568	0.556
1,4-DMDBT	0.644	0.615	0.589	0.563
1,6-DMDBT	0.631	0.619	0.606	0.595
2,4-DMDBT	0.564	0.528	0.494	0.462
2,6-DMDBT	0.644	0.598	0.557	0.518
2,7-DMDBT	0.300			
2,8-DMDBT	0.390	0.515	0.489	0.464
3,4-DMDBT	0.544	0.532	0.507	0.483

Table 2. Calculated Adsorption Energies, Mayer Bond Orders, Pseudo-First-Order HDS Rate Constants, and Total Electron Density Distributions for Various Sulphur Compounds

Sulphur Compounds	ΔE kcal/mol	Bond order BO	Rate constant h^{-1}	Electron density	
				HOMO	HOMO-5
DBT	-28.9	1.1466	3.84	0.4827	0.3950
2-MDBT	-35.3	1.1538	4.79	0.4510	0.3547
3-MDBT	-35.6	1.1569	4.23	0.4819	0.3615
3,7-DMDBT	-30.7	1.1602	4.69	0.4827	0.3609
4-MDBT	-34.6	1.0301	0.88	0.4591	0.4008
4-EDBT	-35.2	0.9331	0.69	0.4560	0.3134
2,4-DMDBT	-35.1	0.9070	0.89	0.4534	0.3779
2,6-DMDBT	-35.2	0.9822	0.90	0.4453	0.3812
3,6-DMDBT	-35.5	1.0269	0.89	0.4600	0.3887
1,4-DMDBT	-38.3	1.0948	0.75	0.4252	0.3036
3,4-DMDBT	-35.0	0.8982	0.76	0.4619	0.3749
2,4,8-TMDBT	-36.2	0.9387	0.94	0.4413	0.3496
2,4,7-TMDBT	-35.8	0.9736	1.10	0.4546	0.3751
3,4,7-TMDBT	-34.81	1.0762	1.08	0.4544	0.3788
4,6-DMDBT	-21.4	0.7900	0.33	0.4514	0.4134
4-E-6-M DBT	-31.4	0.8682	0.63	0.4396	0.3454
2,4,6-TMDBT	-31.5	0.8815	0.37	0.4411	0.3948
1,4,6-TMDBT	-26.5	0.8542	0.53	0.4141	0.3299

Table 3. Relative Reactivities of DBTs in HDS and ODS Using Supported and Homogeneous Catalysts

k/kDBT	Ni-Mo/Al_2O_3	$Na_3PW_{12}O_{40}/H_2O_2$	$HCOOH/H_2O_2$
DBT	1	1	1
4-MDBT	0.23	0.47	2.1
4,6-MDBT	0.083	0.11	4.9

Table 4. Properties of the Feedstocks Used to Investigate Nitrogen Effect on HDS

Feed	Nitrogen [mg/kg]	Total Base No. [mg KOH/g]	Sulphur [wt %]	Saturates [wt %]	Aromatics [wt %]	Density [kg/L]
LCO	728.5	0.392	1.18	28.7	71.3	0.9353
F3	498.4	0	1.17	29.9	70.1	0.9356
F2	362.5	0	1.18	27.6	72.4	0.9352
F1+F2	170.9	0	1.12	29.8	70.2	0.9281
F1	17.9	0	1.11	29.3	70.7	0.9248

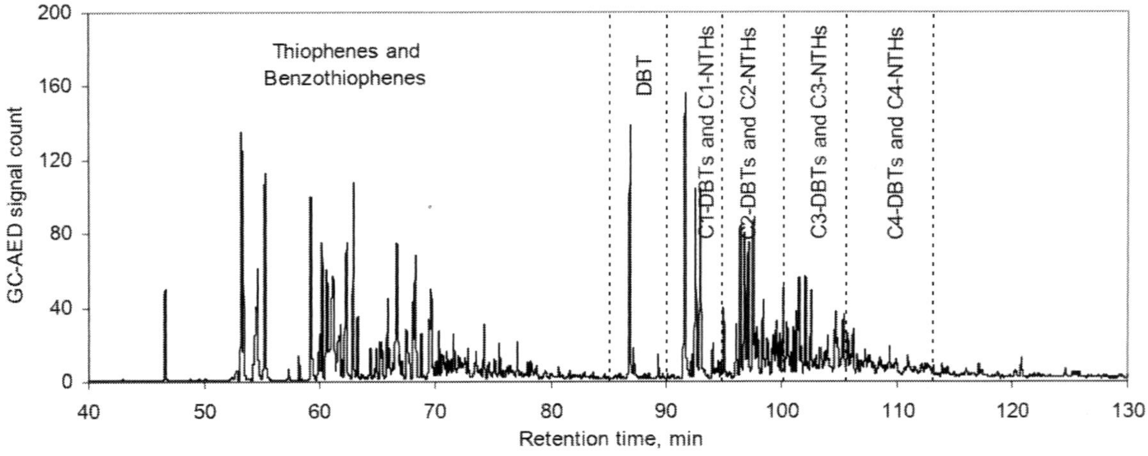

Figure 1a Sulphur chromatogram of the LCO feed (8686 wt ppm of total sulphur content)

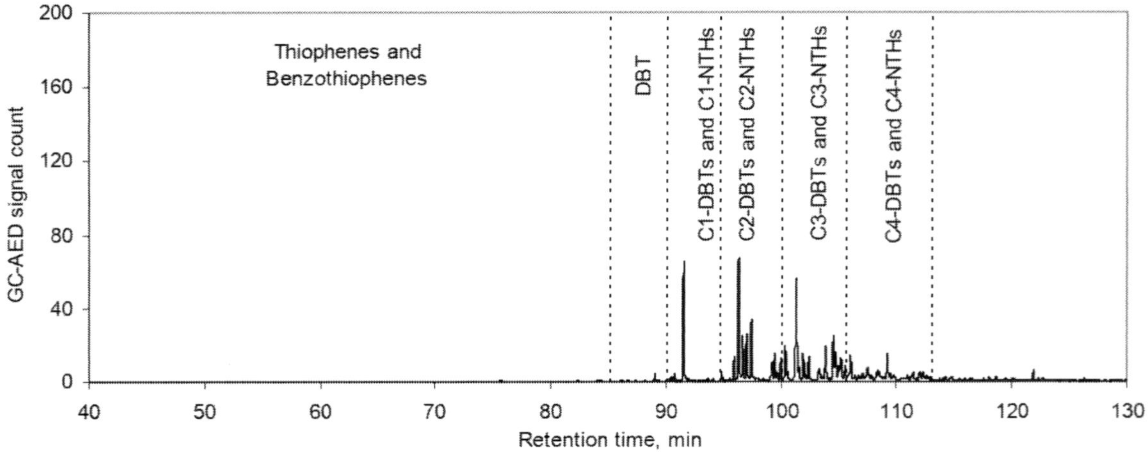

Figure 1b Sulphur chromatogram of effluent at 89.3% total sulphur conversion

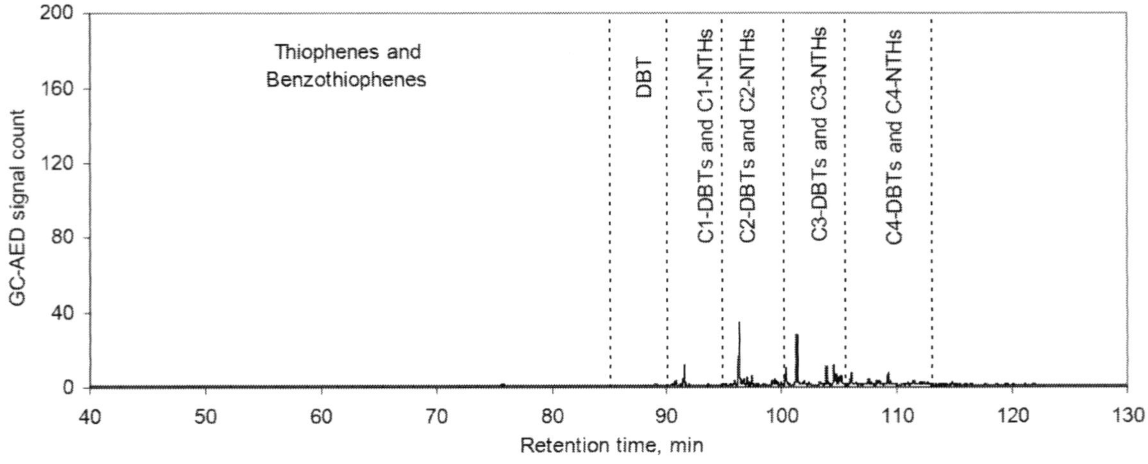

Figure 1c Sulphur chromatogram of effluent at 97.1% total sulphur conversion

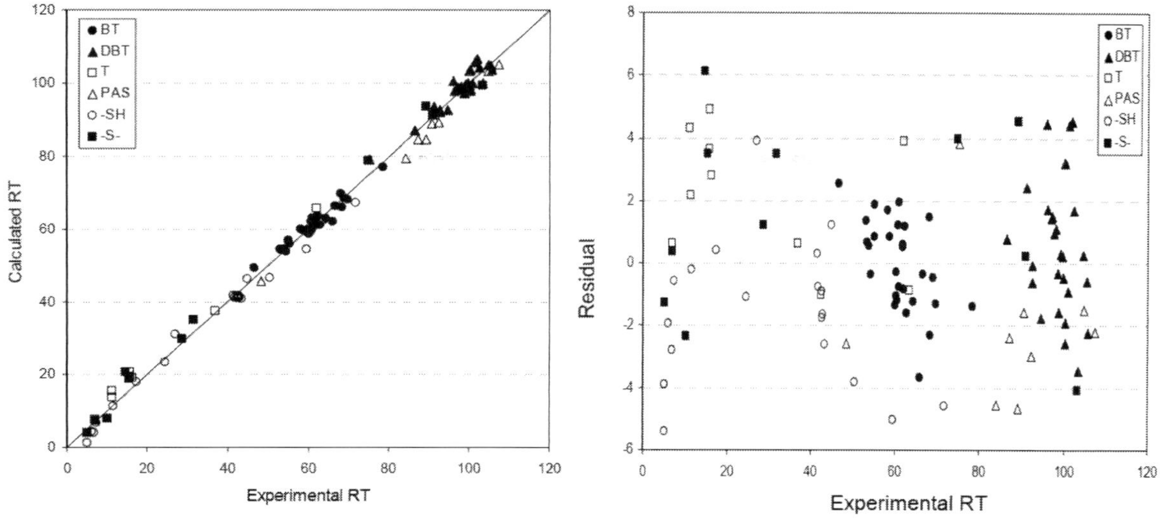

Figure 2a QSPR prediction of GC retention times

$R^2 = 0.9949$, $R^2_{cv} = 0.9937$, $F = 3490$, $s^2 = 6.3745$ (6 descriptors) 114 Sulphur compounds *(19 mercaptans/thiols "-SH", 13 sulphides & disulphides "-S-", 8 thiophenes "T", 29 benzothiophenes "BT", 27 dibenzothiophenes "DBT", 18 polycyclic sulphur aromatics "PAS")*

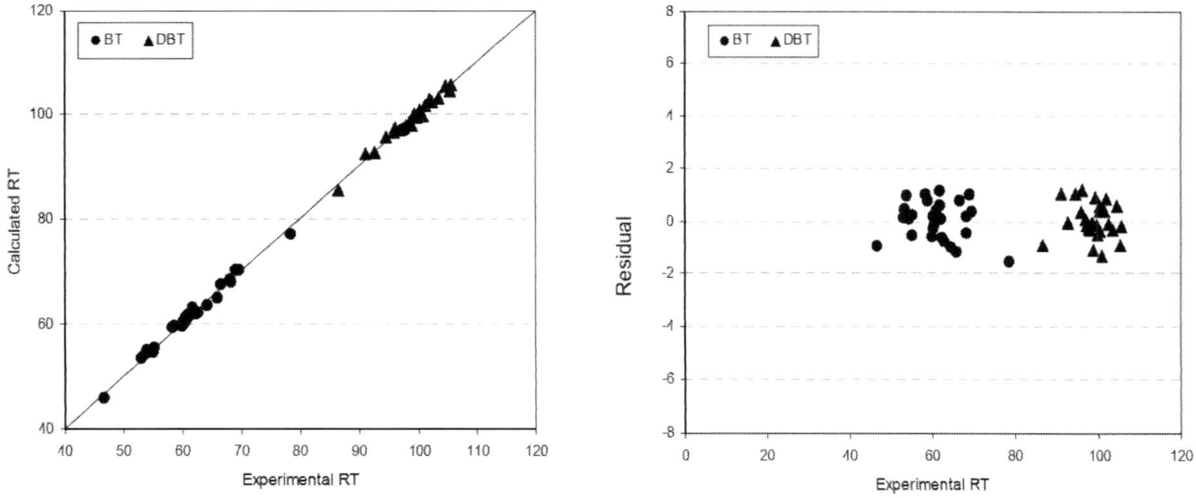

Figure 2b QSPR prediction of GC retention times

$R^2 = 0.9988$, $R^2_{cv} = 0.9984$, $F = 10411$, $s^2 = 0.5083$ (4 descriptors)

56 Sulphur compounds (*29 benzothiophenes "BT", 27 dibenzothiophenes "DBT")*

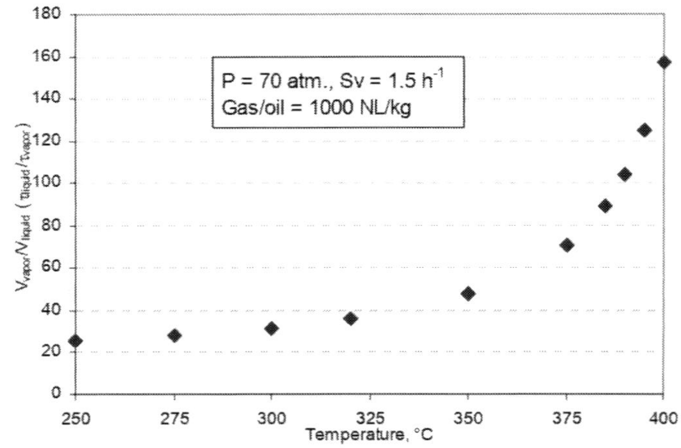

Figure 3 The ratio of the liquid to vapour phase residence times in an LCO hydrotreater

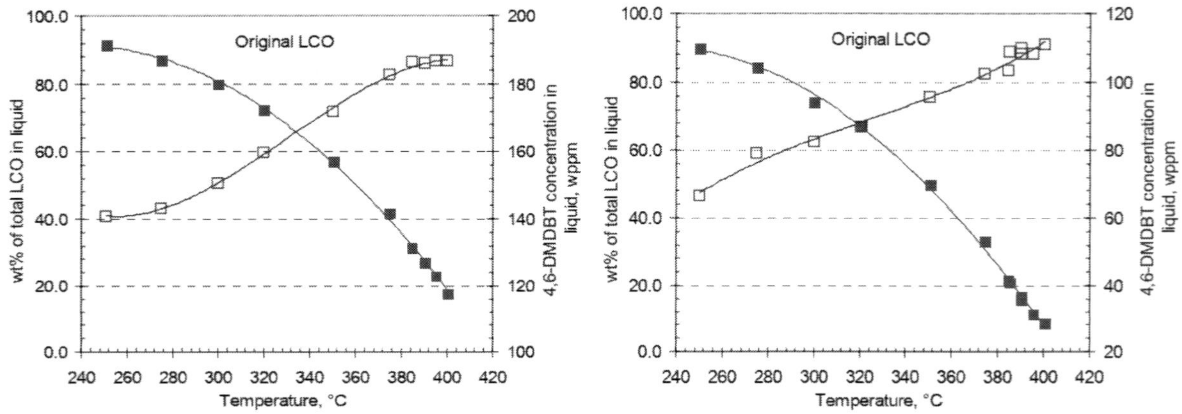

Figure 4 Phase equilibrium at the hydrotreater inlet (LCO) and outlet (hydrotreated LCO)

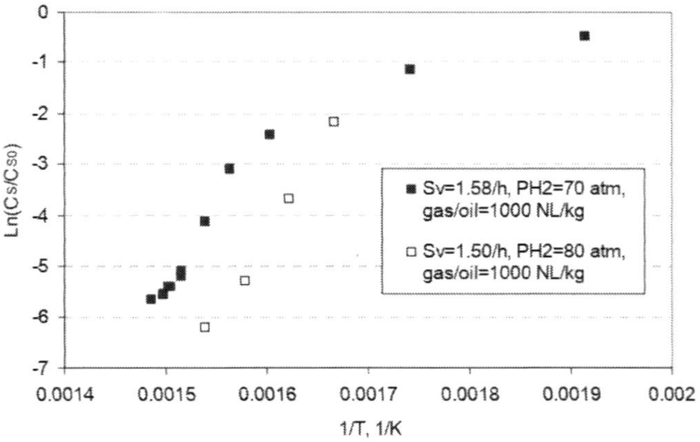

Figure 5 The ratio of effluent and feed sulphur concentrations as function of reciprocal temperature

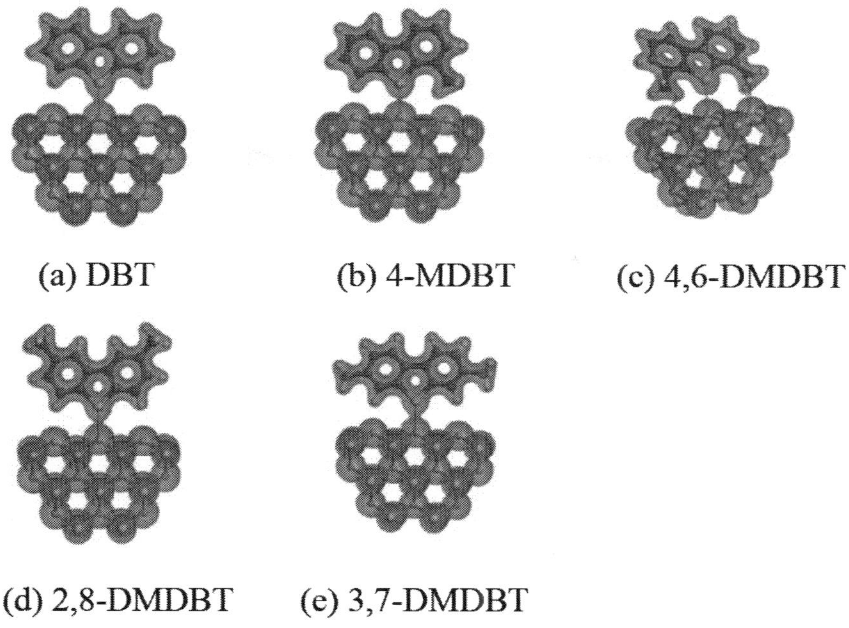

Figure 6 Volumetric plots of total density distribution of different DBTs

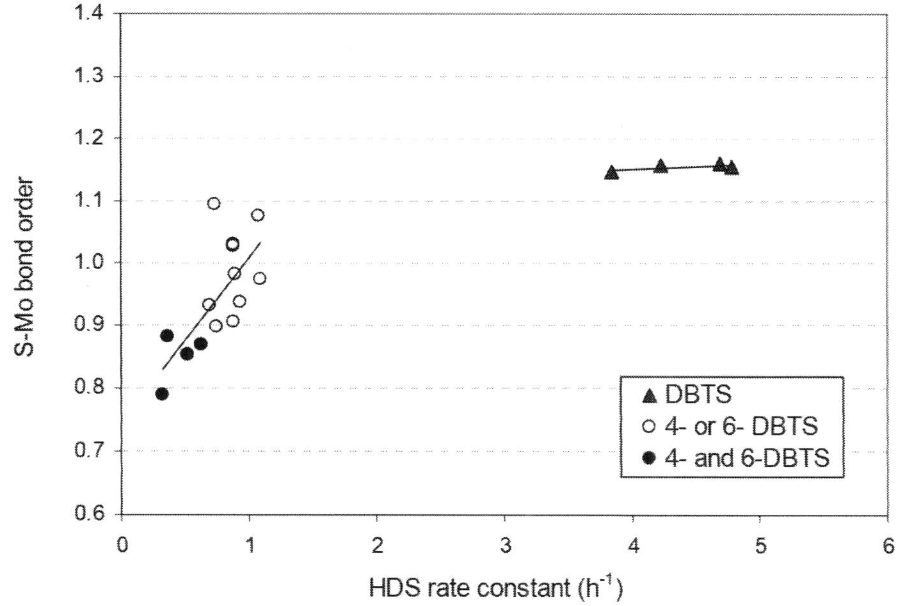

Figure 7 Plots of S-Mo bond order vs. HDS rate constants

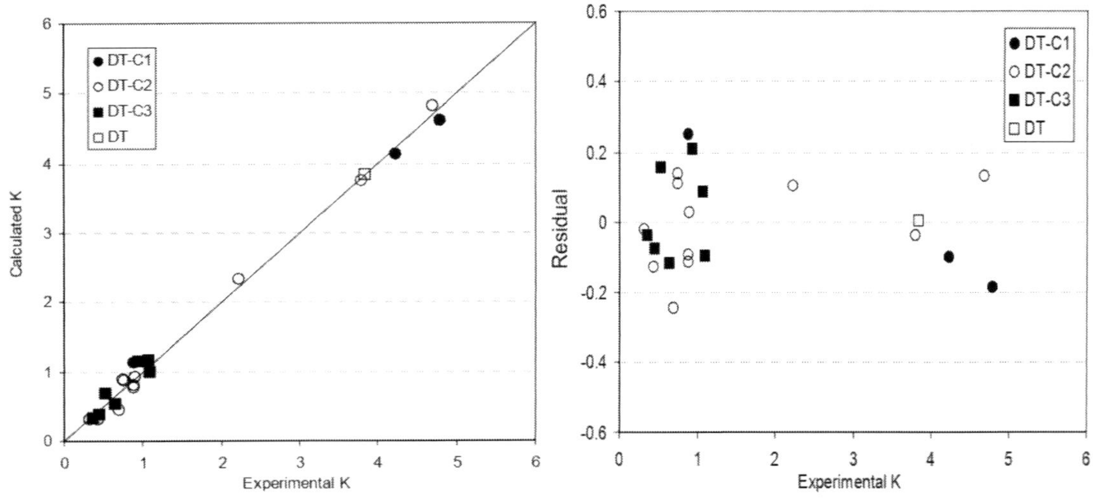

Figure 8 QSPR correlation of the reactivities of 22 DBTs

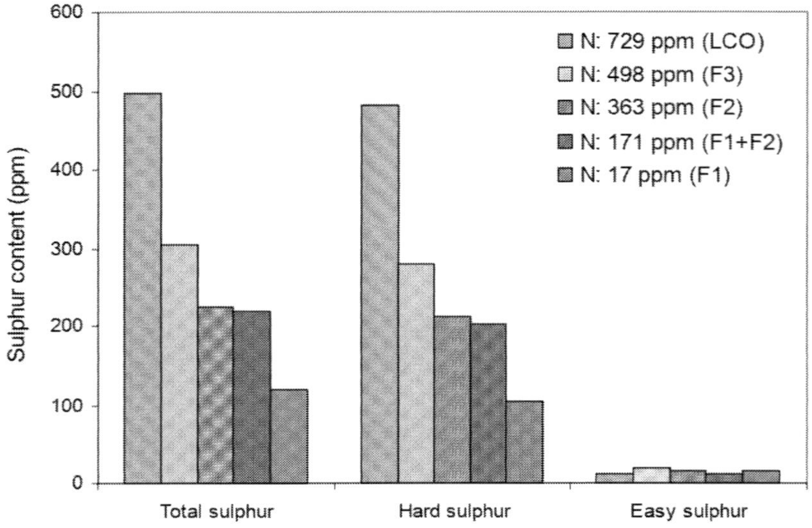

Figure 9 The effect of nitrogen concentration in the feedstock on HDS

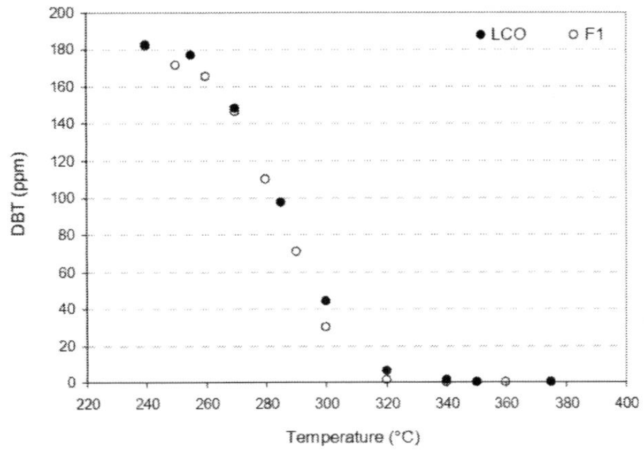

Figure 10a The effect of nitrogen removal on HDS of DBT

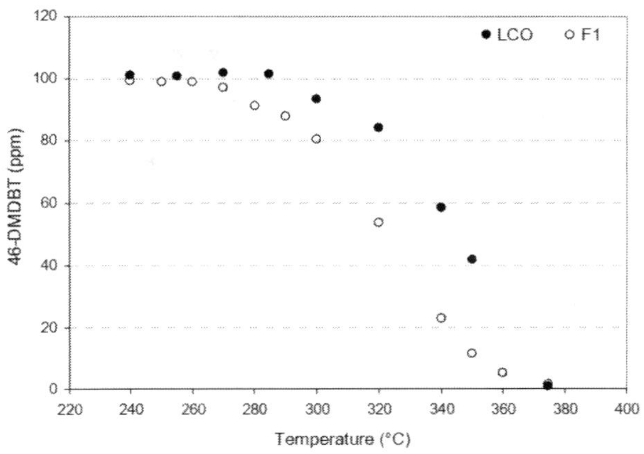

Figure 10b The effect of nitrogen removal on HDS of 4,6-DMDBT

Diesel Strategy Study: Impacts of Future US Diesel Specification Changes on Investment and Operating Costs, H_2 Requirements and CO_2 Emissions

Stephen J. McGovern (Speaker), C.K. Lee and Doug R. Freeman

PetroTech Consultants, 3 Bedford Terrace, Mantua, NJ 08051

Key Words: Clean Diesel, Ultra Low Sulfur Diesel

Prepared for presentation at the AIChE Spring National Meeting 2004

Seventh International Conference on Refinery Processing

April 25-29, 2004, New Orleans, Louisiana

Abstract

In late 2000, the US EPA promulgated new sulfur specifications for on-road diesel fuel, with the maximum sulfur content lowered from 500 ppm to 15 ppm, starting in mid-2006. Additionally, the US EPA has proposed a new sulfur regulation for off-road diesel of 500 ppm in mid-2007 with further reduction to 15 ppm in 2010. Although no new specifications have been proposed for HHO and jet fuel, it is widely expected that the sulfur level in these fuels will also be reduced from their current levels of 2000-5000 ppm.

Although the new diesel fuel regulations set by the US EPA only change the sulfur content of on-road diesel, the state of California and some European governments have also set more stringent limits on other diesel fuel properties such as aromatics, polynuclear aromatics and cetane index or cetane number.

We analyzed the cost to produce ULSD as well as the cost of other potential future distillate fuel regulations based on a comprehensive analysis of the US oil industry. In addition to cost to produce, we also calculated additional hydrogen production requirements and the impact of the regulations on CO2 emissions.

Our analysis shows that the cost to produce ULSD is significantly lower than that reported in other studies; however, it is consistent with the actual cost to produce LSD which required similar refinery modifications. Greenhouse gas emissions will increase as a result of the ULSD regulations.

Introduction

In late 2000, the US EPA promulgated new sulfur specifications for on-road diesel fuel. These regulations required that Ultra Low Sulfur Diesel (ULSD), with a maximum sulfur content of 15 ppm, should be available in mid-2006. As part of the rule making process the US EPA estimated that this new regulation would add 4.5 to 5 cents per gallon to the cost of producing diesel fuel. Furthermore, the US EPA concluded that the refining industry had the time and resources to design and build the facilities that would be required to produce this new lower sulfur diesel fuel.

A number of other studies, commissioned by industry, government or environmental groups concluded that the cost to produce could be significantly higher at 5 to10 cents per gallon. At least one study concluded that the quantity of new high-pressure units that would be required by this regulation, when added to the number of new units that were required to meet low sulfur gasoline regulations, would strain the resources of the construction industry. This would cause refiners to pay a premium for pressure vessels and for skilled welders and that there might even be project delays because of these shortages that would delay the availability of ULSD.

All of these studies appear to have overestimated the severity of the process conditions and hence the capital and operating costs that will be required to produce ULSD.

Although the new diesel fuel regulations set by the US EPA only change the sulfur content of on-road diesel, the state of California and some European governments have also set more stringent limits on other diesel fuel properties such as aromatics, polynuclear aromatics and cetane index or cetane number. At this time, the US EPA sees no need to change any of these specifications; however, 10 years ago they did not foresee the need for the additional reduction in sulfur content that was recently promulgated.

Although on-road diesel fuel accounts for almost half (45%) of the distillate range products that are sold in the US, there are other important distillate products that are not currently subject to the new ULSD regulations. These include kerosene type jet fuel and other kerosene fuels (about 30% of total distillate sales), residential and commercial heating oils (about 10%), farm, construction and military diesel fuels (about 10%) and marine and railroad diesel fuels (just over 5% of total distillate sales). Farm, construction, military, marine and railroad diesel fuels are often collectively called off-highway or off-road diesel fuels although the qualities of each type can be significantly different as will be discussed later.

Current US Distillate Supply and Demand

The US distillate fuels market is rather complex. There are several different grades of distillate products, which are supplied by both domestic refineries and through the import of finished products. The Energy Information Administration (EIA) of the US Department of Energy (DOE) maintains detailed accounting records of petroleum product production, sales and import data by both geographic region and product type. The information discussed in this paper is primarily based on 2001 data.

Most of the distillate products that are consumed in the US are produced in US refineries. Imports of finished distillate products account for less than 9% of the total distillate produced in the US. Over 75% of the imported distillate products come from 3 countries, Canada, Venezuela and the US Virgin Islands and almost 80% of the total imports are into PADD I. Only 25% of the imported distillate is LSD. The rest is higher sulfur distillate. Imports of LSD are less than 5% of current LSD production.

Figure 1 shows historic trends in refinery capacity, crude run and gasoline and distillate products for the US. The raw weekly data available from EIA are smoothed with a 13-point (one quarter) moving average to remove the large weekly fluctuations that occur because of refinery shutdowns, batch product shipments and traders timing product movements based on short term pricing trends.

Although total crude capacity is growing in the US, the number of operating refineries is shrinking.

Figure 1

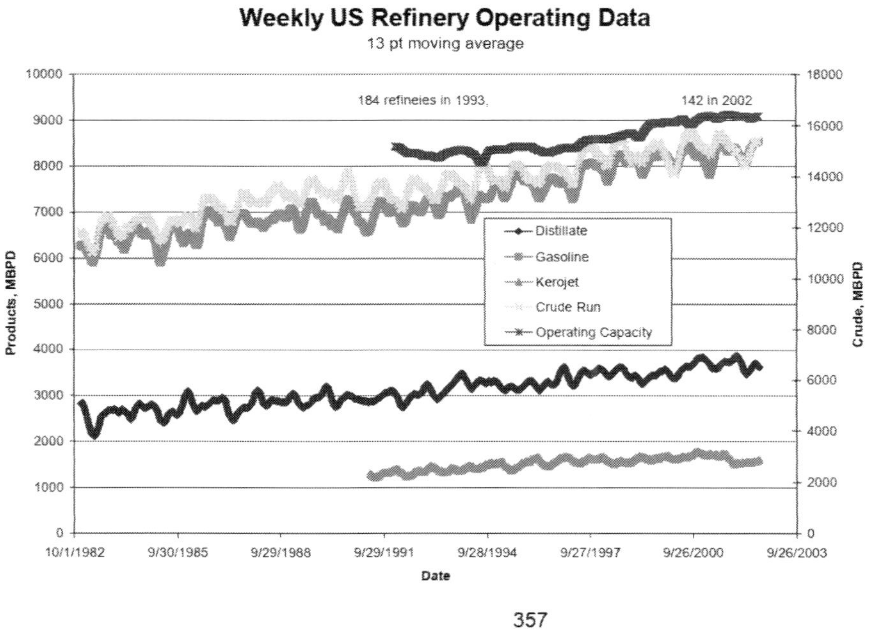

At the beginning of 1993 (prior to the LSD regulations) there were 184 operating refineries with an average crude distillation capacity of about 80 MBPD. At the beginning of 2002, there were only 143, but their average capacity had grown to 115 MBPD. At least 4 times during the past 10 years, the US refining industry was operating near 100% of operable capacity.

In anticipation of the introduction of 500 ppm LSD in 1994, the US refining industry began producing LSD in the second half of 1993. EIA has tracked the production and sale of LSD as a separate product since mid-1993. The production data for LSD and high sulfur diesel are shown in Figure 2. Since its introduction, LSD has experienced steady growth while the production of higher sulfur distillates has declined. Imports of both high and low sulfur distillates remain essentially constant and at relatively low levels. The spike in high sulfur distillate imports in 2001 was partially caused by government purchases to fill the Northeast Heating Oil Reserve. At least part of the growth in LSD and decline in HSD is due to the availability of LSD and a low price difference between HSD and LSD. Most likely, LSD is being substituted for HSD in applications that currently do not require it by law.

The total distillate production rates appear to follow the gasoline production data; however, the heating oil portion of the distillate market (about 10% of total distillate) is highly seasonal, as shown in Figure 3 comparing production and sales data. Distillate storage capacity in the US exceeds 120 million barrels. Much of the seasonal variations in distillate demand are covered by filling or drawing down distillate inventory.

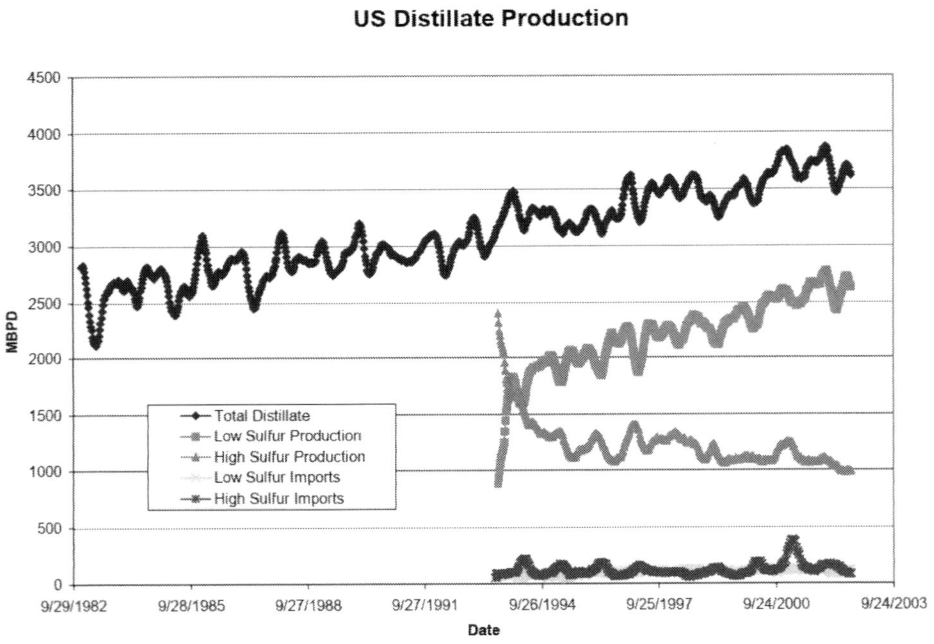

Figure 2

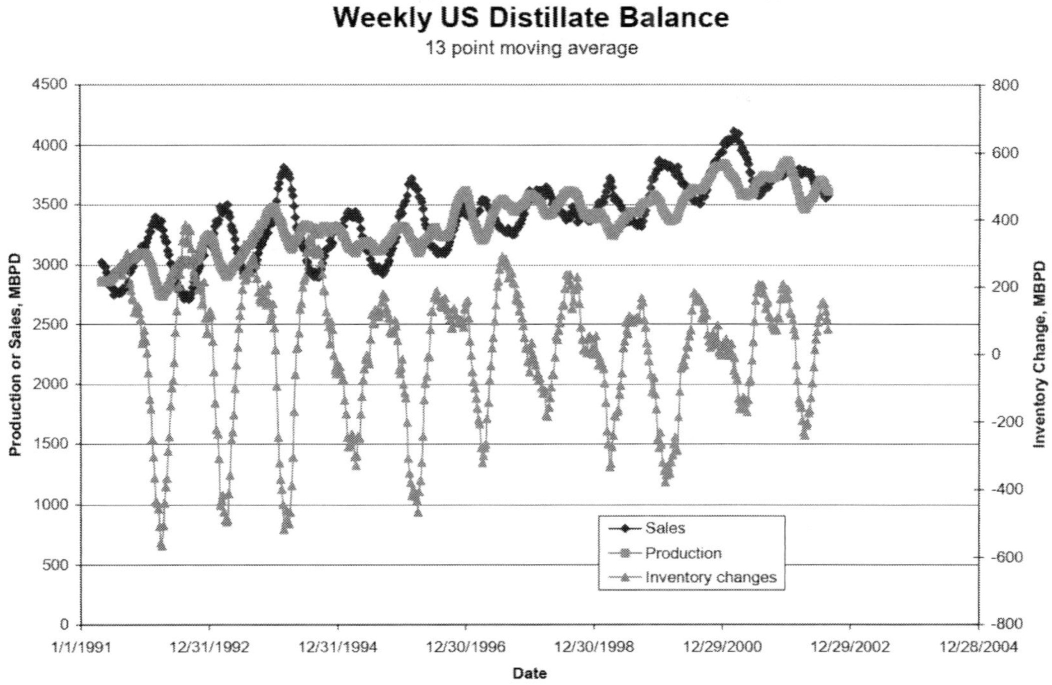

Figure 3

Annual sales of other distillate products (Figure 4) show that LSD is by far the largest single type of distillate product sold in the US. Furthermore, on-road and off-road diesel are the only categories that show net growth during the period from 1996 to 2000, the last full year for which detailed adjusted data are available.

Figure 4

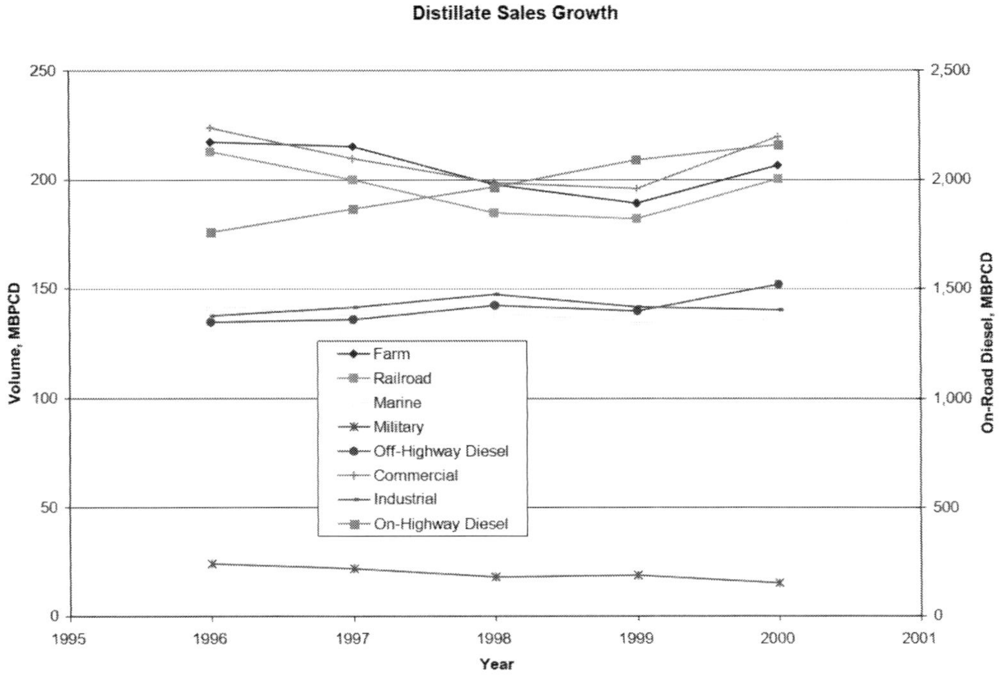

Analysis of US Refining Industry

The US refining industry is growing in capacity, but shrinking in the number of operating refineries. Mergers and acquisitions have also decreased the number of companies that operate refineries in the US. The refineries that continue to remain viable are either larger and more complex or have local supply and demand advantages because of their locations.

As of Jan 1, 2002 there were 143 operating refineries with a total crude distillation capacity of 16,565,000 BPCD. During 2001, the total crude run averaged 15,000 MBPCD, giving a 91.4 % utilization rate. Based on total crude capacity and the number of refineries, the average US refinery capacity is just over 115,000 BPD. More than 75% of the US crude capacity is in the 56 largest refineries. There are 87 small refineries in the US, with capacities less than 114 MBPD. 25 of these refineries are specialty refineries that do not even have naphtha reformers and produce HFO, asphalt, lubes or jet fuel as their primary (only) product. The rest of the crude is then sold to other refiners. 9 have naphtha reformers, but no vacuum gaso oil (VGO) or heavier conversion capabilities. These refineries sell long resid as well as G&D. There are also 53 small, "complex" refineries, 17 in PADD IV and Washington State and 36 in other parts of the US. These primarily serve local markets, many times refining local crudes, and currently have a significant product transportation advantage because there are often no product pipelines into their regions.

There are three basic levels of refinery complexity:

1) The simplest refineries are those that do not have a catalytic naphtha reformer. These are mostly small, specialty refineries that produce asphalt, lubes, HFO or jet fuel for local markets. They have few, if any catalytic units and depend on the quality of local crudes or their remote locations for their profitability. They distill the crude to recover products with high local demand and then sell the rest of the products as intermediates to other refineries. These specialty refineries account for very little of the on-road diesel that is currently produced, except in Alaska where the on-road diesel does not have to meet the current 500 ppm sulfur spec. The specialty refineries most likely will not install facilities to produce ULSD. Only one of these 25 refineries has a distillate hydrotreater that uses pipeline hydrogen. The average crude capacity of these specialty refineries is 18 MBPD and they account for less than 3% of the total US crude distillation capacity and an even smaller fraction of the G&D production.
2) The next level of complexity is the simple topping/reforming refinery. These 11 refineries have a naphtha reformer for gasoline production and a distillate hydrotreater that uses the hydrogen from the reformer to hydrotreat the distillate streams. They do not catalytically upgrade the VGO; however, 3 produce lubes from the VGO and 4 produce asphalt from the resid. These refineries have an average crude distillation capacity of 36 MBPD and are located in all 5 PADD regions.
3) There are 106 "complex" refineries that have at least some form of catalytic VGO upgrading. 100 of these use FCC units to upgrade the VGO while 6 have VGO hydrocrackers. Most of these refineries also have some form of vacuum resid upgrading. These complex refineries produce cracked distillates that must be desulfurized in addition to the virgin distillates that the simpler refineries must treat. VGO and resid conversion greatly increase the amount of sulfur that must be removed from the gasoline and distillate pools.

The 106 complex refineries in the US account for 93.6% of the total crude distillation capacity in the US and can be further subdivided according to the method that they use to convert or dispose of the vacuum residue from the crude. The resid disposition methods are listed below.

	Total Resid Capacity, MBPD	Percent of Crude	Number of Units
Coking	2,092	13.6	51
Resid HDT/HDC	580	3.8	11
Asphalt	406	2.6	33
HFO(from sales data)	721	4.7	21+

The capacities and number of units for the different resid disposition methods are from the two refining surveys. The percent of crude capacity column is based on the total stated resid unit capacity for the US divided by the total crude capacity for the US. The number of units is really the number of refineries that have that type of unit. The refinery surveys no longer list the capacities for each individual unit in every refinery. They only list the total capacity for each different type of unit. If a refinery has two cokers, for example, it lists the total capacity for both cokers.

None of the refinery surveys list the refineries that produce heavy residual fuel oil (HFO). The total capacity is from residual fuels sales data, while the number of refineries is based on the refineries that have VGO conversion, but no stated resid conversion unit. There are 21 such refineries. Some of these refineries feed the resid to their FCC unit to convert it. Some refineries use more than one method of resid disposition. For example, refineries that crack resid in their FCC unit also sell some HFO to dispose of the unconverted FCC bottoms stream. Some refineries sell asphalt and coke a portion of the resid, while others hydrotreat or hydrocrack the resid before converting it in a coker or FCC. This results in some double counting of resid disposition.

The major types of complex refinery configurations are show below.

Configuration	Crude Capacity, MBPD	# of Refineries	Average Refinery capacity, MBPSD
Coking, FCC	6,695	36	186
FCC, no resid unit	2,096	21	100
Asphalt, FCC	2,000	22	91
Coking, Asphalt, FCC	1,932	10	193
Coking, Resid HDT/HDC, FCC	1,112	4	278
Resid HDT/HDC, FCC	1,075	7	154
VGO HDC, no FCC	600	6	100

The Coking, FCC refinery configuration is by far the most common configuration in the US, accounting for over 40% of the total US crude distillation capacity. Adding in the next three categories represents over 75% of the total US crude capacity and almost 85% of the complex refineries.

The following table lists the average kerosene and heavier downstream unit capacities for the 106 complex refineries in the US grouped by refinery type. There are six different types of kerosene and heavier hydroprocessing units listed in the surveys, four hydrotreating units and 2 hydrocracking units. The hydrotreating units are classified by the feed type: kerosene, diesel, VGO or resid. The HDC units are classified either as distillate, which could be diesel range or VGO, and resid. In general, the units processing heavier feeds operate at higher pressure and with more catalyst.

	Coking, Asphalt, FCC	Coking, Resid HDP, FCC	Resid HDP, FCC	Coking, FCC	Asphalt, FCC	FCC, No Resid Unit	VGO HDC, No FCC
Kero HDT	6.4	8.1	12.2	7.3	15.4	3.2	0
Diesel HDT	25.1	18.8	16.9	19.2	13.9	14.8	7.8
VGO HDT	20.6	8.7	4.0	16.4	12.0	2.1	3.3
Distillate HDC	7.3	10.3	3.8	7.6	2.3	2.3	24.5
FCC	32.8	39.1	39.8	37.1	35.8	41.6	-
Resid HDT	-	6.7	19.4	-	-	-	-
Resid HDC	-	8.2	12.8	-	-	-	-
Coking	19.4	19.5	-	20.9	-	-	20.6
Asphalt/HFO	8.8	-	-	-	11.8	-	0.2
Total Distillate & Heavier HDP	59.3	67.0	69.2	50.4	43.5	22.5	35.6
Average Crude Capacity, TBD	193.2	278.0	153.6	188.4	90.9	99.8	100.0
Number of Refineries	10	4	7	35	22	21	6
Total Crude Capacity, TBD	1,932	1,112	1,075	6,595	2,000	2,096	600

Those refineries that hydroprocess resid have a higher total distillate and heavier hydroprocessing capacity than those that do not. They also have higher resid processing unit capacities because they either hydrotreat long resid as FCC feed or partially convert short resid and coke the bottoms from the resid HDC unit. This causes some double counting of resid disposition capabilities.

The average coking capacity of refineries with cokers is very consistent at 19.4 to 20.9 % of crude capacity. FCC capacities are also relatively consistent between 32.8 and 41.6% of crude. The largest FCC capacity is for refineries without a resid conversion unit, indicating that some resid is being cracked in the FCC. The differences in total hydroprocessing capacity and total resid conversion capacity reflect the types of crudes that the refineries are capable of running.

Refinery Investments Required for Future Regulations

Our analyses of the US distillate market and the US refinery configurations and hydrotreating requirements allowed us to develop algorithms that could be used to evaluate each individual refinery's capability to meet the potential future distillate regulations. These algorithms used the refinery configuration to predict the total distillate production from each refinery. The capability to produce 500 ppm sulfur diesel was determined from the distillate hydroprocessing capacity. This approach matched both the total distillate production and the amount of distillate that is sold without being processed in a distillate hydrotreater to within 2%.

There are six different combinations of future potential distillate regulations that will require additional investment in the refineries. The regulatory scenarios that require new hydrotreating capacity are as follows:

1. Highway Diesel to 15 ppm sulfur. No changes to other products or other diesel specs.
2. Scenario 1, plus off-road diesel and jet fuel to 500 ppm sulfur. No other changes to any product.
3. Scenario 2 plus Home Heating Oil to 500 ppm.
4. All distillate products to 15 ppm sulfur.
5. Diesel Aromatics to 20% vol
6. Diesel Aromatics to 10% vol

In developing the refinery modifications that were required to meet the future distillate fuels regulations, we established the following ground rules:

- No reduction in kerosene or diesel production volumes. These are the growth products. The sales of other distillate products have been declining and some small reductions in these lower value products were allowed if necessary.

- Minimum 1 year HDS unit cycle length. Although some refiners currently run longer and shorter cycles, one year was felt to be a reasonable minimum cycle length.
- Minimize new unit capacity. From our experience, we felt that most existing distillate HDS units could be expanded by up to 20% at a lower cost than building a new unit. Furthermore, informal discussions with many refiners indicate that most distillate HDS units do not have state of the art reactor internals. Upgrading the reactor internals on these units would increase their effective catalyst volume by at least a factor of 2 and in some extreme cases up to a factor of five. Upgrading reactor internals and using higher activity catalyst is a very low cost method for upgrading a diesel hydrotreater that is currently producing 500 ppm.

All capital cost estimates are based on second quarter 2002 US gulf coast construction costs escalated to mid-2005 at 2 % per annum. The capital cost estimates for an existing unit upgraded to ULSD capabilities are based on a recent unit upgrade to ULSD capabilities that we designed. This was a 45 MBPD unit that added a new reactor, feed effluent exchanges, HPS and pump and compressor modification. The cost was just over $20 MM with start-up in early 2002. These modifications also increased capacity by 17% in addition to giving the unit the capability to produce 10ppm sulfur diesel. To be conservative, all revamp costs for ULSD were scaled off of a $30MM revamp cost for a 45 MBPD unit using a 0.6 capacity factor.

New unit costs were estimated based on a simple HDS unit design without product ffractionation. Major equipment costs were estimated for a 30 TBD unit and then total project costs were estimated using multipliers to the total equipment costs. The multiplication factors that were used are given below:

	Percent of Bare Equipment Costs
Bare Equipment	100
Direct Construction Costs	140
Total Indirect Costs	225
Offsite Costs	100
Total Project Cost	565

This procedure gave a cost of about $50MM for a new 30MBPD unit for ULSD production. The costs for other units of different sizes were scaled from the 30 MBPD unit cost using a 0.6 factor. Cost estimates for other HDS unit types, i.e. Arosat units or units to treat railroad and marine diesel were estimated in a similar way.

The cost of a new hydrogen plant was not included in the capital estimates. The incremental hydrogen was assumed to be purchased from a hydrogen pipeline system and its cost was included in the operating costs. Purchasing pipeline hydrogen, or hydrogen from a third party plant has become much more cost effective and reliable than the refinery building its own hydrogen plant.

There are three major HDS unit operating costs: 1) fuel gas for the process heater and hydrogen production, 2) electricity for the pumps, compressors and air coolers and 3) catalyst replacement. The other costs, such as, cooling water, wash water, amine circulation and regeneration and steam all total less than 0.1 cents per gallon of distillate and therefore are insignificant.

Since the probability of accurately predicting future energy costs is very low, we have used the average gulf coast prices for the last five years. These gave a cost for electricity of 4.5 cents per KWH and a natural gas cost of $3.2/Mcf (20.4 $/FOE).

Scenario 1 – Highway Diesel to 15 ppm Sulfur

Our evaluation of the US refining industry indicated that essentially all of the current on-road LSD (plus some additional distillate products) is currently produced via distillate hydroprocessing units. Many of these units could be upgraded to produce ULSD without building new distillate desulfurization capacity. Via survey of the refining industry, the US EPA has estimated that 80% of the refiners will produce ULSD by modifying existing units and that only 20% will build new units.

We have briefly investigated the ability to produce ULSD by running lower sulfur crude. Although this might be an option for a few small refineries, there is insufficient low sulfur crude in the world to make this a viable industry wide option. Also, if more refineries adopt this option, the spread between low and high sulfur crudes will increase, making it more attractive to invest in the facilities required to process the higher sulfur crudes.

There is a small amount of heavy distillate and LCO that is currently hydrotreated and blended into LSD. Some refiners will probably install facilities to improve the fractionation of these streams to make them easier to desulfurize. Any project to improve fractionation will most likely also include facilities to increase distillation capacity to improve the return on the total investment. There could also be some minor optimization of feed components to the HDS units to move the more difficult to desulfurize streams into the higher sulfur products. This option is only available to refineries that currently produce higher sulfur products. Stream reallocation is only a short-term solution since the market for higher sulfur distillates is declining and will probably be reduced by future distillate regulations.

As previously shown, essentially all of the existing kerosene and diesel HDS capacity must be upgraded to switch all of the current LSD production to ULSD production. In all refinery configurations, hydrotreated kerosene is blended into the LSD pool. A breakdown of the current kerosene and diesel HDS units by size is shown below:

Capacity Range, MBPD	Assumed Average Capacity, MBPD	Number of Kerosene HDS Units	Number of Diesel HDS Units
0-10	7.5	11	22
10-20	15	15	20
20-30	25	12	13
30-40	35	4	16
40-50	45	3	16
50-60	55	1	3
60+	70	4	7
Total		50	97
Total Capacity, MBPD		1,217	2,725

The total investment required to upgrade all of these units to ULSD capabilities is $3.1 billion dollars. Incremental operating costs are 0.7 cents/gal. When the total investment is amortized over 15 years at 15% ARR and added to the incremental operating costs, then the total incremental cost to produce ULSD is 1.55 cents/gal. This cost does not include any additional distribution facilities or product downgrade that might occur during shipment to the final point of sale. The total amount of incremental hydrogen required to produce ULSD was estimated to be 400 MMSCFD. This is about a 40% increase in the current hydrogen pipeline system in the US. Hydrogen consumption accounts for about 0.38 cents per gallon of the incremental cost to produce ULSD.

The maximum investment case for ULSD production would be to install new HDS units at each refinery to reprocess the existing LSD into ULSD. Assuming the same size distribution of units gives a total investment of $6.2 billion and a total cost to produce ULSD of 2.45 cents/gal. Since the true investment scenario is somewhere between these two extremes (80% revamps and 100% new units), the incremental cost to produce ULSD is between 1.55 and 2.45 cents/gallon of ULSD.

This cost is significantly lower than any other previous study for several reasons:

- Recent experience has shown that ULSD can be produced at much less severe conditions than previous studies assumed and with lower hydrogen consumption.
- Actual revamp costs were used to estimate the cost of other revamps.
- Any costs for additional distribution facilities and product downgrade were not included.

Scenario 2 – Other Distillates, Except HHO, to 500 ppm Sulfur

A previous section showed that current LSD production exceeds highway diesel sales. We have assumed that all of the current producers of LSD will upgrade their facilities to produce ULSD. In fact, some will actually expand their production facilities as part of the modifications for ULSD. A reasonable method for estimating the new facilities that are required to desulfurize the other distillate streams is to calculate the total distillate that currently enters the distillate pool without being processed in a distillate HDS unit. This includes virgin distillates as well as some portion of the distillates produced from VGO and resid HDP units. Although a few high severity VGO HDS units are able to produce distillate side draws with less than 500 ppm sulfur, most VGO and resid HDS units produce distillates that exceed 500 ppm sulfur and their distillate must be reprocessed to meet the lower sulfur spec. Some current distillate capacity is also used to treat higher sulfur streams down to the 0.2 to 0.5 wt% sulfur levels of marine and railroad diesel and home heating oil (HHO).

In Scenario 1, we modified all existing hydrotreaters from 500 ppm sulfur product to 15 ppm sulfur product. Not all of the expanded HDS capacity is currently producing 500 ppm. Some is desulfurizing high sulfur distillate to 0.2 to 0.5 wt% sulfur products. The same modifications that upgraded a unit from 500 to 15 PPM sulfur would allow a unit that was currently producing 0.5 wt% sulfur to now make 500 ppm sulfur. Thus part of the investment required to lower the sulfur of other distillate streams was made in Scenario 1. The incremental operating cost will be higher for reducing sulfur from 5000 to 500 ppm, however, because these higher sulfur products currently contain more cracked gas oils and will consume more hydrogen in reducing their sulfur from 5000 to 500 ppm.

Currently there are about 40 refineries that produce about 1.25 MMBPD of untreated distillates for distillate sales. To meet the Scenario 2 specs, 33 of these refineries will have to build a total of about 950,000 BPD of new grassroots HDS capacity to process the untreated distillate to 500 ppm. Although about half of the distillate that needs to be treated is kerosene, most refiners will probably choose to install a higher pressure HDS unit and shift existing HDS capacity to the less severe kerosene service.

Marine and Railroad diesel (the other half of the high sulfur product that will be desulfurized in scenario 2) contains a high percentage of cracked distillates. We have estimated that the new HDS units will consume an average of about 600 SCFB of hydrogen or a total incremental hydrogen consumption of 580 MMSCFD. The total incremental cost to produce is 4.5 cents/gallon. 1.9 cents is due to the total capital investment of 1.6 billion dollars and 2.6 cents is due to hydrogen consumption.

Scenario 3 – HHO to 500 ppm

After upgrading marine and railroad diesel and jet fuel to 500 ppm sulfur, 7 refineries will still be producing about 315 MBPD of high sulfur home heating oil. These are essentially all in PADDs 1 and 3. A relatively large percentage of the current HHO is untreated virgin gas oil and LCO. Reducing the sulfur content of these materials will consume less hydrogen than reducing the sulfur content of the marine and railroad diesels. The total incremental hydrogen consumption for this scenario would be 95 MMSCFD. The total cost to reduce the HHO sulfur would be 3.3 cents/gallon, with about 1.6 cents/gallon attributable to the capital investment of $0.5 billion.

Scenario 4 – All Products to 15 ppm

For this scenario all of the new HDS capacity that was built to produce 500 ppm sulfur distillates (about 1,250 MBPD) must now be upgraded to produce 15 ppm sulfur distillates. The unit modifications will be similar to those that were applied to upgrade the highway diesel units; however, these streams are more aromatic than the highway diesel and will consume more hydrogen. The total incremental hydrogen consumption for this scenario would be 190 MMSCFD. The total cost to reduce the HHO sulfur would be 1.5 cents/gallon, with about 0.8 cents/gallon attributable to the capital investment of $0.9 billion.

Scenario 5 – Highway Diesel to 20% Aromatics

We have estimated that the current highway diesel pool is about 26 % volume aromatics. On average, the 20% aromatics spec could be met by shifting about half of the LCO that is currently in the highway diesel pool into other distillate products with very little additional cost. Although this shift is theoretically possible "on average", some refineries process crudes (such as heavy Canadian or Venezuelan crudes) where the virgin distillate approaches or even exceeds the 20% aromatics spec. These refineries would most likely have to add distillate arosat units to meet the 20% aromatics spec. Since the US imports about 2.5 MMBPD of crude from these countries (16% of total crude run), we have estimated that about 15% of the total highway diesel pool (or 340 MBPD) must be processed through a diesel arosat unit to meet the 20% aromatics spec.

Production of 20% aromatics diesel fuel would require an investment of about $0.8 billion for 11 or 12 new arosat units. The total incremental cost to produce ULSD with 20% aromatics would be about 0.6 cents per gallon; however, the total $0.8 billion investment would be made by only 11 or 12 refineries. The cost to produce for these refineries would increase by almost 5 cent/gallon while the other refiners would only have marginal additional costs due to stream reallocation. The total incremental hydrogen consumption would be about 150 MMSCFD.

Scenario 6 – All Highway Diesel to 10 vol % Aromatics

To meet the 10% aromatics spec, essentially all of the current LSD production must be sent to an arosat unit to reduce its aromatics concentration from the interim level of 20% to the new spec of 10% aromatics. This would require the construction of almost 100 new arosat units at a cost of about $5.8 billion dollars. The incremental cost to produce the 10% aromatics diesel would be about 4.2 cents/gallon and would consume an incremental 530 MMSCFD of hydrogen.

Impact of Future Regulations on CO_2 Emission

Based on the refinery modification scenarios that we developed in the previous sections, we were able to estimate the increase in total industry wide CO_2 emissions that would result from the implementation of the different distillate fuel regulations.

CO2 emission from the refining and combustion of distillate fuels currently exceeds 2.5 million tons per day, with over 95% coming from the combustion of the fuels themselves. Although increasing the hydrogen content of distillate fuels via hydrotreating would appear to reduce CO2 emissions, the total CO2 emission actually increases via hydrotreating, as shown below:

Increase in CO2 Emissions, Tons per day

	From Distillate Fuel Combustion	From Distillate Hydroprocessing	Total	Percent Increase
Scenario 1	792	9607	10400	0.41
Scenario 2	871	14160	15032	0.59
Scenario 3	284	2383	2667	0.10
Scenario 4	362	4549	4910	0.19
Scenario 5	4635	3670	8305	0.33
Scenario 6	15539	14034	29572	1.16
Total	22483	48403	70887	2.78

The average change in diesel fuel energy content due to reducing sulfur content from 500 to 15 ppm (scenario 1) is less than 0.1% and smaller than the current variations in diesel fuel energy content produced in different refineries. The largest reduction in energy content is from reducing diesel fuel aromatics to 10% (scenario 6). This specification change also gives the largest overall increase in CO2 emissions.

Other Observations

The new ULSD regulations will eventually cause the shutdown of many small refineries. The larger refineries will install the equipment that is required to make ULSD. These projects will most likely also include facilities that will increase the crude processing and distillate production capabilities of the refinery. A minor 10% increase in the distillate production capabilities of the 56 largest refineries (ie those with crude capacities >114 MBPD) can replace the distillate currently produced by the 39 smallest complex refineries.

Although there were some local shortages of LSD during the first year after its introduction, supply quickly exceeded demand. During the first year, the price difference between high and low sulfur diesel was as high as 7 cent/gallon on the gulf coast, but has averaged about 1.6 cents per gallon since then, much lower than the predicted price spread. The lower-than-predicted price difference is due to both an exaggerated estimate of the cost to produce and over building of capacity in the industry because of the high predicted price differential.

Major Conclusions

- Current LSD supply exceeds on-road diesel use by about 20%. This excess supply is most likely being consumed by off-road diesel users. All current LSD users will expect to shift to ULSD when it is available.

- The only true high sulfur distillate products currently being sold are: jet fuel, railroad and marine diesel and home heating oil.
- There is insufficient low sulfur crude available to make crude substitution a viable method of producing ULSD.
- A post-audit of the introduction of LSD in 1993 showed that the price differential between LSD and home heating oil has averaged less than 2 cent/gallon, significantly below the predicted cost to produce LSD of 3-6 cents/gallon estimate by EPA and others prior to the regulation.
- Capacity expansions by the large refineries as part of their ULSD strategies will result in the closure of some small refineries.

Paper 93a

A WEI-PRATER KINETIC MODEL FOR XYLENE ISOMERIZATION OVER THE BP AMOCO HSDE CATALYST

Jeffrey A. Amelse ** and Lisa M. Pinsent

bp Amoco Chemical Company
150 West Warrenville Road
Naperville , IL 60566
AmelseJA@BP.com

** To whom correspondence should be sent

Keywords: kinetic, model, xylene, isomerization, paraxylene, Wei, Prater

Prepared for Presentation at the 2004 Spring National Meeting, New Orleans, LA, April 25-29
UNPUBLISHED

Abstract

The Wei-Prater method decouples the differential equations that describe the kinetics of reversible mono-molecular reactions. For the isomerization of the three xylene isomers, the reaction paths towards equilibrium are constrained to a reaction triangle. A linear transformation moves the problem from the natural 3D concentration coordinate system in which three species react to a new 2D coordinate system. In the transformed system two new species (linear combinations of the xylene isomers) react independently and decay towards zero concentration. The problem of finding six rate constants in the natural coordinate system becomes one of finding two straight-line reaction paths (two of the eigenvectors of the rate constant matrix), the equilibrium point (the third eigenvector), and the decay constants of the transformed species.

Previously, Amelse (1993) had shown how to very accurately determine the equilibrium distribution of the xylene isomers via the intersection of the slow straight-line reaction paths for shape selective and non-shape selective molecular sieves. In this paper, we will show how one can determine the equilibrium distribution by adjusting the position of the equilibrium point along the slow path to improve the fit to curved reaction paths. The equilibrium distribution so determined is slightly different than that reported previously. A kinetic model will be presented for the bp Amoco Chemical Company HSDE shape selective xylene isomerization catalyst.

Introduction

Para-xylene (pX) is a bulk industrial chemical intermediate. It is oxidized and purified to Purified Terephthalic Acid (PTA). PTA is reacted with ethylene glycol to produce polyester, which has a number of uses, including fiber for clothing, film, tire cord, and plastic beverage bottles. World capacity for pX was about 24 million metric tons per annum in 2003. bp Amoco Chemical Company (bp) is one of the largest producers of pX.

The primary feedstock for pX is mixed xylenes isolated from refinery reformate. pX units based on reformate xylenes are generally comprised of three sections: isomerization, fractionation, and pX separation. pX is separated from mixed xylenes by one of two methods: low temperature crystallization or via selective adsorption on a molecular sieve. bp uses its own crystallization process. The bp process isomerization and fractionation sections are highly heat integrated. Thus, bp pX units are very energy efficient.

The xylene isomers in the fresh reformate mixed xylenes feed are near equilibrium. Thus, they are sent to the pX separation section directly or via the fractionation section. After pX removal, the pX separation section reject is sent to the isomerization section, where the xylenes are isomerized catalytically to near equilibrium. The catalyst must also convert ethylbenzene (EB) and P&N's present in the feed to lighter or heavier byproducts to prevent their build-up in the recycle loop, since these components cannot be separated from pX by distillation.

All bp pX units use modifications of the Amoco (now bp) pX crystallization process, and bp xylene isomerization catalysts. bp has developed several versions of its xylene isomerization catalyst over the years. bp currently uses its High Selectivity DeEthylation (HSDE) that comprises its patented medium pore borosilicate molecular sieve AMS-1B in all of its units. The HSDE catalyst converts EB present at levels of 12-20 wt% in typical reformate primarily via reaction with hydrogen to produce benzene and ethane (EB hydrodeethylation). Amelse (1987 and 1988) has shown the important role that shape selectivity provided by medium pore molecular sieves plays in allowing high selectivity for hydrodeethylation, in suppressing xylene loss via methyl transfer, and in suppressing certain C10 aromatic isomers formed by ethyl transfer.

The bp HSDE catalyst exhibits high activity for isomerization of the xylene isomers, and extremely low xylene loss. It also exhibits high activity for the conversion of P&N's. Thus, is ideal for use with unextracted mixed xylenes feedstocks coming from medium-high pressure reformers. The bp DC2 unit routinely processes mixed xylenes with a P&N content in excess of 6 wt%. The catalyst is insensitive to sulfur, an added bonus.

Wei (1982) showed that a model he had developed twenty years earlier (Wei (1962 a and b)) for diffusion limited isomerization of a 3-component system having unequal diffusivities adequately describes enhanced para-selectivity during isomerization. This model is an extension of the classic treatment of the kinetics of reversible mono-molecular reactions developed by Wei and Prater (1962).

The accepted carbonium ion mechanism for xylene isomerization over acid catalysts indicates that methyl groups jump one carbon atom at a time around the aromatic ring. This means that there is no direct conversion of oX to pX. Rather, oX is first isomerized to mX, and mX is then isomerized to pX. However, diffusion limitation of xylene isomerization and unequal isomer diffusivities leads to apparent direct conversion of oX to pX, as explained by Wei (1982).

Amelse (1993) examined the parametric sensitivity of the Wei model to show under what conditions enhanced para-selectivity will occur during xylene isomerization. More importantly, Amelse (1993) showed that an extension of the Wei model could be used to determine the equilibrium distribution of the xylene isomers with extreme accuracy. The distribution calculated from the published equilibrium data (Stull, et al. (1987)) available at that time was shown to be in error. This prompted new measurements of the thermodynamic properties of the xylene isomers (Chirico, et al. (1997 a-e). While the equilibrium distribution of the xylene isomers calculated from the revised thermodynamic expressions was now in agreement with the Amelse (1993) experimental value within the accuracy of those expressions, the imprecision of the calculated value is too high to be of great use to pX producers. An imprecision of several tenths of a weight percent of pX can mean a difference of several percent plant capacity.

Accurate knowledge of the equilibrium distribution of the xylenes is important. The older literature is full of claims of isomerization to greater than 100% pX approach-to-equilibrium (PATE). When researchers see a PATE of 104%, they are tempted to try for 120%. Old reports of greater than 100% PATE are due to the use of the wrong equilibrium distribution. It is not possible to ***isomerize*** to greater than true 100% PATE. However, it is possible to achieve a pX concentration of greater than equilibrium when producing xylenes via reactions such as toluene alkylation or disproportionation (Young, et al., (1982)).

Amelse (1993) showed that the equilibrium distribution of xylenes can be very accurately determined as the intersection of the slow straight-line reaction paths for a shape selective medium pore molecular sieve and a non-shape selective large pore molecular sieve. (More on straight-line reaction paths below.) In this paper, we will show how one can determine the equilibrium distribution by adjusting the position of the point along the slow path to improve the fit to curved reaction paths. The equilibrium distribution so determined is slightly different than that reported previously. A kinetic model will be presented for the bp Amoco Chemical Company HSDE shape selective xylene isomerization catalyst.

Experimental

Experimental data were obtained via a conventional fixed-bed hood-scale pilot plant, fitted with a 0.5 inch OD stainless steel reactor loaded with 1.5 gm of bp HSDE catalyst. All data were obtained at 250 psig (1.72 MPa gauge), and an H2/Hc ratio of 1.5. PATE was varied by varying WHSV. Most data were obtained at 400 °C. However, an additional curved path was obtained at 382 °C to determine activation energies. Products were condensed and feed and effluent liquid samples were analyzed by capillary gas chromatography (FID detector) using a 60 m x 0.25 micron ID Supelcowax 10 column. This column is capable of giving baseline separation of the xylene isomers. Data analysis was via a custom FOTRAN program that follows Wei and Prater (1962). Triangular plots were constructed with EXCEL. 3D plots were constructed via the "3D Grapher" software available from Roman Lab Software: http://www.romanlab.com/3dg/.

The Wei-Prater Method

Details of the Wei-Prater method are provided in the classic article by Wei and Prater (1962). It is a long article. However, it is very well written and includes step-by-step instructions and many examples.

The Wei-Prater method decouples the differential equations that describe the pseudo mono-molecular diffusion-disguised isomerization of the three xylene isomers, as shown in Figure 1.

Wei-Prater Method

- **Coupled System (3-D coordinates)**

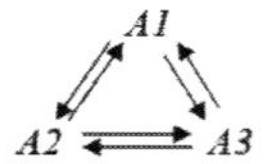

- **Rate equation:**

$$\begin{bmatrix} da_1/dt \\ da_2/dt \\ da_3/dt \end{bmatrix} = \begin{bmatrix} -(k_{21}+k_{31}) & k_{12} & k_{13} \\ k_{21} & -(k_{12}+k_{32}) & k_{23} \\ k_{31} & k_{32} & -(k_{13}+k_{23}) \end{bmatrix} \begin{bmatrix} a_1 \\ a_2 \\ a_3 \end{bmatrix}$$

- **Solution:**

$$a_1 = c_{10} + c_{11}e^{-\lambda_1 t} + c_{12}e^{-\lambda_2 t}$$
$$a_2 = c_{20} + c_{21}e^{-\lambda_1 t} + c_{22}e^{-\lambda_2 t}$$
$$a_3 = c_{30} + c_{31}e^{-\lambda_1 t} + c_{32}e^{-\lambda_2 t}$$

- **Uncoupled System (2-D coordinates)**

B_0 *does not react*

$$B_1 \xrightarrow{-\lambda_1} 0$$
$$B_2 \xrightarrow{-\lambda_2} 0$$

- **Rate equation:**

$$\begin{bmatrix} db_0/dt \\ db_1/dt \\ db_2/dt \end{bmatrix} = \begin{bmatrix} 0 & 0 & 0 \\ 0 & -\lambda_1 & 0 \\ 0 & 0 & -\lambda_2 \end{bmatrix} \begin{bmatrix} b_0 \\ b_1 \\ b_2 \end{bmatrix}$$

- **Solution:**

$$b_0 = b_{00}$$
$$b_1 = b_{10}e^{-\lambda_1 t}$$
$$b_2 = b_{20}e^{-\lambda_2 t}$$

Figure 1. Wei-Prater Method

The solution to the uncoupled system lies in the 3D natural coordinate space with the mole fractions of the 3 xylene isomers lying in the direction of the three coordinate axes. However, two constraints place the solution onto a 2D reaction triangle.

The first constraint is that the sum of the mole fractions is equal to one:

$X_{oX} + X_{mX} + X_{pX} = 1$

This constrains the solution to a plane. The second constraint is that all mole fractions are greater than zero:

$X_i > 0$

This constrains the solution to a reaction triangle sub-region lying on the plane. Thus, the problem drops from 3D to 2D. Moving to the 2D system involves a linear transformation from the natural coordinate system involving mole fractions to a new coordinate system involving fictitious species b_0, b_1, b_2, which are linear combinations of the mole fractions. The geometric relationship between the two coordinate systems is illustrated in Figure 2.

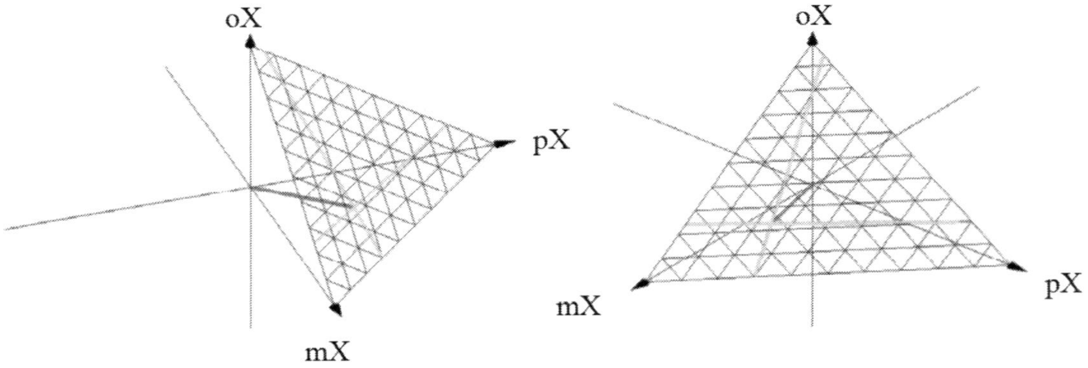

Figure 2. Two Views of the Relationship Between the Natural and Transformed Coordinate Systems

Any point on the reaction triangle can be viewed as the sum of three vectors in the direction of the three transformed coordinates, b_0, b_1, and b_2, as shown in Figure 3.

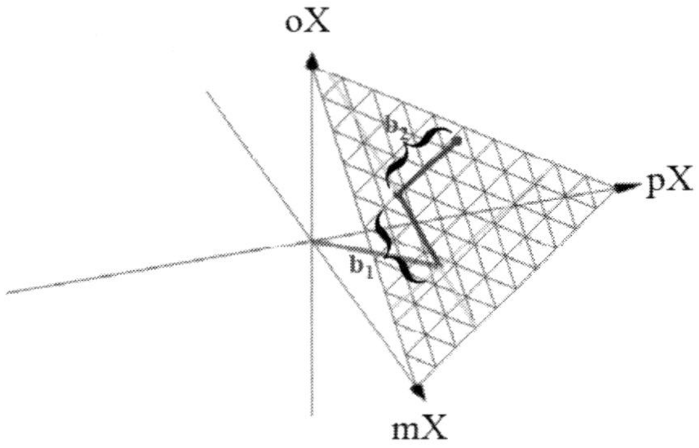

Figure 3. A Point on the Reaction Triangle Expressed as the Sum of the 3 *b* vectors

The solution at any point in time is:

$b_0 = b_{00}$

$b_1 = b_{10}\, e^{-\lambda_1 t}$

$b_2 = b_{20}\, e^{-\lambda_2 t}$

where b_{00}, b_{10}, and b_{20} are the concentrations in the transformed coordinate system at time zero.

b_0 points from the origin in the 3D coordinate system to the equilibrium point. This vector puts the solution onto the reaction triangle. The b_1 and b_2 vectors can be considered as departures from the equilibrium point in the direction of b_{10} and b_{20} respectively. These departures decay exponentially to zero. Thus, as time approaches infinity, the solution points to the equilibrium point.

A starting composition that deviates from equilibrium in the direction of b_1 or b_2 with no component in the other direction will isomerize towards equilibrium along a straight-line path that points in these directions. Isomerization of a starting composition with both b_1 and b_2 components will trace a unique curved path towards equilibrium. Usually, one of the time constants (λ_1 and λ_2) is larger than the other, and by convention this is assigned to b_2. Thus, b_2 decays faster than b_1, and as time becomes large, the solution for isomerization of a starting composition that does not lie exactly in the direction of b_2, will always approach equilibrium along the direction of b_1.

The problem of finding six rate constants in the Natural coordinate system becomes a problem of finding the direction of three vectors, b_0, b_1, *and* b_2, and two time constants λ_1 and λ_2. These vectors turn out to be eigenvectors of the matrix of rate constants in the Natural coordinate system, and the time constants are eigenvalues. As noted above, b_0 points the equilibrium point. The direction of b_1 can be found by taking data close to equilibrium. b_2 can then be constructed knowing b_1 and b_2 by applying orthogonality relationships, as shown by Wei and Prater (1962).

Prior to the Wei-Prater method (and still to this day), some have attempted to obtain relative rate constants in the Natural coordinate system by isomerizing pure components at differential conversion. Wei and Prater (1962) have discussed the pitfalls of determining rate constants by extrapolating data to zero conversion. Often the solution obtained in this manner does not approach true equilibrium.

The equilibrium point can be determined by operating at long residence time. The problem is determining how long is long enough. Wei (1982) showed that for isomerization in the presence of diffusion limitation with unequal diffusivities of the components, the b_1 and b_2 eigenvectors rotate about the equilibrium point in the plane of the reaction triangle. Amelse (1993) showed that the equilibrium point can be determined very accurately from the intersection of the slow straight line reaction paths for a shape selective and a non-shape selective molecular sieve. In this paper, we will show how to improve the accuracy of the equilibrium distribution by moving the equilibrium point along the direction of the slow straight-line reaction path to improve the model fit to two curved-line reaction paths.

The development of a kinetic model via the method of Wei and Prater (1962) is summarized by the following steps:

1. Determine the equilibrium point, which determines b_0.

2. Experimentally determine the position of the slow straight-line reaction path, b_1.

3. Calculate the position of the fast straight-line reaction path, b_2, from b_0 and b_1 by applying the orthogonality relationships and construct the **X** matrix as described in Wei and Prater (1962).

4. Transform the **a** vector of mole fractions in the Natural coordinate system to the **b** coordinate system via the equation:

   $$\mathbf{b} = \mathbf{X}^{-1}\,\mathbf{a}$$

5. Obtain the relative decay constants in the B-system by plotting ln b_1 versus ln b_2:

   $$\ln b_1 = \lambda_1/\lambda_2 \ln b_2 + \text{constant}$$

6. Obtain the rate constant matrix in the Natural coordinate system via:

   $$\mathbf{K} = \mathbf{X}\,\Lambda\,\mathbf{X}^{-1}$$

Results and Discussion

Our prior work (Amelse (1993)) showed that the slow straight-line reaction path for catalysts comprising shape selective molecular sieves intersects the reaction triangle near 95% oX and 5% pX. The position of this path was refined by taking new data for a feed containing 94.20% wt% oX, 4.98 wt% pX, 0.54 wt% mX, and 0.23 wt% cumene impurity. WHSV was varied to obtain reactor effluent compositions close to equilibrium (from 23.16% to 23.99% pX amongst the xylenes). The slow straight-line path for these data is shown in Figure 4. The coordinates for this plot place mX at the origin of an x-y plot, pX at the point (10,0), and oX at the point (10*cos (60), 10*sin(60)). The correlation coefficient, R^2, for this plot is 0.9990.

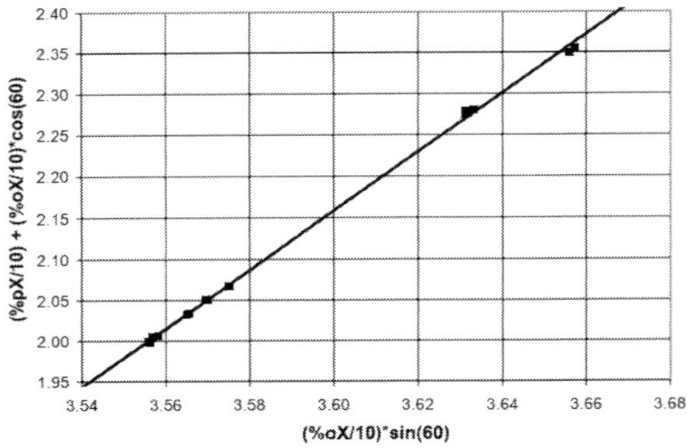

Figure 4. Slow Straight Line Path for HSDE Catalyst at 400 °C

Two curved paths were obtained with a feed containing 49.65 wt% pX, 0.32 wt% mX, 49.87 wt% pX, and 0.12 wt% cumene impurity; and a feed containing 0.08 wt% pX, 49.97 wt% mX, 49.78 wt% oX, and 0.12 wt% cumene impurity.

Using the equilibrium composition at 400 °C published by Amelse (1993) of 23.99% pX, 53.21% mX, and 22.80% oX, the experimentally determined position of the straight-line path, and the position of the fast of the fast path calculated via the procedure outlined in Wei and Prater (1962), a complete model was constructed. Once a complete model is constructed, curved paths for any feed composition can be constructed.

Using our previously published equilibrium composition, the model is just a bit off as shown in Figure 5. The experimental points lie slightly below the calculated curved paths.

The model can be improved by moving the equilibrium point a bit along the slow straight-line path. The equation for the straight-line path is:

y = -10.7413 + 3.5832 * x

where:

y = (%oX/10) * sin(60)

x = (%pX/10) + (%oX/10) * cos(60)

The improved model is shown in Figure 6.

The best fit of the experimental data is obtained, within about +/- 0.02 wt% pX, using an equilibrium distribution at 400 °C of:

24.100% pX, 53.148 wt% mX, and 22.7518 wt% oX

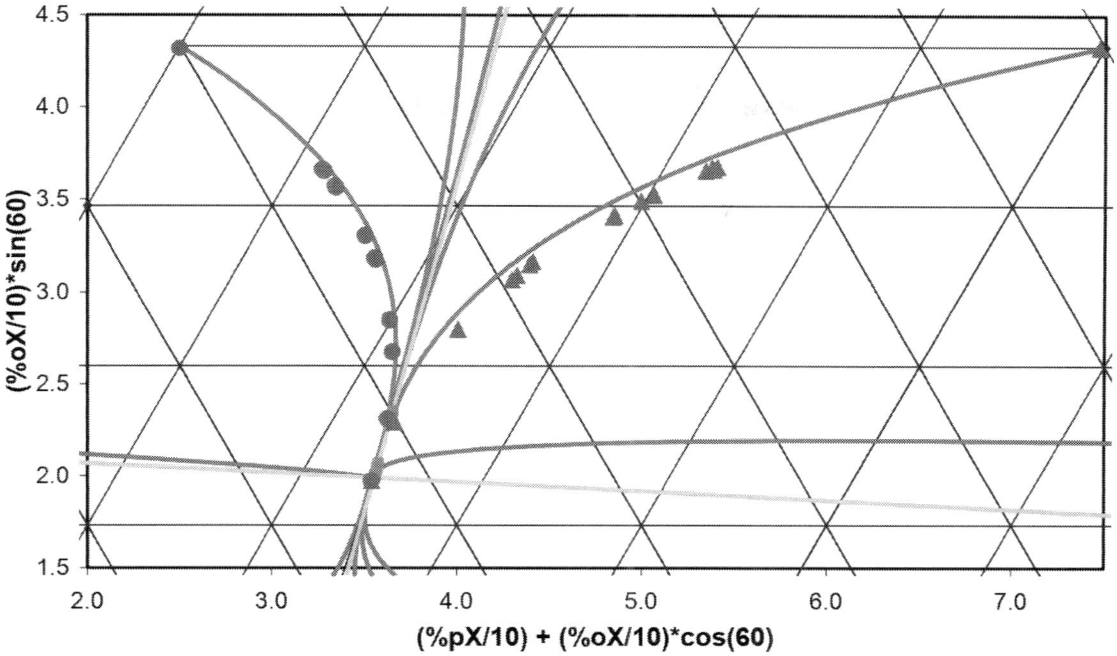

Figure 5. Model based on equilibrium distribution of Amelse (1993)

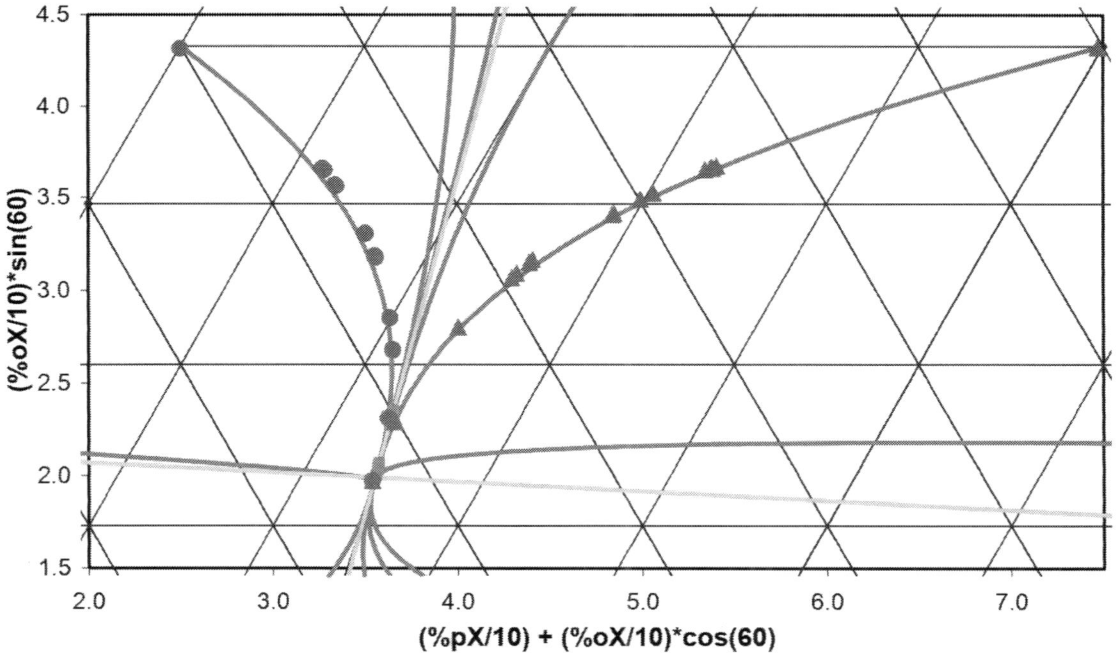

Figure 6. Improved Model Based on an Equilibrium Distribution of 24.100% pX, 53.148 wt% mX, and 22.7518 wt% oX

The **X** matrix and $\mathbf{X}^{-1}$ matrix for the improved model are:

```
X MATRIX
   0.2275179    0.7165806   -0.2275179
   0.5314821   -0.5314821   -3.8426368
   0.2410000   -0.1850985    4.0701547


 X INVERSE MATRIX
   1.0000000    1.0000000    1.0000000
   1.0747297   -0.3412320   -0.2620811
  -0.0103360   -0.0747297    0.1745607
```

The relative time constants are:

λ_1 = -0.4194
λ_2 = -1.0000

The relative rate constant matrix in the Natural coordinate system, **K**, is:

```
RELATIVE RATE CONSTANT MATRIX:

     -0.5313045        0.1397070        0.1934836
      0.3263557       -0.5931559        1.0000000
      0.2049489        0.4534489       -1.1934836
```

where

$$K = \begin{bmatrix} -(k_{21}+k_{31}) & k_{12} & k_{13} \\ k_{21} & -(k_{12}+k_{32}) & k_{23} \\ k_{31} & k_{32} & -(k_{13}+k_{23}) \end{bmatrix}$$

As noted above, the direct conversion of oX to pX cannot occur. oX must pass through mX on its way to pX. Thus, k_{13} and k_{31} should be equal to zero. However, in the presence of diffusion limitation within the medium pore molecular sieve crystals, and with unequal diffusivities, the diffusion-disguised rate constant matrix appears to indicate that direct oX to pX conversion is not only possible, but appreciable. This is due to the fact that the diffusivity of pX is much higher than that of the other isomers.

Conclusions

A kinetic model is reported for isomerizing the xylene isomers over the bp Amoco Chemical Company HSDE xylene isomerization catalyst. This catalyst is comprised of bp's proprietary medium pore borosilicate molecular sieve, AMS-1B. The diffusion-disguised rate constant matrix shows appreciable apparent direct conversion of oX to pX, in contrast to the accepted step-by-step mechanism for isomerization that has oX going through mX on its way to pX. This is due to diffusion limitation within the molecular sieve crystallites, and a diffusivity of pX that is much higher than that for the other isomers.

To obtain the best fit to experimental data, the equilibrium point was moved slightly along the slow straight-line reaction path relative to that reported previously by Amelse (1993). The following equilibrium distribution of the xylene isomers at 400 °C is believed to be accurate within about +/-0.02 wt% pX:

24.040 %pX, 52.976 %mX, 22.983 %oX.

Acknowledgement

The authors would like to thank Mr. Mark Pinderski for showing Ms. Pinsent how to safely operate the pilot plant, and for overseeing the experimental effort, while she was a Co-op student at bp.

REFERENCES

Amelse, J. A., Studies in Surf. Sci. Catal., Vol. 38, J. W. Ward, ed., p. 165 (1987).

Amelse, J. A., 1988 Summer Natl. AIChE meeting, Denver, CO, Aug. 21-24 (1988).

Amelse, J.A., Proc. 9th Internatl. Zeolite Conf., Montreal, 1992, Eds. R. von Ballmoos, et. Al., Butterworth-Heinemann, p.457 (1993).

Chirico, R. D., Knipmeyer, S. E., Nguyen, A., and Steel, W. V., J. Chem. Eng. Data, **42**, 248 (1997a).

Chirico, R. D., Knipmeyer, S. E., Nguyen, A., Reynolds, J. W., and Steel, W. V., J. Chem. Eng. Data, **42**, 475 (1997b).

Chirico, R. D., Knipmeyer, S. E., Nguyen, A., Cowell, A. B., Reynolds, J. W., and Steel, W. V., J. Chem. Eng. Data, **42**, 758 (1997c).

Chirico, R. D., Knipmeyer, S. E., Nguyen, A., and Steel, W. V., J. Chem. Eng. Data, **42**, 773 (1997d).

Chirico, R. D., and Steele, W. V., J. Chem. Eng. Data, **42**, 784 (1997e).

Young, L.B., Butter, S. A., and Kaeding, W. W., J. Catal., 76, 418 (1982).

Wei, J., J. Catal., **1**, 526 (1962a).

Wei, J., J. Catal., **1**, 538 (1962b).

Wei, J., and Prater, C. D., Adv. Catal., **13**, 203 (1962).

Wei, J. Catal., **76**, 433 (1982).

Benefits of implementing APC on a Styrene Plant

Venky Venkataraman, BP – Oil, Texas City, USA
Clay Shafer, BP – Chemicals, Texas City, USA

Introduction

This paper describes the benefits of implementing, advanced control (using DMCplus model-based predictive control) at the Texas City Styrene Plant.

From a technical perspective, the control has decreased the plant energy usage, reduced byproduct polymer formation, increased unit capacity, smoothed process operations generally, and reduced the frequency of process alarms. Operator acceptance has been good.

Background

The Vision 2003 Project strategy was comprised of several smaller projects, each justified individually. This advanced controls project was one aspect of the Vision 2003 strategy. Others included upgrading instruments from pneumatic to electronic, adding positioners to valves, upgrading the safety system, and upgrading some control equipment. These other projects were enablers for the advanced control project.

DMCplus is the technology chosen for the advanced control. It is a linear controller and is designed to control the process to a constrained optimum, using Linear Programming techniques.

Model-based controllers, whether linear as here or nonlinear, make control moves to maximize the profitability of an operation by controlling some set of variables by manipulating a set of variables such that the value of an objective function (e.g. profit) is maximized. Because the controller has a model of the process interactions, it makes timely compensatory moves of the right magnitude to control the process tightly. These controllers will tend to push the process to a constrained optimum.

Process Description

General Process Description

In the de-hydrogenation reactor section, ethyl-benzene is dehydrogenated to styrene in 3 radial bed reactors. The main reaction, which is highly endothermic, is carried out in the vapor phase over a catalyst in the presence of steam. The reactor effluent is then condensed to separate crude styrene from water in the condensation subsection. Crude styrene is distilled in the refining section to provide a styrene monomer with purity greater than 99.9% weight.

Dehydrogenation Reactor section description

The dehydrogenation reactor subsection is shown in figure 1. The purpose of the reactors subsection is to heat the EB feed and dilution steam to the reactor temperatures in the super-heater, to dehydrogenate the ethyl-benzene to styrene and hydrogen in the three-stage reactor system, and to recover the heat from the high temperature reactor effluent in a waste heat recovery system.

Since 2 moles are formed from one mole of ethyl-benzene is dehydrogenated to styrene and hydrogen in the dehydrogenation reactors, thermodynamics dictates that the lowest possible pressure in the reactor will shift the equilibrium to form more styrene. The steam added to the EB serves several purposes:

1. Provide the heat needed to sustain the highly endothermic reaction
2. Dilute the reactants in products and thus shift the equilibrium in favor of styrene
3. Remove coke from the catalyst and maintain active state of catalyst.

Along with styrene and hydrogen, byproducts such as benzene, toluene, paraffin's, olefins, CO and CO2 are also formed. The reactions result in fairly high styrene selectivity and high EB conversion.

Styrene Refining section process description

The styrene refining subsection is shown in figure 2. Crude styrene from the dehydrogenation reactors contains ethyl-benzene, toluene, benzene and other impurities. To obtain the desired product purity 99.9 wt%, crude styrene is refined in a series of distillation columns. Low temperatures, vacuum and polymerization inhibitors are used in these columns to minimize polymerization.

The first column in the refining section is the benzene toluene column, in this column the benzene and toluene are sent overhead while the bottoms product contains styrene, EB and less volatile components.
The Styrene and EB are separated in the two Ethyl-benzene recycle columns. In these columns the EB is sent overhead and recycled to the dehydrogenation reactors while the styrene and heavy components in the bottom are sent to the styrene-finishing column. In styrene finishing column, styrene monomer is sent overhead with a product purity of 99.9%.

Schematics of the Dehydrogenation Reactor and Refining sections are shown in Figures 1 and 2.

Advanced Control Project Scope

The scope of this project included building a process control model for the Dehydrogenation reactors and furnace, one for the Styrene Refining section, and one for the condensate stripping tower. Supporting inferential models were built for the reactor conversion and for the tower analyzers. Control model sizes ended up being 8 MV x 34 CV x 9 FF for the Dehydrogenation section and 12 MV x 51 CV x 27 FF for the Styrene Refining section[1].

Schematics of the Dehydrogenation Reactor and Refining sections are shown in Figures 1 and 2.

[1] CV is a Controlled Variable. MV is a Manipulated Variable (a variable manipulated to control a CV. FF is a Feedforward variable, a variable which this controller has no control over and to which it must simply respond.

Dehydrogenation Reactor Section Control Strategy

- Maximize feedrate until furnace D-coil gassing occurs or until steam pressure backs out condensate to waste heat boiler. Constraints inferred by valve positions.

Minimize steam to hydrocarbon ratio (SHCR) subject to minimum limit specified.

- Reduce reactor temperatures to minimum limit specified.
- Implement inferred calculation of conversion.

Figure 1 depicts this section of the process.

Styrene Refining Control Strategy

The Styrene section is made up of four distillation towers as shown in Figure 2. The two EB/Styrene splitters are operated in parallel. Each tower has its own sub-controller to allow a single tower control to be turned off.

- AS-4201 (Benzene Toluene Column)
  - Feed-rate is adjusted to match dehydrogenation rates.
  - Reflux is adjusted to keep overhead EB within specification.
  - Bottoms toluene concentration is controlled by bottoms temperature.
  - Bottoms level is included in the control model.
- AS-4202 (Ethyl Benzene Recycle Column)
  - This tower is based loaded at a constant feed-rate.
  - EB in bottoms is pushed to its high limit.
- AF-202 (Ethyl Benzene Recycle Column)
  - Feed-rate to this tower changes when AS-4201 changes.
  - This tower is moved to compensate for EB changes in AS-4202.
  - EB in the final product is pushed to its high limit.
- All tower GCs were modeled.
  - The GCs are continuously used by Aspen IQ to update analyzer predictor biases.
  - Both the GCs and the analyzer predictors are used by DMCplus.
    - DMCplus uses the predicted value when the GC is out of service and the GC when the predictor is not working well.
    - The predictors have less dead time than the GCs, allowing DMCplus to correct disturbances faster.

Performance

Dehydrogenation Section

Inferred conversion implemented and DMC originally successfully controlled conversion, but it caused reactor inlet temperature swings not appropriate for furnace stability. As a result, conversion is used only an indication to help offline optimization of the process.

Maximum rates are obtained when both the feed and condensate valves are at their maximum constraints. The plant has reached sustained feed-rates of 4% higher compared to traditional maximum.

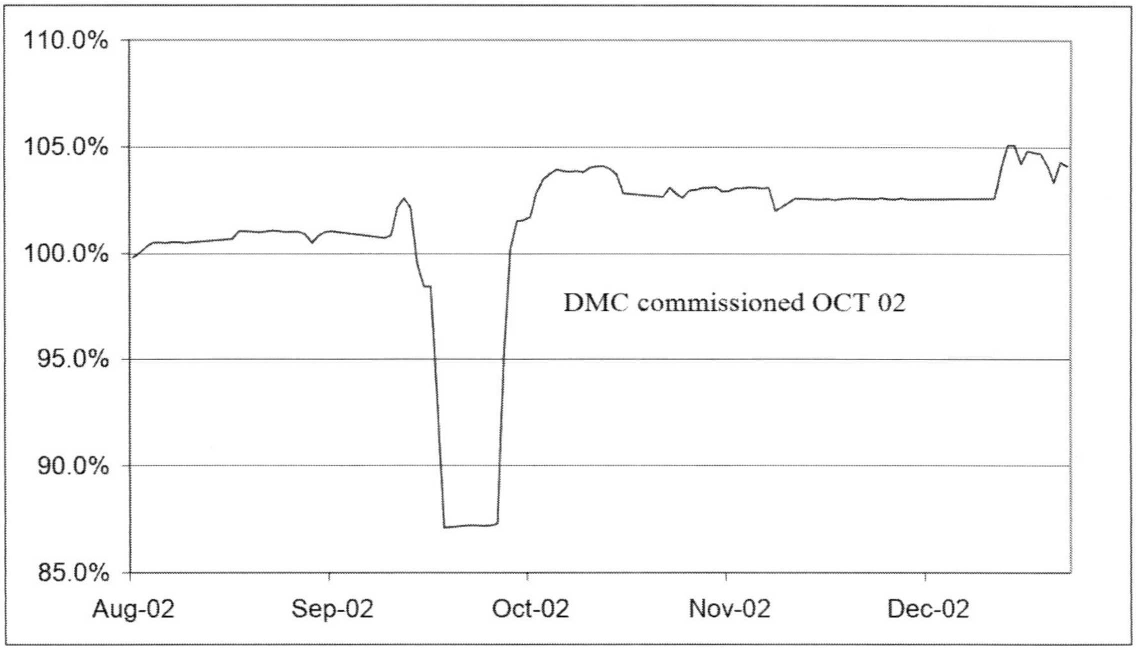

The controls can handle large rate changes. A large turndown from 104 % to 81% was tested and found acceptable. The controller provides a gentler ramping of reactor conditions, causing less shock to the catalyst.

Furnace firing is much more stable, leading to significantly fewer draft and oxygen alarms, which has allowed much tighter adjustment of excess oxygen. At the end of the catalyst life, DMCplus is maximizing reactor temperature while maintaining furnace coil outlet temperature within safe operating limits. This maximizes optimization while protecting the furnace tubes.

Reactor Temperatures Before DMC

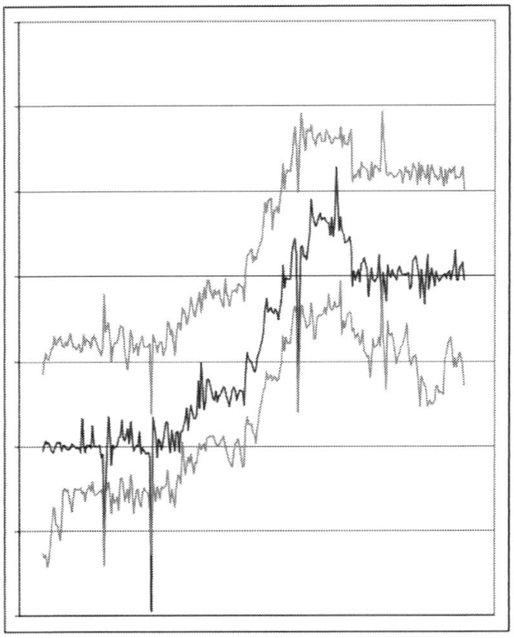

Reactor Temperatures After DMC

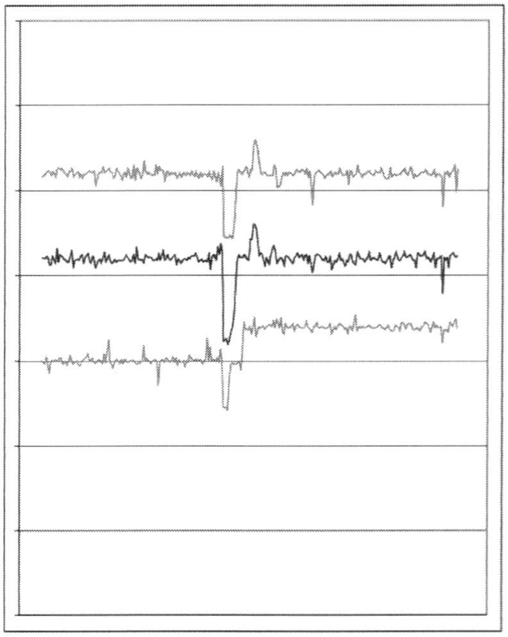

5 Deg F/ Div.

The SHCR has been reduced by 4 - 5 % (mass of steam per mass of hydrocarbon). Steam flows are balanced to reduce energy cost. Having DMC is helping in optimization runs to study the effects of lower SHCR on catalyst activity and selectivity to ensure low cost production, and yet maximize the catalyst life.

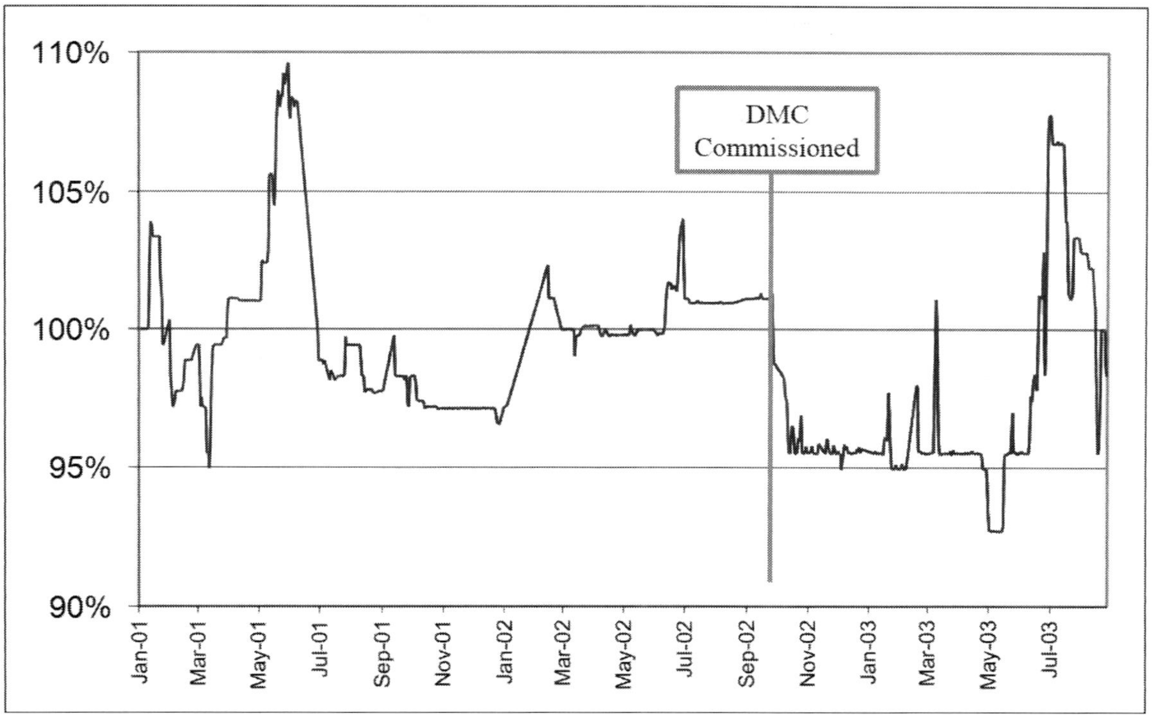

Styrene Refining

As stated in the controller scope discussion, DMCplus can use either GC data or predicted analyzer values. Currently an operator or engineer must make the decision to use one versus the other.

EB in the final product is pushed to its high limit. Tower bottoms temperatures are pushed to their low limit in all three towers. Together, these moves reduce unit energy requirements. In the benefit calculations for saving steam energy, an economic value of zero was taken for saving steam if steam venting to atmosphere was taking place.

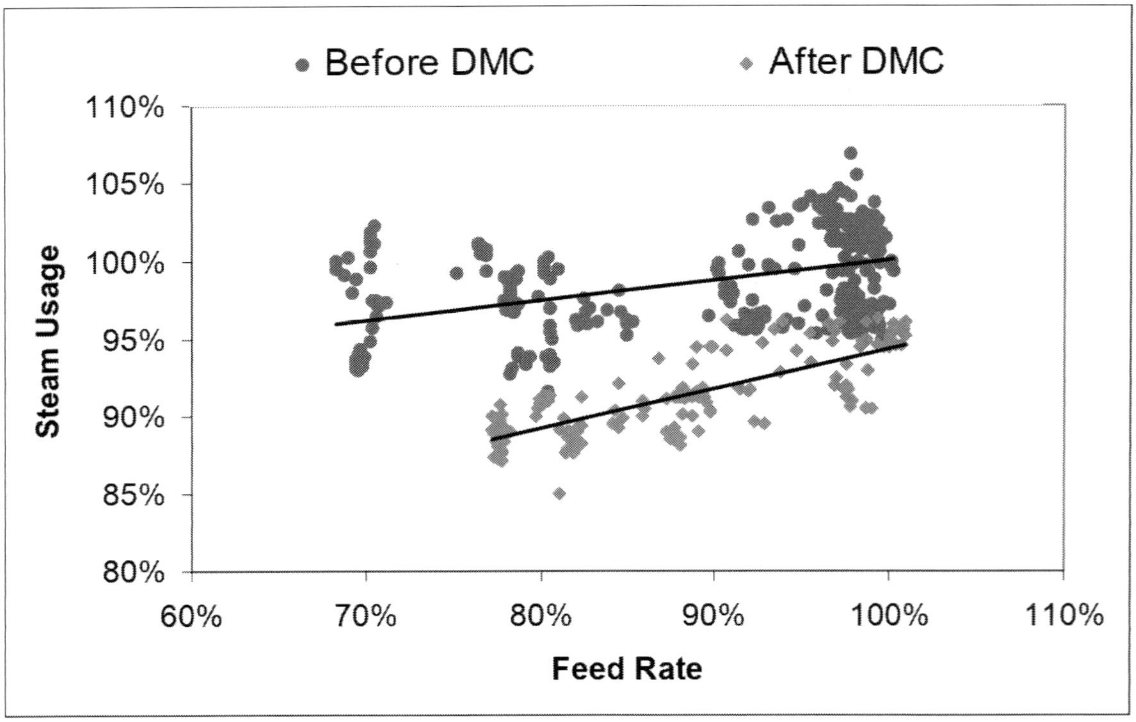

Reduced AS-4201 bottoms temperature reduced steam usage 15-23%. Its tighter control of EB in the overhead provided stability important for the success of a new toluene export stream. Control of the crude styrene tank level provided matching feed-rates with the dehydrogenation production rate and provided smooth transitions during rate changes.

Reduced AS-4202 bottoms temperature reduced steam usage from 4-12%. Its increased stability has helped remove swings from AF-202.

AS-4202 benefits have been limited due to several mechanical issues.

Reduced AF-202 bottoms temperature has led to steam savings of 4-10%. It has reduced polymer production. This in turn has provided an opportunity to better optimize inhibitor usage. And the moving bottoms temperature has provided data to build a polymer model that is now being used for inhibitor optimization.

AF-202 tower is operating more stably than before. Its improved disturbance rejection has allowed operation closer to the EB specification.

General

Operator acceptance has been high. DMCplus is only turned off during major unit upsets or during maintenance periods.

Current unit problems are limiting production. DMCplus however is maximizing feed-rate against constraints, and feed-rate capability is up 4% compared to earlier.

A tool that would help Operations improves unit profitability would be one that properly calculates the lost opportunity cost due to constraints entered. Although shadow prices are available, these are only as good as the linear model of the plant allows. The plant being nonlinear, some of these shadow prices cannot be correct (for example the effect of recycling styrene to the EB section). The models generally need to be updated to reflect the current operating regions and equipment condition.

Operators noted that fewer alarms were generated following the controller commissioning, particularly around the furnace operation, but no alarm data were saved to make a quantitative estimate of the reduction in alarms. With the unit being controlled closer to constraints than before, there is a need to revisit the alarm limits set generally because some of them are set too close to (now) current operation.

Lessons Learnt and Challenges Faced

1. Develop inferentials early on during the project. It takes time to fine tune the inferentials.
2. Always estimate TTSS of the process by looking at the analyzer, it will reveal the inherent dead times and lags in its response. We found dead times as long as 2.5 hours, which was solved by increasing the sampling pump speed. This was found during the step test, but could have been detected earlier.

Dehydrogenation Section
Figure 1

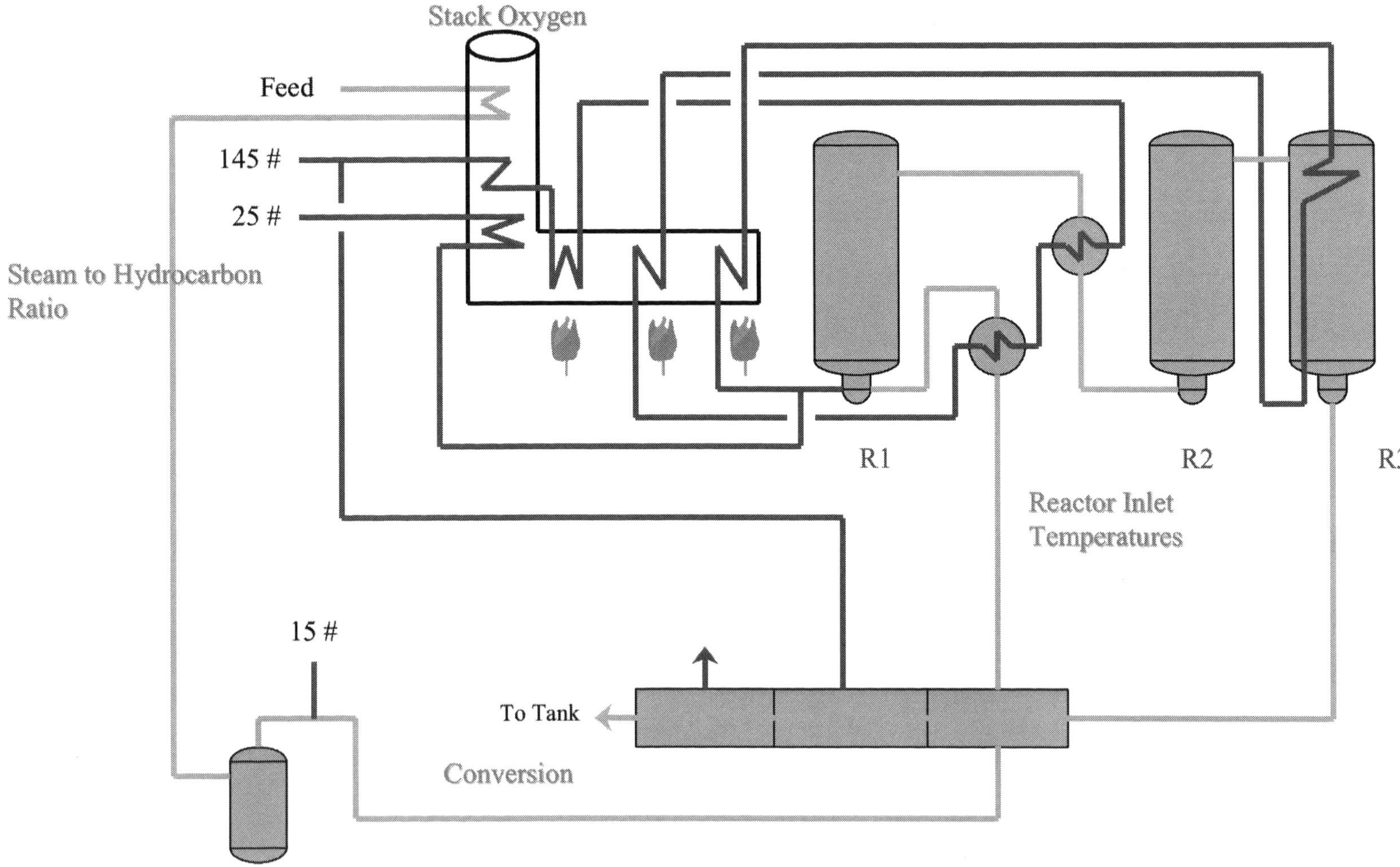

Styrene Refining Section

Figure 2

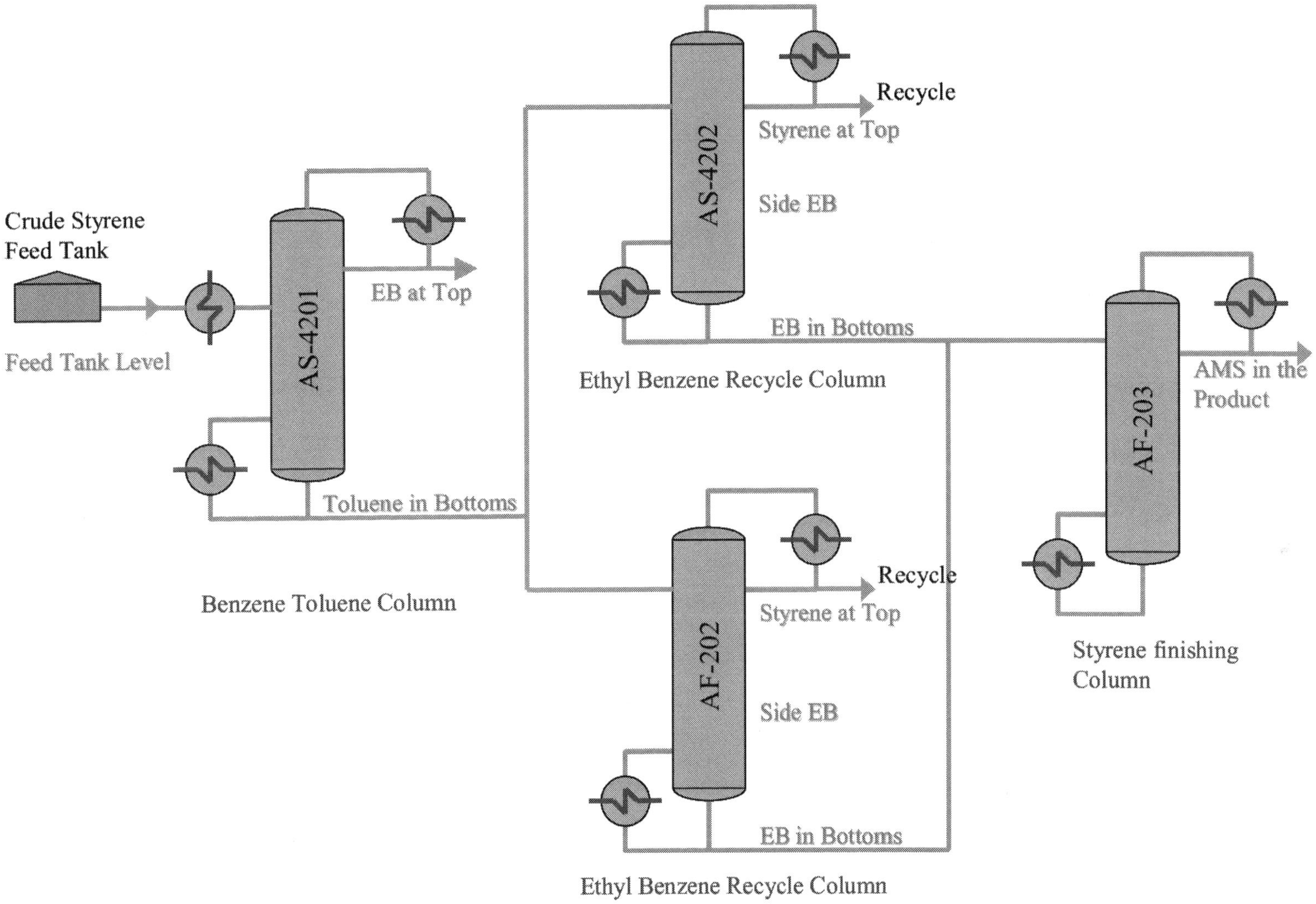

Paper No.93C

Influence of Feed Hydrotreatment on FCC Product Aromatics

Andrea de Rezende Pinho, William Richard Gilbert, Ricardo Drolhe Montaury Pimenta and Donizeti Aurelio Silva Belato

PETROBRAS R&D Center, Rio de Janeiro, Brazil

Key words: FCC, Aromatics, hydrotreated feedstock.

Prepared for Presentation at the AIChE Spring National Meeting, April 26, 2004, New Orleans, LA, 6th Topical Conference on Refinery Processing

January/2004

Unpublished

Abstract:

Three FCC feeds derived from Brazilian Marlim based crudes were hydrotreated at varying severities to evaluate the cost effectiveness of pre-treatment in the new scenario of very low sulfur fuels: pure Marlim Vacuum Gas Oil, a blend of 80% Marlim Vacuum Gas Oil plus 20% Coker Vacuum Gas Oil Marlim and a blend of 65% Vacuum Gas Oil Marlim plus 35% Atmospheric Residue.

The aromaticity of the Gasoline cut was analyzed by PIANO GC, while the aromaticity of the Light Cycle Oil and Slurry Oil cuts were analyzed by SFC (Supercritical Fluid Chromatography).

Hydrotreatment of the feed is clearly unable to decrease Gasoline aromatics. Actually, aromatics seem to increase with the severity of the hydrotreatment in some cases.

The predominant family in the Light Cycle Oil cut are the di-aromatics, as expected, which is considerably higher than the other fractions: around 60wt% of the total. For the hydrotreated feeds the concentration of aromatics in the LCO range varied from 80 up to 90% wt.

Surprisingly, a substantial amount of saturates remains in the Slurry, from 27% up to 45%. The data also clearly demonstrates that the hydrotreatment increases the saturate level in the Slurry. After saturates, the most important hydrocarbon family group in this cut are the tri-aromatics.

Introduction

Brazilian oil industry has traditionally relied on the FCC (Fluid Catalytic Cracking) as the chief conversion process in its refineries. With the discovery, in the 1980s, of the oil fields in the Campos Basin, light imported crudes were progressively substituted for Brazilian heavy crudes. At the same time, the market for fuel oil started to decline, being displaced by natural gas. There was a dramatic demand for extra conversion and the FCC had a crucial role in the change, being adapted to process increasingly heavier feedstocks. In this scenario, the prospect of using FCC feed hydrotreatment was never considered a profitable alternative. The heavier than average FCC feedstocks processed by Petrobras, including in many cases substantial amounts of Vacuum Residue, would have demanded special hydrotreating technology, which to this day has not achieved a strong position in the refining industry. Also, the large capital expenditure, high operating costs and long payback cycles outweighed the potential benefits of feed hydrotreating.

Since the late 1990s, the refining industry worldwide has met the new challenge of having to drastically improve the quality of the fuel produced, particularly in the case of sulfur content which is being pushed to extremely low levels, in an effort to improve air quality. The environmental pressures are not limited to fuel quality, and are directly affecting the refinery, which is required to reduce SOx and NOx emissions to the atmosphere, to which the FCC is the major contributor. Feed hydrotreating is an obvious solution to both problems. Hence, the economic attractiveness of FCC feed hydrotreating in Brazil had to be reappraised leading Petrobras to conduct a thorough study on the effects of feed pretreatment on FCC product quality [6], which produced an extensive database in which the question of feed quality effect on aromatic production could be carefully examined over the whole range of the FCC product slate.

FCC aromatics could have several sources [3, 4, 8]: they may come from side-chain cracking or dealkylation reactions of higher boiling range aromatics already in the feed, from naphtene dehydrogenation or from linear precursors, via olefin cyclization followed by dehydrogenation. Controlling the hydrogen transfer character of the catalyst does influence the aromatic content of the product to some extent. For instance, changes to the catalyst that affect gasoline aromatics will be key factors in controlling the product slate, maximizing gasoline yield or light olefins [5]. The feed quality effect is not so clear and would depend on which of the already mentioned mechanisms of aromatic formation would predominate.

The aromatic content of the FCC product is important in different ways. In the gasoline range, aromatics boost octane. However, the question of gasoline aromatics may be expected to change in the next decades, particularly in the context of "dieselization" seen Europe, where oil companies and automakers predict that there will be a market for a new fuel, with the boiling range of gasoline and the high cetane of diesel, as required by the new HCCI (Homogeneous Charge Compression Ignition) engines [1].

In the LCO (Light Cycle Oil) range, aromatics, typically the predominant hydrocarbons, may be seen as a problem if the LCO is sent to the hydrotreater, before blending in the diesel pool. LCO aromatics are the hardest products to hydrotreat, require a large amount of hydrogen and have very low cetane numbers. In the Slurry Oil cut, aromatics may be seen as a desirable product, because they contribute to the BMCI (U.S. Bureau of Mines Correlation Index) of Aromatic Residues, the starting material for Carbon Black production, a higher added value alternative than fuel oil.

Experimental

Three different FCC feedstocks, representing the typical range of feedstocks processed in Petrobras FCC units, were hydrotreated at different severity levels by adjusting WHSV (Weight Hourly Space Velocity):

- Feedstock A: Marlim Vacuum Gasoil (VGO);
- Feedstock B: 20% Coker Heavy Gasoil + 80% Marlim VGO;
- Feedstock C: 65% Marlim Vacuum Gasoil (VGO) + 35% Atmosferic Residue (ATR).

Table 1 presents the properties of the FCC feedstocks used in the study. The numeric index is related to the severity of the hydrotreatment, where the zero represents the untreated feedstock (A0, B0 and C0) and the highest number the most severe level of hydrotreatment (A2, B4, C3). The rate of catalyst deactivation when hydrotreating of the VGO-ATR mixtures was twice that of the other feedstocks, so mixtures of the hydrotreated gasoils and the ATR as such were also evaluated in case refiners would choose this option to guaranty reasonable Hydrotreater turnaraounds. These feedstocks are represented by the "+R" sufix.

Table 1 – Properties of the FCC feedstocks

	°API	KUOP	RCR wt%	B.Nitrog mg/kg	Sulfur mg/kg	C/H Ratio	Saturates SFC wt%	MonoAro SFC wt%	DiAro SFC wt%	TriAro SFC wt%
A0	16.6	11.60	0.98	1437	7840	0.566	38.8	15.5	14.5	7.1
A1	20.9	11.82	0.23	344	168	0.543	53.3	20.6	13.8	3.7
A2	21.6	11.84	0.21	203	77	0.539	55.2	20.3	12.7	3.6
A1+R	20.1	11.79	2.23	655	2150	0.546	53.8	16.2	11.2	3.6
B0	16.4	11.43	0.86	1790	8730	0.568	39.2	15.0	13.9	7.5
B1	20.7	11.75	0.24	552	299	0.545	49.0	19.4	15.2	3.3
B2	21.2	11.76	0.22	345	157	0.544	51.7	19.5	14.6	4.3
B3	21.8	11.78	0.19	264	103	0.538	53.3	21.9	13.3	3.3
B4	22.5	11.81	0.15	121	47	0.538	55.8	20.1	12.2	3.6
B1+R	21.0	11.76	1.69	546	1560	0.552	53.8	16.2	11.2	3.6
C0	16.8	11.66	2.84	1501	7010	0.559	39.4	13.7	13.4	6.6
C1	20.3	11.82	1.56	707	597	0.544	49.6	18.1	13.3	4.0
C2	20.6	11.84	1.50	615	464	0.538	52.0	17.5	11.4	3.3
C3	21.9	12.05	1.18	295	203	0.533	56.3	18.5	11.4	3.2

The 14 feedstocks were processed in a DCR circulating pilot Riser [9], totalising 23 pilot riser runs. Pilot Riser conditions were: 540°C Reactor Temperature, 720°C Regenerator Temperature, 1000g/h feed flow. Products were collected for characterization at Catalyst to Oil ratios (CTO), 7 and 11. An equilibrium catalyst with high rare earth (2.47 wt%), typical of Petrobras refineries, was used in the tests with 168 m^2/g surface area, 2800 wt ppm nickel, 2000 wt ppm Vanadium and 71 wt% MAT (Micro Activity Test). TBP (True Boiling Point) distillation was used to fractionate the liquid product into Gasoline (IBP-220°C), LCO (220°C-344°C) and Slurry Oil (344°C+). The aromaticity of the GLN cut was analysed by PIANIO, while the aromaticity of the LCO and Slurry cuts was analyzed by SFC (Supercritical Fluid Chromatography), which fractionates the hydrocarbons into saturates (all non aromatic hydrocarbons), mono, di, tri and poli-aromatics.

Results and Discussion

Product aromaticity was verified in two ways: as aromatic concentration in the specific cuts, i.e. gasoline, LCO and Slurry; and as aromatic yields relative to the feed. Figures 1 to 3 show the composition of Gasoline, LCO and Slurry cuts, respectively, obtained at constant catalyst to oil ratio (CTO) of 7. The hydrotreatment of the feed was clearly unable to decrease aromatics concentration in these cuts, as can be seen for families A, B and C. Actually, in the case of Gasoline, aromatics seem to increase with the severity of the hydrotreatment for family C (compare C0, the untreated feed, against C3, the highest hydrotreatment level). Di-aromatics is the predominant fraction in the LCO cut, followed by mono-aromatics. The di-aromatics concentration increases with the hydrotreatment. Hence, hydrotreatment of the FCC does not contribute to improve the cetane number of LCO.

Surprisingly, the predominant fraction in the Slurry turned out to be the saturate hydrocarbons, ranging from 27% to 45% of the total. Also, feed hydrotreating increases the saturate concentration in the Slurry, which is the opposite of what happens in the LCO and Gasoline cuts.

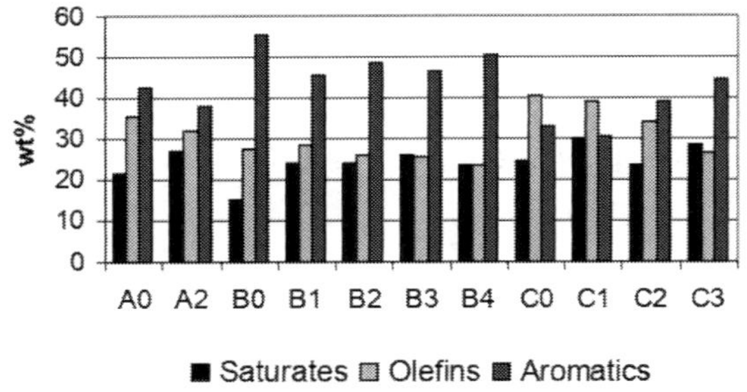

Figure 1 – Gasoline Composition by PIANIO

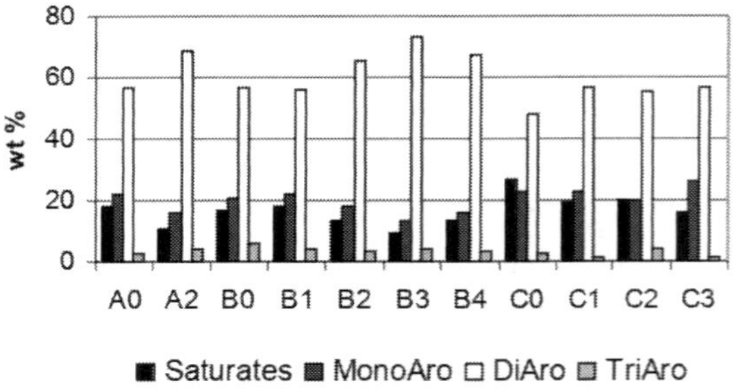

Figure 2 – LCO Composition by SFC

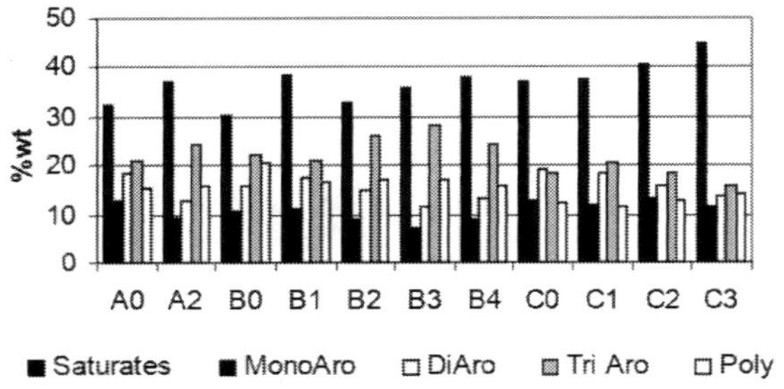

Figure 3 – Slurry Composition by SFC

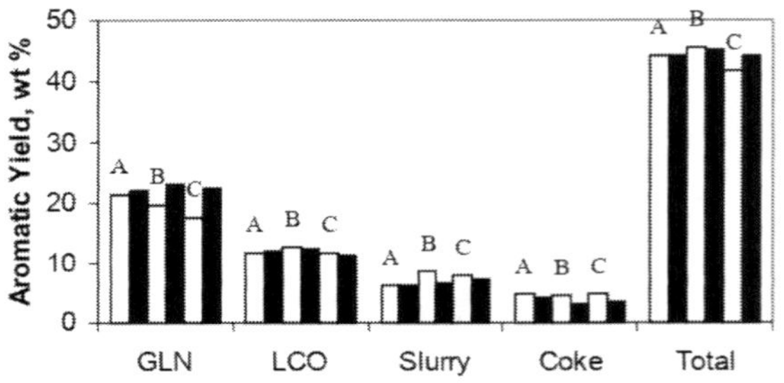

Figure 4 – Aromatics Yield in GLN, LCO, Slurry and Coke for low (□) and high (■) severity hydrotreatment of feeds A, B and C.

Figure 4 shows the effect of hydrotreatment on the aromatics yield in GLN, LCO, Slurry and Coke at constant conversion (defined as 100% - LCO – Slurry) for each of the three feedstock families: A, B and C, where the bars represent the aromaticity for a low and a high severity hydrotreatment. Aromaticity of the coke was assumed to be 100%. Hydrotreating severity actually increases the yield of gasoline aromatics. The yield of LCO aromatics doesn't change much and there are small decreases in the Slurry range. The sum of aromatics in the four groups didn't change at all.

To better visualize the effect of the process variables on the experimental results statistical techniques of Principle Component Analysis and multivariate correlation were applied to the data developing a model that could isolate the effect of individual variables. The CTO and 14 different feedstock properties were considered as possible independent variables and 47 product results (yield profiles plus selected product properties) were considered as the experiment dependent variables. The Principal Component Analysis [2] was applied for data compression, before the 23 x 47 output matrix could be modelled as a function of the 14 x 14 input matrix, reducing the number of independent variables to 4 (3 feed quality principal components and the CTO) and the dependent variables to 6 (Conversion, Coke yield and 4 product principal components). Cumulative variance for the first four feed property principal components was: 81.5%, 93.0%, 96.1% and 98.0%; while for the product properties it was: 49.6%, 63.7%, 73.1% and 78.8%. Principle component analysis and multivariate correlation were performed with the help of the SatSoft Inc., Statistica 6.1 package.

In experiments with commercial feedstocks, as in this case, it is often difficult to manipulate one feed property at a time and isolate its effects on the results. By extracting the principal components of the feedstock properties, it was possible to identify two major sources of variability, responsible for 93% of the variance in the 14 feeds tested: the first Principal Component (PC1) correlated very well with the Feed Conversion Factor [7] (the feed theoretical conversion in FCC) and the Second Principal Component (PC2) with the Carbon Residue of the feed (RCR). The predominant effect of the hydrogenation of one of the parent feeds would be to reduce the value of PC1 or improve Conversion Factor (move to the left on the PC1 vs PC2 plot – figure 5), adding atmospheric residue to one of the feedstocks would reduce PC2 (increase RCR and move down in the PC1 vs PC2 plot). Of course, other uncontrolled effects also got superimposed on these changes, for instance, the atmospheric residue used in the study had a higher Conversion Factor than the parent gasoils, so the addition of the residue would also produce a small improvement in cracking performance of the mixture (decrease in the value of PC1). Also, the hydrotreating would reduce the RCR of the feedstock, increasing the value of PC2. The resulting moves on the PC1 versus PC2 graph are indicated in figure 5 by the top arrow for hydrotreating and the bottom arrow for residue addition.

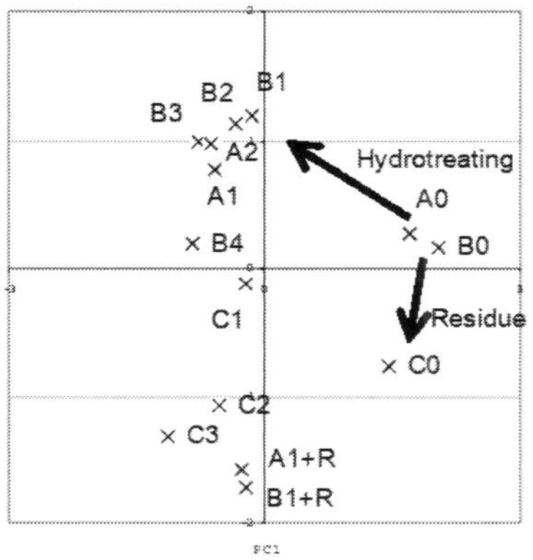

Figure 5 – Principal component plot for the feed properties.

The degree of variability and the randomness induced by experimental error of the pilot unit cracking results was much greater than for the feed quality, indicated by the large number of principle components needed to describe the data set (low cumulative variance for the first four principle components). When an empirical model based on the multivariate correlation of the results as a function of the independent variables was attempted, reasonable correlations could only be found for the first four principle components of the result matrix, and it was also necessary to develop individual correlations for conversion and Coke yield, since the prediction based on the principal component models alone was not accurate enough. Tables 2 and 3 present the average error and correlation coefficient between predicted and observed results for the model.

Table 2 – Pilot unit product yield prediction model based on the feed properties and CTO.

	Convers. wt%	DryGas wt%	LPG wt%	Gasoline wt%	LCO wt%	Slurry wt%	Coke wt%
Error	2.7	0.1	0.3	0.7	0.1	0.5	0.1
Error %	3.8	3.9	1.7	1.4	0.7	3.7	2.5
R^2	0.95	0.60	0.80	0.68	0.74	0.80	0.80
Average	72.5	3.4	17.1	47.5	14.0	13.5	4.4
Max/Min	78.3/55.4	5.9/2.2	22.5/10.8	53.6/35.7	16.5/12.8	28.1/8.6	6.5/2.3

Table 3 – Pilot unit Gasoline, LCO and Slurry oil selected properties prediction.

	GLN			LCO				Slurry			
	Sat. wt%	Olefi wt%	Arom wt%	Sat. wt%	Mon wt%	Di wt%	Tri wt%	Sat wt%	Mon wt%	Di wt%	Tri wt%
Error	0.66	0.68	0.99	0.76	0.49	1.08	0.28	8.1	5.5	1.1	1.7
Error %	2.5	2.2	2.4	4.7	2.5	1.8	7.1	21.8	51.8	7.5	7.7
R^2	0.63	0.63	0.58	0.60	0.64	0.69	0.08	0.75	0.0	0.52	0.21
Average	27.1	31.3	39.2	14.8	18.5	62.8	3.8	35.8	10.1	15.0	23.5
Max/Min	34/16	39/25	47/29	27/8	26/13	74/47	7/1	45/27	14/7	19/11	28/16

The effect of the feed hydrotreating and atmospheric residue addition on the feed properties principal components was already discussed. With the help of the model, it is now possible to investigate the effect of the two first feedstock principal components on the pilot unit cracking results (figures 6 and 7). For the sake of clarity, the principal components were substituted for the Feed Conversion Factor (PC1) and the Carbon Residue (PC2). The yield profiles and the hydrocarbon composition of the gasoline, LCO and Slurry cuts were calculated at three different CTOs (5, 7.5 and 10), for varying Feed Conversion Factors (PC1) in figure 6, the same thing was done for the Carbon Residue in figure 7.

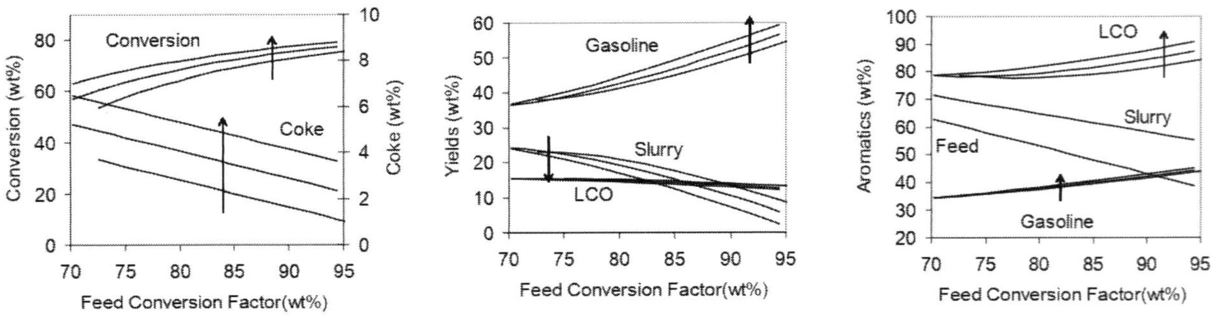

Figure 6 – Model predicted effects of the change of the Feed Conversion Factor on the FCC products. From left to right: (A) Conversion and Coke yield; (B) Gasoline, LCO and Slurry yields; (C) Feed, Gasoline, LCO and Slurry aromatic content. Vertical arrows show the effect of increasing CTO (between 5, 7.5 and 10).

The increase in Feed Conversion Factor (decreasing PC1) produced the expected increase in Conversion and reduction in Coke yield (figure 6 left). The effect of CTO on Conversion was more pronounced at low Conversion Factors, while the effect on Coke was linear. Gasoline yield goes up and Slurry down with both CTO and Conversion Factor while the LCO yields change very little with either (figure 6 middle).

The increasing aromatic content in the Gasoline and LCO fractions at the same time that Feed aromatic content was decreasing (figure 6 right) seems to show that most of the aromatics in FCC are being formed from saturate species in the feedstock and the contribution of feed aromatics is minor. Since Gasoline aromatics and Gasoline yield increased at the same time, the production of Gasoline range aromatics on a feed basis increased even further, as can be also seen in figure 5.

Changing the feed Carbon Residue (PC2) over the data base range affected the cracking results much less than the Feed Factor (PC1). Changes in most product yields, except for Coke was very small, and there was a small decrease in aromatic content in the Gasoline, LCO and Slurry cuts with increasing RCR (figure 7).

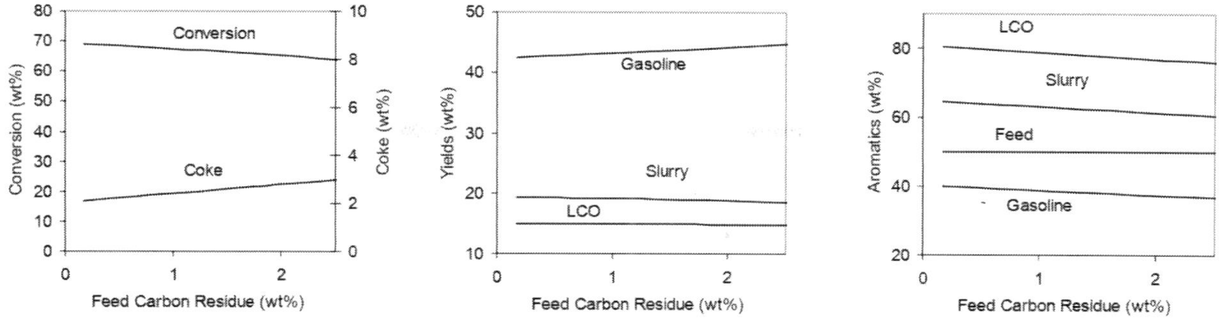

Figure 7 – Model predicted effects of the change of the Feed Carbon residue on the FCC products. From left to right: (A) Conversion and Coke yield, (B) Gasoline, LCO and Slurry yields, (C) Feed, Gasoline, LCO and Slurry aromatic content.

Conclusions

Hydrotreatment of typical Petrobras FCC feedstocks at different severities and the addition of the Atmospheric Residue to hydrotreated feedstocks produced changes in the feedstock properties that could be reduced to two Principal Components of feed composition Matrix. The two Principal Components could be interpreted as the Feed Conversion Factor and the Feed Carbon Residue. While hydrotreating would produce a large improvement of the Feed Factor and a smaller reduction of the Carbon Residue relative to the untreated feed, Atmospheric Residue addition would produce a large increase in the Carbon Residue and a small change in the Conversion Factor.

Gasoline and LCO aromatics increased with increasing feed quality, while the aromaticity of the Slurry oil decreased. These results indicate that, with high zeolite and high rare earth catalysts typical of Petrobras refineries, a great part of the aromatic species recovered in the gasoline and LCO fractions are produced from saturated species in the feed, either naphtenes or most probably by the cyclization and dehydrogenation of olefin intermediates. The aromatic content of the Slurry oil, contrary to expectations, decreased with improving feed quality, leading us to the conclusion that it is composed to a large extent of unconverted hydrocarbons in the feedstock.

References

1. Ashley, S., Scientific American, June, 74-78 (2001).
2. Beebe, K.R. et al., Chemometrics: A Pratical Guide, Wiley and Sons, New York (1998).
3. Christensens, G. et al., Chem.Eng.Sci, 54, 2753-2764 (1999).
4. Nace, D. M., Ind.Eng.Chem.Pro.Res.Dev., 8 (1), 31-38 (1969).
5. O'Connor, P. et al., 17th WPC, Block 2, Forum 13, Rio de Janeiro, (2002).
6. Pimenta, R. D. M., Belato, D.A.S., Gilbert, W.R. Effect of heavy feedstocks pretreatment on FCC yields and product quality, 226th ACS National Meeting, New York (2003).
7. Torem, M.A., 226th ACS Nat.Meeting, Chicago (1993).
8. van Klink, A. J. E. M et al. in H.J. Lovink and L.A. Pine (Editors), Proceedings, Symposium on the Hydrocarbon Chemistry of FCC Naphtha Formation, ACS Div. Pet. Chem. Inc., ACS, Miami, Sept. 1989, Technip, Paris (1990).
9. Young, G.W. et al., W. Simulating Commercial FCC Yields with the Davison Circulating Riser Pilot Unit. NPRA Annual Meeting, AM-88-52 (1988).

Embedded APC Tools Dramatically Lower Implementation Costs: A Refinery Case Study

by

Pete Sharpe
Emerson Process Management

Chris Kominar
Ergon West Virginia, Inc.

Prepared for presentation at the

**AIChE 2004 Spring National Conference
New Orleans, LA**

Session TA012

Embedded APC Tools Dramatically Lower Implementation Costs: A Refinery Case Study

Abstract

Ergon West Virginia Inc. recently installed a new digital control system (DCS) on part of their Newell, WV lube oil refinery. This paper describes the application of embedded Advanced Process Control (APC) tools on the atmospheric crude and vacuum columns in the refinery. The project included the use of two Model Predictive Controllers (MPC) along with neural-net inferential sensor technology to stabilize control of the critical product properties in the columns. Utilizing embedded APC technologies allowed Ergon to implement the project for a fraction of a traditional APC project costs while still achieving significant reduction in variability and capturing most of the potential benefits from this technology.

Introduction

Ergon operates a 20,000 BPD lube oil refinery in Newell, WV which has recently completed the second phase of a control system modernization project for a number of process units. As part of the project, a modern DeltaV DCS from Emerson Process Management was installed and most of the associated field devices were upgraded to smart instruments using Foundation Fieldbus technology. Phase II conversion included the Crude Unit, Propane Deresining Unit, Vacuum Fractionation Unit, and Sour Water Unit. Several more phases of the control modernization project are planned over the next few years to complete the remaining refinery process units.

The total size of the Phase II project included a total of around 580 I/O, of which approximately two thirds were Foundation Fieldbus (FF) with one third conventional I/O. For those loops that are on FF, the actual control algorithm is executed in the device rather than the DeltaV controllers.

History has shown that today's modern digital control systems, with their ability to incorporate Advanced Process Control (APC) functions, can deliver substantial performance improvements over traditional control systems. With new APC tools that are embedded in the DCS platform, the incremental cost to deploy these strategies has dropped dramatically as compared to older DCS systems. Nevertheless, many control system replacement projects do little more than replicate the previous control logic. Consequently, much of the potential benefits are never realized or at least postponed for several years.

Ergon had a different approach for their control modernization project. As part of the project engineering, the control system vendor reviewed Ergon's regulatory control strategies and recommended enhancements that would improve the operation of the units and take advantage of some of the features available in the new DCS. In addition, the engineering team reviewed the process operating objectives as well as the incentives and economics and identified areas that would benefit from APC applications. This review was incorporated into the front-end engineering for the project and the enhanced regulatory controls and APC functions were planned from the outset.

This paper describes the use of the embedded APC tools to achieve incremental improvements in the operation of the Crude and Vacuum Distillation units at the refinery.

Process Description

The Atmospheric Crude / Vacuum Distillation Unit (CDU) at the Newell refinery represents a typical process configuration for a lube refinery. Crude is preheated in a number of product and pumparound exchangers before entering a flash tower that removes light components from the crude feed. The crude is further heated in a 2-pass heater that fires refinery fuel gas to control a combined coil outlet temperature. The atmospheric column produces a straight run naphtha (SRN) product overhead, and three side-draw products: kerosene, heavy kerosene and atmospheric gas oil (AGO), which go through strippers before going to storage tanks. Two pumparound systems remove heat from the tower and preheat crude. The overhead vapors are condensed in a set of fin-fan exchangers. A simplified process flow diagram of the Atmospheric Column is provided in Figure 1 below.

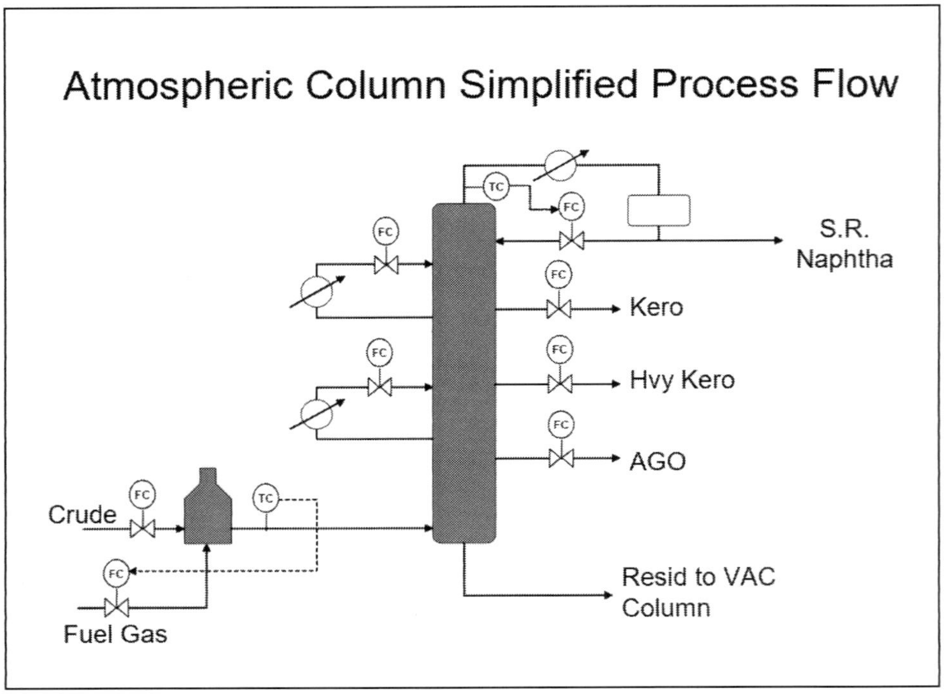

Figure 1

Atmospheric resid is taken from the bottoms of the atmospheric crude tower and fed directly to the Vacuum Tower heater. This is a 2-pass heater, also fired primarily with fuel gas. The vacuum column uses a steam ejector system that reduces pressure to nominally 50-70mm Hg. Three sidedraw products are removed from the column and sent to their respective strippers. The side draw products are Vacuum Gas Oil (VGO), wax distillate and heavy wax distillate. The top section of the column is packing with a total draw tray for the VGO product. Part of the VGO stream is cooled in fin-fan exchangers and returned to the top of the tower as cold reflux and part bypasses the exchangers and is returned as hot reflux. A portion of the heavy wax stream is also returned to the tower as hot reflux for the bottom section of the column. A simplified process flow diagram for the Vacuum Column is provided in Figure 2 below.

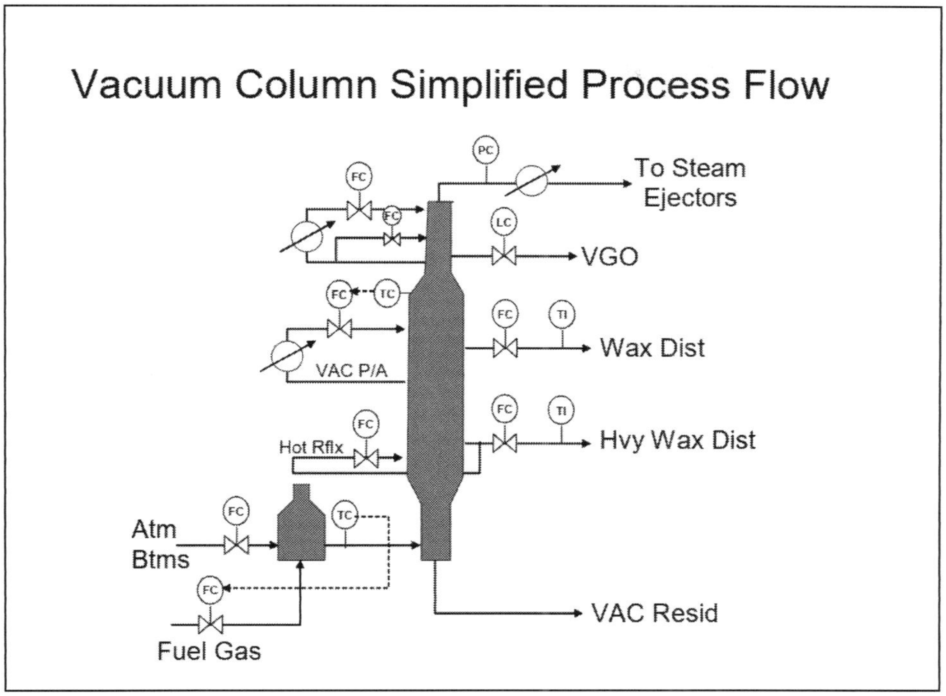

Figure 2

The Newell CDU typically runs a consistent slate of Appalachian crudes chosen to maximize production of the valuable lube stock components. Key product qualities are the endpoints of the various side draws that are sold directly as finished products, processed in downstream units or blended into final product grades. Lab samples of the key products are taken every 4 hours and the column yields are adjusted accordingly,

The CDU is rarely the refinery bottleneck, so limited incentives exist for increased capacity. The main process disturbances are caused by switching crude feed tanks and weather related upsets.

Advanced Control Strategies

The APC functions were designed to make use of the embedded tools within the DCS, primarily using Model Predictive Control (MPC) and Neural Net blocks. The control objectives for both columns are:

- Stabilize the column and control against desired product quality targets
- Maintain the unit within specified constraints and process limits
- Increase yields of the more valuable products up to their key quality specifications.
- Minimize upsets during crude switches and inclement weather.

These objectives are met with a combination of MPC blocks that manipulate the reflox flow and product draw flows to control the tower temperature profiles. Neural-net inferential sensor applications provide real-time indicators for the key product qualities. Pressure Compensated Temperature (PCT) calculations are used to account for the effect of pressure fluctuations on the tower temperatures and further stabilize the key product quality variables.

For the Atmospheric Column, the MPC block is configured with the following variables:

Variable	Description
MV1	Reflux Flow
MV2	Kero Draw
MV3	Heavy Kero Draw
MV4	AGO Draw
CV1	Tower Overhead Temp
CV2	Kero Draw Temp
CV3	Heavy Kero Draw Temp
CV4	AGO Draw Temp
DV1	Reflux Temperature
DV2	Heater Outlet Temperature
DV3	Total Charge Flow
CNST	None

Where:
MV = Manipulated Variable
CV = Controlled Variable
DV = Disturbance Variable
CNST = Constraint Variable

A Neural block is used to predict naphtha and AGO end points in real-time to provide operator guidance. Inputs for the neural block include the tower tray temperatures, pressures, product yields, heater outlet temperature, stripping steam and reflux rates. The inferential predictions may eventually replace their corresponding PCT as controlled variables as sufficient plant data is accumulated to obtain a robust and accurate prediction.

For the Vacuum Column, the MPC block controls three controlled variables, essentially the tray temperatures (PCT's) in the column that were found to respond best to changes in the product draw rates. The Vacuum Column controller includes the following variables:

Variable	Description
MV1	T-102 Pump Around Flow
MV2	Wax Distillate Flow
MV3	Heavy Wax Distillate Flow
CV1	VGO Chimney Vapor PC Temp
CV2	Wax Distillate PC Temp (Tray 11)
CV3	Heavy Wax PC Temp (Tray 7)
DV1	Pump Around Temperature

Variable	Description
DV2	Heater Outlet Temperature
DV3	Total Charge Flow
CNST1	Wax Distillate Draw Temp

A Neural block is used to calculate the wax distillate 95% point, primarily for operator guidance. Inputs for this block include the tower tray temperatures, pressures, product yields, heater outlet temperature, stripping steam and reflux rates.

A New Project Methodology

Due to the embedded nature of the APC tools, the application of advanced control is much easier and most activities happen at the plant on the actual operating system. The engineering required for implementing, staging and testing the solution has been greatly reduced. The project started with a 3-day kickoff meeting at the plant to confirm the APC objectives, design the control strategies and perform preliminary step tests on the units to capture basic process dynamics and constraints. A brief Functional Design Specification (FDS) document was prepared which defined the advanced control block configuration, special calculations and testing procedures. The advanced control functions were configured directly from this document, eliminating the need for a separate Detailed Design phase. Configuration of the complete advanced control applications for both columns, including the custom PCT calculations, MPC and Neural blocks as well as modifications to the operator graphics was completed in less than a day.

The configuration of the Atmospheric Column MPC block is shown in Figure 3 below. This configuration is done graphically using the Control Studio environment, in much the same way as a PID or other control block is configured in the system. The embedded APC tools automatically set up the history module as well as the regulatory control mode checks, watchdog timers, failure mode switching and other functions required for a closed-loop control system.

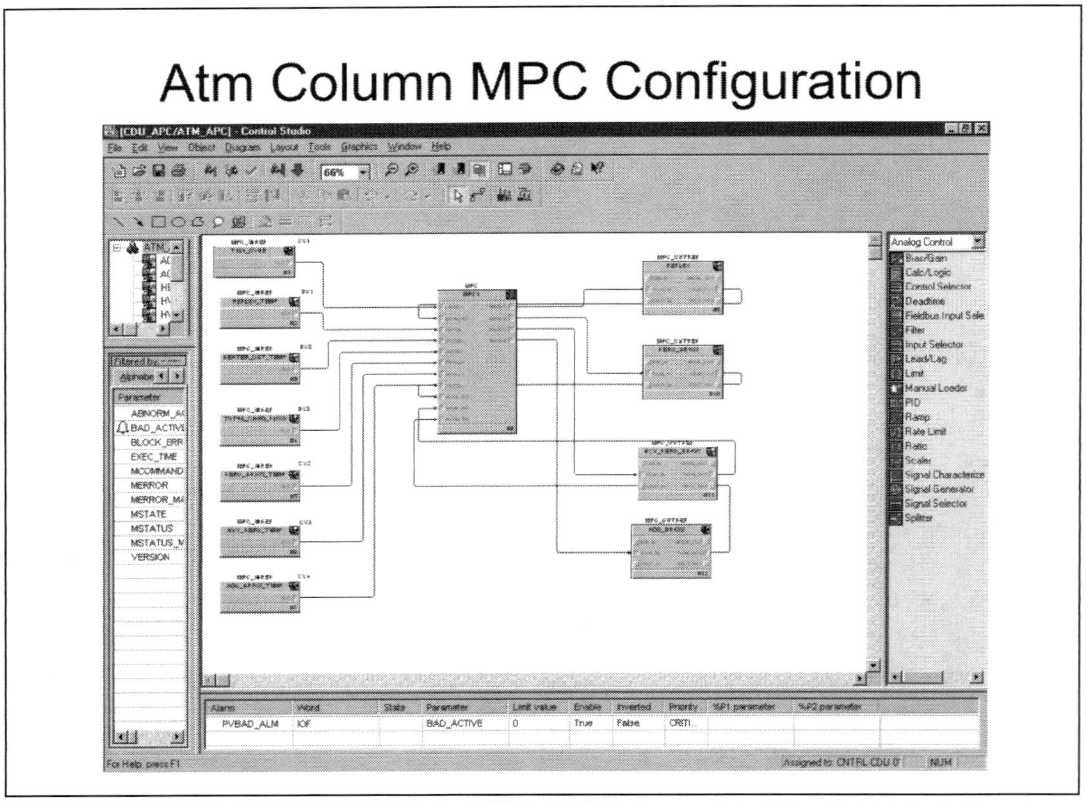

Figure 3

Once the configuration is complete, the modules are downloaded to a controller or an application station and the automated step test tools are used to step the manipulated variables. The MPC automated step test tool uses a "pseudo-random binary signal" (PRBS) approach to step tests which moves all variables around their current setpoint in a special test pattern. The MV's can be stepped individually or at the same time, depending on the complexity of the process and the experience of the user. An example step test setup and execution screens for the Atm Column AGO product is shown in Figure 4 below.

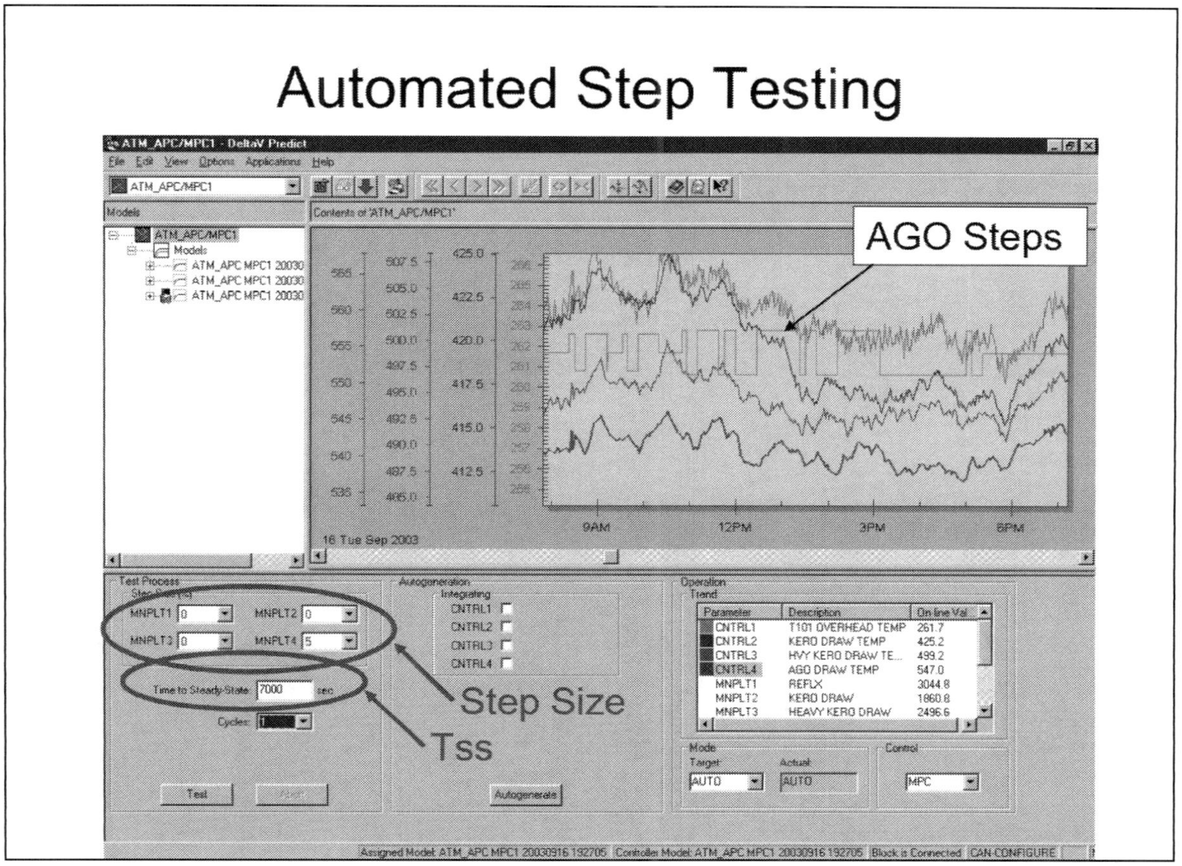

Figure 4

To initiate a step test, only the step size for each MV and the time-to-steady-state (Tss) needs to be entered. Pressing the "Test" button starts stepping the MV setpoints in an automated pattern around their current point. At any time, the operator can change the setpoints of any manipulated controller without affecting the test.

As the step tests are completed, the built-in model identification package is run against historical data to fit the models and validate their results. The final models obtained for the Atmospheric Column are shown in Figure 5 below. This figure shows the final step response curves observed for a 1% change in the MV or DV. From this view it is apparent that the top reflux flow (MV1), reflux temperature (DV1) and heater outlet temperature (DV2) had a strong impact on all four of the tower tray temperatures, thus leading to the interactive nature of the crude fractionation control problem.

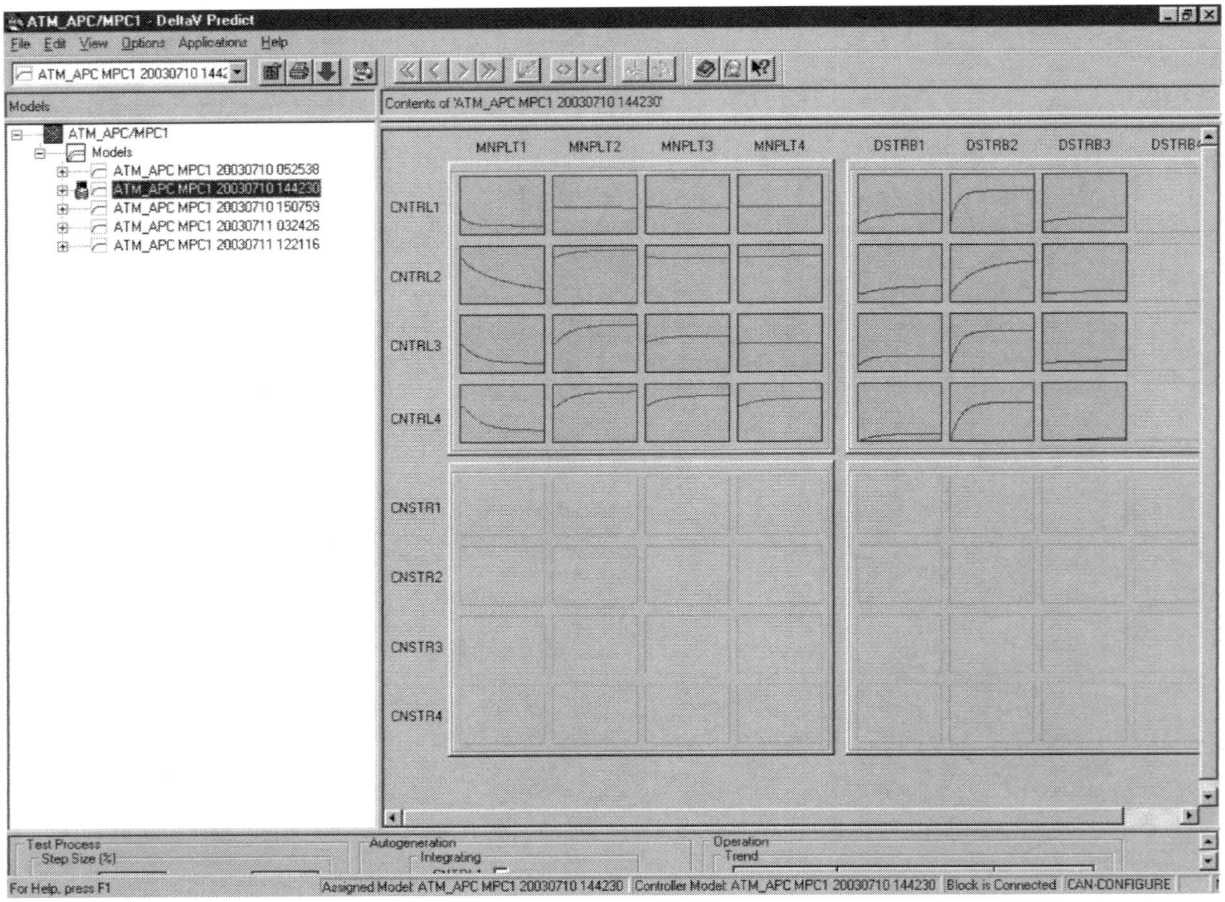

Figure 5
Atmospheric Crude Column Models

Models are validated against original or selected process data by clicking the "Validate" button and selecting a timeframe. The validation results from Ergon, showing actual vs. predicted are shown in Figure 6 for each of the 4 CV's. The observed offsets between some of the actual and predicted values that can be seen in these figures are not an issue as long as the process gain and time constants are reasonably correct. These offsets represent biases caused by "unmeasured disturbances" that can be corrected for on each execution of the controller. Basically, these step tests have yielded some fairly good models that explain most of the variation seen for each of the CV predictions over the step period.

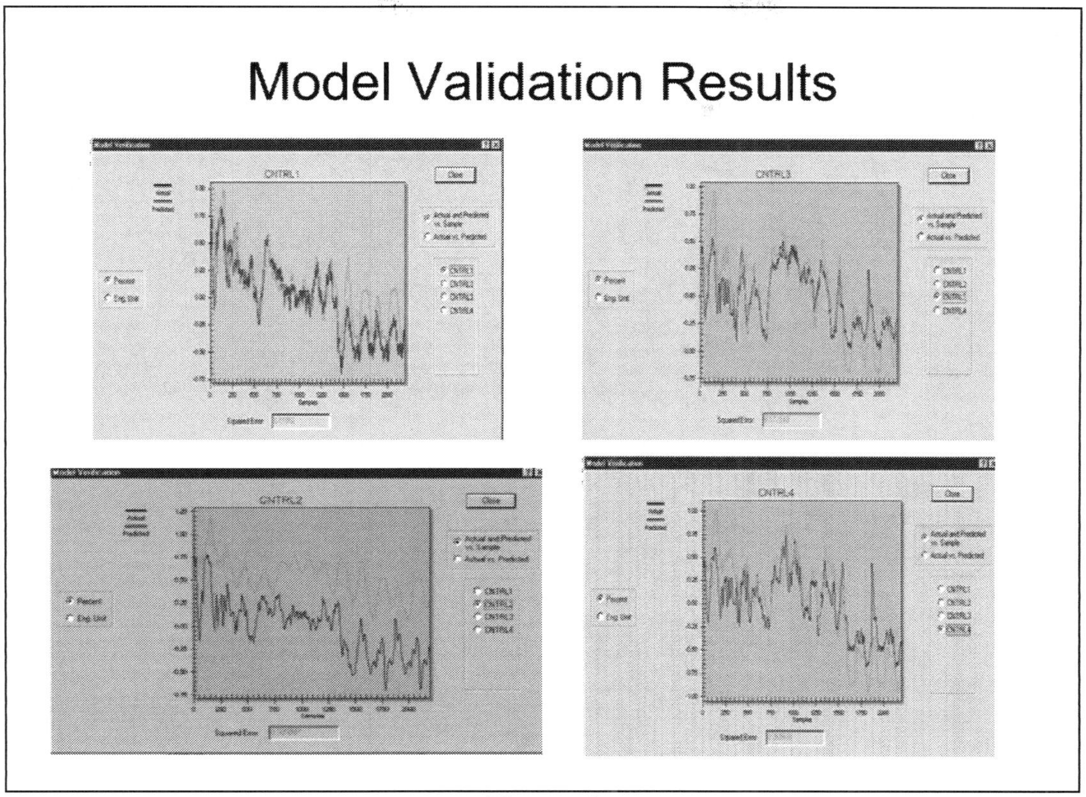

Figure 6

While waiting for the step tests to complete, all of the neural blocks were trained using process and laboratory data stored in the refinery data historian system. The built-in Neural Tools provide the ability to set a delay value for each input. This delay is calculated automatically to obtain the best correlation between the input and output variables, but ultimately needs adjustment based on process knowledge. For each predicted variable, a reasonable fit was obtained in the initial training set, so these calculations were put on-line and the lab update functions enabled. Figure 7 shows the Neural training results for the naphtha endpoint as an example.

Neural Fit of Naphtha EP

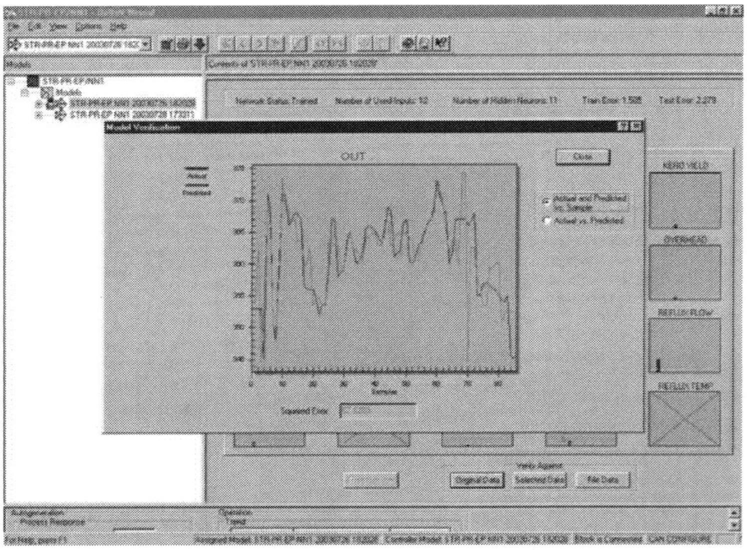

Figure 7

The operator interface to the APC functions is provided through the DCS console graphics. Key calculated variables, neural net predictions and the controller targets can be viewed through the main process overview displays. Detailed information on the MPC blocks is provided by standard MPC Operate displays that are automatically configured from the block configuration. These displays include the following five main windows as shown in Figure 8 below for the Atmospheric Column:

- Controller Mode Window
- Trend Window
- Controller Status Window
- Variable Faceplace Displays
- Detailed Faceplate

All operator interaction can be accomplished through these screens including changing controller modes, setpoints and limits.

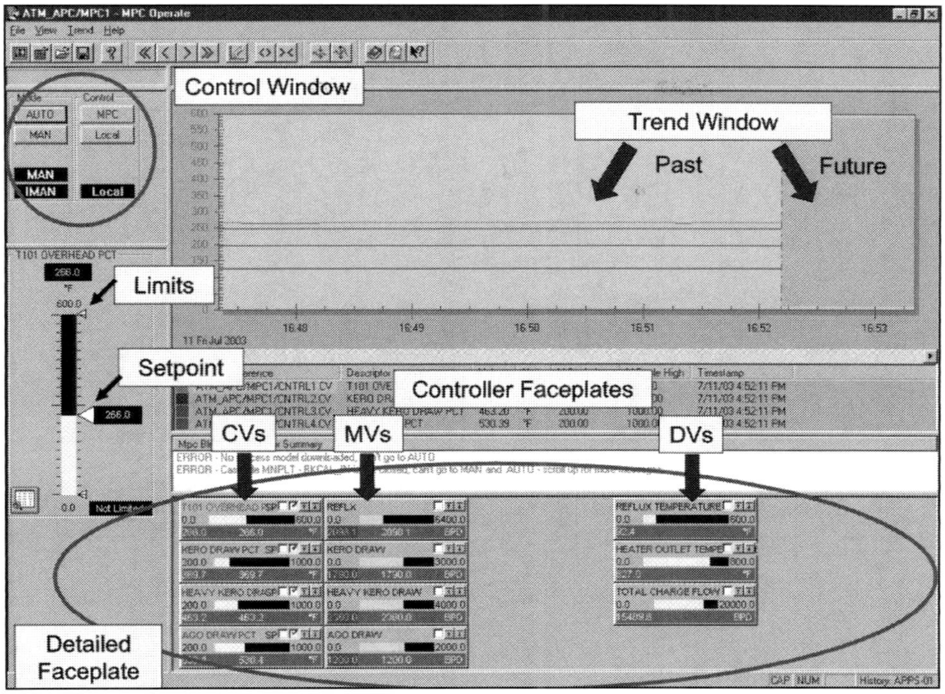

Figure 8
MPC Operate Display

Overall, the step tests, model identification and validation activities for both the Atmospheric and Vacuum Columns were completed in slightly less than ten days. The advanced control functions were commissioned in a week in July, 2003 a few days after the step tests were completed. Within two days, both controllers were on and controlling temperatures as designed. Subsequent changes to the controller design were implemented primarily by Ergon's own control staff and the final Site Acceptance was completed in August, 2003. The following table shows the calendar time associated with each project phase.

Activity	Timeframe
Functional Design Specification	3 weeks
Application Configuration	< 1 day
Step Tests & Model Identification	10 days
Commissioning	1 week
Site Acceptance Test	3 days

Results

Overall, operator acceptance of the new technology has been excellent, with closed-loop time essentially at 100% since it was commissioned. The controllers are able to hold the tower temperatures, for the most part, to within less than a degree from their setpoints are shown in Figure 9 below. Some excursions may occur when switching crudes, but the operators usually leave the MPC blocks in control and allow it to recover. Similar results were observed for the Vacuum Column.

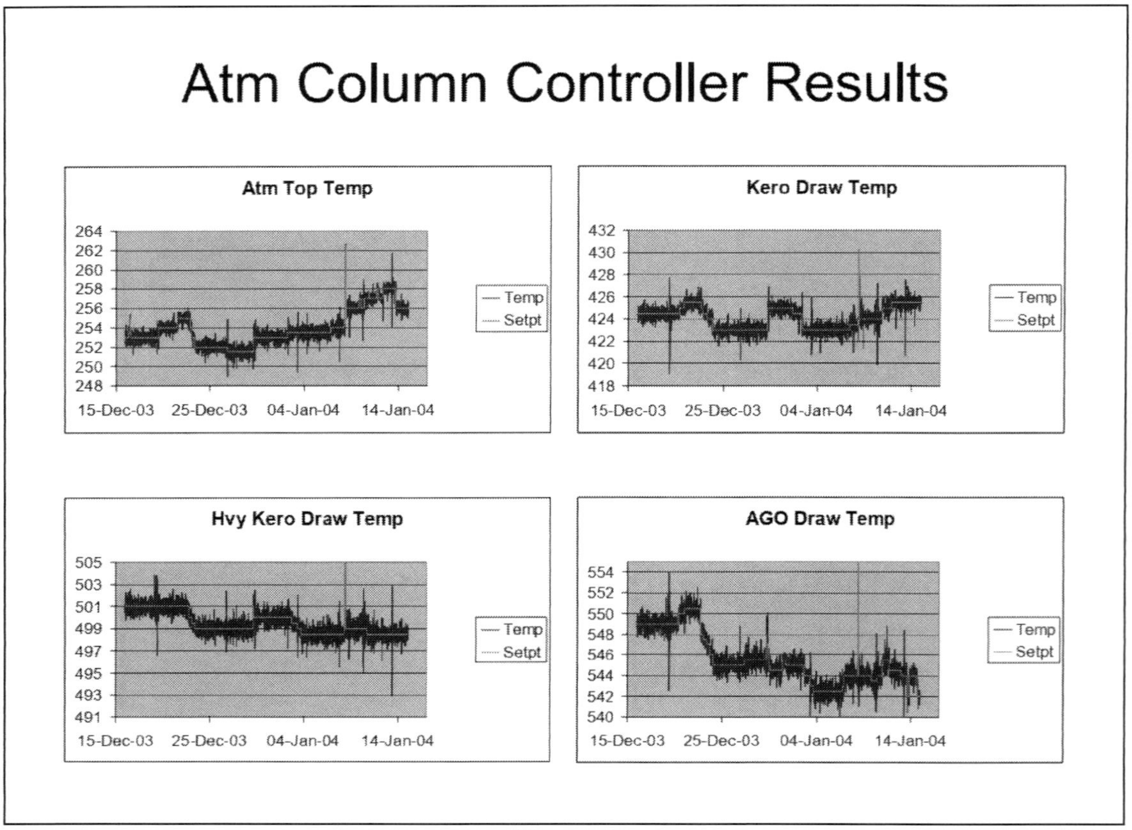

Figure 9

A comparison of process data collected before and after commissioning indicates a reduction in standard deviations for all quality variables as well as the tray temperatures. This data is provided in the table below.

	Before		After		Reduction in Std. Dev.
	Average	St. Dev.	Average	St. Dev.	
Atm Column					
SR NAPHTHA EP	353.69	5.40	339.79	4.53	16.0%
AGO EP	633.35	10.54	626.95	3.54	66.4%
OVERHEAD TEMP	263.67	3.55	253.87	1.84	48.1%
KERO TRAY TEMP	419.56	4.64	424.11	1.06	77.0%
HVY KERO TRAY TEMP	494.73	4.53	499.46	1.11	75.6%
AGO TRAY TEMP	531.13	4.52	545.57	2.45	45.8%
Vacuum Column					
WAX DIST EP	934.36	12.28	936.88	3.40	72.3%
VGO CHIMNEY TEMP	358.58	7.32	353.17	1.89	74.2%
WAX VAP TEMP	542.47	5.99	424.15	1.09	81.8%

The reduction in standard deviation of the key product qualities resulting from the new control strategy has allowed the refinery to reduce the operating target ranges set by the planning group from 10 to 6 deg. F. The improved process performance not only improves the quality of the products delivered to Ergon's customers, but allows the refinery planners to set targets that increase the overall refinery operating margins. As an example, a 4 degree F change in the Heavy Naphtha/Kero cutpoint shifts the yields about 400 BPD, which at a $2/Bbl price differential generates $280,000 in profit.

Conclusions

In the past, high implementation costs and special technical skills required to implement and maintain advanced control applications have relegated these technologies to the largest, most sophisticated refineries or petrochemical plants, which have sufficient incentives to justify such an investment. With the advent of embedded APC tools, that is no longer true today. These APC functions are now available to any process control engineer, who can learn to use them with only a small amount of training. APC applications are now feasible for even small process units with limited benefit potential. Simple regulatory functions like loop decouplers or overrides can be implemented easily and cost effectively using the built-in MPC blocks. Neural-net inferential sensor technology can be configured and tested with just a few mouse clicks. For more than 50 years, control engineers have been making do with PID, feedforward or lead/lag algorithms and high/low selectors because that is all that came with a basic control system. It is now time for APC tools to find their way into a typical control engineer's toolbox, to be used along with all the other basic control functions he or she uses on a daily basis.

Controlling Crude Oil Blend Using Scheduling Tools

Dhaval Dave and Nan Zhang
Department of Process Integration, UMIST,
Manchester, M60 1QD, United Kingdom.

Abstract

The profit of a refinery now a day depends upon how refinery responds to crisis-making situations and dynamic market conditions. It is very important to exploit short-term market opportunities such as low-cost availability of crude oil in the spot market, buying/trading cargoes and so on. Such short-term decisions are always coupled with inventory crisis and most importantly risk of running unstable operations. In this paper, we propose optimum scheduling based method to control crude oil blend composition and to ensure stable performance of crude distillation unit during dynamic situations such as above. An index independent of time and property has been defined on the scale of 0-100 to represent compatibility of different crude oil when they mix. Both continuous and batch blending operations are considered in the scheduling model. Consideration of batch blending generated more profit and provided feasible schedules where continuous blending models failed. Overall problem is a large scale mixed integer nonlinear programming (MINLP) problem, which is solved using MILP-NLP decomposition method. Several disturbances such as late arrival of cargoes are considered in order to demonstrate the capabilities of this algorithm.

1. Introduction

The first disturbance to any refinery can be a change in crude oil quality or a variation in the crude oil supply. Also, many large refineries that have a variety of down-stream processes buy 6-8 or more number of large varieties of crude oils depending upon the market conditions. A variety in downstream processes offers large operational window and hence enables the refinery to accommodate various crude oils. Smart storage and blending procedure can accommodate sweet to sour and light to heavier crude oils and hence provides more flexibility in crude oil purchase decisions. It is not always very straightforward to make a purchase decision. For example, consider the case when some amount of crude oil is available cheaply in the market for instant delivery. If refiners buy such crude oil, savings in terms of raw material cost is obtained. This also generates the risk of running unstable operations, product property constraint violation and exceeding storage limitations. On the other hand, if refiners do not purchase the crude oil, refinery can be operated with predetermined stable operations but the opportunity to save raw material cost has been lost. Similarly, delaying and selling crude oil cargo can generate extra profit sometimes but it is also associated with risk of unstable operations.

In literature scheduling approach is popular in order to optimize blending decisions while controlling refinery constraints. Both discrete time (Shah (1996), Lee *et al.* (1996), Wenkai *et al.*(2003) , Kelly and Mann (2003a, 2003b) and continuous time (Jia *et al.* (2003)) mathematical formulations are available for crude oil blend scheduling problems. Continuous time formulation requires less number of discrete variables since it avoids the discrete variables for not-happening tasks. Continuous time formulations are still not favorite because of nonlinear formulation, large optimality gaps and computational time (Pinto *et al.*, 2000).

Crude oil blend scheduling problem turns bilinear when property values in the tank are calculated. Lee *et al.* (1996), Shah (1996) and Jia *et al.* (2003) controlled the property values in the tank by assuming property as a component and controlling its flow and volume. This relaxation avoids bilinearities but property values in the tank are not controlled tightly. Wenkai *et al.* (2003) proposed iterative MILP-NLP solution method for solving MINLP problem considering bilinearities due to property calculation. Recently, Kelly and Mann (2003a, 2003b) provided detailed features of crude oil blend scheduling problem and mentioned the substantial benefits offered by optimum scheduling. They also separated quantity and logistic MILP problem from quality based NLP problem and solved the later problem with successive linear programming (SLP) techniques. Kelly and Mann (2003b) with transferline restrictions first modeled batch blending operation of different crude oils in the charging tanks. These restrictions included were, only one component can flow into the transferline at a time and time window for blending in transferline was restricted to arbitrarily chosen value 20hr.

In summary, most of the research effort has been made for mathematical formulation of short-term crude oil unloading, tank inventory management and crude distillation unit charging schedules. As reported by Kelly and Mann (2003b), scheduling tools should be used for important tactical decisions such as buying/trading crude oil cargoes and its application should not only be limited to create optimum time-sequence of various activities. In the present work, an index independent of time and property of crude oil has been developed, called Blend Compatibility Index (BCI) for evaluation of crude oil, in order to make decisions such as buying or trading of cargoes. Also both continuous and batch blending is considered for crude oil blending in charging tanks using discrete time formulation. Present method for batch blending does not require transferline restrictions and hence it generates better quality schedules. Also, the objective of optimization process is to maximize the profit rather than cost minimization. The main motivation is to reduce computational time by avoiding waiting for small cost improvements. Modification in MILP and NLP iterative solution procedure proposed by Wenkai *et al* (2003) has been made for the case when NLP fails to generate feasible schedule for feasible MILP solution. Finally, case studies are performed to demonstrate the applications of scheduling model for handling crisis situations and making tactical corporate decisions.

2. Crude oil Blending System

Generally, crude oil purchase deals are of two kinds: (i) long-term contracts and (ii) short-term contracts. Long term contracts are made well in advance, delivery schedules are generally fixed and planning procedure (single period) is used to make the decisions. This procedure also sets production targets and various operating strategies. On the other hand, Short-term contracts are generally made in the spot market. Spot market is defined as a market in which crude oil or its products are traded for immediate delivery at the prevailing price. Since the deal is valid only for the short time period, many constraints such as product demands and production mode are not subjected to vary. Scheduling tools can be applied for evaluating short-term contracts. Crude oil from pipeline or cargo is first stored into tank farm storage tanks. Crude oil from storage tanks are then blended into the charging tanks. In some refineries, there is no clear distinction between storage tank and charging tanks; different cargoes upload the crude oil in storage tanks and blending occurs over a period of time. Blending of different crude oil in charging tank is tightly controlled because of several reasons (design of distillation column, crude oil compatibility, market constraints, downstream processes requirements and so on).

Crude oil properties such as API gravity, overall sulfur, Watson K factor, TBP curve (product distribution) etc. are maintained in charging tank. Some times individual product properties such as kerosene pour point and diesel Cetane number are also controlled in this charging tank. Assigning appropriate weight factors to these properties, in an optimization process can accommodate different priorities. For example, in Table (3), sulfur is the most crucial property for crude oil blending operation (CF has the highest value). Scheduling problem has an added dimension – time, that requires crude oil evaluation index which is independent of it. For this purpose Blend Compatibility Index (BCI) is defined on the scale of 0-100, 100 value signifies that all the property constraints are satisfied during scheduling time horizon and value less than 100 denotes for constraints violation. Sample calculation procedure for BCI is as follows:

Let the time horizon be 30 days (T_{days}). Assume all the properties are within the range except crude oil gravity that is out of range for 5 days (N_{days}). Then the blend compliance index for API is given by $BCI_{API} = CF_{API}\text{-}N_{days}(CF_{API}/T_{days}) = 10 - 5(10/30) = 8.33$. BCI for all other properties are same as their criticality factors (CF). Hence overall BCI of crude oil = 90 +8.33 =98.33.

In order to maintain the quality of a blend by providing enough time for mixing, some operational constraints such as loading and unloading of a tank can not be allowed simultaneously, are imposed. Once the desired quality of blend is achieved in the charging tank, it is can feed to crude distillation unit (CDU). In the CDU, crude oil blend is separated in to six or seven straight-run fractions such as LPG, naphtha, kerosene, diesel, atmospheric residue, etc. Atmospheric residue is then processed in vacuum distillation unit (VDU) in order to provide light and heavy vacuum gas oil and vacuum residue. The operations of VDU and CDU are highly interlinked and atmospheric residue from CDU is generally sent to VDU without storing. Although the basic boiling ranges of these fractions are more or less determined by the crude oil assay (e.g. 180-350°F for naphtha, 400-550°F for kerosene, and 580-650°F for light gas oil, etc.), some overlaps might exists in between the two adjacent fractions. For instance, the fraction with boiling range 350-400°F might either belong to the naphtha or the kerosene fraction. The qualities of this overlapped fraction that are cut into naphtha and kerosene are determined by the cutpoint between naphtha and kerosene. While the cutpoint varies, not only the yield but also the properties of naphtha and kerosene (sulfur, aromatics, flash point, etc) change. Hence cut point scheduling is an important optimization decisions in present scheduling framework.

Blending processes are of two types: (1) continuous blending and (2) Batch blending. In continuous blending process, simultaneous mixing of different crude oils is achieved in charging tank and property constraints are imposed at every time. Sometimes, because of large number of crude oil to be processed and a limited availability of storage tanks, both the functions storing and blending are performed in one tank, as shown in Figure (2) – this is known as batch blending. No feasible schedules can be obtained with continuous blending (where property constraints are applied at each time period), because the tank having sour crude oil needs to wait for sweet crude oil cargo to match the blend quality requirements. Batch blending is applied to model such situations where the required components are mixed in a destination tank sequentially and the same recipe/composition can be achieved while the tank is charging. In general, batch blending offers wider operating range with higher integration between charging decisions and quality constraints. Tighter quality constraints can generate large changeover (of charging tanks to charge CDU), and this must be avoided in order to overcome operability and control problems. A novel uniform discrete time MINLP model for this scheduling problem is as follows.

3. Mathematical formulation

3A) Objective function Profit = product revenue – waiting cost – unloading cost – inventory cost for storage and charging tanks – quality constraint violation cost – change over cost.

Subject to:

3B) Crude oil selection and unloading operations

Crude oil vessels carry different crude oils in discrete sizes and transportation can be made through different modes.

Equations for specifying crude oil cargo arrival time $Tarr_V$, unloading start time Ts_V and unloading finish time Tf_V.

Various unloading constraints are specified (such as speed of unloading, number of ships that can unload simultaneously, in which tank crude oil vessel should unload and so on), depending upon marine tank farm and shipment management team.

3C) Material balance for vessels

Volume of crude oil in vessel v at time t = initial volume of crude oil in vessel v – crude oil transferred from vessel v to storage tanks up to time t.

3D) Material balance for storage tanks

Volume of crude oil in storage tank i at time t = initial volume of crude oil in storage tank i + crude oil transferred from vessels to storage tank i up to time t – crude oil transferred from storage tank i to charging tank j up to time t.

In order to maintain quality, operating rule on crude oil transfer from storage tank i to charging tank j at time t have been imposed. No crude oil is allowed to transfer when charging tank is feeding to CDU.

Amount of material delivered to the storage tanks = initial volume in vessel.

Capacity equations for storage tanks are specified.

3E) Material balance for charging tanks

Crude oil blend in charging tank j at time t = initial volume of blend in charging tank j + crude oil transferred from storage tanks to tank j till time t – crude oil blend charged to CDU up to time t.

Pump capacities provide limitation to charging.

Capacity equations for charging tanks are specified.

CDU should always be charged with minimum amount of crude oil all the time.

3F) Controlling properties in the charging tank

Amount of crude oil key components such as sulfur, specific gravity, K value and so on are calculated. Volume of component k in charging tank j at time t = initial component k in charging tank j + component k in crude oil transferred from storage tanks to charging tank j up to time t – component k in crude oil mix j transferred to CDUs up to time t.

CDU products such as naphtha, kerosene, etc. and the key components of these products such as sulfur and aromatics content in naphtha etc. are treated as sub-key components of crude oil and similar equations are applied.

Equation for calculating flow rates of key components from storage tanks to charging tanks should be included.

Volume of key components within the storage tank and the outgoing flow rate of key components from the charging tank are constrained. Equations for continuous blending operations:

$$Fcdu_{j,l,t}Mink_{j,l,k} \leq fcdu_{j,k,l,t} \leq Fcdu_{j,l,t}Maxk_{j,l,k} \quad \forall j=1,..,Nj; k=1,..,Nk; l=1,..,Nl; t=1,..,Nt \tag{1}$$

$$Vb_{j,t}Mink_{j,k} \leq Vbk_{j,k,t} \leq Vb_{j,t}Maxk_{j,k} \quad \forall j=1,..,Nj; k=1,..,Nk; t=1,..,Nt \tag{2}$$

Relaxation Equation (3) and Equation (4) replace Equation (1) and (2) for the case of batch blending operations:

$$(Vb_{j,t}Mink_{j,l,k} - Vbk_{j,k,l,t}) \leq M\sum_{l}(1-D_{j,l,t}) + S\_lo_{j,k,t} \quad \forall j=1,..,Nj; k=1,..,Nk; t=1,..,Nt \tag{3}$$

$$(Vb_{j,t}Maxk_{j,l,k} - Vbk_{j,k,l,t}) \geq M\sum_{l}(1-D_{j,l,t}) + S\_up_{j,k,t} \quad \forall j=1,..,Nj; k=1,..,Nk; t=1,..,Nt \tag{4}$$

Overall sum of CDU products should be equal to the amount of crude oil charged to the CDU.

3G) Operating rules for crude oil charging

Charging tank j can charge at most one CDU at any time t.
CDU i must be operated throughout scheduling horizon.
Equations for calculating feed changeover in charging from tank j to j` in CDU l.

3H) Cut point management

Two equations are used to model cutpoint changes by swing-cut concept. Equation (5) calculates the flowrate (F_p) of each product on the basis of weight percentage (wt_p) of total crude oil flow F. In Equation (6), the total swing-cut flowrate, which counts for the percentage of the crude oil flow (ws_pF) is the sum of the amount going to the product above (F^-_p) and amount going to the product below (F^+_p). This two equations are applied at every time period.

$$F_{p,t} = F_t wt_{p,t} + F^+_{t,p-1} + F^-_{t,p} \quad \forall t = 1,...,Nt; p = 1,...,Np \tag{5}$$

$$ws_p F_t = F^+_{p,t} + F^-_{p,t} \quad \forall t = 1,...,Nt; p = 1,...,Np \tag{6}$$

3I) Blend Compliance Index (BCI)

Slack variables used in equation (4) and (5) are then used to calculate the properties that have been violated and Blend Compliance Index (BCI).

3J) Demand Equations

Total production should meet the market and internal demand of individual CDU products.

3K) Nonlinear Equations

The composition of key component within the charging tank and the composition of key component within the flow coming out of charging tank should be equal.

$$\frac{Vbkey_{j,k,t}}{Vb_{j,t}} = \frac{fcdu_{j,k,l,t}}{Fcdu_{j,l,t}} \quad \forall j=1,....,Nj; k=1,....,Nk; l=1,....,Nl; t=1,....,Nt \tag{7}$$

For cut-fractions composition of key component is assumed constant throughout the fraction.

4. Solution Strategy

Equations (3A) to (3K) represent a large MINLP problem, which was solved using decomposition method as shown in Figure (1). Firstly, mixed integer linear programming (MILP) problem is solved which does not consider the fact that while the tank is charging, the key components composition with in the tank and key component composition of the material coming out of the tank are same (equation 3K).

Hence, the solution of MILP needs to be updated by applying nonlinear programming (NLP), which incorporates this fact (by equation 3K). The main reason for such an iterative decomposition is to avoid nonlinearity in first combinatorial problem. The solution of MILP provides good starting point for small size NLP, which is then easier to solve.

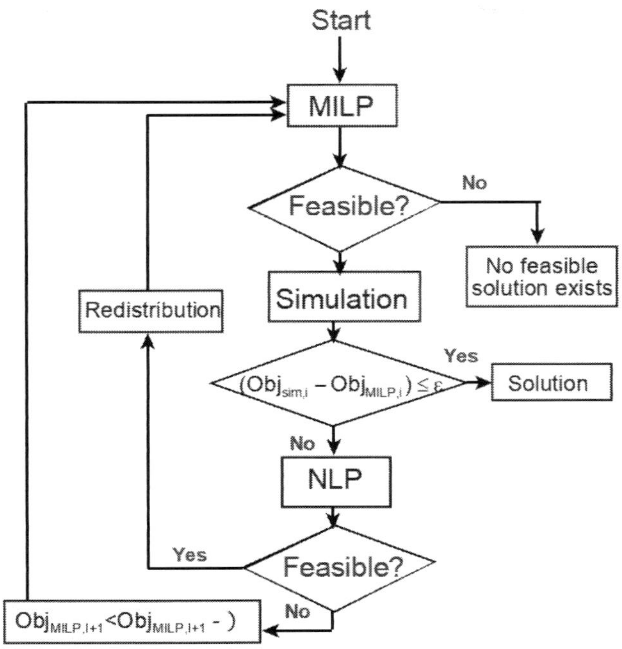

Figure (1) MINLP problem solution strategy

5. Case studies

In order to demonstrate the performance of above approach, three case studies are made. Case study (a) shows batch blending generates a feasible schedule, while continuous blending assumption provides no feasible crude oil blend schedule. Case study (b) deals with crisis making situation and provides the impact of late arrival of crude oil cargo to the refinery and case study (c) shows application of crude oil scheduling model to tactical decisions such as purchase of crude oil.

Case study (a) Batch blending vs. continuous blending

As shown in Figure (2) depicts crude oil systems for this case study where blending of different crude oils occurs over a period of time in storage tanks. There are 4 crude oil vessels, 4 storage tanks and 1 distillation column in this system; crude oil vessel and storage tank information is provided in Table (1) and Table (2) respectively.

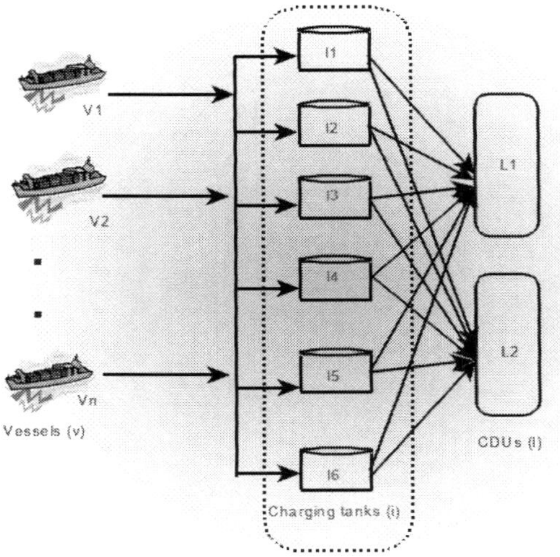

Figure (2) Crude oil system for case study (a)

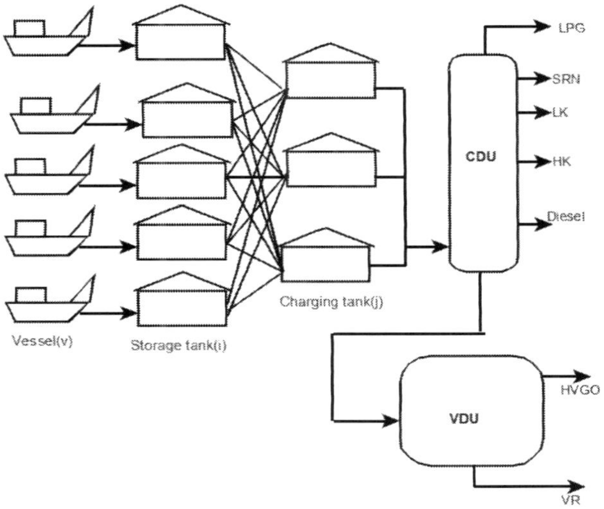

Figure (3) Crude oil system for case study (b)

Table (1) Vessel information for case study (a)

Initial volume	Arrival time	Key component K_1	Key component K_2	Key component K_3
100	1	0.885	2.25	11.81
150	2	0.862	1.8	11.76
140	5	0.864	1.99	11.779
100	7	0.865	1.98	11.778

Table (2) Storage tank information for case study (a)

Initial volume	Minimum volume	Maximum volume	Initial composition of key components	Maximum composition of key components	Mainimum composition of key components
120	10	350	(0.865,1.9,11.77)	(0.865,2.0,11.78)	(0.85,1.8,11.76)
70	10	350	(0.862,2.0,11.78)	(0.865,2.0,11.78)	(0.85,1.8,11.76)
75	10	350	(0.864,1.99,11.779)	(0.865,2.0,11.78)	(0.85,1.8,11.76)
90	10	350	(0.865,1.98,11.775)	(0.865,2.0,11.78)	(0.85,1.8,11.76)

Continuous blending assumptions (equation (1) and (2)) generated no feasible schedule for time horizon 10 days, using GAMS (Brooke *et al.*, 1992). Batch blending formulations (equation (3) and (4)) provided a feasible schedule with minimum objective function value 1172.25. Charging schedules, inventory profile and key component (specific gravity) composition profile is shown in Figure (4), (5) and (6) respectively.

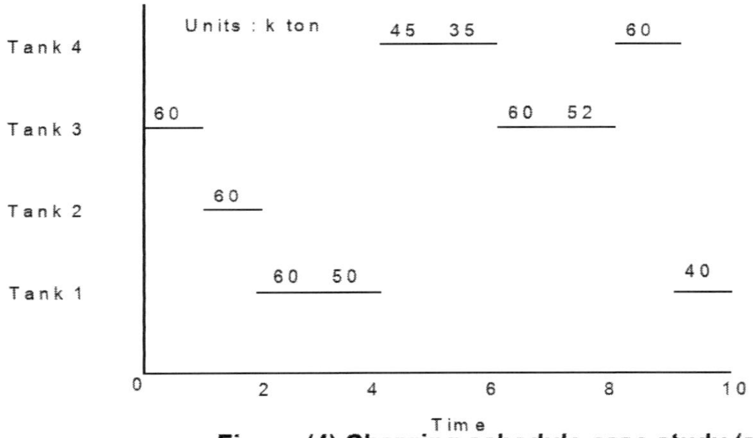

Figure (4) Charging schedule case study (a)

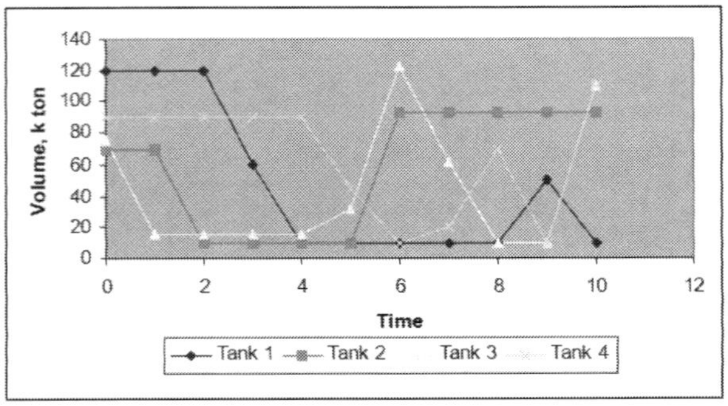

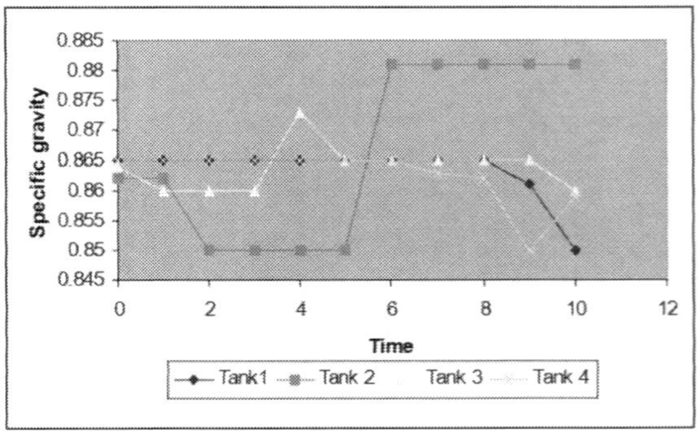

Figure (5) Inventory profile case study (a)

Figure (6) Specific gravity profile for case study (a)

Case study (b) Late arrival of cargo

Figure (3) represents crude oil system for this case study. There are 5 vessels carrying Arabian Heavy, Ratawi, Iran Heavy, Dubai and Kuwait crude oils with sizes 250 k ton each and arrival time 1st, 13th, 22nd, 9th and 5th day of the month. Table (3) shows feasible operation range and criticality factor for properties for this case. MINLP solution provides charging schedule as shown in Figure (7) and Figure (8) represents need for solving quality based NLP problem. MILP solution over-estimates the profit with relaxed quality constraint (BCI achieved = 100), where as simulation provides realistic picture with nonlinear quality constraints (BCI = 78.33). Nonlinear programming problem (NLP) brings back property values within bounds (BCI = 100).

In above case, if Ratawi crude cargo arrives 3 days late, maximum achievable BCI value (after optimization) is 81. Since Ratawi crude is sour (sulfur content 4.07 wt%) and heavy (LPG fraction ~0 wt%), late arrival produces sweeter and lighter blend. As shown in Table (4), minimum sulfur constraint is violated for 15 days and maximum LPG constraint is violated for 6 days. Hence, refiners can delay Ratawi cargo if they have sufficient LPG handling facility (sulfur is not a main issue here as crude is getting sweeter). Similarly, more such 'what-if' scenarios can be studied to gain more operational insights.

Case study (c) Integrated crude oil supply decisions

In scheduling framework mentioned in section 3, tactical decisions such as which crude oil to buy, size of the order and ship arrival time is added in order to make optimum integrated crude oil supply decisions while satisfying operating constraints and market demands. Table (5) provides the list of crude oil in spot market, their qualities and initial volume in the refinery. Integrated optimization reported 10.01 MM$/month profit compared to 9.87 MM$/month profit for the base case. BCI for integrated decision was 100 and supply decisions are mentioned in Table (6).

Table (3) Feasible operation range

Property	Target range	Criticality factor (CF)
API gravity	29.3 – 30.1	10
Overall sulfur, wt%	2.70-2.85	30
K value	11.90-11.92	20
LPG max, wt%	1.7	20
VR max, wt%	25.6	20

Table (4) Blend compliance after optimization

Property	Criticalilty factor, CF	Property violation no. of days, N_{days}	$BCI_{property}$	Maximum Deviation (%)
API gravity	10	0	10	-
Overall sulfur, wt%	30	15	15	18
K value	20	0	20	-
LPG max, wt%	20	6	16	24
VR max, wt%	20	0	20	-
Overall BCI			**81**	

Table (5) Available Crude oils and its properties

Name	Gravity	Sulphur	LN	HN	ATF	AGO	HVGO	HNS	HVGOS	ATF-PP	Cost	Initial Vol.
CrudeA	0.854	0.0045	0.099	0.235	0.082	0.185	0.149	8.00E-05	0.012	-38	118	80
CrudeB	0.8239	0.017	0.1372	0.1131	0.2468	0.1377	0.2934	3.00E-06	0.0038	-54	125	115
CrudeC	0.8576	0.014	0.07	0.1436	0.2124	0.1437	0.278	0.0003	0.02	-27	109	78
CrudeD	0.8647	0.0252	0.0556	0.1263	0.18	0.1147	0.2533	0.0009	0.0312	-43	100	70

Table (6) Integrated crude oil supply decisions

Vessel	Crude type	Size (k ton)	Arrival time (th day)
V_1	A	100	27
V_2	B	150	17
V_3	C	0	-
V_4	D	0	-
V_5	B	100	26
V_6	A	0	-

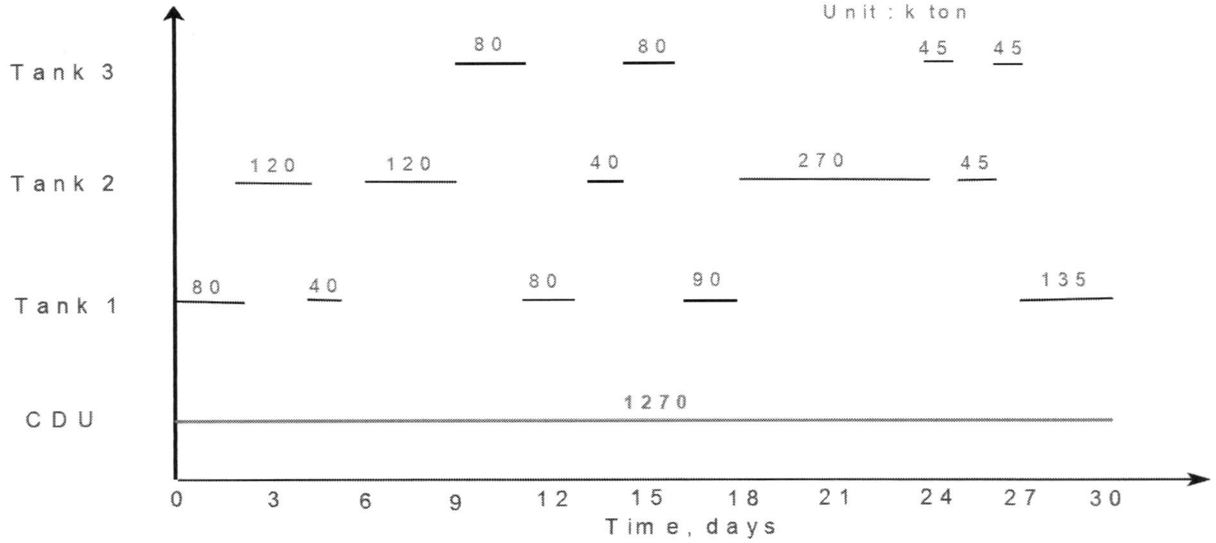

Figure (7) Charging schedule for case study (b)

Figure (8) Property value for MILP, simulation and NLP runs respectively (constraint violation)

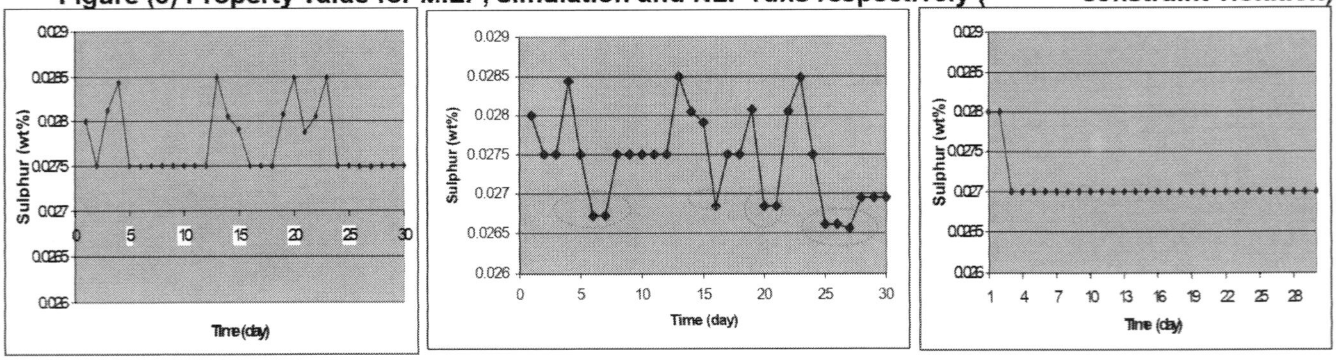

6. Conclusions

In this paper, scheduling based approach is used to control crude oil blend composition and to make tactical crude oil purchase decisions. Efficient batch blending operation is considered without transfer line restrictions that generates more profit and provides feasible schedules where existing continuous blending model fails. Case studies show that this procedure provides tactical insights and it can be applied to make decisions such as buying/trading crude oil cargoes and for optimizing current operations.

Notation

Indices

i	Storage tanks
j	Charging tanks
k	Key components
l	Distillation units
p	Products from CDU
t	discrete time points
v	Crude oil vessels

Parameters

$Maxk_{j,k}$	Maximum composition of key compoent k in tank j that feeds distillation unit l
$Mink_{j,k}$	Minimum composition of key component k in tank j that feeds distillation unit l

Variables

$Fcdu_{j,l,t}$	Flowrate to CDU l from charging tank j at time t
$Fcdup_{j,l,p,t}$	Flowrate of product p from CDU l which is charged by tank j at time t
$Fs_{p,l,t}$	Flowrate of sulfur in product p from CDU l at time t
$S\_lo_{j,k,t}$	Slack variable for equation (3)
$S\_up_{j,k,t}$	Slack variable for equation (4)
$Tfcdup_{l,p,t}$	Total flowrate of product p from CDU l at time t
$Vb_{j,t}$	Volume of tank j at time t
$Vbk_{j,k,t}$	Volume of key component k in tank j at time t

Literature Cited

Brooke, A.; Kendrick, D.; Meeraus, A. GAMS- A User's Guide, release 2.25; The Scientific Press: San Francisco, CA, **1992**.

Jia,Z.;Ierapetritou, M.;Kelly, J. Refinry Short-Term Scheduling Using Continuous Time Formulation: Crude-Oil Operations. *Ind. Eng. Chem. Res.* **2003**, 42(13). 3085.

Kelly, J.D.; Mann,L. Crude Oil Blend Scheduling Optimization: an Application with Multimillion Dollar Benefits – Part I. *Hydrocarbon Process.* **2003a**, 82(6), 47.

Kelly, J.D.; Mann,L. Crude Oil Blend Scheduling Optimization: an Application with Multimillion Dollar Benefits – Part II. *Hydrocarbon Process.* **2003b**, 82(7), 72.

Lee, H.; Pinto, J.M.; Grossmann,I.E.; Park,S. Mixed-Integer Linear Programming Model for Refinery Short-Term Scheduling of Crude Oil Unloading with Inventory Management. *Ind. Eng. Chem. Res.* **1996**, 35, 1630.

Pinto,J.M.; Joly,M.; Moro,L.F.L. Planning and Scheduling Models for Refinery Operations. *Comput.Chem.Eng.* **2000**. 24, 2259.

Shah, N. Mathematical Programming Techniques for Crude oil Scheduling. *Comput. Chem. Eng.* **1996**, 20, S1227.

Wenkai, L.; Hui,C.W.; Hua,B.; Tong,Z. Scheduling Crude oil Unloading, Storage and Processing. *Ind. Eng. Chem. Res.* **2002**, 41,6723.

Online Tuning Procedures for Crude Distillation Unit Model

Dhaval J. Dave, Murtuza Z. Dabhiya, S.V.K. Satyadev, Saibal Ganguly and Deoki N. Saraf
Process Control Laboratory, Department of Chemical Engineering, Indian Institute of Technology, Kanpur – 208016.

Abstract

Crude distillation unit (CDU) is an important refinery operation and its product properties play an important role in control and optimization. Various properties such as specific gravity, flash point and pour point are function of product composition, which is calculated by simulator requiring crude oil composition (in terms of True Boiling Point –TBP curve and average specific gravity) and process parameters. The exact feed TBP is often unavailable during the plant operation. Stratification of raw crude oil into layers in the large tank farm sections cause severe operating problems in terms of the stability of the column. Also, steady state simulator equations are based on ideal conditions where it is assumed that phase equilibrium is attained on each stage, but it is seldom achievable in real life owing to a variety of column non-idealities. Moreover, no single thermodynamic model can successfully predict the thermodynamic properties in the entire range of boiling points with the same accuracy. These reasons lead to deviation of the simulator results from the plant data. The present study aims to provide CDU simulator tuning methods for online applications such as inferential control and supervisory optimization. In-house developed tuning methods are described and performance of tuned simulator package with refinery data is compared.

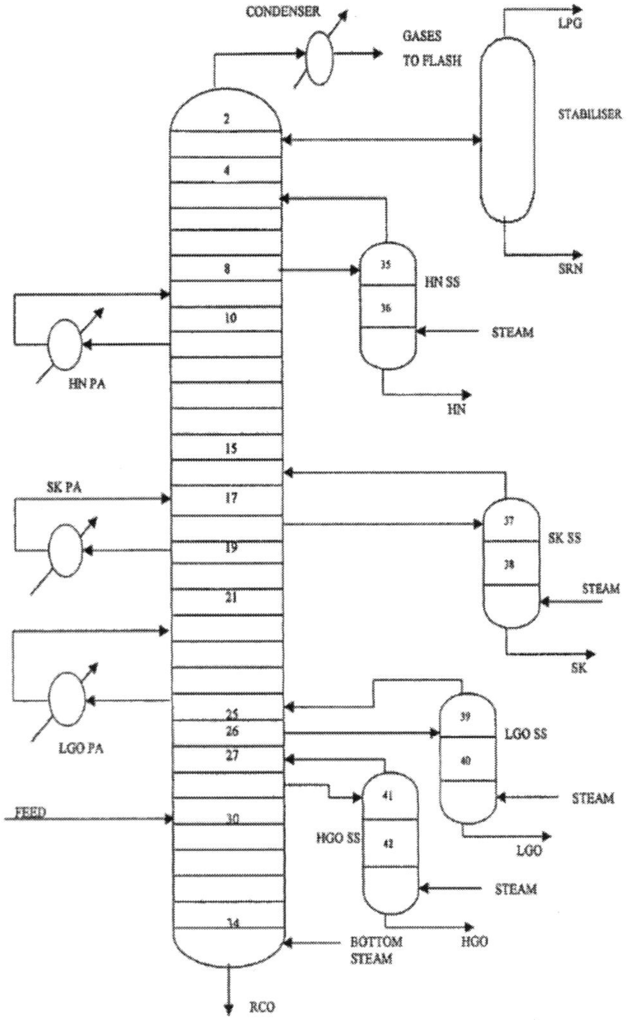

Figure (1) Schematic diagram of CDU

1. Introduction

A crude distillation unit (CDU) processes entire crude oil entering in the refinery. Figure (1) shows a schematic diagram of a unit having various pump arounds and side strippers for products. Stringent environmental laws and highly competitive market imposes tighter quality specification of CDU products. All these products are complex mixtures of hydrocarbons and it is not possible to characterize them in terms of individual components. Moreover, it is often sufficient to characterize these products in terms of certain gross properties such as Reid Vapor Pressure for volatile products, Flash Point for light distillates, Pour Point for heavy distillates etc. Unfortunately no suitable hardware sensors are available in the market which could be used to measure required properties of various petroleum products on-line. The laboratory measurement procedures are tedious and time consuming, and, hence, it is seldom practical to measure these properties more often than once in a shift or even once in 24 h. This essentially necessitates development of "soft sensors" which allow the desired product properties to be estimated on-line since reliable online hardware sensors are not available for field use. The crude oil is characterized in terms of average specific gravity and a True Boiling Point (TBP) curve available from crude assay. The product properties are a strong function of product TBP, which in turn, depends on feed TBP and column operating parameters. Therefore, ability of the package to predict properties depends upon simulator performance and hence accuracy of model equations and input data provided to the simulator.

CDU modeling and simulation is a mature area of research now. Various steady state CDU simulators such as HYSIS are available which calculates product composition with the help of feed TBP, specific gravity and various process data. In order to use such models for online application, it is important to tune the model frequently using online operating data. Although very little information is available in the published literature, stating application of such models and procedures to tune the simulator for online applications. The present study aims to discuss CDU simulator tuning methods for online applications such as inferential control and supervisory optimization. CDU simulator tuning can be made in terms of (i) calculating crude TBP curve (ii) incorporating stage efficiencies as a model tuning parameter and (iii) tuning property package with industrial data. Section 2 describes in-house developed CDU simulator, online and offline calculation of crude TBP is mentioned in section 3. Efficiency tuning procedure is provided in section 4 and section 5 describes property package used. Finally results are provided in section 6 and conclusions are drawn in section 7.

2. Crude Distillation Unit Simulator

A steady state model had been developed in-house for the CDU and that would be referred to as the simulator. A theoretical analog of the actual column was developed for the purpose of simulation. Theoretical analog is defined as one containing the required number of theoretical stages between consecutive product withdrawal positions in order to obtain the same characteristics of the product streams as obtained from the actual column. In practice, finding a theoretical analog is a trial and error procedure. For the column used in the present study, 42 analog stages were required to represent 81 actual stages. Although the analog column used in the present study reasonably approximates the plant behavior, it was found necessary to fine tune the simulator by manipulating efficiencies for better quantitative predictions. The condenser is the first equilibrium stage followed by the rest of the stages numbered from top down. The simulator takes TBP, specific gravity, and flow rate of feed, coil outlet temperature, column pressure at various points, pumparound positions, their flows and return temperatures, side stripper connectivity with main column, product flow rates, reflux flow and its temperature, nature of condenser, steam inputs, efficiencies of the trays etc as inputs.

Crude oil, being a complex mixture of a large number of hydrocarbons, is difficult to analyze in the laboratory. It is, therefore, customary to divide the feed crude oil into a finite number (generally 25 to 40) of pseudo-components. Each pseudo-component is characterized by an average boiling point and an average specific gravity. The rest of the physical and thermodynamic properties are estimated using empirical correlations, equations of state, etc. The simulator uses a C+3 formulation where C is the number of pseudo-components. This implies that for every tray, the liquid phase composition, liquid and vapor volumetric flow rates and temperature are the independent variables, which are C+3 in number. Once the feed is characterized, the feed stream to the column is flashed and the C+3 equations for each of the N trays are simultaneously solved using a modified Newton-Raphson method. The equation solver has been specially developed for the sparse matrix system present in the model and it ensures a fast and robust convergence of the solution within the prescribed accuracy. The thermodynamics used for the CDU simulator is a modification of BK10 adjusted in such a way so as to suit the purpose. The simulator accepts both cold start and warm start options. In the former case, iteration starts at the initial guess and in the latter case, initial calculations begin at the converged results obtained in the previous run (Kumar and Saraf et al., 2001).

3. Feed TBP Calculation Procedure

True Boiling Point (TBP) curve is an important characteristic of crude oil feed and hence it has significant contribution to product yields and compositions. The TBP data that are available from assay of pure crudes do not provide very accurate information on actual feed for several reasons. Firstly, a typical refinery operates with ten to twelve different crudes on regular basis because of economic considerations and availability. The refinery switches crudes frequently, and two crudes get mixed in varying proportion during switch over. Some of the imported crudes are mixtures of different crudes taken from different oil reservoirs. Also in the crude storage tanks incoming crude mixes with the residual crude present which may be of the different type. Lastly, the stratification of the crude in the storage tanks causes the heavier components to concentrate in the bottom section which further alter the composition of the crude being processed as a function of time. Clearly the use of original TBP data as initially established in the laboratory for pure crudes cannot be expected to predict the fractionator performance later when the crude composition may have been changed for above reasons. This leads to erroneous property predictions and hence erroneous control action and loss of profit because of quality give away. The above problem provides a great incentive to develop a methodology to estimate crude TBP for online applications. Following are in-house developed methods to estimate crude TBP.

3.1 Offline Estimation of Crude TBP

The basic idea here is to use the product ASTM temperatures to back calculate feed TBP curve. In the laboratory, usually product ASTMs are measured which can be converted to TBP temperatures using an empirical correlation. Unfortunately product ASTMs are measured occasionally and therefore, this TBP reconciliation procedure can only be used in offline mode whenever complete product ASTM curves are available.

The objective function is formulated as a least squared minimization problem where the difference between the simulator calculated product TBPs and those obtained from laboratory measured product ASTM temperatures is minimized subject to model equations as constraints. An additional constraint arises from the fact that the feed TBP is a monotonically increasing function of the volume percent distilled. i.e. TBP temperature of the 'i' vol.% distilled is always greater than the TBP temperature of 'j' vol.% distilled for all i > j. The TBP reconciliation problem can be mathematically expressed as:

$$\underset{\text{Feed TBP}}{\text{Minimize}} \quad I = \sum_i (\text{Product TBP Measured}_i - \text{Product TBP Calculated}_i)^2$$

Subject to : TBP_j vol% of crude < TBP_i vol% of crude ; if i>j (1)

The success of nonlinear optimization based calculation of feed TBP depends upon the selection of manipulated variables (for example, number of manipulated variables should be more that shapes 0-70 vol% vaporization TBP curve), bounding strategy for manipulated TBP temperatures and selection of product TBP temperatures in objective function. The initial guess is used as laboratory measured TBP curve of crude oil. (Chatterjee and Saraf, 2004)

3.2 Online Estimation of Crude TBP

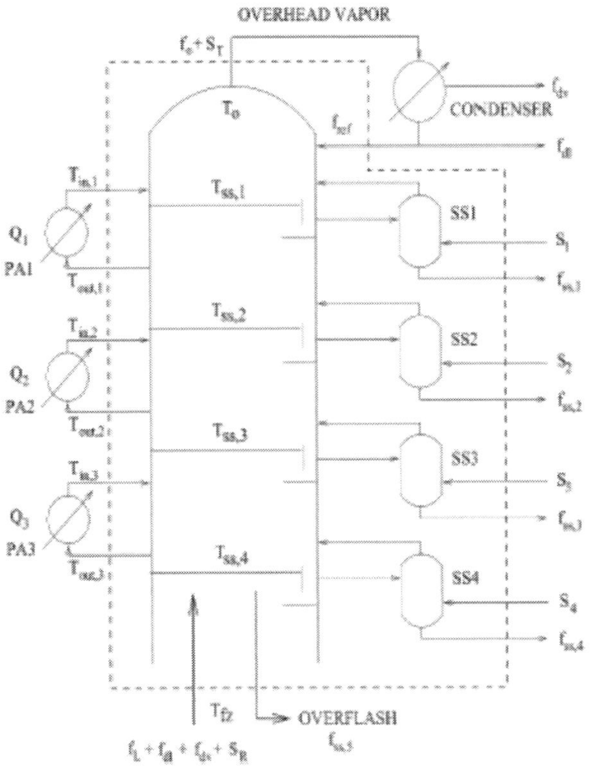

Figure 1a Heat balance around rectifying section

TBP backcalculation model for estimation of feed TBP curve has been developed by the application of heat balance around the crude fractionator's rectifying section similar to that used by Friedman (1985). The procedure provides for the estimation of a set of six points which allows plotting a more realistic TBP curve than a straight line used by Friedman (1985) and Ebbesen (1992). Figure 1a shows the rectifying section of CDU column. The dotted line shows the envelope around which heat balance is performed. At steady state the total heat balance can be written as :

$$H_{in} - H_{out} = 0 \tag{2}$$

The reference conditions for the enthalpy are chosen as the saturated hydrocarbon vapor at the flash zone temperature and pressure. This makes the enthalpy entering the envelope with the vapors from flash zone as zero. Also it is assumed that the enthalpy of the stripping steam at the inlet and outlet conditions are roughly the same.

$$H_{in} = -[H_{ref}^{V} + c_{pl,\,ref}(T_{fz} - T_{ref})]\,f_{ref} \tag{3}$$

The enthalpy leaving the column is the sum of the side stream enthalpies, overflash enthalpy, overhead vapor enthalpy and heat duties of the pumparounds.

$$H_{out} = \sum_{i=1}^{N_{ss}} -[H_i^{V} + c_{pl,\,i}(T_{fz} - T_i)]\,f_{ss,\,i} - c_{pv,\,o}(T_{fz} - T_o)\,f_o + \sum_{m=1}^{N_{pa}} Q_m \tag{4}$$

Where the pumparound heat duty $Q_m = c_{pl,m}(T_{out,m} - T_{in,m})\,f_{p,m}$ (5)

Assuming that the combined liquid product enthalpy can be calculated as if all the liquid products leave the tower at an average temperature, T_{avg}.

$$\sum_{i=1}^{N_{ss}} [H_i^V + c_{pl,i}(T_{fz} - T_i)] f_{ss,i} = (H_{avg}^V + c_{pl,avg}(T_{fz} - T_{avg})) f_L \tag{6}$$

where f_L can be defined as the sum of the total side-stripper products and the overflash stream. The mass average property, for example (T_{avg}) is given as

$$T_{avg} = \frac{\sum_{i=1}^{N_{ss}} f_{ss,i} T_i}{\sum_{i=1}^{N_{ss}} f_{ss,i}} \tag{7}$$

Similarly the other mass average properties are defined. The overflash liquid is assumed to be at the flash zone temperature, implying that on the left hand side of equation 6, $T_{i=5} = T_{fz}$, reducing the liquid enthalpy term to zero. Thus for the overflash liquid stream, the only contribution is that from the heat of vaporization, H^V_5. In view of this it is assumed that the overflash enthalpy contribution is very small and hence can be neglected.

$$f_L = \frac{\sum_{m=1}^{N_{pa}} Q_m + f_{ref}[H_{ref}^V + c_{pl,ref}(T_{fz} - T_{ref})] - c_{pv,o}(T_{fz} - T_o) f_o}{H_{avg}^V + c_{pl,avg}(T_{fz} - T_{avg})} \tag{8}$$

It is assumed that f_L responds a lot faster to changes in the vapor load in the column (Ebbesen, 1992). The measured values of top tray temperature, T_o, flash zone temperature, T_{fz}, and side-stripper column draw temperatures, $T_{ss,i}$ (i=1, 2, 3 and 4), correspond to the hydrocarbon product partial pressures in the column at appropriate locations. However, EFV and TBP curves are at atmospheric pressure, hence these temperature have to be corrected for the pressure difference. The partial pressure of the hydrocarbon product vapors is calculated from the actual tower pressure and the molar flow rate of hydrocarbons. The molar flow of each hydrocarbon stream is calculated by dividing mass flow rate by its molecular weight. E.g.

$$F_{dl} = f_{dl} / MW_{dl} \tag{9}$$

The partial pressure of the hydrocarbon vapors in the overhead is given by Watkins (1981) as

$$p_{top} = P_{top} \left(\frac{F_{ref} + F_{dl}}{F_o + S_T} \right) \tag{10}$$

The partial pressure of the hydrocarbon vapors at the flash zone is given by

$$p_{fz} = P_{fz} \left(\frac{F_L + F_{dl}}{F_L + F_{dl} + F_{dv} + S_R} \right) \tag{11}$$

The partial pressure of the hydrocarbon product vapors above the draw tray for the ith side stripper is given by

$$p_{ss,i} = P_{ss,i} \left(\frac{L_{r,i-1}}{L_{r,i-1} + V_{ss,i} - F_{ss,i-1}} \right); \; i = 1 \ldots N_{ss} \tag{12}$$

where $L_{r,i-1}$ is the hydrocarbon liquid reflux flow to the ith side stripper column draw tray (liquid flow from the tray above the draw tray) and $V_{ss,i}$ is the total hydrocarbon vapors and steam flow leaving the ith side stripper column draw tray. The flow rate of the previous side-stripper product ($F_{ss,i-1}$) is neglected from the total hydrocarbon vapor leaving the draw tray. The reason behind this is that the product vapor, $F_{ss,i-1}$, is near its critical temperature as it leaves the draw tray under consideration and hence, it is assumed to have no effect on partial pressure. But the product vapors which are to be withdrawn from the side-strippers above the previous side-stripper, i.e. $F_{ss,i-1}$, are above their critical temperatures when they are at the temperature of this draw tray. Thus, like the steam, they behave as non-condensibles and lower the boiling point of the product in accordance with Dalton's law of partial pressures. For the calculation of hydrocarbon partial pressure at the first side-stripper draw tray, liquid distillate (corresponds to $F_{ss,i-1}$ for this case) must be subtracted from the total hydrocarbon vapors leaving the draw tray. The vapor distillate still behaves as non-condensible and must be accounted for.

The tray temperatures measured from the column are corrected to atmospheric pressure using Clasius - Clayperon equation. Thus the top temperature, T_o, when corrected corresponds to the dew point of the overhead vapors (liquid distillate), T_{oc}, while the draw tray temperatures, $T_{ss,i}$ (i=1, 2, 3 and 4) when corrected correspond to the bubble points of unstripped liquid hydrocarbons on the respective draw trays, $T_{ssc,i}$ (i=1, 2, 3 and 4). Also the correlated flash zone temperature, T_{fz}, correspond to the bubble point of the overflash liquid, T_{fzc}, in the flash zone. Thus these temperatures form the cut points on the EFV curve corresponding to the respective cumulative volume percent distilled.

The above determined EFV temperatures are correlated to TBP temperatures of feed using empirical relations. It was found that the thermodynamic properties, required in equation 8, were sensitive to the composition of the hydrocarbons. Thus first the simulator was used with initially available laboratory TBP for generating those thermodynamic properties and then model was used to backcalculate six points on TBP curve. The six points generated were joined with straight lines. As in the backcalcultion, it is not possible to predict the initial and final points on TBP curve, these were taken to be the same as in the initially available laboratory TBP curve. After backcalcultion the simulator was operated with backcalculated TBP curve to generate the product TBPs. (Dave and Saraf *et al.*, 2003)

4. Online Efficiency Tuning

The simulator equations are based on ideal conditions where it is assumed that phase equilibrium is attained on each stage, but it is seldom achievable in real life owing to a variety of column non-idealities. Moreover, no single thermodynamic model can successfully predict the thermodynamic properties in the entire range of boiling points with the same accuracy. These reasons lead to deviation of the simulator results from the plant data. Murphree tray efficiencies are traditionally mentioned in the literature to account for deviation from the equilibrium. However, there is no easy way to calculate these efficiencies and certainly not on-line. Moreover these efficiencies do not guarantee a perfect match with the plant performance. Certain tuning factors were, therefore, associated with each stage to enable the simulator results match the available measured plant data. These tuning factors are introduced in the equilibrium relations in the same way as the point efficiencies (Treybal, 1981) and can, therefore, be referred to as point efficiencies and are defined as :

$$y_{i,j} = K_{i,j} \times \eta_{i,j} \times x_{i,j} \qquad (13)$$

Here $\eta_{i,j}$ is the point efficiency or the tuning parameter for jth component on ith stage. It is further assumed that this factor is same for all the components at all locations on a single stage. This implies that only one η_i is used for each stage and referred to as stage efficiency.

These stage efficiencies or tuning parameters can be obtained by solving the following nonlinear programming problem (Dave and Saraf *et.al.*, 2003).

$$I = \text{Min}\ [\Sigma\ (\text{Column parameters}_{\text{measured}} - \text{Column parameters}_{\text{calc}})^2]$$

$$\text{Subject to}: \quad \eta_{I,}\ \text{Model equations and}$$

$$\eta_i^{\,minm} \leq \eta_i \leq \eta_i^{\,maxm} \qquad i=1, 2, \ldots, p \qquad (14)$$

Column Parameters Used

1. Distillate flow rate, (m^3/h) , 2. Top tray temperature, (°C), 3. HN pumparound draw tray temperature, (°C), 4. SK pumparound draw tray temperature, (°C), 5. LGO pumparound draw tray temperature, (°C)*, 6. Flash zone temperature, (°C), 7. HN side-stripper draw tray temperature, (°C), 8. SK side-stripper draw tray temperature, (°C), 9. LGO side-stripper draw tray temperature, (°C)*, 10. HGO side-stripper draw tray temperature, (°C), 11. Bottom tray temperature, (°C)

* Both temperatures correspond to the same tray

Design Variables

If each stage is assigned efficiency, the optimization problem will have 42 manipulative variables. However, it can be argued that for a reasonable solution, the number of manipulative variables should be less than the column parameters and certainly should not exceed this number. In view of only 10 measured variables being available, it was decided to estimate only 8 efficiencies. The rest of the efficiencies were assigned by linear interpolation with a view to obtain a smooth temperature profile. Thus tuning of the simulator became an eight variable constrained optimization problem with least square objective function as given in equation 19. The equality constraints are the nonlinear algebraic equations of the model used for the simulation of CDU with the equilibrium relation as modified in equation 13.

5. Property Package Tuning

The refiners are concerned about improving the economics as long as products are within the specifications and plant operating conditions are within allowable limits. Several products obtained from CDU, such as distillate, heavy naphtha (HN), superior kerosene/ aviation turbine fuel (SK/ATF), light gas oil (LGO), heavy gas oil (HGO) and residual crude oil (RCO) are obtained through top condenser, side strippers and bottom. Each of these products are characterized by one or more properties such as: Reid Vapor Pressure (RVP) for distillate and HN, flash point for HN and SK/ATF, end point for SK/ATF, recovery at 366°C for HGO and RCO. These product properties are strong functions of product composition achieved from simulator (such as product TBPs, molecular weight, specific gravity and so on). Empirical correlations are developed with large number of product properties data obtained from laboratory and the coefficients for these correlation equations are tuned.

6. Results

In the present study, three different crudes, one each from Persian Gulf, Indian origin and Arabian origin have been considered and are denoted as Crude – 1, 2 and 3 respectively in further discussion. The TBP data of these crudes are available from the crude assays. From Figures 2 it is clearly seen that the backcalculated feed TBP matches well with the reconciled TBP. The maximum observed deviation in TBP temperatures for all the cases is less than 8 K, which is 1.28 % with the average deviation being less than 1%. It was observed that the match is closer in the middle region of the TBP curve where the six TBP points were backcalcuted but not in the initial and final regions. The larger deviation of TBP temperatures at lower and higher volume percent distilled is due to the fact that the TBP curve is highly non-linear in these regions but the points are joined by linear interpolation. An attempt was made to join these points by cubic spline method, which aggravated the problem by taking arbitrary excursions in between the calculated points and in the two end regions. The region from 70 to 100 percent distilled contributes to RCO which is not of much economic interest; hence this product is not usually required to meet any stringent properties. Moreover, accurate TBP data are seldom available for this region. However, the 0 to 20 volume percent distilled region is of significant importance since product properties of light distillates depend heavily on this part of the curve. It should be possible to calculate one additional point in this region if plant data about fraction of vapor product in the distillate is known. However, in the present work this could not be done because of lack of the required information.

After backcalculating the feed TBP, the exercise of tuning the simulator by efficiencies is undertaken to match the performance of the simulator with that of plant. Figure 5 show comparison of the calculated temperature profiles before and after TBP back calculation and efficiency tuning. From these figure it is observed that the temperature profile obtained without TBP backcalculation and efficiency tuning deviates substantially from the measured values, at times by as much as 30°C. The match, however, improves significantly after TBP backcalculation and incorporation of the efficiencies. Table 1 presents the comparison of the predicted properties before and after tuning along with the laboratory data for Crude 1. In actual practice, all the product properties are not measured at a given instant, so an average value of some of these have been included in the last column of this table. The properties of distillate (density and RVP) have been compared with the corresponding properties of straight run naphtha and hence cannot be expected to match. It can be seen that after tuning, the simulator predicted properties match well with the laboratory results. The large deviation in the properties before tuning is due to the fact that the simulator predicts improper distillate flow rate, thus shifting the cut points in TBP curve. This makes the products lighter or heavier increasing the deviations in the properties. In the tuning process the distillate rate gets improved upon along with the temperature profile of the column thus improving the prediction. Similar results were obtained for the other two crudes and are therefore, not included.

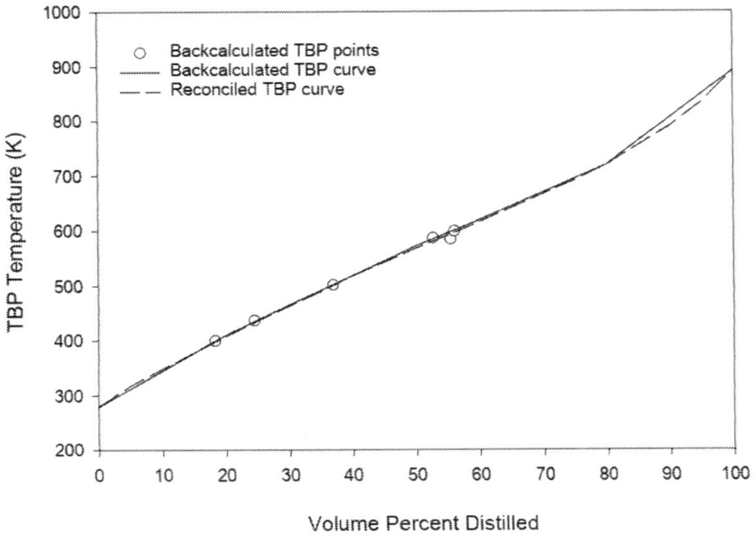

Figure 2 Comparison of backcalculated feed TBP curve and reconciled feed TBP curve for Crude - 1

Table 1: Comparison of product properties before and after tuning the simulator for Crude – 1

Product Properties	CDU Simulator			Measured values
	Before TBP back calculation	After TBP back calculation	After TBP back calculation and Tuning	
Top distillate density, (gm/cm^3)	0.600	0.732	0.732	0.740*
Top distillate RVP, (atm)	0.726	0.539	0.531	0.440*
HN density, (gm/cm^3)	0.770	0.780	0.783	0.785
HN RVP, (atm)	0.079	0.050	0.046	0.040†
HN Flash point, (°C)	2.4	10.3	11.3	11.5†
SK density, (gm/cm^3)	0.801	0.811	0.813	0.815
SK Flash point, (°C)	32.5	42.3	42.8	43.0
SK End point, (°C)	236.0	249.1	249.6	240.0
LGO density, (gm/ cm^3)	0.846	0.857	0.858	0.860†
LGO Flash point, (°C)	56.1	67.1	67.5	74.0†
LGO Pour point, (°C)	-19.5	-17.3	- 3.2	NR
HGO density, (gm/cm^3)	0.875	0.888	0.889	0.883
HGO Flash point, (°C)	74.6	93.9	93.8	96.4
HGO Pour point, (°C)	-11.9	- 5.1	-4.4	NR
HGO Recovery at 366 °C, (%)	79.16	73.82	73.27	88.0†
RCO density, (gm/cm^3)	0.963	0.962	0.969	0.970
RCO Recovery at 366 °C, (%)	13.25	12.75	12.74	18.00†

NR : Not reported in experimental set of values

* : Properties pertaining to SRN

† : represents average value

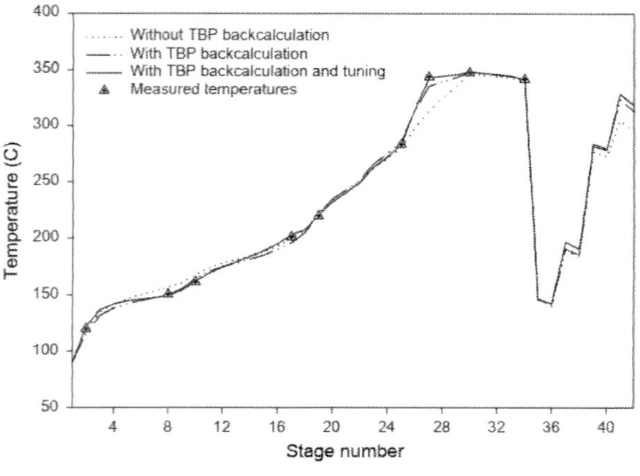

Figure 5 Comparison of calculated temperature profile before and after tuning with the measured values for Crude - 1

Conclusions

The properties calculated by steady state simulators deviate from actual properties because of several reasons. Various CDU simulator tuning procedures to match product properties with industrial data are mentioned. Online and offline crude TBP calculation along with tuning of CDU simulator with efficiencies improves the performance of package. The backcalculation model has been tested for different crudes and it has been observed that the deviation between the reconciled and backcalculated feed TBP is small, and hence validating the model within the required accuracy. Tuning of the simulator (with tray efficiencies) has been achieved by minimizing the error between calculated and measured column parameters. It has been observed that the predicted product properties match reasonably well with those measured in the laboratory. The backcalculation model coupled with efficiency calculation can be used to tune the simulator for on-line optimization, to improve overall CDU performance. For effective feedback control of CDU to ensure product quality control it is necessary to predict product properties using soft sensors online. The model based sensors depend on the simulator output and in order that the properties are accurately determined the simulator must be tuned using online operating data. The existence of steady state has been assumed in the development of the simulator as well as the backcalculation procedure and it is, therefore, necessary to ensure that the plant is operating under steady state conditions before proceeding to use these models.

Reference:

Dave, D.J.; Dabhiya, M.Z.; Satyadev, S.V.K. ; Ganguly,S. and Saraf, D.N. "Online tuning of a steady state crude distillation unit", J. Process Control, **13**, pp 267–282 (2003).

Chatterjee, T. and Saraf, D.N. "On-line estimation of product properties for crude distillation units", J. Process Control, **14**, pp 61-77 (2004).

Ebbesen, L.E. Main fractionator crude switch control", European Symposium on Computer Aided Process Engineering –1, *Computers and Chemical Engg.*, **16**, pp S165 – S177 (1992).

Friedman, Y.Z. Control of crude fractionator product qualities during feedstock changes by use of a simplified heat balance, Paper presented at American Control Conference, pp 1084 – 1089 (1985).

Kumar, V., Sharma, A., Choudhury, I.R., Ganguly, S. and Saraf, D.N. A crude distillation unit model suitable for online applications, *Fuel Process Tech.* in press (2001).

Treybal, R.E. Mass-Transfer Operations, McGraw-Hill Book Company, Singapore, Third edition (1981).

Watkins, R.N. Petroleum Refinery Distillation, Gulf Publication Company, Houston, Second edition (1981).

DETERMINATION OF THE CONCENTRATION OF LUBE OIL IN PARAFFINS USING NIR SPECTROSCOPY

PHILIPPE IWANSKI, LUCIANO MONTANARI and ROBERTO GIARDINO

EniTecnologie, v. F. Maritano, 26, 20097 San Donato Milanese (Mi), Italy. Fax +39-02-52036347 / +39-02-52056612.

Key words: NIR, paraffin, lube oil, PLSR.

Prepared for presentation at the 2004 Spring National Meeting, New Orleans, LA, April 25-29
7th Conference on Refinery Processing
Session TA012 Application of Control to Refining

December 2003
Unpublished

ABSTRACT

Near infrared (NIR) spectroscopy and multivariate analysis were applied to estimate the oil content in paraffin streams. The technique could be applied to continuously monitor the slack wax oil content in a base oil manufacturing plant, thus providing a fast feedback to solvent dewaxing operations and maximisation of the yield of base oil. Its application to control the hard wax production quality can also be investigated. The regression models, based on industrial hard and slack waxes, were validated using cross-validation methods. Good correlations between NIR spectroscopic data and oil concentrations were obtained: the average prediction error (SDEP) in the determination of oil concentrations down to 1% Wt. was close to ± 0.09 % .

INTRODUCTION

Slack waxes are the by-product of the dewaxing process of lube oils from vacuum distillation of crude oil and deasphalted vacuum residue. They are made up of a combination of mainly saturated hydrocarbons with higher melting (linear paraffins) and low melting (iso-paraffins and entrapped lube oil) components.

Slack waxes can contain varying amounts of oil (usually ranging from 1 to 20% Wt. wt) and must be de-oiled to produce hard waxes. The oil content of slack waxes mainly depends on filtration and washing operation efficiency which in turn depend on operating conditions.

In hard waxes the oil content affects significantly several of their properties (strength, hardness, flexibility , melting point, etc.) that can be important depending upon their ultimate use. The oil content of hard waxes is usually lower than 1% Wt. wt.

The chemical properties and structure of petroleum products can be well described through spectroscopic techniques. In particular near infrared spectroscopy (NIR) combined with multivariate data analysis offers many interesting perspectives in the advanced control of petroleum processes due to high speed computations and reliability [1-4]. Moreover, the use of fiber optics allows it to be easily used on-line. The basis of the process involves establishing relationships (models) between absorbances in the near infrared and chemical-physical characteristics of the products.

The most studied application in the petroleum field is gasoline and gasoil quality prediction, mainly for efficient control of fuel blending plants, but applications to other petroleum products and processes have also been described [5-9].

In this work the NIR/PLSR (Partial Least-Squares regression) technique is applied to estimate the oil content in paraffin streams. A feasibility study has been reported in a previous work[9] using laboratory mixtures of base oil in paraffin as calibration samples. The present study has been, instead, carried out on industrial samples of slack and hard waxes.

The technique could be applied to continuously monitor slack wax oil content in a base oil manufacturing plant which is currently determined using laboratory spot analysis, i.e. once a day, according to ASTM D721 method. A fast feedback will allow to adjust in real time the dewaxing operating conditions in order to minimize the oil content in slack waxes and, consequently, to increase the dewaxed base oil yield. A more accurate control will also allow continuous operating of the unit at its maximum capacity without losing oil trapped with the removed waxes. The technique could also be applied to control the de-oiling/fractionation section of the hard wax production cycle in order to avoid production out of the desired specification.

EXPERIMENTAL SECTION

Three calibration sets, corresponding to different paraffin cuts, were used to test the system NIR/PLSR. Two sets have oil contents between 0.3 % Wt. and 1.5 % Wt. . One set has a higher concentration range:

- 33 paraffin samples (Slack wax from heavy fraction) with high oil contents from 2.4 to 16.5% Wt.
- 15 paraffin samples (Hard wax from medium fraction – melting point 133/137 °F) with low oil contents between 0.2% Wt. and 0.6% Wt. .
- 23 paraffin samples (Hard wax from heavy fraction – melting point 145/148 °F) with low oil contents between 0.5% Wt. and 1.5% Wt. .

The oil content of petroleum waxes has been determined according to ASTM D721 method. The sample is dissolved in methyl ethyl ketone. The resulting solution is cooled to –32°C to precipitate the wax and filtered. The oil content of the filtrate is determined by evaporating the methyl ethyl ketone and weighing the residue. The repeatibility of the reference method is ± 0.06 units (% Wt.) plus 8% of the measured value.

Near-infrared transmission spectra were recorded between 9200 and 6200 cm^{-1} (1100 - 1600 nm) at 4 cm^{-1} resolution with a Bruker Equinox 55 Fourier transform (FT-NIR) spectrometer. The spectrometer was configured with a tungsten-halogen lamp as the light source and a PbS detector. Two 2.5 m long optical fibers connected the spectrometer to the heated steel measurement cell with 2 cm optical pathlength. The cell temperature was regulated at 90°C with a band heater and a thermocouple connected to a temperature controller. The samples were melted at 90°C and carefully stirred before introduction in the measurement cell. The cell was cleaned with n-heptane between measurements. Sixteen spectra were averaged for each sample. A reference spectrum (air) was recorded before each measurement. The spectra were baseline corrected and normalized according to the mean normalization method. The typical spectra of the samples are presented in figure 1.

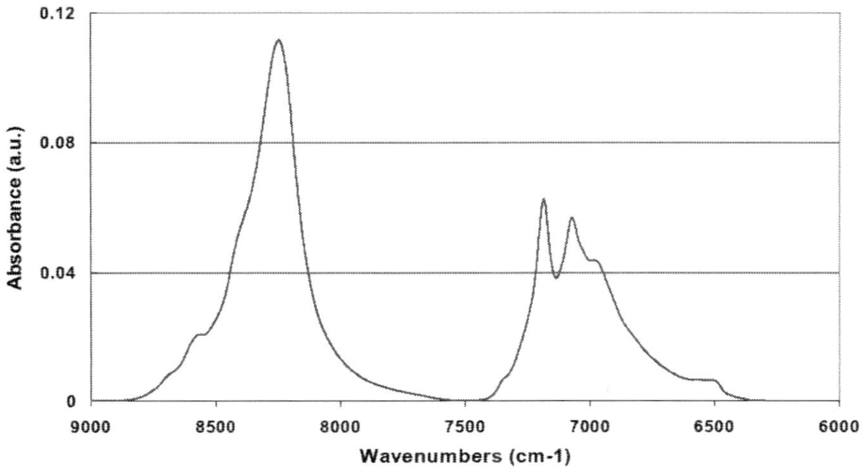

Figure 1. NIR spectra of paraffin samples with oil: 23 samples Hard wax (heavy fraction)

The absorption bands in the range 9000-7500 cm^{-1} correspond to the second overtone of the fundamental stretching vibrations of the C – H bonds in hydrocarbons. The bands in the 7500 - 6500 cm^{-1} region are combination bands (stretching C-H + bending C-H in methyl and methylene groups).

PLS methods were used to compute predictive models of base oil content in paraffin using a program developed by CAMO (Trondheim, Norway): the UNSCRAMBLER version 7.0.
The absorbance values in the NIR spectra of the *n* calibration samples were used as independent variables (X). The oil content of the samples was the dependent variable (Y). The relationship between the NIR spectrum of a mixture and its oil content is given by the equation :

$$\%\ \text{Wt. oil} = \sum_{i=1}^{n} [B_i . A(\lambda_i)]$$

where $A(\lambda_i)$ is the absorbance at wavelength λ_i and the B_i are the regression coefficients calculated with the PLS algorithm. The predictive ability of the models was evaluated through the standard deviation of the error of prediction (SDEP) value , defined as follows [10]:

$$\text{SDEP} = \sqrt{\frac{\sum_{i=1}^{n}\left(Y_{i,ref} - Y_{i,calc}\right)^2}{n}}$$

where $Y_{i,ref}$ and $Y_{i,calc}$ are, respectively, the weighed and the calculated oil content in the ith sample. The calculated Y were obtained by leave-one-out (LOO) method or by grouped cross-validation method [10].

RESULTS AND DISCUSSION

Repeatability of the NIR measurements

Two samples of hard wax with different oil content (0.3% Wt. and 1.2% Wt.) were repeatedly measured in order to estimate the NIR statistics. The standard deviation and the confidence have been estimated at two wavelengths (8250 cm^{-1} and 7180 cm^{-1}) of the spectrum corresponding to different absorbance intensities. An estimation of the practical discrimination between the samples is given by a comparison of the averages. A Student test at a confidence level of 95% [$t_{0.05(2),10} = 2.228$] has shown that the observed averages are significantly different in both the spectroscopic regions ($t_{8250\ cm-1} = 29.95$ and $t_{7180\ cm-1} = 5.31$). In order to see the influence of the NIR measurement on the precision of the actual predicted oil percentage, a statistical estimation has been made on the values obtained with the cross validation routine of the PLS regression step. The repeated measurements were included in the calibration set one at a time. The results are reported in table 1.

Sample **1.2% Wt.**	**Predicted % Wt. (Latent variable n. 1)**	Sample **0.3% Wt.**	**Predicted % Wt. (Latent variable n. 1)**
1.2_1	0.873	0.3_1	0.299
1.2_2	0.925	0.3_2	0.227
1.2_3	0.881	0.3_3	0.384
1.2_4	0.889	0.3_4	0.375
1.2_5	0.981	0.3_5	0.246
1.2_6	0.911	0.3_6	0.262
average	0.91		0.30
Standard deviation	0.036		0.067
confidence	0.03		0.05

Table 1. NIR measurements statistics.

There is a systematic error of 0.3 units for the sample with higher oil content (1.2% Wt.) but the accuracy is depending also on the reference method and on the goodness of fit of the regression model. The precision, which, instead, depends only on the NIR measurement, is good in both cases. The confidence is ± 0.03 units for the sample at 1.2% Wt. and ± 0.05 units for the 0.3 % Wt. sample. This value was used to represent the error bars in the plots of predicted versus reference values.

Multivariate regression models

Classification based on Principal Component Analysis (PCA) was made on the complete data set. It can be seen (figure 2) that the paraffin grade has a relevant influence on the classification, comparable to the oil content. In this case a regression of the oil percentages on the principal components derived from the complete data set does not give a good fit for the low oil values.

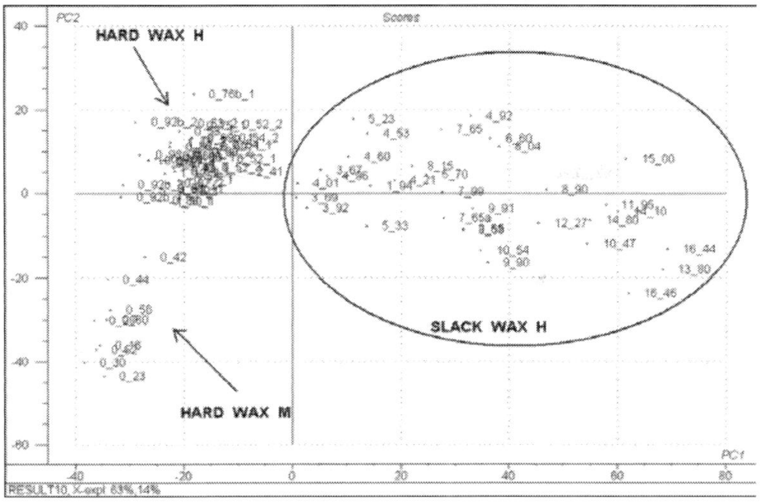

Figure 2. PCA analysis of paraffin samples

Because of this relevant influence of the paraffinic fractions, the regression models were elaborated considering the sample sets separately.
A good linear regression model of the oil concentration on the first 3 latent variables was obtained with the samples of slack wax (heavy fraction). A plot of the calculated values versus the reference values of the oil content is reported in figure 3.

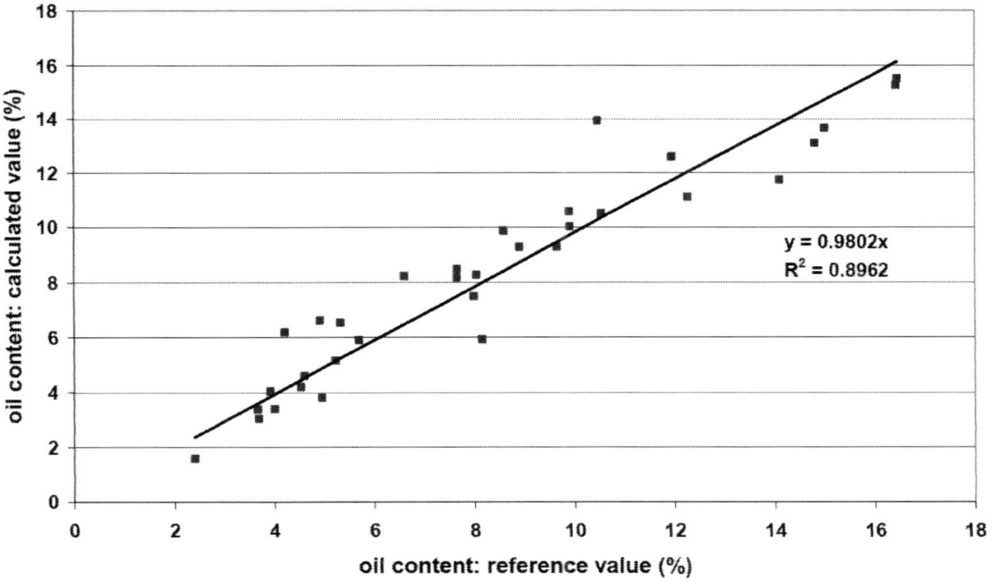

Figure 3. Prediction of oil content in the slack wax samples (heavy fraction).
Plot of the predicted values (leave-one-out cross-validation) versus reference values.

The error of prediction (SDEP), calculated with the leave-one-out (LOO) cross validation method, is better than 1.3 % Wt. . Model validation has also been checked through iterations of a segmented cross-validation (11 groups) and resulted in SDEP values ranging from 1.2 to 1.3 % Wt. , pointing out a stable model. The regression coefficients corresponding to the model with three latent variables are plotted together with a rescaled spectrum in figure 4.

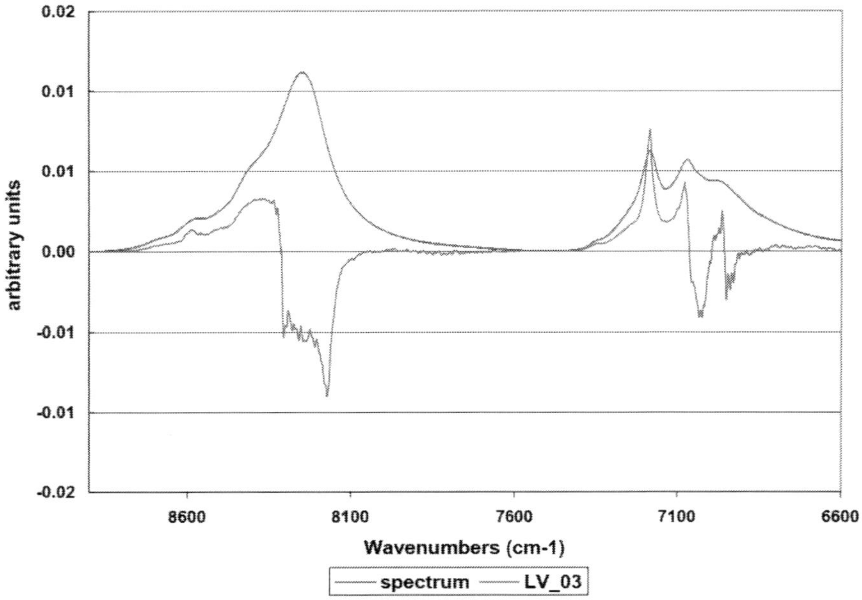

Figure 4. NIR spectrum and regression coefficients (slack wax)

The SDEP value appears to be somewhat high when compared to the value of 0.31 % Wt. reported in the feasibility study [9]. This result can be explained by the fact that the calibration samples used in the previous work had oil concentration values obtained by precise weighing during the laboratory preparation of the mixtures. In the present work, instead, the oil content determination is actually affected by a much larger error due to the reference method accuracy and precision limits.

In the case of the hard wax (heavy fraction) samples, the regression was made on the first latent variable and the validation (LOO) gave a SDEP value of 0.09 % Wt. in oil concentration but pointed out the presence of outlying samples at the highest oil content in the set (1.4 % Wt. and 1.5% Wt.). The application of more severe segmented cross-validations (9 and 6 groups), after the exclusion of the outliers, gave SDEP values in the range 0.08 to 0.1 % Wt., yet indicating a stable model (figure 5).

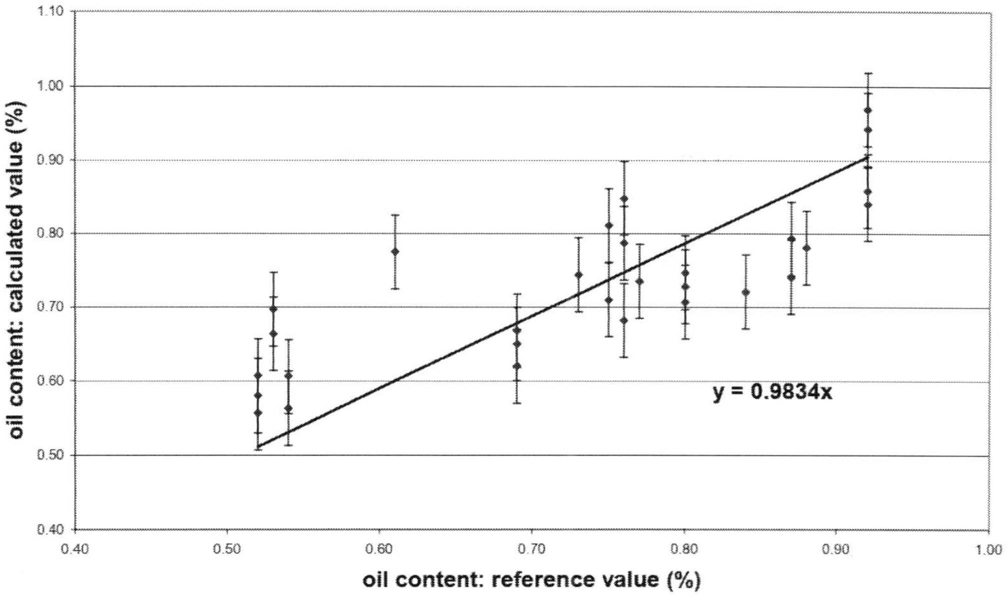

Figure 5. Prediction of oil content in the hard wax samples (heavy fraction). Plot of the predicted values (leave-one-out cross-validation) versus reference values

The regression model for the hard wax (medium fraction) sample set was also built up on the first latent variable. The cross-validation (LOO and segmented) resulted in SDEP values in the range 0.07 to 0.09 % Wt. after the exclusion of 2 outliers.

The presence of a higher number of outliers in the low oil concentration samples can be caused by the more difficult discrimination between oil and paraffin in the narrow range of oil percentages and by the intrinsic errors of the reference method for oil determination.

CONCLUSIONS

Although the absorption and combination bands of the various types of hydrocarbon C-H bonds are broad overlapping features in the near-infrared region of the spectrum, a multivariate statistical method such as PLS can be applied to near-infrared spectra to predict the content of oil in a paraffin sample. The advantage of the NIR/PLS approach is its speed and simplicity compared to more traditional methods. Moreover, the use of fiber optics allows it be used on-line.

The standard deviation of the residuals of prediction (1.3 % Wt. for the slack wax in the range 2 - 16% Wt. and 0.09 % Wt. for the hard waxes in the range 0.2 – 1 % Wt.) is good compared to the repeatibility of the ASTM method. The stability of the models was checked with different cross-validation methods.

BIBLIOGRAPHY

1. Welch, W. T.; Bain, M.L.; Maggart, S.M.; May, J.M. Experience leads to accurate design of analysis systems. *Oil & Gas Journal*, **48**, 51 (1992)
2. Maggard, S. Process for predicting properties of multi-component fluid blends. WO Patent 17392-A1 (1994)
3. Michael, J.H. Method for prediction of physical properties data of hydrocarbon products. WO Patent 17391-A1 (1994)
4. Espinosa, A.; Lambert, D. ; Martens, A; Ventron, G. Method for the direct determination of physical properties of hydrocarbon products. EP Patent 0304232-A2 (1988)

5. Lambert, D.; Descales, B.; Bages, S.; Bellet, S.; Llinas, J.R.; Loublier, M.; Maury, J.P.; Martens, A. Optimize steam cracking with online NIR analysis. *Hydrocarbon Processing*, **74** (12), 103 (1995)
6. Espinosa, A.; Lambert, D.; Valleur, M. Use NIR technology to optimize plant operations. *Hydrocarbon Processing*, **74** (2), 86 (1995)
7. Chimenti, R.J.L.; Halpern, G.M. Method and system for controlling and optimizing isomerization processes. US Patent 5404015 (1995)
8. Faraci, G.; Giardino, R.; Janis, B.; Iwanski, P. Procedimento per produrre basi lubrificanti raffinate. IT Patent 1269182 (1997)
9. Iwanski P., Giardino R. and Janis B.- Dewaxing Process Control Using Nir Spectroscopy - Industrial & Engineering Chemistry Research - **38**, 571-574 (1999)
10. Cruciani, G; Baroni, M..; Clementi, S.; Costantino, G.; Riganelli, D.; Skagerberg, B. Predictive ability of regression models. Part I. *J. Chemometrics*, **6**, 335 (1992).

FOUNDATION FIELDBUS IN AN ON/OFF WORLD

by

Jim Beals
Emerson Process Management

Prepared for Presentation at the

AIChE 2004 Spring National Conference
New Orleans, LA

Session TA012
Paper 94f

Abstract
You're the first to the planning meeting for the company's next big project, new DCS, new piping, new wire, and new smart devices with digital communications everywhere. Shortly after you arrive, the new Plant Manager walks in with her management team, Engineering, Operations, Instrumentation, Maintenance, and Purchasing, to name a few. The purpose of the meeting, to decide which communication protocol (or bus) will be used, where it will be used, why it will be used, and cost. You're up first so you take a deep breath, smile, just a little and dazzle them with brilliance, especially when it comes to the discrete valves and Foundation Fieldbus, thanks to the last AIChE meeting you attended.

This paper is about several companies and the groups of individuals that went through this process and why they chose to implement Foundation FieldBus on their On/Off valves.

Kickoff

I walked into the conference room a few minutes early and utter a small, but audible, "Hello?" as I walk through the door. Nobody responds, good, that means I am early. I have everything I need for the meeting, I know what I am talking about, and feel very confident that I can "Dazzle them with brilliance." However, I have also heard the stories: the table pounding "exclamation point", the time he broke his glasses when he threw them across the room, the Vice President that was "wasting time" and told to come back when he had something substantial. To make it worse, he was right every time, he is always right, just ask anybody. His "bedside manner" may not be the best but he is one of the best in our industry and he is certainly the best in our company.

I think to myself "I have an hour, if I get past the first 15 minutes I should be OK."

I hear some noise outside and lean back to see down the corridor, here they come, Richard Dai our President and his 15 (maybe 20) person entourage, including my boss, his boss, and who knows how many more layers of bosses and assistants. I am not sure why this presentation has all this attention, guess I shouldn't have pushed so hard that we can actually save resources, especially money, by spending some extra up front.

My boss introduces me to Richard; we exchange a few pleasantries before we are interrupted by one of his assistants. It's my break to allow me to get to the front of the room so I am ready when the meeting starts, and it will start on time, they always do.

"Ladies and gentlemen, it's 11:00 o'clock, let's begin" and so it begins.

Don't worry, I won't give you a blow-by-blow of the meeting here, just some of the facts. I told them who I was and what we were going to discuss. I asked them to let me finish my presentation before we get into too many questions, as most of their questions would be answered as part of my presentation (that got an interesting look from my boss). I immediately hit them with the main reason they were here, money. We are going to spend money up front to save money down the road. This is not new to them that is why we are installing the system, however, it is new to them at the device or instrument level.

I stepped back a little and brought them up to date on the project. We are installing a new control system for our plant and at the same time automating all the processes. This in and of itself is a large investment. To go from manual / conventional control (we have some traditional 4-20 and pneumatic controls in place) to full automation is a big step. However, when we look at the details we find that the incremental spend due to automation is relatively small. We have analog control and monitoring, but being a batch plant, we have mostly on/off control. The justification for using Foundation FieldBus on the analog part of the plant has already been made. Now we are taking some of that justification and using it here.

Direct Cost

When we decided to implement a communications protocol (often referred to as a bus) in the plant, we discussed the cost savings due to wiring, panels, I/O, installation, commissioning, etc.. The savings are basically the same no matter what protocol we use. Although some of the differences between protocols does affect how much we save when you look at the overall implementation. We'll discuss this more in a minute.

Since we are taking the opportunity to upgrade valves and transmitters as part of this automation project, that element is constant. We need a valve and actuator no matter which protocol we implement. We need the chosen transmitters, as well as indicators, switches, etc. and they are independent of the communications.

The cost difference is best seen in the Valve package. The valve and the actuator are still required. The actuator vendor may change due to our chosen path based on what is available in the marketplace; however, the cost to us is basically the same. The valve controller, the control and feedback part of the valve package, is where we see the cost difference. This cost difference will be basically the same independent of the size of the valve package. Although on a percentage basis, it is higher on the smaller valves than the larger ones.

A traditional implementation with solenoid and switches (and other bits and pieces) costs us about the same as an implementation using a simple communication protocols. We also get about the same information from either implementation, very little. We have a control (output) function and two position (input) indications. Depending on the bus or traditional implementation, we may get a couple of simple status bits to indicate the health of the wire to the solenoid and maybe the switches, but that is basically all.

With a more robust protocol like Foundation FieldBus, we can get more information. Notice I said "can," as in we can but we don't have to. If we want to control and monitor our valve package with just the basic information we "can" or we can use other information available like temperature, health, position, & diagnostics to name a few.

This brings us to our next area of discussion. Note when you are reading through the next sections how every area has a cost factor associated with it. Although not usually identified, it is there.

Information

As noted above, with Foundation Fieldbus we have more information about our processes and about the instrumentation that makes up the processes at our disposal. Things like actual valve position in an on/off valve, the internal temperature of the instrument, how many times the valve, actuator, controller and/or set point have changed direction (i.e. cycled), what to do if there is a fault, who last calibrated the instrument when and where, and so on.

There is also the expectation factor. For example when our operators and maintenance staff start working with the analog devices and the information that is available from them, they will start to expect the same information from the other devices as well.

This becomes even more obvious when we look at our host system and the overall system implementation. Don't misunderstand, this isn't bad, but every device on our system looks the same once it is commissioned. What I mean by this is that from an operator's perspective an AS-i device looks the same as a Foundation FieldBus device at the operator interface level. This isn't a problem until the operator tries to bring up temperature or cycles on an AS-i device and can't.

Along with the expectation factor is the discovery factor. We are implementing something we haven't really done before so we don't know what to expect. We have had the training, however, until we actually use the equipment and see what it can do, we don't know what else we will want it to do. Try to remember when you got your first word processor on your PC. You knew it could do typing, formatting, printing, but then you found styles, and the different fonts, and colors, and different ways of printing, page layouts, cut & paste, and....you get the picture. Now when you use a different PC or application you "expect" the items you "discovered" to be present. When they are not, it takes longer for you to do what you are trying to do.

This brings us to the next area of discussion:

Training

Although many of the communication protocols are similar and at a high level, the training is the same, when looking at the details, they are different and require a different skill base depending on the protocol.

When we implement different communication protocols, we now have to make sure we have someone trained to work with each protocol, and, based on the number of units we are installing, we will need 2 or 3 people trained in each protocol. We also need the training from 2 different perspectives, operator and maintenance. The operator cares more about the information and what he sees on his console. The technician also wants information, although it is usually different than what the operator needs as well as the physical information.

When we use just one or two protocols, we can train everyone at the same time and they can backup each other. In fact, we can bring in our system vendor and get the training on site using our equipment. When we have multiple protocols, we have to send people to schools and learn on generic information.

As noted above, the technicians require information when it comes to protocols, and the protocol determines the information and features available for maintenance, which brings us to our next topic of discussion:

Maintenance

Maintenance is composed of several pieces, the training aspect above is one of them. When working with protocols, the big differences are at what is known as the physical layer; the wiring, the signal, power, connections, and I/O. These vary from one protocol to another and require a different knowledge base when working with the protocol. For example, AS-i uses a different voltage than Foundation Fieldbus and they have different waveforms when signaling. AS-i communications is high speed therefore it can only be used in relatively short distances before needing additional equipment. Also since it is high speed, the wiring used and implemented becomes more critical. The technicians need to be trained on these nuances.

Another piece of the maintenance pie are the tools used to support the protocol. Not only are they different (you can't use a Foundation Fieldbus handheld on an AS-i segment), but how they operate is also different (there's that training thing again). Each protocol has one or more tool used to support it and although some of the tools are the same (Multi-meter and oscilloscope) many are different (bus monitor, hand-held, and PC tools).

One other piece to consider is what we have to keep in stock in order to support the instruments, in other words spares. By using fewer protocols, we have to stock fewer spares. This is important when you consider the pieces to a control system: the actual instrument, power supplies for the protocol, repeaters, wiring, connectors, and so on. It can get large quickly. When it gets too large, we have to hire someone to manage the inventory.

Another resource, that brings us to our next topic:

Resources

Depending on the protocol chosen we can actually reduce the number of people required to do specific tasks. With the increased information from Foundation Fieldbus, we can allow certain areas of the plant to operate without anyone on site, in other words, unmanned or with one or two people instead of several.

This is also true in maintenance. The additional information from the instruments allows diagnostics to be performed at the maintenance station. This allows the technicians to know what is wrong before they go out into the plant. They take the tools and spares they need to fix the problem and don't have to make multiple trips. They also spend less time troubleshooting at the instrument. In some instances, they can even fix the problem from the station and not go in the field at all.

Diagnostics are what allow this to work. With the new diagnostics being offered by vendors, the technician can even plan for a shutdown based on the diagnostics that are done while the units are running the process. With this knowledge, he or she can have the right spares and tools on hand to make the shutdown go smoother.

We can even save resources in operations. Since the plant is automated, an operator can now control more of the plant because the process is providing better control with the additional information it receives. Some of the control systems that include their own instruments take advantage of their knowledge of the instrument and the information available and integrate that into the overall system providing a more robust picture of the plant and its processes. With local consoles allowing the operator access to the system from different areas of the plant, they can do their job better.

These are some good examples of being able to do more with less. This doesn't include savings in engineering and other parts of the plant due to the more robust protocol.

Conclusion

Although I had several more areas of discussion (engineering, procurement, and vendor support options for example), the end was near. As I said at the beginning, you start on time and you end on time. I had presented several detail slides on cost savings as it applies to our plant, which Richard pulled from the presentation binder and added to his file. I did not provide the information here because as they say "your results may vary."

I ended by asking for questions…the room was silent. I thanked everyone for attending and let them know where I can be reached if they have any further questions. There were questions after the meeting, not a lot, but a few, in fact Richard called to discuss a few things and to learn more.

The meeting ended with direction to proceed with Foundation Fieldbus as the primary protocol. We sat down with the vendor and procurement to make sure everything would go smoothly, which compared to our last project they did. We sent some staff off for initial training to help with the design, and even convinced the vendor to supply a small system with instruments to allow us to learn.

We did not implement only one protocol, however, the need for a second protocol was driven by the plant needs and not solely by price. A large part of the on/off devices are Foundation FieldBus including switches and indicators. We also saved money when looking at the overall plant. We spent a little more in one place to save a lot more in another.

Improving Refinery Performance: Process and Control Information from Step Testing

by

Tim Olsen
Emerson Process Management
Tim.Olsen@emersonprocess.com

and

Jim Dunbar
Emerson Process Management
Jim.Dunbar@emersonprocess.com

Prepared for Presentation at the

**AIChE 2004 Spring National Conference
New Orleans, LA**

**Session TA012
Paper 94g**

Abstract

Step testing is typically only used during advanced control implementation to understand the effects of one process variable on other process variables. Not only can step testing prove valuable in understanding interactions between process variables when implementing model predictive control (MPC), it can also be used to identify control valve deficiencies, understand overall process dynamics, and assist with tuning control loops. This paper will demonstrate how step testing helped improve process control and process performance around a crude desalter, hydrocracker hydrogen quench, and gasoline blender.

Introduction

This paper is intended to encourage process control engineers to understand the dynamics of their processes and use this information to improve process control performance. Step testing is a simple and inexpensive tool that every control engineer should utilize.

A step test is when you place a controller in manual and change the output a small percentage so as not to significantly upset the process and then observe the response in the process variable. As shown on the right in Figure 1, the response can be the most common type, a first-order plus dead time response. There are many types of responses that could be observed such as integrating, second order response, inverse response and others. This paper will address the first-order plus dead time response.

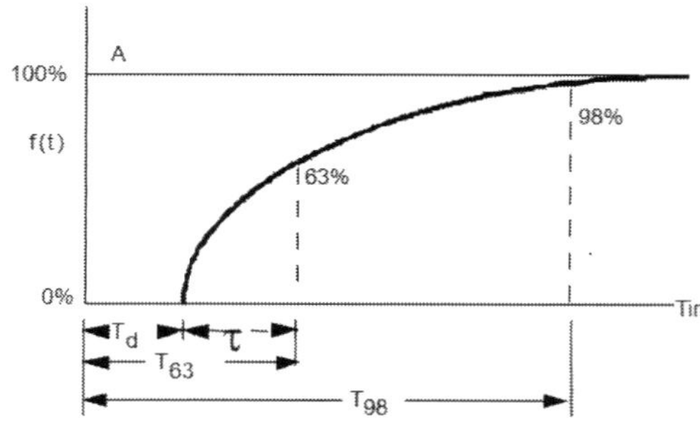

Figure 1: First order with deadtime response

Traditional Step Testing and Model Predictive Control (MPC)

In order to implement MPC on a process, one must first go through a series of moves on each manipulated variable to determine the effects on all controlled variables and the time-to-steady-state. The manipulated variables should be moved in both positive and negative directions from the current operating point to observe the process response. The data collected in the manipulated variables are fit to algorithms in order to generate a dynamic model that best represents the current process operation around the tested operating conditions (design conditions versus turndown rates).

Control Valve Deficiencies

Before beginning any process control optimization program or implementing MPC, one should first look at field device performance and deficiencies. A simple and inexpensive first phase is to execute step tests on each manipulated variable to determine the performance of the final control element. The size of the output step should be large enough to observe the response above any process noise that may be present, yet small enough to minimize process upsets.

To test the performance of a control valve, put the controller in manual and begin with small steps in one direction. Ensure you understand the process before executing step tests to prevent unwanted process upsets (for example, do not reduce hydrogen quench flow to cause a hydrocracker reactor temperature runaway; start with increasing hydrogen and then changing direction back to normal hydrogen use). Start with small steps such as 0.25%, 0.50% or 1% depending on sensitivity of the process. If you do not see a change in flow/temperature/pressure, continue making step changes until you do see a change. It is not uncommon to see control loops requiring an output change in excess of 2% to see changes from the final control element movement (Figure 2 above shows a limit cycle due to a sticky valve). Under-performing field devices add process constraints that cannot be tuned out or compensated with MPC.

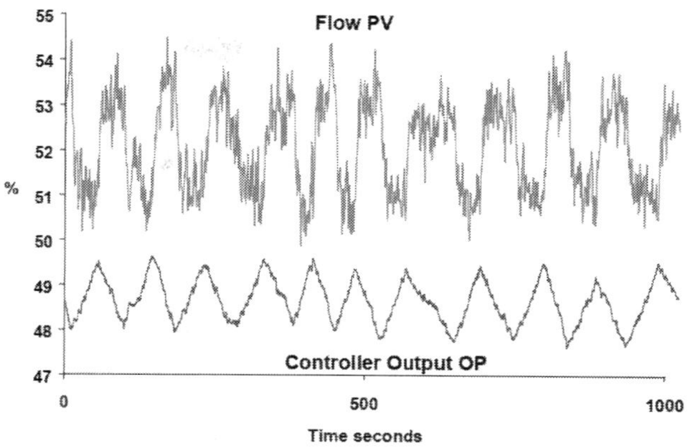

Figure 2: Limit cycle in a flow loop due to a sticky valve

Overall Process Dynamics and Assistance in Tuning Loops

Step tests can be a useful tool in determining overall process dynamics and time-to-steady-state. This information can be used in MPC and to assist in non-MPC tuning. The primary process dynamic information attained from a step test is steady-state process gain, dead time, and the time constant. Typically these parameters are an average of several step tests around the normal operating range.

The steady-state gain can be calculated by taking the measured process response before and after the step test and then dividing by the output change. The time constant Tau is the time required to reach 63.2% of the final process change.

The dead time is the amount of time after the step test is made until the process actually begins to move. See Figure 3.

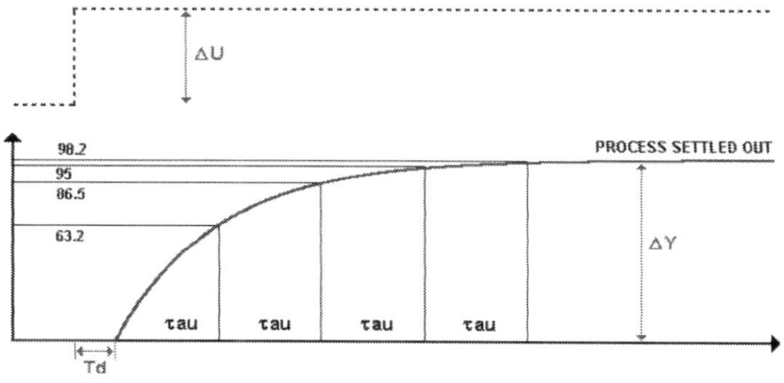

Figure 3: Typical first order step response

Once an engineer understands the process dynamics, model-based tuning such as Lambda Tuning can be applied to set the speed of response for each controller based on operating objectives and priorities. "Lambda Tuning" refers to tuning methods where the control loop speed of response is a selectable tuning parameter; the closed loop time constant is referred to as "Lambda" (λ). Lambda Tuning originated with Dahlin [1] in 1968; it is based on the same IMC theory as MPC [2, 3]. It is model-based and uses pole-zero cancellation to achieve the desired closed loop performance. A recommended Lambda starting point to ensure robustness is three times the larger of Tau or Deadtime. The time for the loop to reach setpoint is approximately four times the selected Lambda value.

Self-Regulating Process (First-order with deadtime model, Classical PID) [3, 4]

τ = time constant (from step test)
T_d = dead time (from step test)
K_p = steady state process gain (from step test)
λ = closed loop time constant (Lambda – user defined speed of response)

Controller Settings

$T_r = \tau$
$K_c = (1 / K_p) * T_r / (\lambda + T_d)$

Crude Desalter (Decoupling Loop Interactions without MPC)

For a refiner, a difficult task is to tune the flow and pressure controllers around a desalter so that they do not fight one another. These controllers have long time constants and long delay times before action sees the measured variable. Tuning-by-feel often results in de-tuning one controller to the point it is similar to being in manual operation. By understanding the process dynamics of each control loop and the overall process operating objectives, one can ensure the loops can be tuned to maximize loop performance while minimizing the tendency of the two loops to fight each other in closed loop control.

Here are the dynamics for one refiner's desalter.

	PT Downstream Pressure	FT Upstream Flow
PVC Pressure Control Valve	Gain 0.87 %/% Delay 15 sec. Tau 20 sec.	Gain 0.24 %/% Delay 15 sec. Tau 30 sec.
FVC Flow Control Valve	Gain -0.4 %/% Delay 11 sec. Tau 30 sec.	Gain 0.2 %/% Delay 25 sec. Tau 35 sec.

One refiner had difficult interactions between the pressure and flow controller around their desalter that resulted in a pressure change of –7 psig from setpoint when the flow increased 1% from normal. This was an operation issue when decreasing flow rate since the pressure increased and lifted the relief valve on the desalter vessel. A standard procedure for the board operator was to manually manipulate the pressure control valve when reducing feed rates to avoid lifting the relief valve. Figure 4 shows the 20-psig pressure change related to a 3% increase in feed rate. The tuning was changed with a 30-second Lambda value for the pressure controller and a 240-second Lambda value for the flow controller. The end result is displayed in Figure 5 with a 3% change in feed rate, yet the pressure only changed 6-psig (a 67% reduction in variation from setpoint). Flow rates could be changed and the pressure controller could remain in automatic without lifting the relief valve.

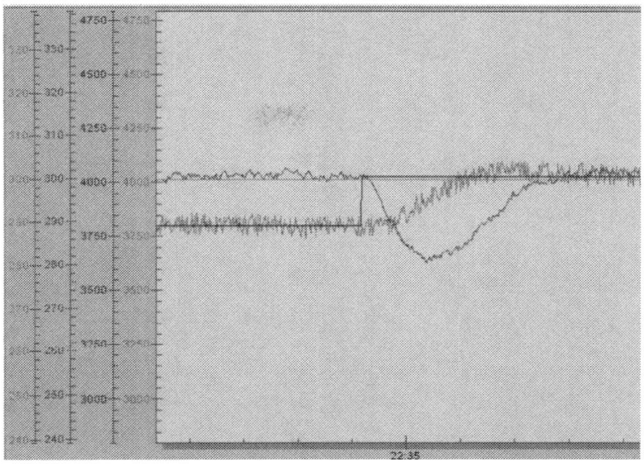

Figure 4: Desalter Flow and Pressure Interactions

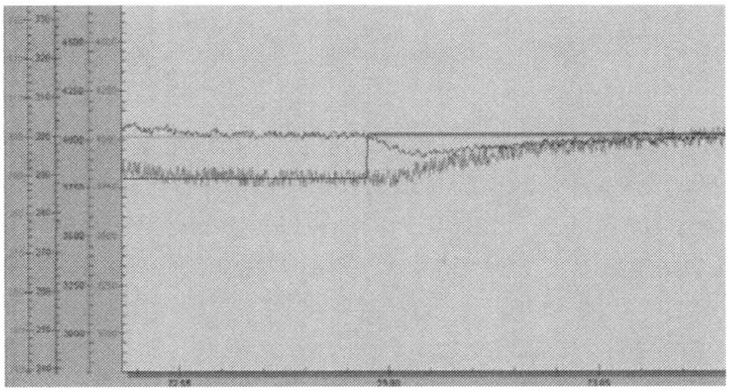

Figure 5: Desalter Flow and Pressure De-Coupled

A simple rule used in cascade control loops is to have the inner control loop tuned 5-10 times faster than the outer loop. The same rule is used to decouple interacting control loops not used in MPC. The most important control parameter, the desalter pressure, will be tuned 5-10 times faster than the crude flow controller.

Hydrocracker Hydrogen Quench (Detecting Hardware Deficiencies)

A refiner was experiencing +/-3 °F control on a reactor bed temperature control via a hydrogen quench. No matter what tuning was used, the performance of the controller could best control within +/-3 °F. As shown in Figure 6, the problem was not the tuning but the resolution of the control valve.

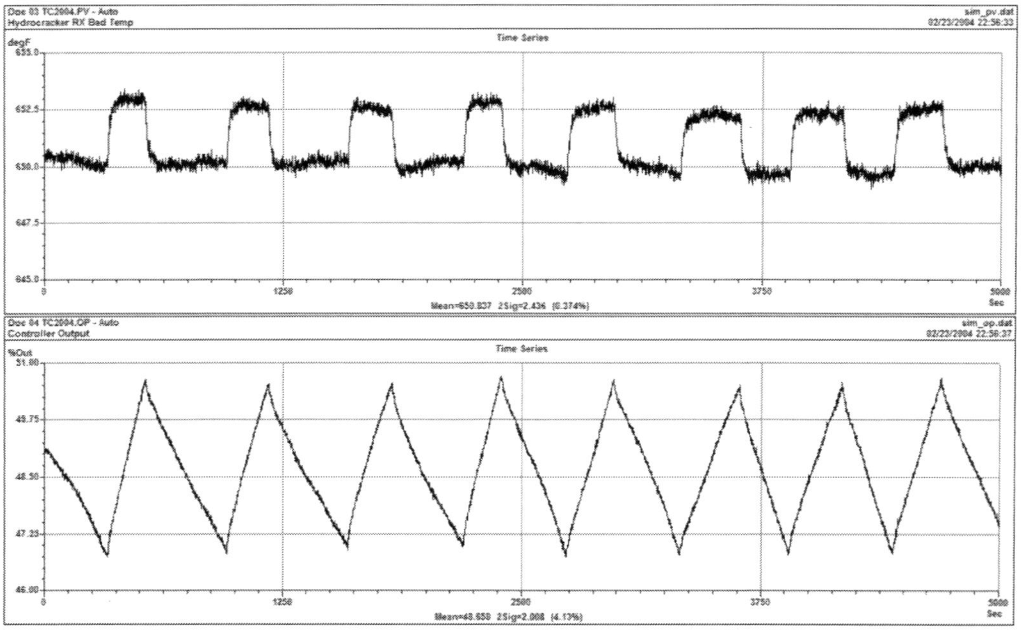

Figure 6: Limit cycle in Rx Bed Quench Temperature due to 3% valve stiction

The hydrogen quench control valve was repaired with less friction packing and a new two-stage positioner which resulted in much tighter control around the hydrocracker reactor bed temperature. The controller was re-tuned and the temperature control is observed below in Figure 7. The temperature set point was increased slightly with the improvement in control without additional hydrogen required.

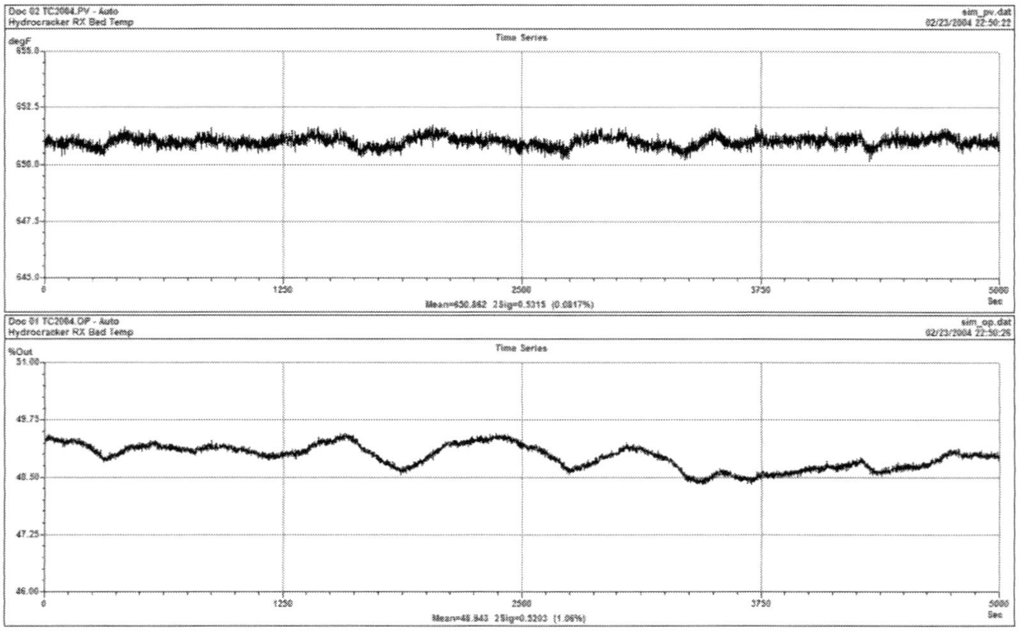

Figure 7: Reactor Bed Quench Temperature Control after improving valve performance

Gasoline Blending (Making Component Dynamics Perform the Same)

Many refiners use fudge factors on their gasoline blends based on experience to account for non-linear octane blending. These compensating factors only work well if no additional field device deficiencies or additional non-linear behaviors exist for each blend component controller. In this application, it is desirable for all of the blend components to have the same dynamic response to maintain a consistent blend even when the components are ramped up or down.

For example, here is one refiner's gasoline component blend control system along with a description of each component dynamics (determined from step testing):

Component	Line Size	Valve Size	Valve Type	Gain	Dead Time (sec.)	Time Constant (sec.)
n-Butane	3	3	Equal %	0.95	2.0	2.2
Isomerate	3	3	Linear	0.57	2.5	2.7
Reformate	4	3	Linear	0.73	3.0	3.0
Cat Gas	6	4	Equal %	0.87	2.3	3.3
Alkylate	3	3	Equal %	0.36	3.2	3.7
Back Pressure	6	6	Equal %	-1.10	2.3	2.7

Tight control of the back pressure controller is desirable to maintain a consistent upstream pressure on the blend flow valves which will stabilize the individual flow controllers. Applying Lambda Tuning, the Back Pressure controller will be tuned to a Lambda of 8 seconds (or 3 times the open loop time constant). To minimize interaction between the backpressure and the flow controllers, it is necessary to tune the flow controllers at least 5 times slower than the pressure controller. Here, each of the flow controllers will be tuned to the same Lambda of 40 seconds (5 times slower than the back pressure controller).

As a result of step testing, finding control element deficiencies and fixing them, understanding the process dynamics, and applying the operation objectives to determine desired speed of response for each controller, the primary benefits for this refiner were as follows:

- 80% reduction in the number of off-spec blends (already had a blend optimization package implemented)
- Cost avoidance related to additional lab samples for off-spec blends
- Better inventory management (less off-spec taking up storage)
- Demurrage avoidance (again due to off-spec blends)
- Octane target shifted from +0.2 above specification to +0.1
- Repeatable process control operation to learn non-linear octane blending characteristics
- Total octane and RVP giveaway reduced from approximately $250K/month to less than $50K/month.

Conclusions

Process control engineers have the ability to understand more about the dynamics of their processes and then use this information to improve process control performance. Step testing is a simple and inexpensive tool that every control engineer can utilize. It can be used to identify control valve deficiencies, understand overall process dynamics, and assist with tuning non-MPC control loops.

REFERENCES

1. Dahlin E.B., *Designing and Tuning Digital Controllers,* Instr and Cont Syst, 41 (6), 77, 1968.
2. Morari M. and Zafiriou E., *Robust Process Control,* Prentice Hall, 1989.
3. Chien I-Lung and Fruehauf P.S., *Consider IMC Tuning to Improve Controller Performance*, Hydrocarbon Processing, Oct. 1990.
4. J. Martin, A.B. Corripio and C.L. Smith, *How to Select Controller Modes and Tuning Parameters from Simple Process Models*, ISA Transactions, Vol 15, April 1976, pp 314-319.

Paper No. 95a

Improvement of Product Phenol Quality via New Effective Method for Removal of Hydroxyacetone

V.M. Zakoshansky, I.I. Vasilyeva, Y.N. Koshelev

ILLA International, LLC.
4756 Doncaster Court, Long Grove, IL 60047-6929, USA

Prepared for Presentation at the Third International Conference on Aromatics and Derivative
AICHE Spring National Meeting, New Orleans, April 25-29, 2004.

This article uncovers the study of a process for hydroxyacetone removal from cumene hydroperoxide cleavage products through extraction of hydroxyacetone from an acid-catalyzed cumene hydroperoxide cleavage products and subsequent chemical conversion of hydroxyacetone into condensation and oxidation products in aqueous sodium sulfate solution in the presence of alkaline catalysts and oxidizing agents. This study has established the kinetic mechanisms of hydroxyacetone conversion in aqueous salt solution in the presence of alkaline catalysts and oxidizing agents. The flow diagram of the proposed hydroxyacetone removal process is presented. It is proved that the novel method as applied along with downstream catalytic phenol purification from other impurities on zeolite catalyst is superior for production of high quality phenol and acetone.

Product quality values are considered to be some of the most essential characteristics of the existing combined phenol and acetone process. The phenol product quality is directly associated with impurities contained therein, some of which such as hydroxyacetone (HA) and 2-methylbenzofuran (2MBF) formed from HA have an adverse impact on color. It is the phenol color that accounts to a great extent for commercial qualities of phenol-based plastics and fibers.

In a conventional cumene to phenol process, 2MBF is formed from the reaction of HA and phenol condensation occurring generally on acid catalyst, which forces utilization of a very energy-incentive step of phenol-water azeotrope distillation to separate 2MBF from phenol. Given high energy consumption and inadequate efficiency of this method, there is an immediate need to find another technique which would effectively overcome the HA-2MBF removal problem.

In view of the fact that it is impossible to totally avoid HA formation at the cumene hydroperoxide (CHP) cleavage stage, as supported by earlier studies described in reference [1], and it is also essentially impossible to remove HA by chemical conversion directly in CHP cleavage product medium without huge losses of target products, there arises an urgency for new uncomplicated and economical approaches to remove HA from CHP cleavage products.

HA is known to be a chemical compound unstable to thermal treatment and decomposed at a boiling point of 145°C. Nevertheless, it would be wrong to think that HA can be totally removed from the products of acid-catalyzed CHP cleavage via just heating.

Our tests demonstrated that HA turns out rather stable in the medium of CHP cleavage products directly or well-diluted aqueous solutions or aqueous salt solutions containing primarily sodium sulfate at a concentration of 2 wt.% through 19 wt.% or crude phenol solutions, in the absence of catalyst, so minor variations in HA concentration proceed at a very slow rate even at very high temperatures of up to 300°C. Table 1 below represents HA conversion in an aqueous salt solution containing 18 wt.% sodium sulfate, 1 wt.% acetone, and 0.7 wt.% phenol at a various temperature and contact time.

Table 1

Contact Time, hr	HA Conversion, %		
	150°C	200°C	300°C
1	2.5	8.1	65.6
2	4.6	21.4	81.7
3	6.1	32.5	89.3

It follows from the data of Table 1 that the temperature of 150°C secures a negligibly low HA conversion even at a long time of the solution retention in reactor. At a high temperature of 300°C, an HA conversion of 80-90% is reached for 2 to 3 hours, but steam requirement to heat up the aqueous salt solution to such hot conditions and, the more importantly, inevitable obstruction of heat exchanging equipment with polymerization products certainly rule out any use of thermal (non-catalytic) method for reducing HA concentration on commercial scale.

When carrying out the reaction without salts, the conditions are even worse, forcing a three-fold reduced HA conversion. Thermal decomposition of HA in aqueous salt solution or aqueous solution causes formation of an impurity in an amount of about 25% relative to the HA decomposed. We found experimentally that this impurity totally carries over to CHP cleavage products during further extraction of aqueous/aqueous-salt solutions. Therefore, both the low HA conversion due to thermal decomposition and the formation of extra undesired products adversely affecting product qualities discredit utilization of thermal method of HA conversion on commercial scale.

In addition, the calculations prove that in this case, the use of thermal treatment of the streams in commercial practice would certainly be uneconomical.

The above considerations make it evident that there is no alternative to catalytic method of HA conversion when applied to commercial conditions.

Catalytic HA conversion (on alkaline catalysts) may be contemplated in two optional occurrences:
(1) Conversion of HA directly in CHP cleavage products medium; or
(2) Conversion of HA separated from CHP cleavage products.

We studied both options. The study revealed the first option to have no prospects since HA conversion goes together with conversion of target phenol and acetone products which is intolerable for commercial process on the grounds of chemical losses of target products and formation of undesired impurities.

In contrast, the second option where HA is first separated from CHP cleavage products through extraction by aqueous salt solution and then subjected to conversion in a special-purpose reactor proved its efficiency [2,4].

Because of the distinctions of all existing technology implementations of phenol-acetone production process, the section of sulfuric acid neutralization by alkaline agents such as sodium hydroxide and/or sodium phenate includes a continuously recirculating aqueous salt phase which contacts with organic products from CHP cleavage reactor. Therefore, there exists a hypothetically potential extractant for HA which, however, actually fails to act as an extractant because its HA concentration reaches the solubility limit and, hence, HA is not extracted from the organics coming to this stage. In order to secure efficient continuous occurrence of HA extraction, the authors of this article suggested that the HA conversion to condensation and oxidation products contained in the aqueous salt phase be carried out in an additional reactor at special conditions which included use of alkaline catalysts, oxidizing agent, higher temperature, etc., and that the aqueous salt phase devoid of HA and most aldehydes be recycled to the extraction stage [2, 3,4]. It was found experimentally that, along with HA, a significant amount of aldehydes is extracted to aqueous salt solution and their conversion in the additional reactor at the special conditions is high enough to efficiently remove aldehydes from CHP cleavage products and thus, in addition to HA removal issue, promote overcoming of the acetone product upgrading problem.

In fact, the development of the process for HA removal from CHP cleavage products was based on three principles [2, 3]:

- Withdrawal of HA from CHP cleavage products by aqueous salt solution at extraction stage;
- Subsequent conversion of HA and carbonyl compounds under the impact of alkaline catalysts and oxidizing agent in the aqueous salt solution, accomplished in additional reactor; and
- Return of HA-free aqueous salt solution to extraction stage.

The study particularly focused on critical operating conditions of HA and aldehyde conversion in aqueous salt phase, so the following purposes were pursued:

- to thoroughly examine the process of HA and aldehyde extraction by aqueous salt phase;
- to find the upper allowable limit of process temperature to minimize utilities consumption and avoid an adverse impact of high temperature, i.e. occurrence of undesired side reactions;
- to find the critical values of alkaline catalyst concentration which accounts for pH value of reaction medium (aqueous salt phase); and
- to find potential ways to speed up the HA and aldehyde conversion rate due to the use of different oxidizing agents such as air, hydrogen peroxide, etc.

Experimental

The process of HA conversion in aqueous sodium sulfate solution containing alkali was studied both in batch mode (in static glass or metal reactors equipped with thermostatic heating jackets and stirring bars) and on a continuous flow-through lab unit which simulated commercial process conditions.

Process Chemistry

One of possible mechanisms of HA conversion in aqueous salt solution is an intermolecular aldol-like condensation. Bimolecular interactions of the following patterns may take place as possible ways of HA chemical conversions:

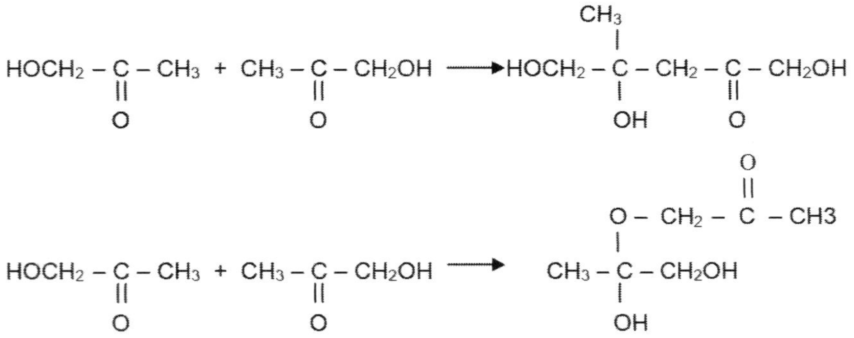

Another possible mechanism of HA chemical conversion may be induced by oxidation reactions which occur as oxidizing agents are introduced to reaction medium (aqueous salt phase). We previously proved in reference [3,4] that the HA and aldehyde oxidation reactions actually occur.

There is only a small group of oxidizing agents which may be suitable for HA conversion in aqueous salt solution on commercial scale. This group includes air, common inorganic oxidizers such as hydrogen peroxide, and also permanganates and chromates. Our study explicitly proved that none of these oxidizing agents, except air, can be used in commercial practice for a number of reasons. The first reason is that these oxidizing agents are classified as "severe" and unselective to an organic substrate to be oxidized. Consequently, their application to the discussed process causes oxidation of not only HA, but also acetone present in the aqueous salt solution. As a result, a crowd of explosive acetone peroxides emerge in the solution. The second reason is huge consumption of oxidizer reactants required to achieve specified HA conversion so that the application of such oxidizers becomes uneconomical.

HA oxidation by air oxygen can be performed easily enough, with formation of carboxylic acids:

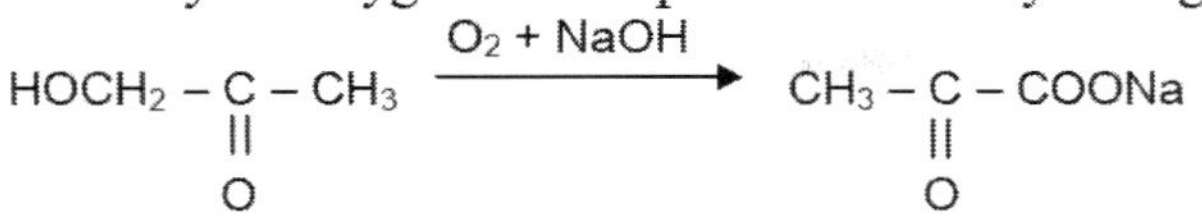

It should be noted, however, that oxidation of organic compounds by air oxygen follows, as our earlier studies demonstrated, a radical pattern and slows down sharply when inhibitors are introduced. In the present case, HA oxidation in aqueous sodium sulfate solution is restrained by phenol which partly enters the aqueous salt solution from CHP cleavage products at HA extraction stage. Phenol is commonly known to always act a strong inhibitor of a radical-chain oxidation process. It was earlier demonstrated in reference [3,4] that the only effective way to overcome this restraint is phenol conversion to sodium phenate due to alkali addition to aqueous salt phase (or organics). Therefore, HA oxidation by air oxygen in aqueous salt solution after HA extraction from acid-catalyzed CHP cleavage products cannot be implemented otherwise than in alkaline medium having a pH value in far excess of 7.

HA oxidation process in alkaline aqueous salt solution occurs rapidly, in contrast to alkaline HA condensation reactions, is irreversible and forms carboxylic acid salts. In the present case, oxidative HA conversion dominates, providing complete conversion of HA.

HA and Aldehyde Extraction Stage

Graphs in Figure 1 and Figure 2 below represent the HA partition between organic and aqueous salt phases obtained for different process operating conditions.
It was found that the HA extraction degree predictably grows with higher water to organics volumetric ratio (Figure 2). However, the HA extraction degree somewhat falls unexpectedly as the HA content in reaction products coming to extraction rises (Figure 2). This can possibly be explained by the fact that HA concentration in CHP cleavage products does not exceed 0.2 wt.%, and in this case the extraction process is described by the laws of ultimately dilute solutions.

Extraction steps may number from one to five, and this is strictly and exclusively defined by specific producer requirements to final product quality. For instance, utilization of one-step extraction leads to a three-fold lower HA concentration in organics. In two-step extraction by aqueous sodium sulfate solution, with an organics to water weight ratio of 1:3, the HA concentration in CHP cleavage products reduces from about 0.2 wt% to about 0.02 wt%, whereas three-step extraction decreases this value from 0.2 wt% to about 0.005 wt%. However, these operating values of HA extraction stage can be achieved only when the HA concentration in the aqueous salt solution contacting the organics at step 1 is 30 ppm or less (preferably 1 to 2 ppm). Accordingly, the HA conversion in the reactor should be at least 98% so as to achieve the required degree of CHP cleavage product purification for HA at the stage of HA extraction by aqueous salt solution. Therefore, the efficiency of HA extraction from organics is rigidly affected by the value of HA conversion in the reactor.

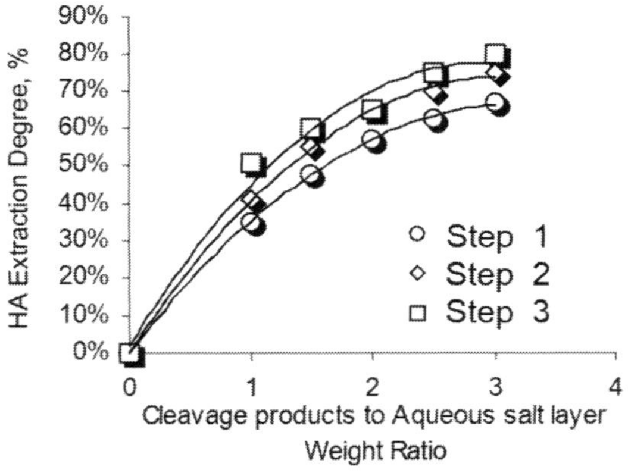

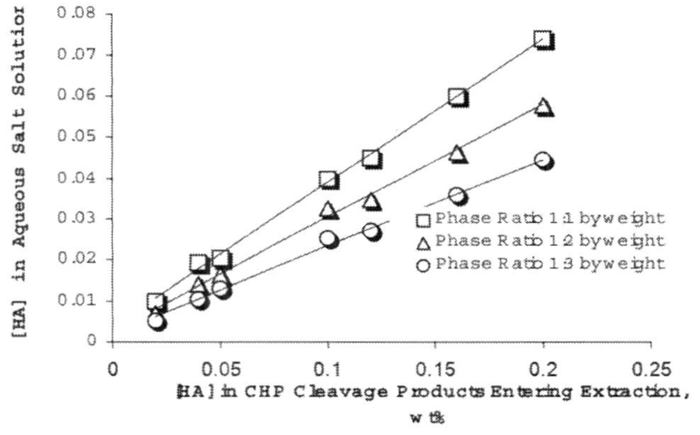

Fig.1. Degree of HA extraction from CHP cleavage products vs. initial HA concentration and cleavage product to aqueous salt solution weight ratio

Fig.2. HA concentration in aqueous sodium sulfate solution vs. initial HA concentration and cleavage product to aqueous sodium sulfate solution weight ratio

1. Catalytic HA Conversion without Oxidizing Agent

During the study, kinetic mechanisms of HA conversion at variable process parameters were examined. Figure 3a through Figure 3d below represent the kinetic curves of HA consumption in a 19% aqueous sodium sulfate solution at variable temperature, air pressure, caustic concentration, or phenol concentration.
The figures indicate that faster HA conversion is promoted by elevation of temperature, caustic concentration, or air pressure. On the other hand, higher phenol concentration sharply inhibits HA conversion, and even addition of equimolar caustic (pH>8) would provide a very slow rate of the process occurrence (see Figure 3a):

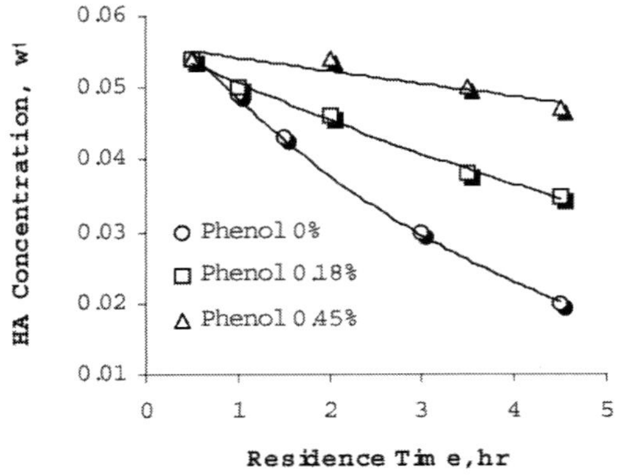

Fig.3a. Kinetic curves of HA consumption at variable phenol concentration in aqueous salt solution (temperature 65°C; NaOH concentration 0.23 wt.%)

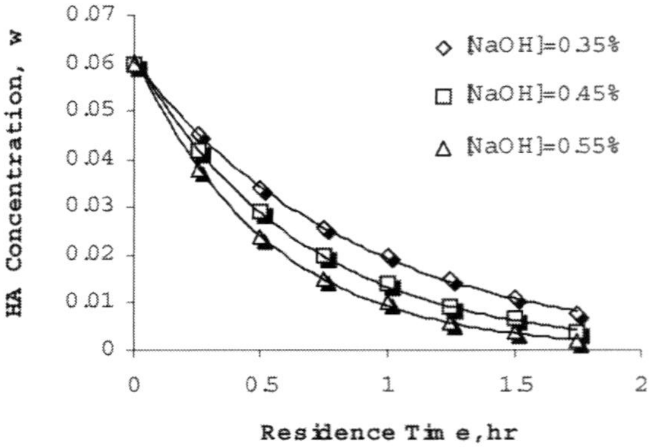

Fig.3b. Kinetic curves of HA consumption at variable NaOH concentration in aqueous salt solution (temperature 80°C)

2. Catalytic HA Conversion with Air as Oxidizing Agent

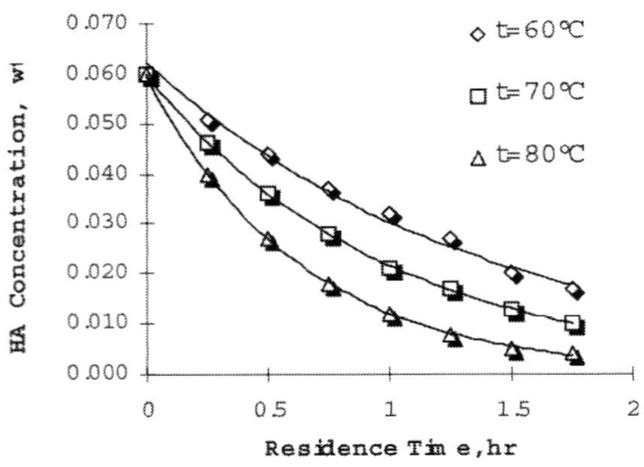

Fig.3c. Kinetic curves of HA consumption at variable temperature (gauge pressure 1 bar; NaOH concentration 0.45 wt.%)

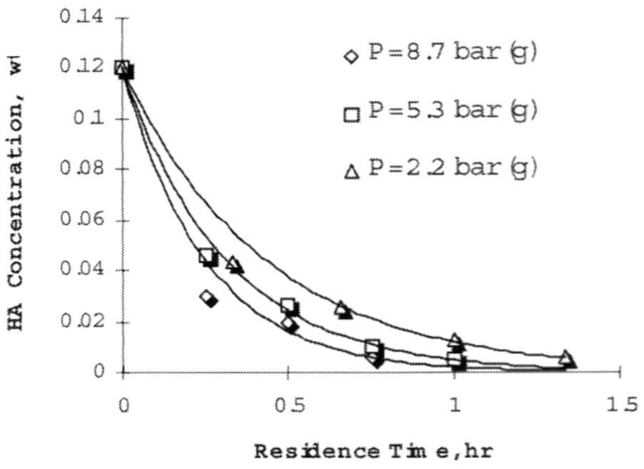

Fig.3d. Kinetic curves of HA consumption at variable gauge pressure (temperature 80°C; NaOH concentration 0.3 wt.%)

Despite the fact that the temperature rise favors the increase in HA conversion rate, a temperature rise above 100°C is inadvisable as it would speed up undesired side reactions of tar formation and induce unreasonable consumption of target products and steam.

The study showed that higher temperature predictably speeds up the HA conversion rate. While examining potential for HA conversion reactor minimization due to temperature elevation, it was explicitly proved that the temperature limit cannot be higher than 130-150°C. The temperature of 130°C initiates and 150°C notably accelerates undesirable deep condensation reactions resulting in very heavy obstructions of both the reactor and heat exchangers and, hence, preventing further process operation. Besides, 1% acetone present in aqueous salt phase causes an almost complete loss of acetone. Another serious aggravation is that high reactor temperatures of 130°C or more require huge steam consumption to heat up the aqueous salt phase, which is in far excess of the steam consumption currently required by commercial method of water-extractive distillation. Therefore, the application of the given method for HA removal from CHP cleavage products is uneconomical..

The kinetic mechanisms of oxidative conversion of hydroxyacetone in aqueous salt solution occurring in the presence of caustic in flow-through mode were studied using a unit schematically illustrated in Figure 1 above. Air addition to the reactor was widely ranged, but the air flow was not found to have any impact on the HA conversion rate.

The quantitative relationships of oxidative conversion of hydroxyacetone in aqueous salt solution in the presence of caustic under air pressure were determined, which allowed elaboration of the general flow diagram of the commercial process called "HAC Process" as represented in Figure 4 below.

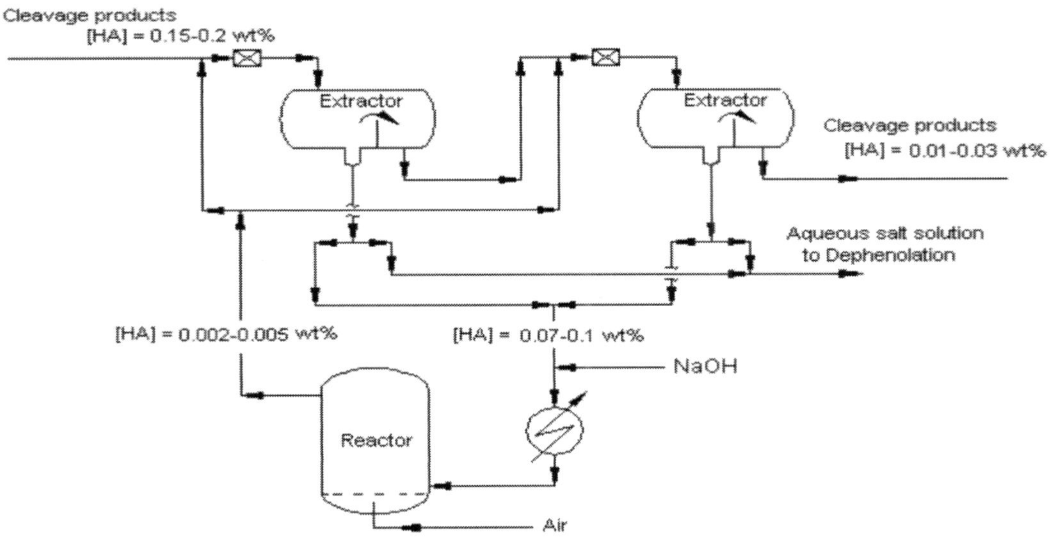

Fig. 4. General flow diagram of commercial process for extraction and conversion of HA and carbonyl compounds in aqueous salt solution (HAC Process)

Capital costs for HAC Process implementation are 2 times less than for alternative methods for the resolving HA and 2MBF problems.

It is important for commercial practice that HA oxidative conversion speeds up under the influence of aldehydes. The aldehydes themselves convert into salts of respective carboxylic acids. This article does not present kinetic data for aldehyde conversions since those conversions only refer to side positive effects of the HAC Process and their effect on phenol quality is much slighter than that of HA.

The presence of aldehydes has a negative impact on acetone product quality, and they are typically subjected to alkaline condensation in fractionation columns so as to avoid carryover to final product. The HAC Process allows a 50% relat. reduction in aldehydes.

Thus, the developed HA removal process allows simultaneous upgrading both of phenol and acetone products.

Mesityl oxide, phorone, α-methylstyrene (AMS), and other impurities contained in phenol stream also, along with HA and 2MBF, negatively affect phenol quality. To remove these impurities, commercial practice commonly employs thermal catalytic treatment on sulfo-IERs, which record a continuous run-length of no longer than 1 to 1.5 year and permit no regeneration and require incineration of exhaust catalyst in special burners.

In contrast, natural and synthetic silica-alumina catalysts practiced commercially are almost free of the weaknesses listed above. They feature high thermal stability threshold, higher mechanical strength, sustainability for multiple oxidative regeneration without compromising catalytic properties, and ability to bury exhaust catalyst without adverse environmental impact. With a wide current availability of commercial silica-aluminas identified by specific crystalline structure and silica module that account for their acidity, they have become applicable for phenol purification.

Extensive study was performed to select a catalyst which would provide an adequate purification degree for carbonyl compounds and α-methylstyrene in phenol medium. Such catalyst was eventually found. The crude phenol purification degree for carbonyl compounds and AMS at variable process parameters is presented in Figure 5a and Figure 5b below.

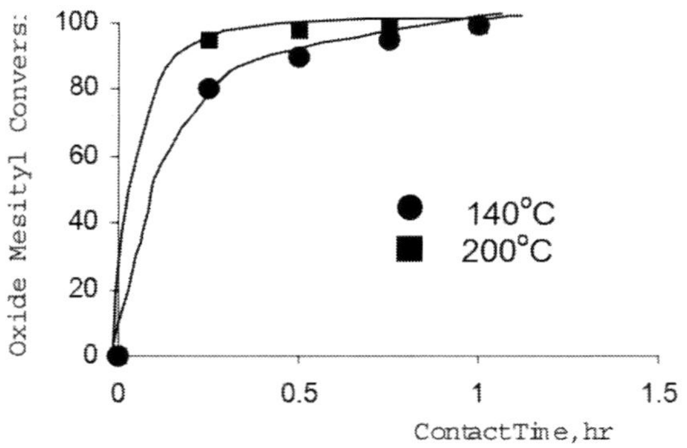

Fig.5a. Mesityl oxide conversion in crude phenol on zeolite catalyst vs. contact time at variable temperature

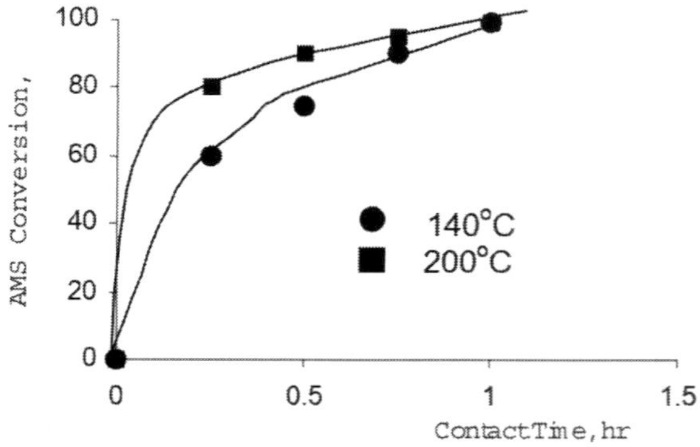

Fig.5b. AMS conversion in crude phenol on zeolite catalyst vs. contact time at variable temperature

The content of total carbonyl-containing impurities as a function of contact time when using sulfo-IER Lewatit vs. zeolite catalyst is presented in the table below.

Contact Time, min	Total Carbonyls, wt%	
	Zeolite (-200°C)	Lewatit (-120°C)
0	0.0619	0.0619
30	0.0161	0.0243
60	0.0153	0.0208
120	0.0134	0.0203
180	0.0105	0.0194
300	0.0099	0.0175

Following commercial operation of the selected zeolite catalyst proved a far better efficiency of phenol purification from aforementioned impurities on zeolite catalyst over various types of IER. With zeolite catalyst applied, the purification degree improves by a factor of 2 to 4. The zeolite catalyst run-length between regenerations is seven years.

The operation of the developed HA and aldehyde extraction method (HAC Process) in combination with downstream catalytic phenol purification on zeolite catalyst allows completely mastering the problem of high quality phenol and acetone production at minimized capital and energy costs:

2MBF	2-5 ppm
Mesityl oxide	< 2 ppm

Total carbonyls	< 50 ppm
Total impurities	< 50 ppm
Color (SAD)	96-97%

Conclusions

1. A novel efficient process for removal of HA and aldehyde impurities from the products of acid-catalyzed CHP cleavage was developed, which is effective in avoiding 2MBF formation at the target product fractionation stages and providing the top quality of phenol and acetone products;

2. A zeolite catalyst for crude phenol purification from carbonyl and unsaturated compounds was selected and tested on commercial scale, which demonstrated a potential for energy and operating cost minimization while producing high quality final products.

References

1. *Vasilyeva I.I., Zakoshansky V.M., Koshelev Y.N., Tchulkov V.P.* Hydroxyacetone formation mechanisms in phenol and acetone production by cumene method // J. Applied Chem. 2000. Vol.73, Issue 12, P.1250.

2. Pat. № 6,066,767 (USA). Method of purifying cumene hydroperoxide decomposition products from hydroxyacetone and from other carbonyls // *V. M. Zakoshansky* et al. 2000.

3. *Vasilyeva I.I., Zakoshansky V.M., Koshelev Y.N.* Hydroxyacetone removal from cumene hydroperoxide cleavage products in phenol and acetone production by cumene method // Oil Refining and Petrochem. 2000. Issue 12, P.507.

4. *Vasilyeva I.I.* Commercial method for high purity phenol production. Abstract of thesis by a Competitor for an Academic Degree of Doctor of Engineering. // St.Petersburg, 2001. P.20.

New Developments in the Sunoco/UOP Phenol Technology

R. J. Schmidt and S.P. Lankton
UOP LLC
25 East Algonquin Rd
Des Plaines, IL 60017

M. Bhinde, S.R. Keenen, and S.M. Gilpin
Sunoco Chemicals
Frankford Plant - Margaret and Bermuda Streets
Philadelphia, PA 19137

Prepared for presentation at the AIChE Spring Meeting, New Orleans, LA, April 25-29, 2004

Abstract

Utilizing the cumene hydroperoxide route to produce phenol, the Sunoco/UOP process remains the process of choice for the production of phenol. Recent design innovations have improved process yields, economics/costs, and process safety. A number of these innovations are highlighted in this article. The Sunoco/UOP process is expected to continue to play a significant role in the future as phenol is a key intermediate feedstock for the growing bis-phenol A and phenolic resins markets.

Keywords: Alkylation, benzene, propylene, cumene, zeolite, phenol, cumene hydroperoxide, acetone, alpha-methylstyrene,

Introduction

The history of the phenol market has been examined in great detail in the literature with much discussion regarding product usage, emerging markets, and process economics over the past two decades [(1)]. More than 90% of the world's phenol production is currently based on the cumene hydroperoxide route. The Sunoco/UOP (formerly AlliedSignal/UOP) phenol process represents the state-of the-art technology for phenol production. It is based on the autocatalytic cumene oxidation and dilute acid cleavage (cumene hydroperoxide decomposition) processing routes. Much of the recent advancement in this technology falls along the lines of improved yield, safety, and economy as a result of specific process and design improvements. The focus of this article is therefore to review the latest improvements to the Sunoco/UOP phenol technology made over the past decade.

Cumene Production

Cumene is the principle feedstock used in the production of phenol and is produced commercially through the alkylation of benzene with propylene over an acid catalyst. Over the years, many different catalysts have been proposed for this alkylation reaction, including boron trifluoride, hydrogen fluoride, aluminum chloride, and phosphoric acid. Cumene processes were originally developed between 1939 and 1945 to meet the demand for high octane aviation gasoline during World War II [(2,3)]. In 2000, about 95% of cumene demand was used as an intermediate for the production of phenol and acetone. A small percentage is used for the production of α-methylstyrene (AMS). The demand for cumene has risen at an average rate of 2–4% per year from 1970 [4,5] to 2003. This trend is expected to continue through at least 2010.

Currently, almost all cumene is produced commercially by using zeolite-based processes. A limited number of remaining units consist of the fixed-bed, kieselguhr-supported phosphoric acid (SPA) catalyst process developed by UOP and the homogeneous $AlCl_3$ and hydrogen chloride catalyst system developed by Monsanto. Two new processes using zeolite-based catalyst systems were developed in the late 1980s. These included technology based on both a conventional fixed-bed system and catalytic distillation system [6-9]. The current state-of-the-art process for cumene production consists of the liquid phase Q-Max™ process offered by UOP based on a beta-zeolite catalyst system [10].

Over the past decade, great progress has been made in improving and optimizing catalyst formulations for improved yields and product quality for cumene alkylation.

Table 1- Improved Cumene Product Quality

	Q-Max	SPA
Cumene Purity, wt%	**99.97**	99.92
Butylbenzene, wt-ppm	**10**	100-200
Bromine Index	**<1**	50-100
Total Nonaromatics, wt-ppm	**10**	300
Ethylbenzene, wt-ppm	20	5-50
n-Propylbenzene, wt-ppm	250-300	200-300

Table 1 shows the typical cumene product quality for the latest Q-Max beta-zeolite technology vs. the older SPA technology. Note that a product purity of 99.97 wt% can be achieved. This is highly desirable for the downstream phenol process due to the reduced levels of impurity precursors leading to undesirable contaminants in the finished phenol and acetone products.

Table 2- Common Cumene Product Impurities

Contaminant	Source	Concern in Downstream Phenol Unit
Nonaromatics	Nonaromatics	Forms acids and other by-products, yield loss
Ethylbenzene	Ethylene, Ethanol	Forms acetaldehyde, acetone contaminant
n-Propylbenzene	Cyclopropane	Forms propionaldehyde, acetone contaminant
Butylbenzenes	Butylenes, Butanols	Resist oxidation, alpha-methyl styrene contaminant
Cymenes	Toluene, Methanol	Form cresols, phenol contaminant
Polyalkylates	Side Reaction	Form alkylphenols, yield loss

Table 2 shows typical cumene feed impurity that lead to undesirable contaminants in the phenol, AMS, and acetone products. The Q-Max process greatly reduces cumene impurities such as butylbenzene and non-aromatics that lead to AMS and phenol impurities. Thus the combination of the Q-Max process and the Sunoco/UOP phenol process results in the highest product quality and highest overall production yields available in the industry.

Phenol Production

The phenol market now spans over three centuries with many production technologies and product uses having been developed over time. 19th century phenol was derived mainly from coal tar and used in the production of products such as disinfectants and salicylic acid. In the early 20th century, phenol was produced primarily using benzene chlorination and sulfonation technologies to support the growing phenolic resin industry after World War I. From the mid-40s through mid-60s, development of the cumene peroxidation process began taking hold such that by about 1990, more than 90% of the world's phenol demand was being satisfied by the cumene route. Refinement of the process over the past decade has been driven by several market forces namely:

(1) the growing bis-phenol A market, driven by the electronics and automotive industries demand for epoxy and polycarbonate resins and
(2) the growing phenolic resin market, driven by housing and furniture industries for composite wood resins.

Also, because competition in these areas for product differentiation and price has become so keen, phenol technology suppliers have been driven to improve production yield, process safety and phenol/acetone product quality to remain competitive. The Sunoco/UOP phenol process represents the most advanced technology for phenol production via the cumene peroxidation route. Recent advances in this technology are highlighted below along with more process details in the following section-by-section analysis.

The Sunoco/UOP Cumene Peroxidation Route to Phenol Production

The Sunoco/UOP Phenol process produces high-purity phenol and acetone by the cumene peroxidation route, using oxygen from air. This process features low-pressure oxidation for improved yield and safety, advanced CHP cleavage for high product selectivity, an innovative direct product neutralization process that minimizes product waste, and an improved, low cost product recovery scheme. The result is a very low cumene feed consumption ratio of 1.31 wt cumene/wt phenol that is achieved without acetone recycle and without tar cracking. The process also produces an ultra high product quality at relatively low capital and operating costs. Extensive commercial experience validates these claims. These improvements are discussed for each of the key major sections of the process.

Overall Process Description/ Chemistry

The main reactions for phenol and acetone production via cumene peroxidation are shown in Figure 1. Both reactions are highly exothermic. Oxidation of cumene to cumene hydroperoxide (CHP) proceeds via a free-radical mechanism that is essentially auto-catalyzed by CHP. The decomposition reaction is catalyzed by strong mineral acid and is highly selective to phenol and acetone. In practice, the many side reactions which take place simultaneously with the above reactions are minimized by optimization of process conditions.

Figure 1 - Main Reactions

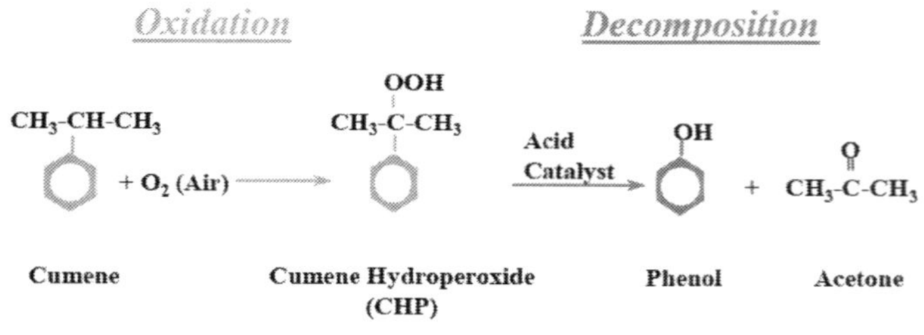

Dimethylphenylcarbinol is the main oxidation by-product, and the DMPC/AMS reactions play a significant role in the plant as shown in Figure 2.

Figure 2 - DMPC / AMS Reactions

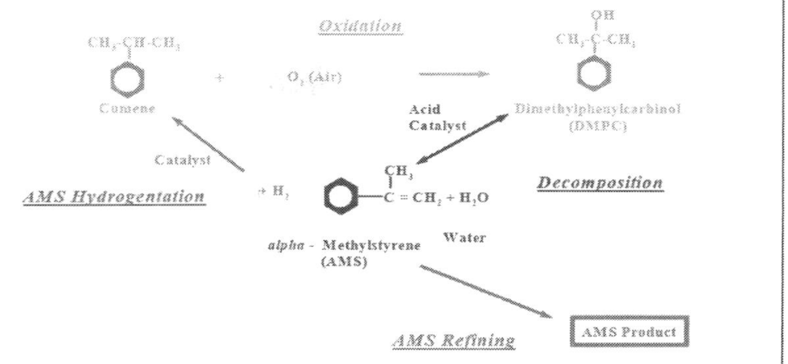

Figure 3 is a block flow diagram showing the major processing steps required to manufacture phenol. The major processing steps include: (1) liquid phase oxidation of cumene to cumene hydroperoxide [CHP], (2) concentration of CHP, (3) acid-catalyzed decomposition of concentrated CHP to phenol and acetone, (4) neutralization of acidic decomposition product, (5) fractionation of the neutralized decomposition product for recovery of acetone, phenol, AMS, and residue, (6) recovery of phenol and the effluent waste water via an extraction process to prepare it for further downstream treatment required to meet effluent quality specifications, and (7) hydrogenation of AMS back to cumene for recycling to synthesis; or, optionally, refining of AMS for sale as a product. The details of each of these main processing steps are discussed as follows highlighting recent technological advances.

Figure 3

Sunoco/UOP Phenol Process

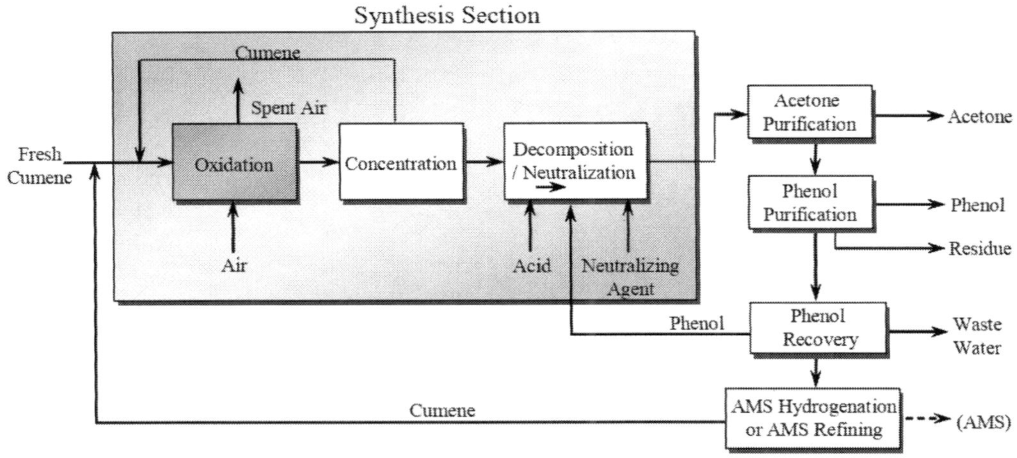

Process Flow and Recent Technology Advances

Oxidation Section Process Flow

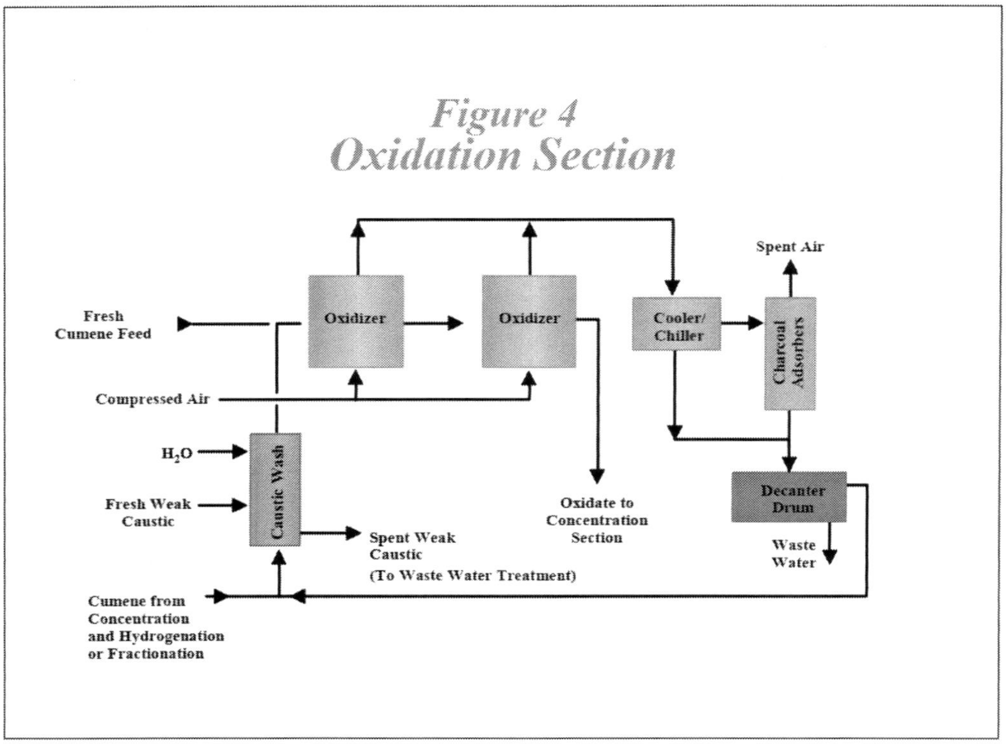

Figure 4 shows a typical series flow for a two oxidation reactor configuration for the low-pressure technology. Five or more reactors can be used in multiple reaction trains depending on the capacity of the unit, location, and processing objectives. This flexibility allows incremental investment staging to meet growing capacity demands.

For large phenol units, it is economical to recover the heat of reaction from the second oxidizer by heat integration with the concentration section. The hot oxidizer circulating liquid stream is used to supply heating to the pre-flash column upper vaporizer. The net oxidate from oxidizer No. 2 (effluent from oxidation section) flows directly to the concentration section.

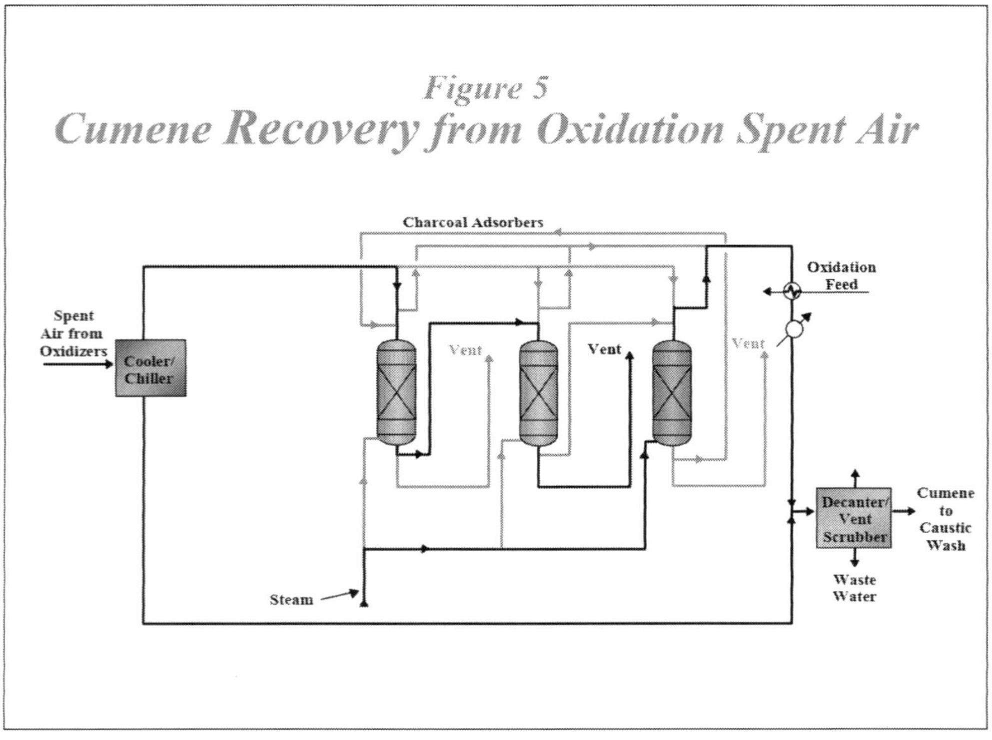

Figure 5 shows the typical cumene recovery system from spent air. Spent air streams from both oxidizers are combined and routed through a water-cooled condenser, a chilled condenser and an entrainment separator for the maximum removal of hydrocarbon and cumene. From the entrainment separator, the air flows to the charcoal absorbers. Two of the adsorbers are always on line in series flow, while the third one is being regenerated with steam.

The cleaned air from the charcoal adsorbers is vented to safe atmospheric disposal. A catalytic incinerator is usually not needed to meet emission limits, but one can be provided if regulations stipulate incineration as the emission control method. The cumene collected by the charcoal adsorbers is recovered by desorption with low-pressure steam followed by condensing the steam and decanting the cumene and water phases. The cumene is then recycled to the feed pre-wash column.

Recent Advances in Oxidation Section Technology

Recent improvements to the oxidation section include: (1) the use of high efficiency charcoal adsorption to recover trace products from spent air, (2) use of an emergency water spray installation and elimination of oxidizer rupture disks, (3) a reduction in oxygen content of vent gas thus reducing air compressor capacity, (4) the elimination of the requirement for caustic scrubbing of fresh feed from the zeolitic cumene unit, (5) use of a dilute caustic wash tower that replaces the feed wash mixer/settler system, (6) the integration of the decanter with the concentration section vacuum system and elimination of the vent gas scrubber, (7) the integration of the feed coalescer into the combined feed surge drum, (8) the use of common spares for oxidizer pumps and emergency coolers. All of these improvements serve to reduce capital and operating costs of the process making it one of the most effective phenol processes available.

With Sunoco/UOP low-pressure oxidation, the air stripping combined with partitioning of the acids to the condensate in the spent air cumene recovery system and the weak-caustic scrubbing of the recycle cumene is so effective that no other method of acid removal is required.

Concentration Section Process Flow

Figure 6
Concentration Section

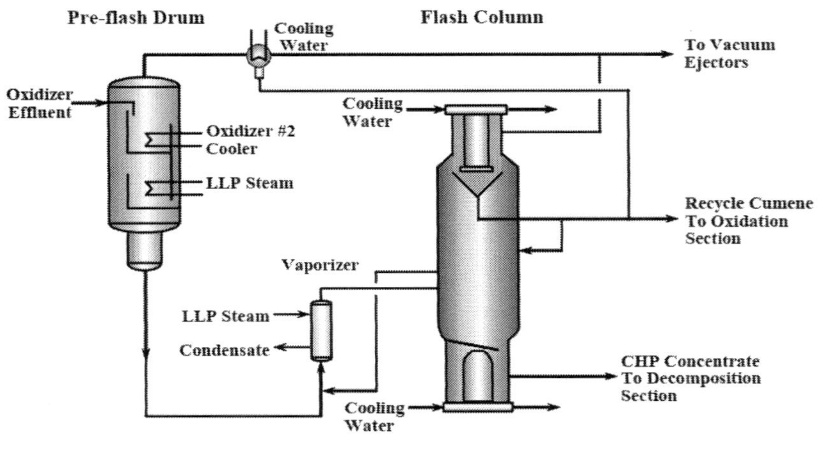

Figure 6 shows an example of the typical flow for the concentration section of the process. The main objective is to concentrate CHP in the oxidizer reactor effluent to a level of about 75 to 85 wt% prior to decomposition to phenol and acetone.

For large phenol units, it is economical to use a two-column concentration system, in which the heat of reaction from oxidizer No. 2 and very low pressure steam are used to vaporize cumene in the first (pre-flash) column, reducing the size of the main flash column. The pre-flash drum and flash column operate under vacuum to minimize the temperature necessary to concentrate the CHP. The vacuum is typically generated by an ejector system. Final CHP concentration is achieved in the flash column vaporizer and flash column, both of which operate under deeper vacuum than the pre-flash drum.

A cumene quench tank is also provided in this section for automatic emergency quenching of various strategic sections of the concentration section if necessary to maintain safe operating temperatures in the event of an incipient CHP decomposition excursion.

Recent Advances in Concentration Section Technology

Recent improvements include: (1) heat integration with oxidation section, (2) a 2-stage concentration section consisting of preflash and flash column, (3) elimination of overhead receivers, (4) use of a Packinox™ welded-plate exchanger in the flash column condenser, and (5) use of power traps instead of level-controlled pumped condensate pots. All of these improvements reduce capital costs for the process.

Decomposition Section Process Flow

The decomposition or cleavage section of the process involves the catalytic decomposition of concentrated CHP in the presence of ppm levels of acid to crude phenol and acetone. The most effective technology for this section is a unique two-step process described in USP 4,358,618[11] by Sifniades/ Allied Corporation patented in 1982. The process involves the use of a backmixed reactor section at low temperature/ higher contact time for the main CHP decomposition step followed by a plug flow dehydration section at higher temperature/ short contact time for conversion of dicumylperoxide (DCP) to AMS. The process represents a breakthrough in AMS yield improvement and, is currently being used by all licensors as the process of choice for modern high yield phenol technology.

Figure 7
Decomposition Section

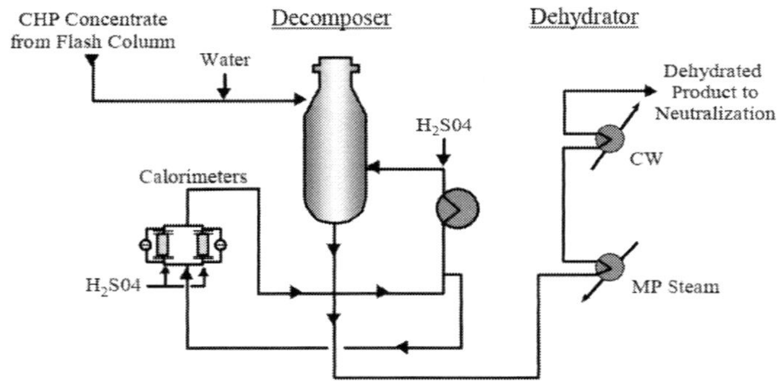

Figure 7 shows an example of the most advanced decomposer technology is the process offered by Sunoco/UOP. It consists of a very simple but elegant drum and loop reactor design where concentrated CHP from the concentration section flows into the decomposer drum, along with a metered amount of water to maintain optimal reaction conditions in the decomposer recycle loop. Sulfuric acid is injected via injection pumps into the loop to provide the catalyst required for the decomposition of CHP to phenol and acetone.

A circulation pump is provided to circulate the content of the decomposer. Sulfuric acid is injected into the circulating stream to such an extent that the decomposition of CHP and dehydration of dimethylphenylcarbinol (DMPC), a key by-product of oxidation reaction, are precisely controlled. The level of unreacted CHP is monitored via calorimeters, to which part of the acid catalyst flow is routed. The effluent from the decomposer is pumped to the dehydrator in which the effluent is heated to a temperature where remaining DMPC is dehydrated and DCP converted to AMS at very high yield. This is a unique advantage of the Sunoco/UOP decomposition technology. The Sunoco/UOP process produces approximately 90% AMS yield from DMPC. This also results in higher phenol yield, thus lower cumene consumption and less residue (e.g., tar) formation.

Recent Advances in Decomposition Section Technology

In addition to very high yields across the decomposition section, the Sunoco/UOP technology offers the following recent improvements: (1) the implementation of advanced process control (APC), (2) a reduction in required recycle rate from 100:1 to 25:1, (3) the elimination of water injection tank and pumps, (4) use of acid totes to eliminate the acid tank dependent on unit size and client preference, and (5) design of the unit for safe containment in most probable relief situations and elimination of the catch tank. The major advantage of these improvements is reduced capital costs and improved process yields and economics. AMS yields as high as 85 to 90% across the decomposition section have been demonstrated making the Sunoco/ UOP technology one of the most selective offerings in the industry today.

Neutralization Section Process Flow and Recent Technology Advances

Figure 8
Direct Diamine Neutralization

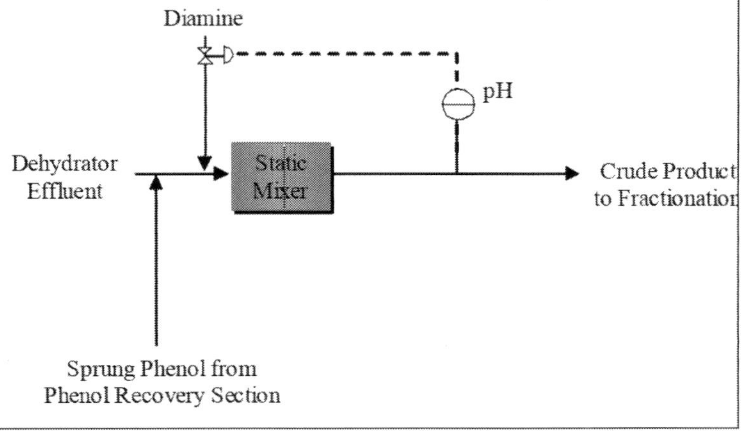

The acid catalyst that is added in the decomposition section must be neutralized to prevent yield loss due to side reactions and protect against corrosion in the fractionation section. The Sunoco/UOP Phenol process uses a novel approach for neutralization: The acid catalyst is neutralized by injecting a stoichiometric amount of a diamine which does not need to be removed from the process, as shown in Figure 8.

The diamine salt is soluble through the distillation train and leaves the unit with the residue stream. The diamine salt does not contribute significantly (if at all) to SO_x or NO_x when the residue is used as boiler fuel. The main advantages of direct diamine neutralization over conventional systems are: (1) a new/simplified design that is easy to operate and reduces capital cost, (2) process uses soluble salts that reduces reboiler fouling and lowers maintenance costs, and (3) does not require water addition for neutralization which in turn lowers waste water production and reduces distillation utilities. By replacing older ion exchange resin technology with the new direct neutralization process, the phenolic waste water production can be reduced by 45% or more at lower capital cost.

Acetone Refining Section Process Flow and Recent Technology Advances

Figure 9
Acetone Refining Section

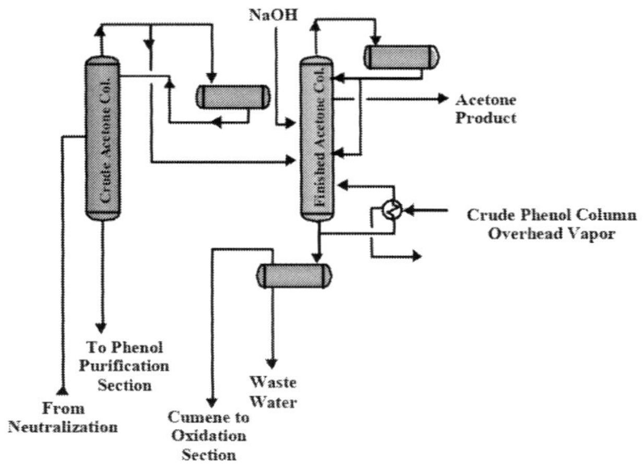

Acetone is a major by-product of the CHP oxidation process for the production of phenol. The overall economics of the process are highly dependent on production of high quality acetone (e.g., 99.7 to 99.9% purity) for sales in the solvents and bis-phenol A markets. Figure 9 shows the typical acetone refining flowscheme for the Sunoco/UOP process. Recent improvements include the use of stabbed-in condensers and elimination of overhead receivers, where appropriate, to save capital cost.

Table 3 shows the typical finished high quality acetone that is achieved in the process. Note that this specification is sufficient to meet or exceed that typically required for bisphenol A manufacture.

Table 3- Typical Acetone Product Quality

Color, Pt-Co	**5**
Specific gravity, 20/20°C	**0.790-0.792**
Total GC impurities, wt-ppm	**250**
Water content, wt%	**0.3 (can design for 0.2)**
Alkalinity, wt-ppm as ammonia	**< 10**
Acidity, as acetic acid, wt-%	**0.001**
Non-volatile matter, g/100 ml	**0.001**
Water solubility	**Passes**
Reducing substances (permanganate test)	**Passes (> 4 hours)**
Distillation range, °C	**0.5°C from start to dry, including 56.1°C**

Phenol Fractionation and Purification Section Process Flow

Figure 10
Phenol Fractionation and Purification Scheme
for AMS Hydrogenation

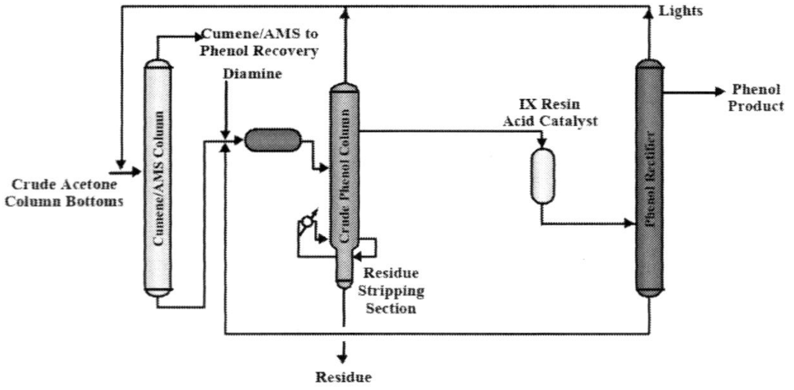

Once the crude phenol has been produced, it must be further fractionated to prepare a finished product that is of sufficient purity to meet downstream user specs. An example of the phenol purification is shown in Figure 10 using AMS hydrogenation as a means of recycling the by-product AMS to maximize phenol production. Note the separate residue stripper column section allows the final stripping of phenol from the residue to be conducted at higher vacuum, which allows both the crude phenol column and residue stripper column to be reboiled with medium-pressure steam. **Thus, no high-pressure steam is required for the phenol plant!**

The crude phenol column has a top pasteurizing section to remove the small amount of light by-products generated during distillation. The main product from the column is taken off as a side-cut and flows to ion exchange (IX) resin treaters, in which the IX resin catalyzes conversion of methylbenzofurans (MBF) and residual AMS to high-boiling components. MBF and AMS are otherwise difficult to remove by distillation. The effluent from the IX resin treaters goes to the phenol rectifier, where the heavy components along with some phenol are distilled to the bottoms and recycled back to the crude phenol column. The phenol rectifier also has a top pasteurizing section for removal of small amounts of light by-products generated during distillation. Phenol product flows by gravity from the rectifier side cut to storage.

Recent Advances in Phenol Fractionation and Purification Section Technology

Figure 11
Phenol Recovery Section

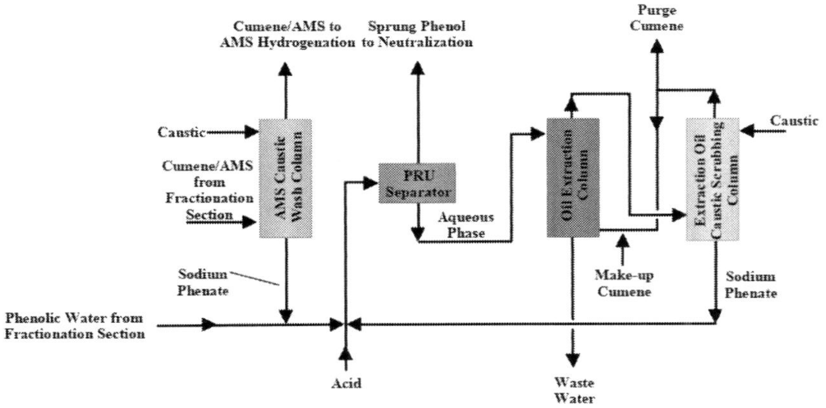

Improvements in the Phenol Fractionation and Purification Section include: (1) replacement of the azeotropic phenol stripper with chemical/resin treating, (2) elimination of the acid neutralizer system, and (3) use of stabbed-in condensers and elimination of overhead receivers where appropriate. The main benefit of these changes has been reduced capital/utilities cost while maintaining the already very high overall phenol product quality. Further, equipment sizes and waste water have been significantly reduced by implementation of the advanced Sunoco/UOP phenol recovery technology shown in Figure 11.

Figure 12 shows the typical fractionation scheme needed to recover high quality AMS. To achieve a high yield of AMS (typically >93% recovery), the crude AMS must first be caustic washed to convert AMS impurities that co-boil with the AMS into heavy by-products that can be removed by fractionation. The washed product is then fractionated to remove cumene and phenol followed by fractionation in the AMS finishing column to remove the heavy by-products.

Figure 12
Alpha-Methylstyrene (AMS) Refining

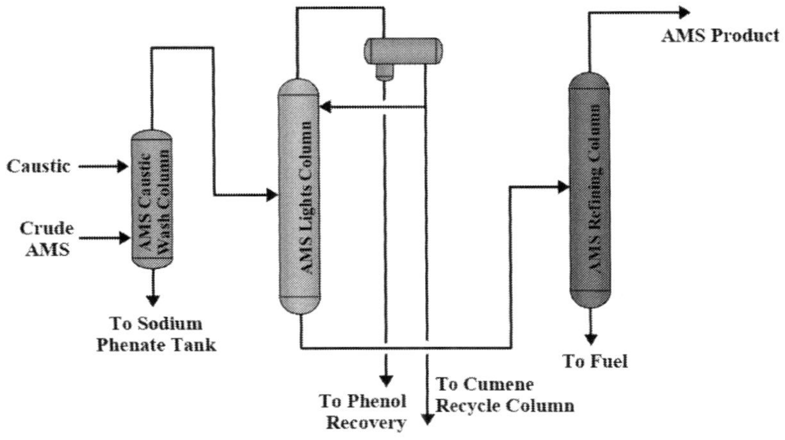

Table 4- Typical AMS Product Quality

	Typical Specifications	Typical Quality
Appearance	Clear, water white liquid	Pass
Odor	AMS, characteristic	Pass
Color, Pt-Co	20, max	10
Specific gravity @ 15.5/15°C	0.912 - 0.915	0.914
AMS content, wt-%	99.0 min	99.2 - 99.6 (up to 99.8 w/ Q-Max)
Sulfur compounds	Free from $H_2S + SO_2$	Pass
Copper corrosion test	No iridescence, gray or black	Pass
Phenol content, ppm	20	5
Inhibitor level, PTBC, ppm	10 - 20	15
Distillation range, 5 to 95%, °C	0.5, including 165.4 °C	0.4 including 165.4

An example of the high quality finished AMS product is shown in Table 4.

Use of an IX resin catalyst that is better at controlling total carbonyls and 2-MBF in the phenol results in a very high purity phenol product in the range of <15ppm total carbonyls and <1ppm 2-MBF. This is compared to claims of <20ppm total carbonyls and <5ppm 2-MBF for some competing cumene hydroperoxide processes.

Table 5- Typical Phenol Product Quality

	High Purity Phenol	Ultra-High Purity Phenol
Color, Pt-Co	5	5
Water, wt-ppm	< 300	< 300
Solidification point, °C	40.8	40.8
Total carbonyls, wt-ppm	< 30	< 15
Total GC impurities, wt-ppm	< 100	< 30
Cresols, wt-ppm	< 50	< 25
2-MBF, wt-ppm	≤ 5	≤ 1

Table 5 shows the typical phenol product quality expected from the process for two different grades of phenol: (1) high purity suitable for most merchant market applications and many bisphenol-A markets and (2) ultra high purity required by the most demanding bisphenol-A and polycarbonates markets. Note that both products can be accommodated in the Sunoco/UOP phenol process by making appropriate adjustments to process operating and fractionation conditions.

AMS Hydrogenation Section Process Flow

The Sunoco/UOP phenol process utilizes AMS hydrogenation technology developed by Huels. The Huels MSHP™ process is a mild hydrogenation process based on a Pd containing catalyst system that operates at moderate pressure. The process achieves nearly complete conversion of AMS with very high selectivity to cumene resulting in a very low overall process cumene/phenol consumption ratio of 1.31 wt./wt.

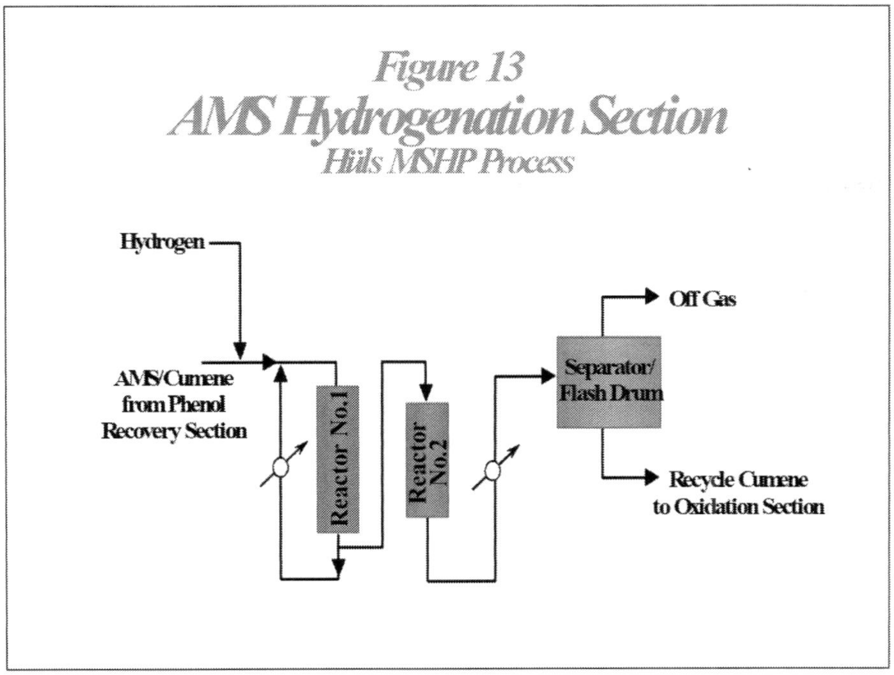

The simple AMS hydrogenation process shown in Figure 13 has been demonstrated to operate without significant catalyst deactivation over multiple years of operation.

Recent advances in the Sunoco/UOP/Huels AMS hydrogenation technology include: (1) elimination of the recycle hydrogen compressor, and (2) elimination of the hydrogen flash drum and pumps. Both of these improvements save capital cost and utilities.

Tar Cracking

Tar cracking of heavy ends produced in the Sunoco/UOP phenol process is no longer required due to the improvement in process yield achieved over the last 10 years. By eliminating tar cracking, along with additional refinements in cumene and phenol fractionation technology, phenol product purity has improved such that total organic impurities have been reduced to a level of <30ppm.

Phenol Process Safety

Safety considerations in the production of phenol and acetone from cumene include design and operating criteria for processing the intermediate CHP. CHP decomposes rapidly to phenol and acetone when exposed to strong acids, even at low temperatures. This reaction is highly exothermic and is the second reaction step in the process. At high temperatures, the rate of CHP decomposition catalyzed by weak acids would also become significant. In addition, CHP reacts with cumene to form dimethylphenylcarbinol. This reaction occurs to some extent under normal conditions in the oxidation, concentration and decomposition sections, but the rate becomes significant at higher temperatures. At still higher temperatures, CHP also decomposes thermally to form acetophenone and methane. CHP decomposition catalyzed by weak acids and the thermal CHP reactions would only become significant from a safety standpoint in the event that heat cannot be removed. In such a case the increasing temperature from the heat of reaction would result in a higher reaction rate, creating the potential for an uncontrolled reaction.

Thus, the availability of heat exchangers, cooling medium, and pumps for cooling CHP mixtures is critical from a safety standpoint. A significant advantage of the Sunoco/UOP low-pressure oxidation technology in addition to high yields is the very mild operating temperature (e.g., typically 82 to 90 °C) required for the process. The lower oxidizer operating temperature translates into a much longer allowable operator response time should intervention be required due to an upset to prevent oxidizer temperatures that are high enough to promote CHP thermal decomposition. The intervention response time may be as long as 24 hours or more to avoid elevated temperatures and high rates of CHP decomposition in the oxidizers. For high-pressure processes offered by other licensors, the response time is much shorter, on the order of only a few hours, to prevent accelerated CHP thermal decomposition due to the higher initial process temperature (e.g., typically 95 to 100 °C or more).

The Sunoco/UOP Phenol process design and operating criteria are based on an industry accepted 10,000-year probability guideline. Safety provisions include emergency coolers and pumps, reliable power supplies, and reliable cooling water supply. Further backup provisions include the ability to use firewater as once-though cooling water, and the capability to use cool cumene to reduce (quench) temperature.

Conclusion

Table 6- Low vs. High Pressure Oxidation

	Sunoco-UOP	High-Pressure Oxidation
Oxidation outlet pressure, kg/cm^2	**0.2**	**5**
Compressor stages	**Single**	**Multiple**
Compression utilities	**Low**	**High (or turbo -expander)**
Number of oxidizers	**2**	**3 - 6**
CHP Selectivity	**95.7**	**~93**
Tar cracking required	**No**	**Yes**
Capital cost including Tar Cracking	**Base**	**Higher**
Cumene/phenol consumption ratio, wt/wt	**1.31**	**>1.35 w/o tar cracking**

A summary of various process and yield improvements is shown in Table 6 for the low pressure Sunoco/UOP process vs. the competitive high pressure technology. Results show that higher yields can be achieved at lower cost with improved process safety as a result of mild oxidation reaction conditions and advanced decomposition and product purification technology employed in the Sunoco/UOP process. With a cumene feed/ phenol product consumption ratio (wt/wt) of 1.31, the process represents the highest yield and lowest cost technology available in the industry today.

REFERENCES

1. M.Bentham, etal., *Process Improvements for a Changing Phenol Market*, DeWitt Petrochemical Review Conference – Houston, Texas,March 19-21, 1991.
2. G. Stefanidakis and J. E. Gwyn, in J. J. McKetta and W. A. Cunningham, eds., *Encyclopedia of Chemical Processing and Design*, Vol. 2, Marcel Dekker, New York, 1977, p. 357.
3. W. Keim and M. Roper, in W. Gerhartz, ed., *Ullmann's Encyclopedia of Industrial Chemistry*, Vol. A1, VCH Verlagsgesellschaft, Weinheim, 1985, p. 185.
4. Z. Sedaglat-Pour, *Cumene*, CEH Data Summary, SRI International, Menlo Park, Calif., Mar. 1989, p. 638.5000A.
5. *Chem. Mark. Rep.* **232**(10), 54 (1987).
6. J. D. Shoemaker and E. M. Jones, Jr., *Cumene by Catalytic Distillation*, 1987 NPRA Annual Meeting, Mar. 29–31, 1987.
7. E. M. Jones, Jr., and J. Mawer, *Cumene by Catalytic Distillation*, AICHE Meeting, New Orleans, La., Apr. 6–10, 1986.
8. W. P. Stadig, *Chem. Process.* **50**(2), 27 (1987).
9. J. D. Shoemaker and E. M. Jones, Jr., *Hydrocarbon Process.* **66**(6), 57 (1987).
10. R. Schmidt, A. Zarchy, and G. Petersen, *New Developments in Cumene and Ethylbenzene Alkylation - Paper 124b*, 1st Annual Aromatic Producers Conference – AIChE Spring Meeting, New Orleans, LA., March 10-14, 2002.
11. U.S.Patent 4,358,618 (November 9,1982), S. Sifniades, A.A.Turnick, F.W. Koff (to Allied Corporation).

Paper No. 95d

Formation of Key By-Products in Phenol Production Process MECHANISM OF *ALPHA*-METHYLSTYRENE DIMER AND *ORTHO-*, *PARA*-CUMYLPHENOLS FORMATION

V.M. Zakoshansky
ILLA International, LLC.
4756 Doncaster Court, Long Grove, IL 60047-6929, USA

Prepared for Presentation at the Third International Conference on Aromatics and Derivative AICHE Spring National Meeting, New Orleans, April 25-29, 2004.

This article presents the study of the kinetics and mechanism of key byproducts formation in phenol production process, specifically α-methylstyrene dimers and cumylphenols. The α-methylstyrene dimer structure was investigated by the vapor-magnetic resonance method. It was proven that these byproducts are formed not under the common concept of protonization of α-methylstyrene double bond, but actually through either dimethyl benzene alcohol (DMBA) hydronium ion formation or acidic particles addition to a DMBA molecule. Methods to reduce the byproduct yield in commercial process are demonstrated.

Commercial process of combined phenol and acetone production features formation of byproducts such as α-methylstyrene dimers and *ortho-*, *para*-cumylphenols at the cumene hydroperoxide cleavage stage, which are critical for the process selectivity.

Except brief information presented in references [1-4], no data on the mechanism of formation and the structure of α-methylstyrene dimers. It is a common opinion that α-methylstyrene (I-AMS) undergoes condensation to form 2,4-diphenyl-4-methylpentene-2 (II, β-dimer) and 1,1,3-trimethyl-3-phenylindane (III, cyclic dimer, indane), although the latter is not formed in the cumene hydroperoxide (IV-CHP) acidic cleavage products, as per authors of reference [1]. The authors of reference [1] show that according to the thermodynamics, the α- to β-isomer ratio shall be less than one, which makes β-isomer dominant in quantity. The authors of the above references do not disclose formation mechanism of α-methylstyrene dimers (D_i) nor uncover their structure but only show the way of dimers formation via α-methylstyrene, which makes one assume the fact of protonization of α-methylstyrene double bond, and identify the conditions to favor formation of dimers II, III, and V.

The general scheme of chemical reactions occurring in the phenol production process, provided by the authors of reference [2], includes a number of products associated with α-methylstyrene conversion reactions, such products comprising 1,3-diphenyl-3-methylpentene-1 and phenylbutenes (cis- and trans-2-phenylbutene-2 and 2-phenylbutene-1). Our studies proved that, given the formation of said products in the course of the acidic cleavage of technical CHP, the amount of such products must be too insignificant to be detected even during the CHP cleavage products concentration. Furthermore, it is not very easy to imagine a reaction occurring between two AMS molecules, wherein one methylbenzene molecule is lost while condensation of the AMS molecules. The same difficulties one has with picturing structural rearrangement of a hypothetically formed carbenium ion A (formula 1), featuring simultaneous migration along the chains of aryl and alkyl groups to form 1,3-diphenyl-3-methylpentene-1. Anyway, we could not detect the latter in the products of the opposing synthesis. Thus, the present study is devoted to formation of only three main α-methylstyrene dimers, formed in a commercial process, which were separated and which structure was disclosed by us.

Experimental

α-Methylstyrene dimers were separated from the byproducts of the acidic CHP cleavage process, or so called "phenol tar". The obtained fraction D_i consisted of three products which weight percentage was about 80 wt.%, 12 wt.%, and 8 wt.%. Similar products were obtained via counter synthesis from dimethyl benzene alcohol (VI-DMBA) employing H_3PO_4 as a catalyst at temperature of about 100°C. The structure of the obtained α-methylstyrene dimers was investigated by the vapor-magnetic resonance method (at 300 MHz) and chromato-mass spectrometry.

Kinetics of D_i and cumylphenols (CP) formation was studied in equimolar phenol and acetone mixture where sulfuric acid was used as a catalyst. CHP was treated according to method of reference [1]. According to the GC analysis, the resulting product contained 99.7 wt.% of CHP and 0.3 wt.% of alcohol VI.

CHP containing either AMS or DMBA was decomposed in a flow-through lab unit with reaction product circulation to the reactor, i.e. under conditions simulating commercial one-stage process of combined phenol and acetone production.

Discussion of Results

I. α-Methylstyrene Dimers Structure

Formula (1) describes conventional concepts of the carbonium ion mechanism of DMBA and AMS conversion reactions occurring under acid catalytic conditions to form carbonium ion which interacts with AMS to form carbenium ion (A):

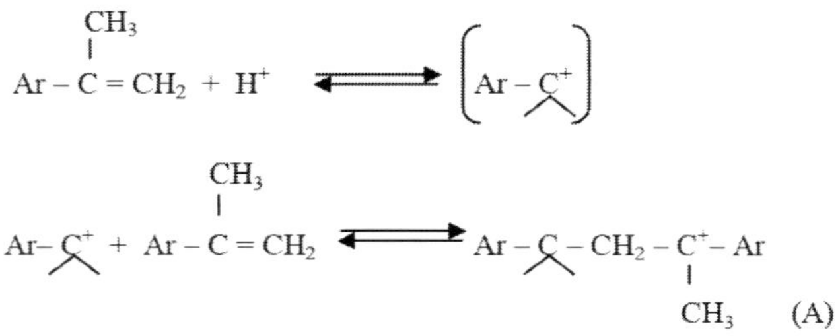

(1)

Stabilization of the formed carbenium ion (A) must result in two isomers of α-methylstyrene dimers such as α- and β-isomers:

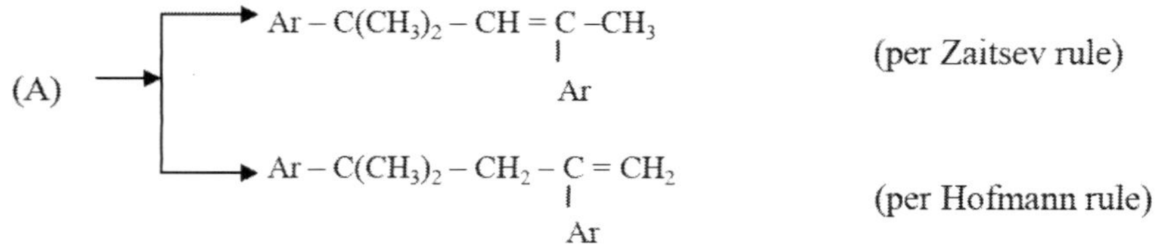

(2)

In addition, it follows from reference [2] that (A) produces a cyclic dimer, indane, which has no double bond. This direction is considered to result from alkylation of the benzene ring in *ortho*-position featured by respective loss of proton. Low probability of the reaction occurrence in this direction must have allowed the authors of reference [1] to exclude potential indane formation under the commercial CHP cleavage process conditions.

However, the study of the products resulting from the decomposition of the VI containing technical CHP revealed nearly total incorrectness of the conclusions made by the authors of references [1] and [2] with respect to the composition and structure of D_i formed in actual commercial process. First, the byproducts of acidic CHP cleavage do not contain any compound formed by the Zaitsev rule, i.e. any β-isomer. Second, the amount of D_i α-isomer in the resultant mixture is about 80% relat. Third, cyclic D_i is proven to be formed under commercial process conditions.

The structure of isomer V was proven by vapor magnetic resonance method where chemical shift of the equivalent methyl groups equals 1.13 ppm (integral intensity of 6 H); methylene protons (intensity of 2 H) produce a singlet signal at δ of 2.78 ppm; geminal vinyl protons produce two signals at 5.04 ppm (hydrogen atom in cis-position with respect to the aromatic ring, $J_{gemin.}$ equals 2.0 Hz) and 4.69 ppm ($J_{gemin.}$ equals 2.0 Hz, J_{allyl} equals 0.8 Hz); aromatic protons produce multiplet within the range of 7.0-7.4 ppm. The obtained specter corresponds to an α-methylstyrene dimer which amount in the produced isomer mixture is about 80% relat., i.e. to 2,4-dephenyl-4-methylpentene-1.

In case the D_i formed had the structure $C_6H_5–C(CH_3)_2–CH=CH–CH_3$, which is attributed to it by the authors of [1] and [2], a proton signal should have been produced in the specter, revealing internal double bond in the area of about 1 ppm, and another proton signal would have revealed carbons having double bonds. Nevertheless, such signals were not detected in the spectrogram neither for the mixture of dimers separated from the commercial phenol process byproducts, nor in the composition of the products resulting from the reaction which was carried out right in the spectrometer cuvette and wherein both AMS and DMBA are used as feedstock.

The data of the chromato-mass spectrometry allow an assumption that the second isomer, separated from the D_i mixture, in an amount of 8% rel. represents a so-called "cyclic" dimer 1,1,3-trimethyl-3-phenylindane (III). No double bond in III was proven by the vapor magnetic resonance method.

In addition, our research revealed that, unlike linear D_i isomers with double bond, compound III is quite insignificantly converted to AMS even given a temperature of 400-450°C and employment of acidic heterogeneous catalyst. Therefore, in addition to the vapor magnetic resonance data, no double bond in compound III structure is supported by its tolerance to the acidic catalyst even at very high temperatures.

The third separated D_i is not a β-dimer as it is assumed by the authors of reference [1] but D_i α-isomer. The α-isomer formation is proven by the fact that in the vapor magnetic resonance specter of the D_i mixture, there is no proton signal revealing an internal double bond in the area of 1 ppm. The exact structure of this D_i was not identified. It can be either 2,5-diphenylhexene-1 (VII) representing a product of addition against the Markovnikov rule, or a trans-isomer of 2,4-diphenyl-4-methylpentene-2 (V). The cis- and trans-conversions of compound V are supported by the fact that with time, acid catalyst impact results in a decrease of the concentration of one of the isomers, while the position of the α-double bond of the D_i in question remains unchanged.

Thus, dimers III, V-cis and V-trans (VII) are formed in the process of the acidic cleavage of CHP containing DMBA.

II. Mechanism of α-Methylstyrene Dimers and Cumylphenols Formation

Correct description of the kinetics of D_i and CP formation requires the exact product (DMBA or AMS) resulting in dimers formation be determined. This is important since the reaction of DMBA to AMS conversion is reversible; we have also found that the equilibrium is achieved very quickly and the equilibrium constant K_e equals 8-12 (depending on the temperature), the equilibrium being shifted towards AMS. Thus, following the conventional carbonium ion mechanism according to which D_i are formed by protonization of AMS molecule [4], it would be natural to expect that in phenol process, where AMS concentration in the reactor is ten times higher than the DMBA equilibrium concentration, the D_i formation rate constant would be in proportion to the AMS concentration. Consequently, the amount of the D_i formed in the reaction products should be proportional to AMS concentration and independent on which compound, AMS or DMBA, the reactor is fed with.

The results of the above examination are given in Table 1. Test run 2 was intentionally carried out at a lower acid concentration unlike run 1, since we proved it before that acidic catalyst activity with respect to all reactions occurring within the phenol process is extremely high at a low water concentration (0.2 wt.% in this case).

Table 1

Acidic Cleavage of CHP Containing DMBA and AMS (*)

No.	Feed Composition, wt.%			Test Conditions		Formed Byproducts Composition, wt.%					
	CHP	DMBA	AMS	H_2SO_4	h_0	AMS	DMBA	D_i	DCP	CP	H_2O
Exp.1	90.8	7.0	0.0	0.035	1.8	2.51	0.61	1.44	0.38	2.42	0.89
Exp.2	90.3	0.3	6.5	0.014	10.0	5.29	0.55	0.39	0.52	0.30	0.20

* Reaction medium is phenol-acetone equimolar mixture; the temperature is 47°C; the reactor is mixing type with the circulation of the obtained products to the reactor; circulation ratio is 17; residence time of the byproducts in the reactor is 34 min; D_i is total AMS dimers.

It follows from the data presented in Table 1 that the amount of the D_i formed is essentially different for the studied cases. In case the reactor is fed with DMBA (run 1), the amount of the formed D_i at the reactor outlet is about 4 times higher vs. the case when the reactor is fed with AMS, despite an about 5 times less acidity value h_0 in run 1.

Produced significant differences in the amount of the D_i formed in runs 1 and 2, 1.44 wt.% and 0.39 wt.% respectively, could be potentially explained by an about 3 times higher H_2SO_4 concentration in run 1, which makes D_i formation follow the common acid catalysis mechanism wherein the reaction rate constants are direct function of the catalyst concentration. If the common acid catalysis took place in the case under study, then the above reactions could hypothetically occur by the conventional carbonium ion mechanism comprising protonization of the AMS double bond followed by the formation and conversion of the formed carbenium ion A.

However, the discovered linear dependence between the D_i formation rate constant and H_2SO_4 concentration value (that is true for phenol-acetone mixture with a constant composition) cannot serve a proof for the reaction occurrence by the common acid catalysis mechanism since there is a linear dependence between the H_0 value and acid concentration in a very wide range. Therefore, it is incorrect to make a selection between two types of mechanisms (common and specific acid catalysis) based on the test runs at variable H_2SO_4 concentration. A selection between common ($k_{Di} \sim [H_2SO_4]$) and specific acid catalysis ($\lg K_{Di} + H_0 = const.$) of D_i formation was based on the studied kinetics of the reaction taking place in a phenol-acetone medium at constant composition and constant H_2SO_4 concentration but at insignificant water additions (from 0.2 wt.% to 1 wt.%). Despite constant acid concentration value, the above conditions provide an about 100 times change in the h_0 value [5]. A slight increase in water concentration was found to cause a very strong reduction in D_i formation rate, which happened to correspond to a decrease in the h_0 acidity value. The above allowed to discard the occurrence of D_i formation reaction by the common acid catalysis mechanism. In fact, the reaction under study follows the dependence $\lg k_{Di} + H_0 = const$, thus being subject to specific acid catalysis laws, which makes the rate of this reaction a function of h_0 acidity value. Therefore, final results of D_i and CP quantitative formation in runs 1 and 2 (Table 1) should be compared by the h_0 acidity value and the reagent (AMS or DMBA) fed to the reactor, i.e. by AMS and DMBA concentration in the reactor, but not by the dependence on the H_2SO_4 concentration in the reaction medium.

As to the change in DMBA concentration in the reactors, in run 1 its concentration varies from 7 wt.% at the reactor inlet to 0.61 wt.% at the reactor outlet with an average actual DMBA concentration of about 3.8 wt.%. In run 2, wherein AMS is fed instead of DMBA, the actual DMBA concentration approaches the equilibrium concentration of 0.55 wt.%. It is evident that in runs 1 and 2, the actual DMBA concentrations ratio changes by 3.8/0.55=6.9 times, which results in proportional change of the amount of the formed D_i (1.44/0.39=3.7).

It follows from the above that it is DMBA and not AMS that is a reactive molecule resulting in AMS dimers formation. In case the term "protonization" is used, produced results allow a conclusion that it is the DMBA molecule that is protonized to form hydronium ion, but not the AMS double bond as it might follow from the common concept of carbonium ion mechanism of acid catalytic reactions. Consequently, it is the DMBA hydronium ion and not the carbonium one that leads to D_i formation in the phenol-acetone medium of the phenol process.

Similar situation is true for cumylphenols formation, when about 7 times reduced actual DMBA concentration in run 2 (vs. run 1) results in a directly proportional decrease (by about 8 times) in CP formation. Thus, it is the reactive molecule of DMBA along with formed hydronium ion or a catalytic complex of DMBA and acid particles, that lead to CP formation.

The same situation, when a reagent's double bond is much less reactive than an alcohol OH-group, was found true for isoamylene alcohol which, condensed with formaldehyde, formed a pyran compound [6]. This reaction occurs in water medium where $H_5O_2^+$ particles with low catalytic activity [7,8] and featuring low h_0 acidity value [9] are present, unlike the aforementioned DMBA conversion at high h_0 acidity value in phenol-acetone medium.

Preferable "protonization" of alcohol OH-group and its higher reactivity compared to the double bond is true not only for organic medium but for water medium as well.

To additionally examine which reagent (AMS or DMBA) is responsible for D_i formation in a commercial phenol process, kinetic runs of DMBA and AMS conversion in the phenol-acetone medium were carried out (see Figures 1 and 2) at the same H_2SO_4 concentration (unlike the runs presented in Table 1).
The trend of the produced kinetic curves of D_i accumulation shows that, given AMS in the feed (Figure 1), D_i start to accumulate only after DMBA has formed from AMS. The D_i accumulation curve is of S-like nature typical of the reactions with the formation of final product through an intermediate one. The test run where the DMBA was in the feed (Figure 2), reveals a totally different type of D_i accumulation vs. the case with AMS in the feed (see Figure 1), since D_i start to accumulate right after acid was introduced into the reaction medium, which is typical of the reactions of the second order. While in the case with AMS in the feed, despite an about 7 times higher initial AMS concentration (about 5 wt.%) vs. the initial DMBA concentration (about 0.7 wt.%) in the test run shown in Figure 2, D_i start to accumulate only when DMBA begins to form.

The obtained results of the kinetic test runs (Figures 1 and 2) prove the dominant part of DMBA in AMS dimers formation as well as no protonization of AMS double bond.

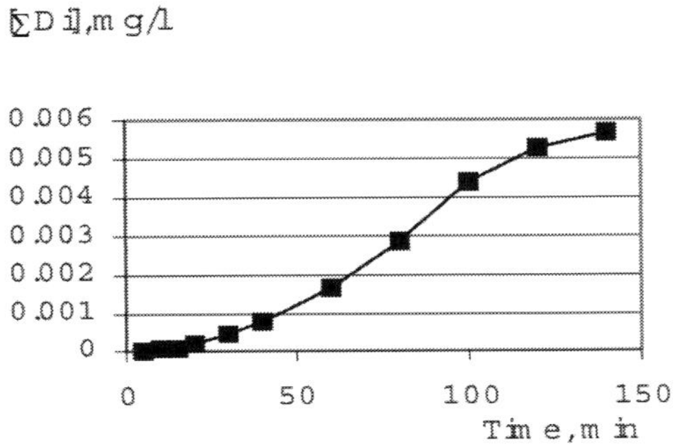

Figure 1. D_i Accumulation with AMS in Feed
t = 25°C; $[H_2SO_4]$ =0.047 wt%
$[AMS]_0$ = 5.43 wt%; $[DMBA]_0$ = 0 wt%

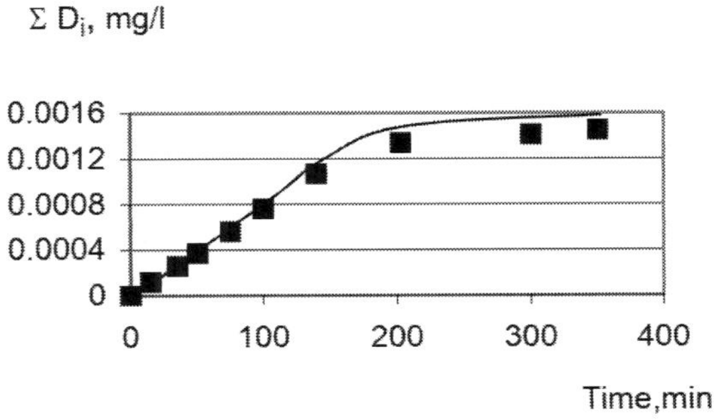

Figure 2. D_i Accumulation with DMBA in Feed
t = 26°C; $[H_2SO_4]$ =0.047 wt%
$[DMBA]_0$ = 0.714 wt%; $[AMS]_0$ = 0 wt%

Given the fact that in phenol-acetone solution, mineral acids do not dissociate to form H^+ but are present as different kinds of acidic particles such as H_3O^+, $H_5O_2^+$, etc. [7, 8], the reversible reactions of AMS and DMBA conversion for the particular case of catalysis via H_3O^+ particles may be illustrated by the following formulas:

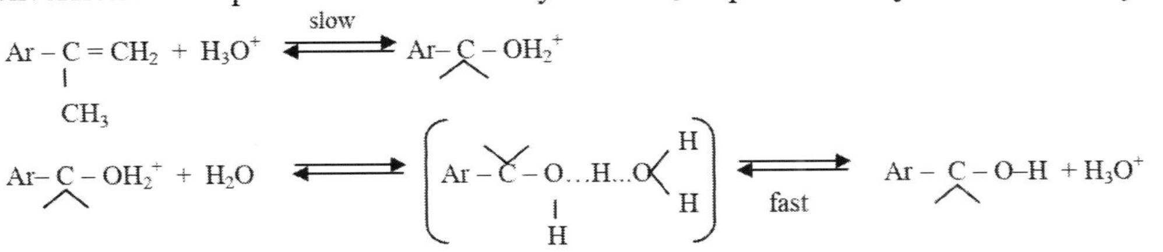

(B)

(3)

(C)

No data can be obtained in support of AMS dimer formation from hydronium ion (B) directly or intermolecular complex (C). It is noteworthy, however, that AMS dimers are formed through (B) or (C) reacting with DMBA molecule rather than AMS. This is the only possible reason for the second order of D_i formation as evidenced from the kinetic tests (see Figure 2). Besides, it was found that the amount of water emerging during the D_i formation reaction corresponds to two moles of converted DMBA. Should the D_i formation reaction have followed the AMS double bond protonization pattern, i.e. two AMS molecules engaged, the water concentration in the reaction medium would have remained constant.

The AMS dimer formation is an essentially irreversible reaction as far as the commercial process of technical CHP cleavage is concerned. The irreversibility of AMS conversion to dimers was proven through durable hold of AMS dimers in a phenol-acetone medium containing 0.07 wt.% of H_2SO_4 at 75°C as no DMBA or AMS was eventually observed in the reaction products.

In view of all the above reasoning, the D_i formation mechanism may be illustrated by formula (4) below:

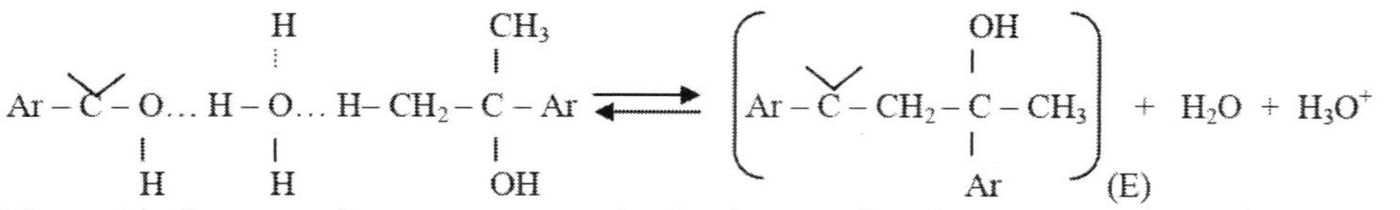

(4)

Thus, the first reaction step occurs in the intermolecular manner through an attack on the DMBA OH group by the acidic H_3O^+ particle and by the CH_3 group of the second DMBA molecule. Consequently, an intermediate complex (D) is formed which is further converted into D_i in the second step represented by formula (5), where the acidic H_3O^+ particle attacks the OH group and the CH_3 group of complex (E) resulting in elimination of the second water molecule and H_3O^+ particle.

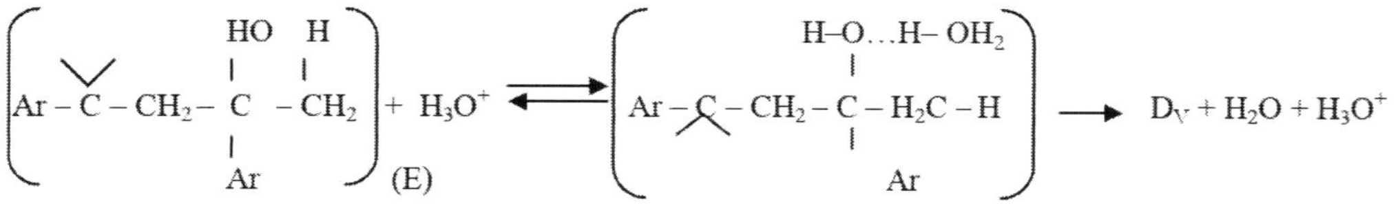

(5)

The first and second steps represented by formulas (4) and (5), respectively, are likely to proceed simultaneously to form 2,4-diphenyl-4-methylpentene-1, i.e. D_V.

A simplified scheme of conversions (4) and (5) appears as follows:

$$2\ \text{DMBA} + 2\ H_3O^+ \longrightarrow H_3C - \underset{Ar}{\overset{CH_3}{C}} - CH_2 - \underset{Ar}{\overset{CH_3}{C}} = CH_2 + 2\ H_2O + 2\ H_3O^+ \qquad (6)$$

Ar- and methyl groups in the D_i formed may be either in a cis- or trans-position, which may be the cause of the formation of two D_i isomers having an α-double bond.

The cyclic dimer 1,1,3-trimethyl-3-phenylindane (III) formation also engages two H_3O^+ particles and two DMBA molecules:

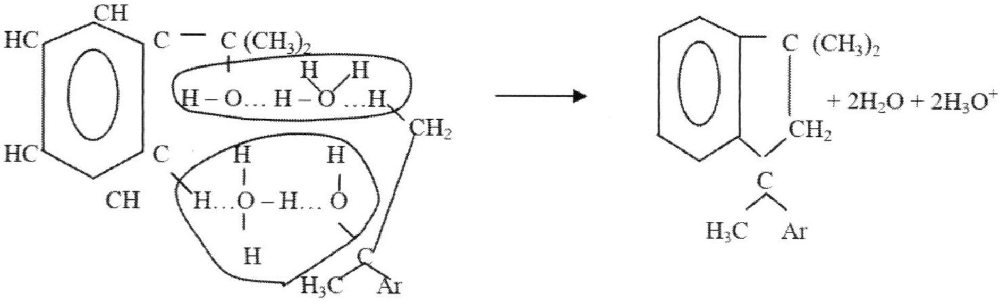

(7)

The influence of reaction medium composition including acetone to phenol mole ratio and water and H_2SO_4 concentration in the phenol-acetone mixture and the influence of temperature were examined. It was found that the isomer quantities and the isomer V to isomer III ratio and the isomer V to isomer VII ratio are affected by all these parameters. The higher the H_2SO_4 concentration and temperature and the lower the acetone to phenol mole ratio (i.e. the more severe conditions are imposed on the process of cleaving DMBA-bearing technical CHP), the greater amount of isomers V, III, and VII is formed in the process. A higher DMBA concentration in technical CHP also leads to a higher yield of isomers V, III, and VII. *Vice versa*, a higher acetone to phenol mole ratio, higher water concentration and lower acid concentration, i.e. milder process conditions, cause a greater isomer V to isomer III ratio and a lower total formation of AMS dimer isomers. Therefore, milder conditions inhibit the indane (III) formation.

The above discussion concerning the AMS dimer structure and isomers ratio should be considered strictly within the process conditions of acid-catalyzed cleavage of DMBA-bearing technical cumene hydroperoxide as a section of phenol/acetone production process.

The structure of dimethyl benzylphenols (cumylphenols) (CP) is commonly known. The general opinion, as expressed in reference [1], is that in commercial phenol/acetone production process, CP is formed both from DMBA and phenol and from AMS condensation with phenol. If the first route is beyond doubt indeed, the second one should be rejected as incorrect, based on the data of Table 1. Even in commercial CP production, with a cumene-AMS hydrocarbon fraction used as the feedstock, a certain amount of water needs to be added to phenol to speed up the CP formation rate. The water does not act as a promoter of acid catalyst, but rather converts AMS entering the reactor into DMBA, which then rapidly reacts with phenol to form CP.

As evidenced from references [10, 11], the phenol-acetone solution containing acid catalyst and water is actually clear of free protons, and it is the free protons that have been commonly believed as responsible for acid-catalyzed reactions occurring by carbonium ion mechanism. It was proven that in the phenol-acetone solution containing acid catalyst and water, there exist acidic particles of various composition such as non-dissociated H_2SO_4, H_3O^+, quasi-ion couple of acetone with H_2SO_4 molecule, and $H_5O_2^+$ [7, 10,11, 12], which in fact catalyze, like in the D_i instance, the CP formation reaction via activating the DMBA OH group. As in cyclic dimer (indane) formation, acidic particles simultaneously attack a carbon atom of the benzene ring to substitute a hydrogen atom and form a new C-C bond between a phenol molecule and a DMBA molecule. The reaction mechanism is described by formula (8):

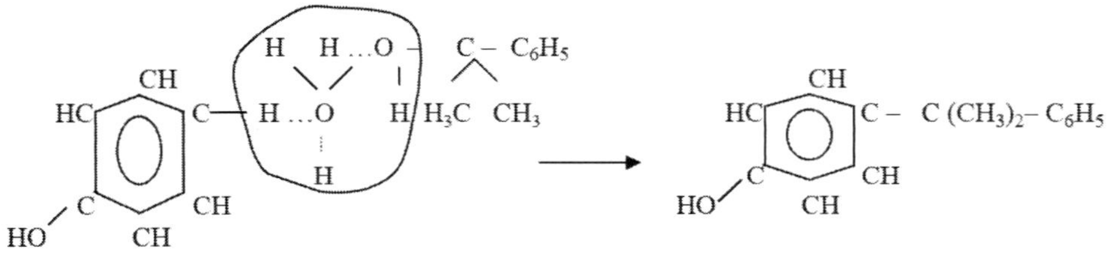

(8)

The research has established that *ortho*- and *para*-CP are formed with a ratio of 6:1 to 10:1. This ratio is determined by the values of acetone to phenol mole ratio and the concentrations of acid catalyst and water in the reaction medium.

The CP formation reaction is described by a first order reaction equation since in the reaction medium of commercial process, a phenol reactant is in a great excess. The CP formation rate constant (k_{CP}), as well as that of CHP cleavage reaction and other reactions of commercial phenol/acetone process, is determined by the environment acidity value (h_0).

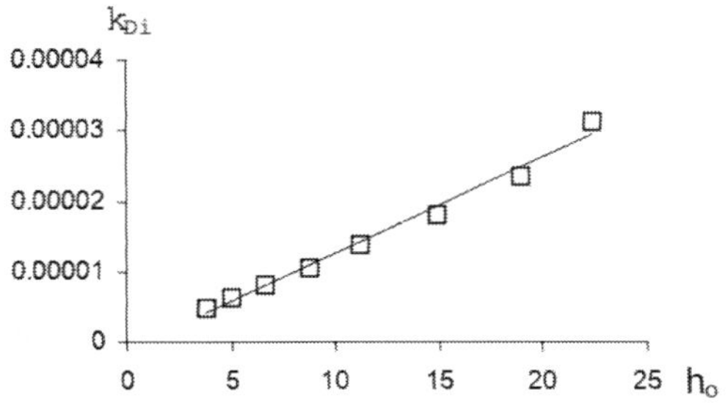

Figure 3. Function $k_{Di} = f(h_0)$

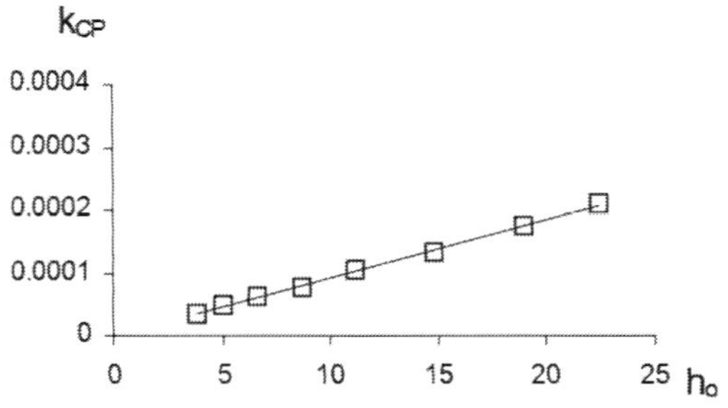

Figure 4. Function $k_{CP} = f(h_0)$

Table 2 below represents the rate constant values for the subject reactions of reversible DMBA conversions and CP/D_i formation and activation parameter values where the h_0 value and the equimolar acetone/phenol ratio in the reaction medium are constant and the water concentration is variable between 0.4 wt.% and 1.0 wt.%. However, any change in the aforementioned species of acidic particles due to variations in acetone/phenol ratio or water concentration, as shown in Table 2, would have a profound impact on the h_0 value and activation parameter values, E and lg A, which may result in gross errors in evaluation of the reaction rate constants when performed without proper correction of the Table 2 data.

Table 2

Reaction	DMBA → AMS	AMS → DMBA	DMBA → CP	DMBA → D_i
Rate Constant	0.47 l/min	0.094 mole/kg x min	0.023 l/min (*)	0.027 l/min (*)
lg A, min^{-1} (**)	5.9 – 8.2	8.8 - 9.2	14.5 –19.2	13.6 –17.4
E_a, kcal/mole x K (**)	8400 - 10700	12600 - 13900	15700 - 24700	13700-17400

* Average value for total CP isomers and D_i isomers, respectively.

** Higher water concentration increases the reaction activation parameter values.

The lower E_a value for the DMBA to AMS conversion compared to the AMS to DMBA case is an evidence of a remarkably greater reactivity of DMBA molecule over AMS. Therefore, DMBA molecule is proven to be responsible for formation of AMS dimers and CP.

An increase in acetone/phenol mole ratio in the reaction medium from 1:1 to 1.5:1 dramatically reduces the D_i/CP formation rate constant values. Therefore, the phenol process occurrence in an acetone-excessive reaction medium will decrease the yield of undesirable byproduct D_i and CP. At the same time, however, acetone excess in the reaction medium will remarkably decrease the CHP cleavage rate constant resulting in a lower CHP conversion and an intolerable increase in CHP concentration in the reactor effluent in terms of process safety. Without prompt computer-aided control of the CHP conversion value, it would be potentially dangerous and even prohibiting to increase the acetone/phenol mole ratio above 1 in the attempt to reduce the D_i/CP yields and improve the AMS yield in phenol process, be it either a one-stage or two-stage process. Moreover, even in commercial process where a prompt computer-aided control is arranged to make sure that CHP conversion is strictly constant, the acetone excess in the reaction medium cannot be varied arbitrarily for the purposes of D_i/CP yields reduction and AMS yield improvement. That is why the statement of Hertzog et al [13] that the CHP concentration in the reactor effluent reduces when extra acetone is added to the reactor is clearly a scientific and technical nonsense. The latter understanding is supported both by commercial practice and scientific research of acidity value h_0 (see Figure 5) and the kinetics of CHP cleavage, DCP hydrolysis, and D_i/CP formation reactions occurring in phenol-acetone medium either for the case of equimolar mixture or acetone mole excess, as shown in Table 3.

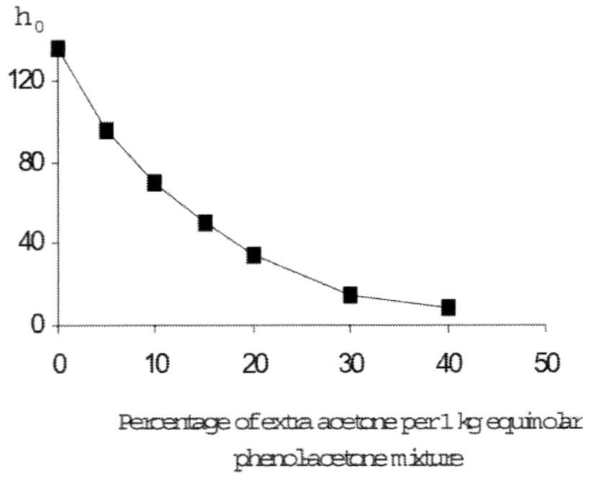

Figure 5. H_2SO_4 Acidity (h_0) as a Function of Extra Acetone Addition to Equimolar Phenol-Acetone Medium
[H_2SO_4], [H_2O], t -°C, [cumene] = const.

A significant decrease in the rate constant values indicated in Table 3 is caused by variation in the composition of acidic particles (non-dissociated H_2SO_4, quasi-ion couple of acetone with H_2SO_4, H_3O^+, or $H_5O_2^+$) with higher acetone concentration in the reaction medium. These particles are distinguished with different catalytic activity, which fact is responsible for h_0 variations, as shown in Figure 5, affected by extra acetone and water addition to the reaction medium. Therefore, the subject reactions of AMS dimers and CP formation are not only affected by H_3O^+ particles, as in formulas (4), (5), (7), and (8), but also by non-dissociated H_2SO_4 and $H_5O_2^+$, which in particular accounts for the variable ratio between the AMS dimers obtained.

Table 3

Extra Acetone, %relat. per 1 kg Equimolar Phenol-Acetone Mixture	Rate Constants x 10^2			
	k_{CHP}, s^{-1}	k_{DCP}, kg x $mole^{-1}$x min^{-1}	k_{Di}, min^{-1}	k_{CP}, min^{-1}
0	21	20	2.3	2.7
5	14	13	-	-
10	9.5	10	1.4	1.8
15	7.3	7.5	-	-
20	6.1	6.0	0.7	0.9

Conclusions

1. **It was found that 2,4-diphenyl-4-methylpentene-1 is formed in commercial phenol/acetone production process. This is an α-isomer which makes the major percentage among the total isomers of α-methylstyrene dimers obtained.**
2. **It was established that dimethyl benzene alcohol, not AMS, is responsible for AMS dimers formation.**
3. **It was exhibited that the commonly recognized concept of AMS dimers formation by carbonium ion mechanism is incorrect. This article presented a pattern including an attack by acidic particles on the DMBA OH group and the carbon atoms of benzene ring and the DMBA CH_3 group, which accounts for formation of both linearly structured unsaturated α-isomers and the double-bond-free cyclic AMS dimer named indane.**

References

1. Kruzhalov B.D., Golovanenko B.N. *Joint phenol and acetone production,* Moscow: Goskhimizdat, 1963.
2. Messina G., Lorenzoni L., Cappellazzo O., Gamba A.// La Chim. e l' Ind. 1983. V.65, No.1. P.10.
3. Levshtein V.A.// Oil Refinery and Petrochemistry series: NTRS, Moscow: CNIITEneftekhim. 1971. No.4. P.33.
4. *Cationic Polymerization*// P. Plesh, Moscow: Mir. 1966. P.264.
5. Zakoshansky V.M., Uriev U.N., Alexandrov A.S., and Idlis G.S.// Journal of Physical Chemistry. 1984. V.58. No.5. P.1265.
6. Zakoshansky V.M., Idlis G.S., Ogorodnikov S.K., Batalin O.E., and Blajin U.M. // Journal of Organic Chemistry. 1975. V.XI. Issue 1. P.152.
7. Zakoshansky V.M.// Catalysis in Industry. 2004 (to be published).
8. Zakoshansky V.M., Artemov A.V., and Antonovsky V.L.// Petroleum and Shale Processing series: Subject Review, Moscow: CNIITEneftekhim, 1988, Issue 6. P.56.
9. Librovich N.B., Mayorov V.D.// Bulletin of USSR Academy of Science, Chemistry series, 1977. No.3. P.684.
10. Kislina I.S., Bushmakin I.G., Sysoyeva S.G., Antonovsky V.L., Zakoshansky V.M., and Vinnik M.I.// Kinetics and Catalysis. 1989. V.30. No.1. P.229.
11. Kislina I.S., Vinnik M.I., Sysoyeva S.G., Svetlichnaya I.V., and Zakoshansky V.M.// Bulletin of USSR Academy of Science, Chemistry series, 1992. No.1. P.72.
12. Mayorov V.D., Kislina I.S., Sysoyeva S.G., Zakoshansky V.M. and Vinnik M.I.// Bulletin of USSR Academy of Science, Chemistry series, 1991. No.5. P.1001.
13. Patent Application 09/865,190 (USA)/ Hertzog R.R., Sifniades S., Fisher W.B.

Paper Number 96a

GLOBAL ENERGY MANAGEMENT SYSTEM

Kirtan Trivedi, Ph D, P.E.
ExxonMobil Research and Engineering,
3225 Gallows Road,
Fairfax, VA 22037

Paper presented at the Spring National AIChE Meeting, April 25-29, 2004

Session: Application of Process Integration tools for Energy Efficient Design

7th International Conference on Refining Processes

UNPUBLISHED

Extended Abstract

ExxonMobil refineries and chemical plants account for over 75 percent of corporate energy consumption and 60 percent of greenhouse gas emissions, so a sharp focus on energy conservation and efficiency is appropriate. Over the past 25 years, we have improved the energy efficiency of these facilities by over 35 percent -- saving the equivalent of 1.8 billion barrels of oil and reducing carbon emissions by over 200 million tonnes.

Significant opportunities also exist to improve energy efficiency and reduce emissions through optimization of existing facilities. We have redoubled efforts in this area with our Global Energy Management System (G-EMS). The G-EMS business model is based on a three-step approach to energy management. First, address base case performance issues by operating existing facilities optimally and efficiently through application of best practices. Second, identify economic investment opportunities above an optimized base for step-change improvement to address structural differences. And third, implement strong management systems to provide rigor and discipline necessary to sustain efficiency gains and drive continuous improvement. Thus far, we have identified opportunities to improve the energy efficiency of our refineries and chemical plants an additional 15-20 percent. At full implementation of G-EMS, potential savings could approach half a billion dollars per year at current energy prices, with an associated reduction in carbon dioxide emissions of nearly 9 million tonnes per year. We expect to capture about half of these benefits with little or no investment.

This presentation discusses the G-EMS system and its application within ExxonMobil refineries and chemical plants

Global Energy Management System

Spring National AIChE Meeting
April 25-29, 2004,

Kirtan Trivedi
ExxonMobil Research and Engineering Company, Fairfax, VA

ExxonMobil 1

ExxonMobil Facts

- Business in nearly 200 countries and territories employing almost 100,000 people worldwide
- Number 2 in Fortune's Global 500
- Revenues of $ 213 billion, assets of over $ 143 billion
- Produce 2.5 million barrels of oil and 10.3 billion cubic feet of natural gas daily
- Ranked among top 3 petrochemical companies
- Revolutionized deep water oil and gas exploration and production
- Among the world's largest independent producers of electricity
- A leader in fuels for advanced engines and fuel cells

ExxonMobil 2

WORLD ENERGY OUTLOOK

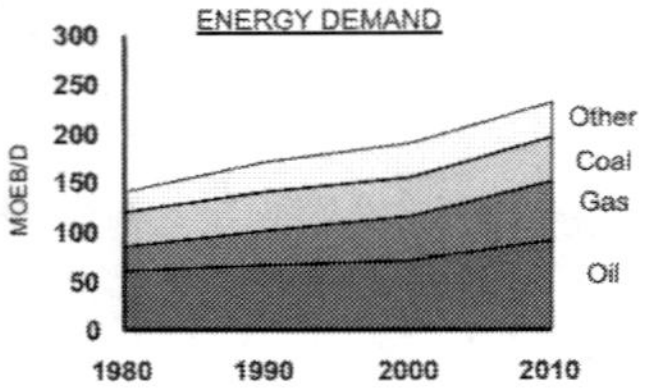

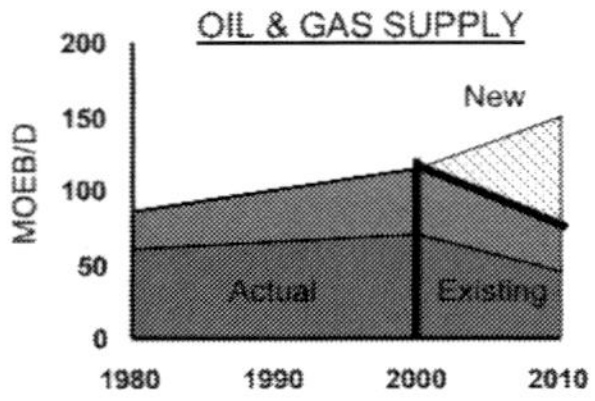

- Oil and natural gas supply 60% of world energy demand
- About 50% of 2010 supply will come from new production
- Balanced approach required to meet the supply challenge

ExxonMobil 3

MEETING THE CHALLENGE

- ExxonMobil engaged on a number of fronts to meet tomorrow's energy needs ...
  - New exploration and production
  - New technologies and improved products
  - Energy conservation and efficiency

ExxonMobil 4

A CORPORATE FOCUS

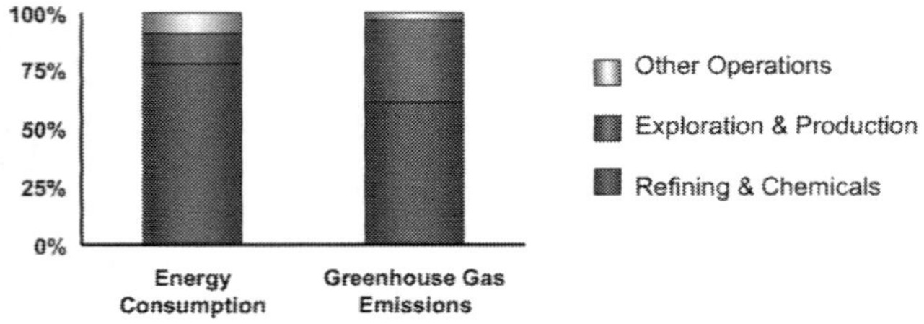

- Refining and Chemicals account for over 75% of energy consumption and 60% of greenhouse gas emissions
- Initial efforts aimed at improving Downstream energy efficiency
- Initiatives also underway in Upstream and Lubes Manufacturing

ExxonMobil 5

IMPROVING ENERGY EFFICIENCY

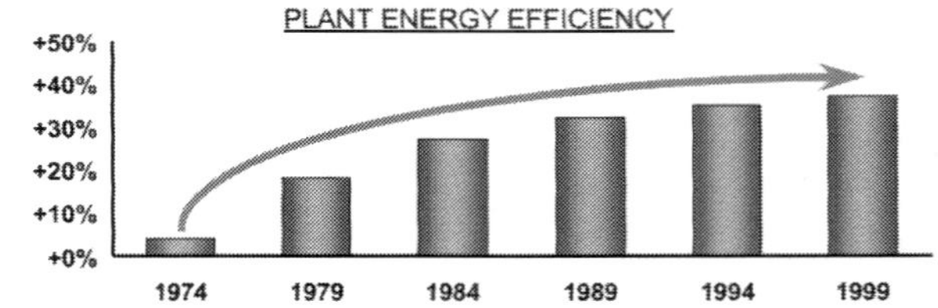

- Plant energy efficiency has improved over 35% since 1973
  - Saved cumulative equivalent of 1.8 billion barrels of crude oil
  - Translates to over 200 million tonne decrease in GHG emissions
- Ongoing initiatives expected to provide additional improvement
  - Additional investment in highly efficient cogeneration facilities
  - Implementation of Global Energy Management System (G-EMS)

ExxonMobil 6

COGENERATION

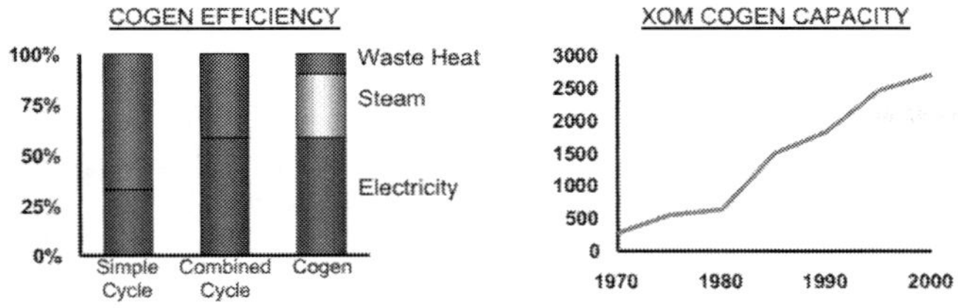

Cogeneration 30-35% more efficient than conventional technology

Facilities at 32 locations produce 2700 megawatts of electricity

- Efficiency gain sufficient to service nearly 1 million U.S. residential households
- GHG emissions reduced nearly 7 million tonnes per year versus alternatives

Additional 970 megawatts planned at 5 sites in U.S. and Canada

- Potential to further reduce GHG emissions more than 2 million tonnes per year when new capacity fully on line

PLANT ENERGY EFFICIENCY

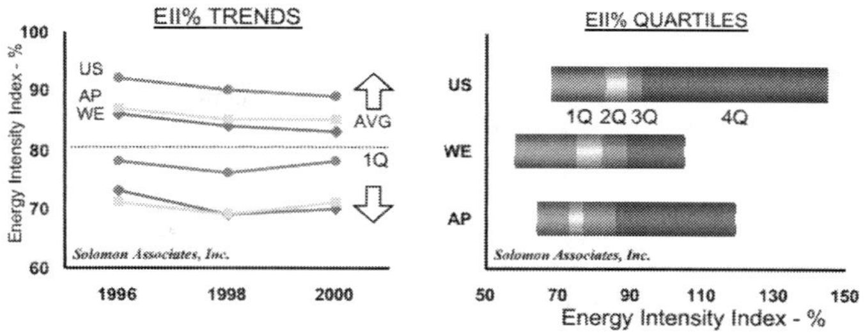

- Industry trend groups improving 1% per year on average
- Top performers appear to have lined-out ... for now
- Data highlights potential opportunity for many plants

A GLOBAL APPROACH

OBJECTIVES

Establish a single, comprehensive global energy management system (G-EMS)

Utilize a common methodology to identify performance gaps, implement closure plans, sustain progress, and continuously improve results

BUSINESS MODEL

Operate existing facilities optimally and efficiently through application of best practices

Identify economic investment opportunities above an optimized base for step-change improvement

Implement strong management systems to sustain progress and drive continuous improvement

G-EMS DESIGN

G-EMS DEVELOPMENT

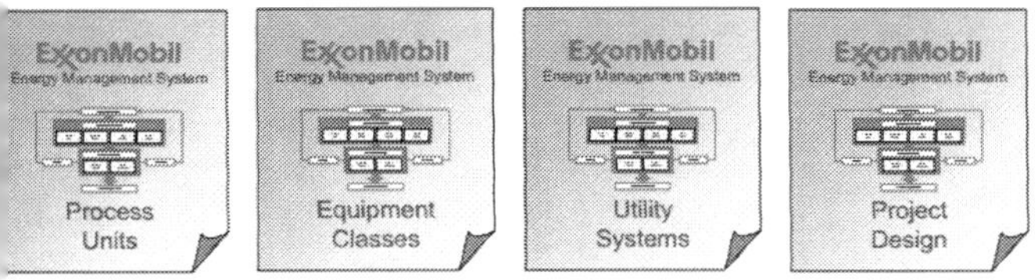

- Best practices documented in 12 volume set of manuals
- Contain 1200 pages and identify over 200 key energy variables
- Cover key process, equipment, and utility operations
- Also address energy efficiency in project design

Identify Best Practices

Provide Performance Measures and "How-to's"

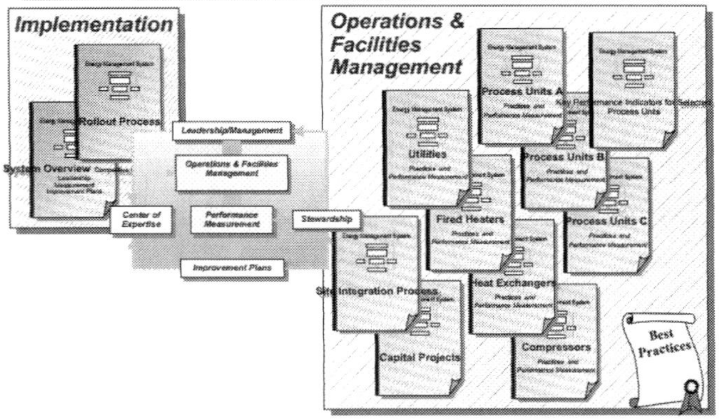

G-EMS MEASUREMENT

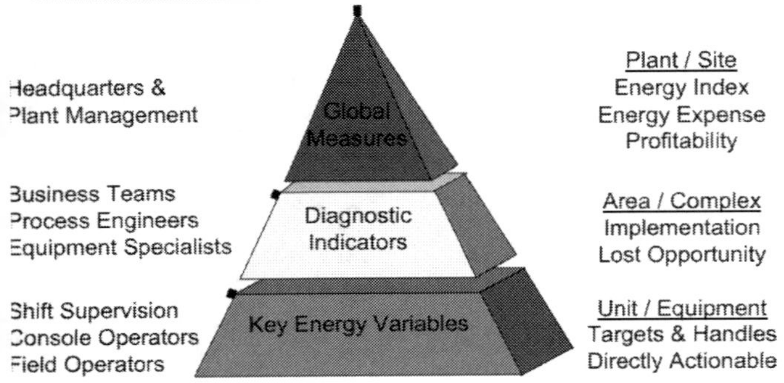

G-EMS ROLLOUT PROCESS

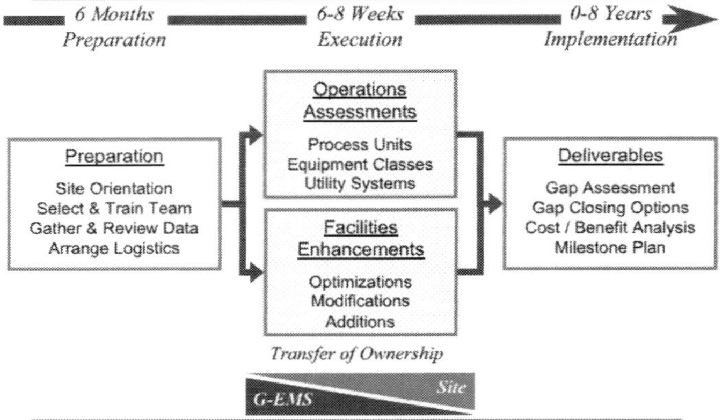

Plan Implementation

Deploy Through Rollouts to Largest Opportunity Sites

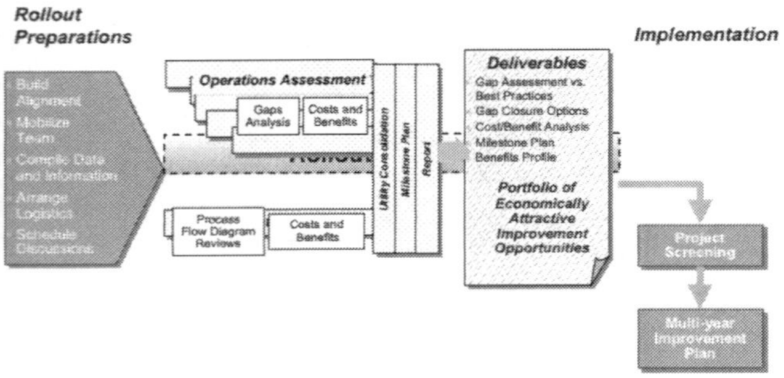

G-EMS DEPLOYMENT PLAN

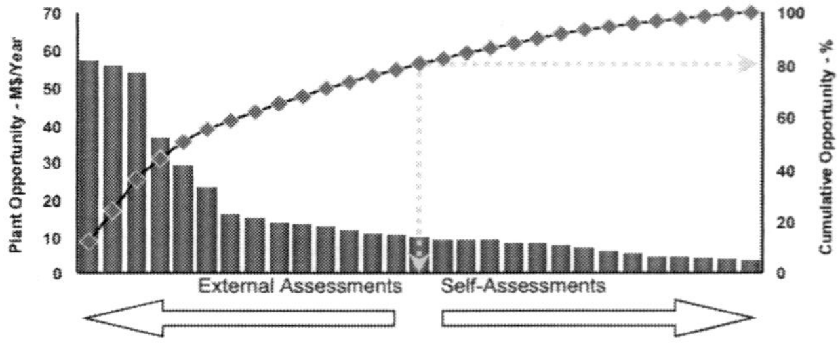

- Rollouts at 15-20 sites address 80% of economic opportunity
- Balance of opportunity addressed via rigorous Self-Assessments
- 13 Rollouts executed to date ... 5 Rollouts planned in 2003

G-EMS ROLLOUT RESULTS

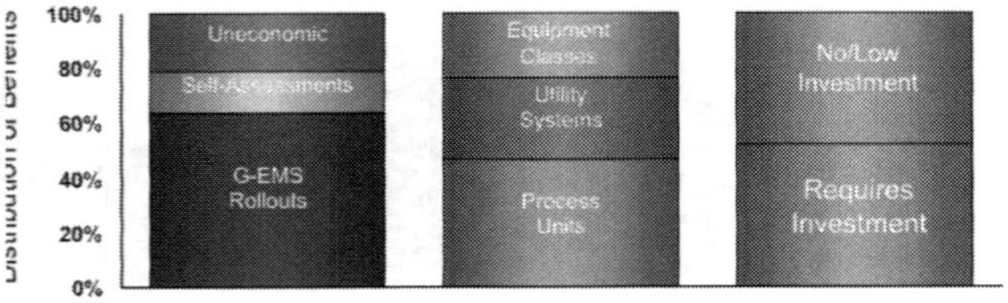

- Rollouts and Self-Assessments address 80% of total opportunity
  - Process Units total nearly 50% -- mainly investment related
  - Utility Systems account for over 30% -- mix of no/low and investment items
  - Equipment Classes total 20% -- many no/low investment options
- Potential savings equal to 15-20% of plant energy bill on average
  - About half of the benefits identified require little or no investment to capture
  - Potential to further reduce GHG emissions nearly 9 million tonnes per year at full implementation

Existing Flow scheme 1/2

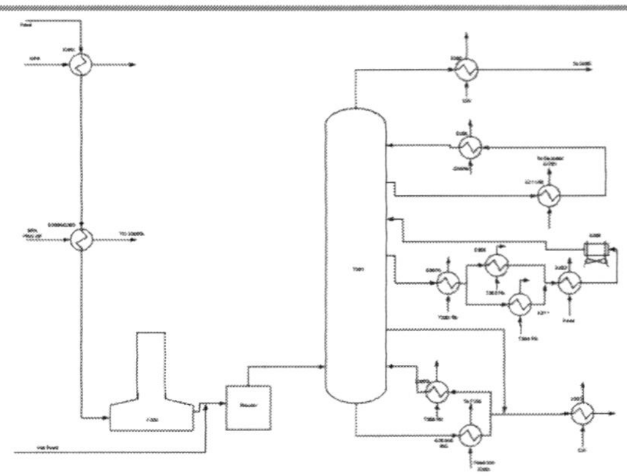

Existing Flow scheme 2/2

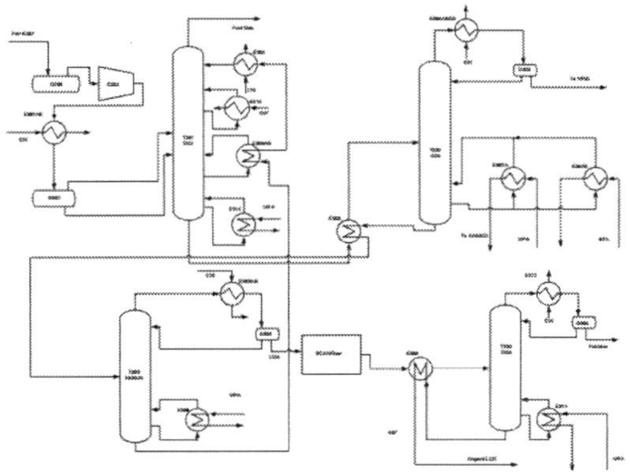

Options

- Two options
  - Generate 600 # steam, shift 125 # from T-301 reboiler to feed preheat
  - Recover waste heat to save 600 # (E-407,E406AB), 125 # (E406C) and fuel (F-202)
- Generate 600 # steam
  - Services on MPA and 125 # steam are moved to TPA (E311 and E314)
  - Additional MPA duty is shifted to pre-heat the feed
  - 125 # steam is shifted from T-301 reboiler to pre-heat the feed
  - BPA feed pre-heat is shifted to generate 600 # steam
  - BFW is heated in the convection section and steam is superheated in the convection section of F202.
  - Superheated 600 # steam is generated with this scheme resulting DCF 33.5 %
  - Additional savings obtained due to capital savings in another grass root project requiring CW

Waste Heat Recovered, MBtu/h	
TPA to CW	16
MPA to Air	18
CFB to CW	4
F202 stack	11
Total	**49**

Flow scheme for 600 # steam generation 1/2

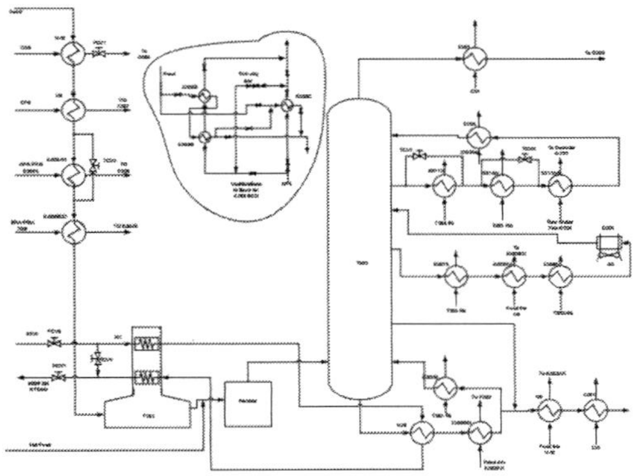

Flow scheme for 600 # steam generation 2/2

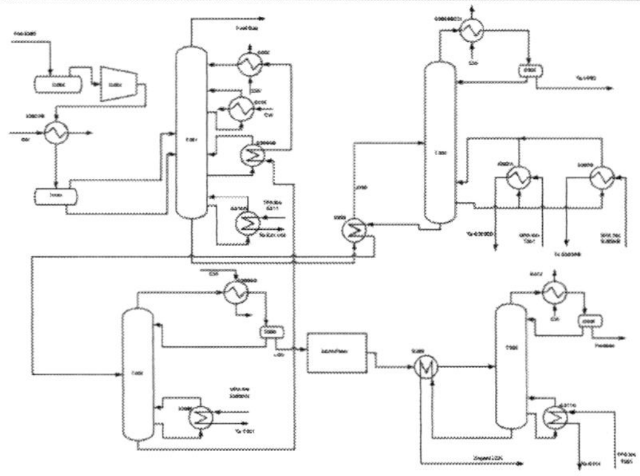

Alternate Waste heat recovery projects

- Three independent projects identified
- Heat feed in the convection section of F-202
  - DCF 47.4 %
- Waste heat recovery from MPA stream currently rejecting heat in Air cooler
  - MPA is used to reboil T404 and other unit feed resulting in saving 600 # and 125 # steam
  - DCF 42.4 %
- Waste heat recovery from TPA
  - TPA is used to reboil T301 resulting in saving 125# steam
  - DCF 39.5 %

Waste Heat Recovered, MBtu/h	
TPA to CW	10
MPA to Air	15
F202 stack	12
Total	**37**

Flow scheme waste heat recovery 1/1

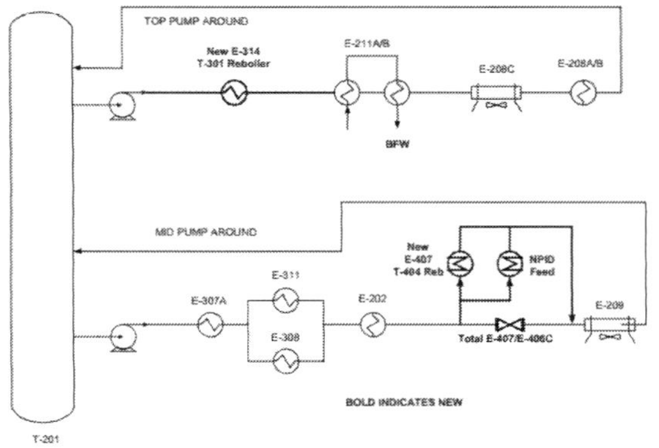

Incremental Economic Analysis

Projects	DCF %
Waste Heat recovery Projects	43
600 # steam generation	33
Incremental for 600 # steam generation project	21

- Steam generation project is recommended as the incremental cost has a 21 % DCF return and the DCF for the whole project is 33.5 %
- Margin credits associated with steam generation
  - 600# steam imbalance during a yearly 21 day Boiler TA requiring FCCU throughput reduction.
- DCF for the steam generation project including margin credits will be 37 %
- Capital savings in alternate grass root project

ExxonMobil

PATH FORWARD

- Potential for climate change poses a legitimate long-term risk
- Path forward must be marked by rational scientific, economic, and technical analysis
- Maintain focus on increasing supplies, developing next generation products and technology, energy conservation and efficiency
- Continue support for scientific and economic research and participate in cooperative programs with industry and government
- Conservation and efficiency -- the first path
  - Develop 970 MW of additional cogeneration capacity
  - Improve refinery/chemical plant efficiency an additional 15-20%
  - Reduce GHG emissions about 11 million tonnes per year
- Benefits consumers, companies, and the environment alike now and well into the future

ExxonMobil

Paper Number: 96d

Using Spreadsheets for Pinch Analysis

Alan P. Rossiter,
President, Rossiter & Associates,
Bellaire, Texas 77401
Tel: 713-660-9503; Fax: 713-669-1987; e-mail: alan@rossiters.org

Prepared for presentation at the AIChE Spring National Meeting,
New Orleans, LA, April 25-29, 2004

Seventh Topical Conference on Refinery Processing

Session: Application of Process Integration Tools for Energy Efficient Design

Unpublished

Using Spreadsheets for Pinch Analysis

Alan P. Rossiter,

Spreadsheets are now the workhorse for day-to-day repetitive calculations, but most of us ignore them for larger, more complex computing requirements. In this paper the author describes the development of a spreadsheet that delivers most of the functionality of mainstream pinch software, and illustrates its use with a practical refinery example.

Introduction

Pinch analysis(1) is a systematic technique for analyzing heat flows through an industrial process, based on fundamental thermodynamics. It is widely used in the energy efficiency area, and also has applications in waste minimization(2), plant debottlenecking and other forms of process improvement(3).

Several major commercial software packages are now available for pinch analysis, as well as older packages that lack the functionality and user-friendliness of current generation of mainstream programs. This paper describes a Microsoft® Excel spreadsheet, developed by the author, which provides pinch targeting and heat exchanger network analysis capabilities similar to those of mainstream commercial programs. A refinery example (FCC Unit) is illustrates the use of the program.

Scope of Software

The most prominent commercial pinch software packages are sophisticated tools, and include capabilities that are not used in most projects(4). The objective of this work was to develop a spreadsheet that provided only the functionality required for a pinch study of a typical process unit. The required capabilities are:

i. composite curves
ii. basic energy targets
iii. pinch identification
iv. grand composite curves
v. multiple utility targets (including segmented utility targets)
vi. quantification of heat crossing pinches in each heat exchanger
vii. grid diagrams

Another important requirement of the spreadsheet was that it should have a simple and user-friendly data entry mechanism. The format adopted to achieve this is a heat exchanger data entry form (Table 1). Here the data for each heat exchanger (equipment name, stream names, heat load and stream temperatures) are entered, using one line for each heat exchanger. Drop down menus are used to facilitate data entry. A similar data entry form is used for utility data (Table 2).

Microsoft® Excel was chosen as the vehicle for the program because of its familiarity in the engineering community, and also because of its compatibility with other Windows software. This compatibility is important, as it greatly facilitates reporting on pinch projects. The spreadsheet was given the name "PinchExcel". Design and development of the program to its present form has taken roughly three man-months.

Table 1: Process (Heat Exchanger) Data Entry Form

First, enter Hot and Cold Streams Names (columns 1 and 2) Then, enter Equipment Number, Name and Duty (columns 3 through 5) and Stream Characteristics for each heat exchanger (columns 6 through 13)
For this data set, temperatures are in °C and heat loads are in MW.

Process Data Form

Error Status: No errors

First, enter Stream Names **Then:** **Enter Stream T, Q and HX Data** ΔT_{min} = **30**

Hot Stream Names:	Cold Stream Names:	Equipment Number	Equipment Name	Duty	Hot Side Stream Name	Hot Side Utility Name	Hot Side Ts	Hot Side Tt	Cold Side Stream Name	Cold Side Utility Name	Cold Side Ts	Cold Side Tt
Bottom Pumparound 1	Tank Feed	E1A	Mixed Feed Prehtr A	11.46	Bottom Pumparound 1	Process	343.0	281.0	Hot Feed	Process	181.6	274.0
Bottom Pumparound 2	Hot Feed	E1B	Mixed Feed Prehtr B	12.28	Bottom Pumparound 2	Process	343.0	281.0	Tank Feed	Process	125.0	274.0
Bottom Pumparound 3	Mixed Feed	H1	Feed Heater	20.34		Fired Heater			Mixed Feed	Process	274.0	360.0
Slurry Product		E2	17 barg Steam Generator	16.88	Bottom Pumparound 3	Process	281.0	232.0		17 barg Steam Gen	152.0	208.1
Heavy Cycle Oil Prod		E3	Slurry Product Cooler	1.82	Slurry Product	Process	343.0	121.0		Air		
Light Cycle Oil		E4	HCO Cooler	1.23	Heavy Cycle Oil Prod	Process	263.0	49.0		Cooling Water		
Mid Pumparound		E5	Tank Feed Prehtr 1	2.05	Light Cycle Oil	Process	202.0	97.0	Tank Feed	Process	51.0	76.0
Top Pumparound		E6	LCO Cooler	1.23	Light Cycle Oil	Process	97.0	49.0		Cooling Water		
Tower Overhead		E7	Tank Feed Prehtr 2	4.04	Mid Pumparound	Process	254.0	193.0	Tank Feed	Process	76.0	125.0
		E8	BFW Preheater 1	0.80	Mid Pumparound	Process	193.0	179.0		17 barg Steam Gen	107.0	154.0
		E9	MPA Cooler	5.77	Mid Pumparound	Process	179.0	77.0		Cooling Water		
		E10	BFW Preheater 2	0.75	Top Pumparound	Process	163.0	157.0		17 barg Steam Gen	107.0	150.0
		E11	TPA Cooler	14.04	Top Pumparound	Process	157.0	49.0		Cooling Water		
		E12	Tower Ovhd Cooler 1	18.17	Tower Overhead	Process	111.0	60.0		Air		
		E13	Tower Ovhd Cooler 2	5.25	Tower Overhead	Process	60.0	37.0		Cooling Water		

Table 2: Utility Data Form

For this data set, temperatures (Ts, Tt) and temperature differences (ΔT) are in °C and segment heat loads (Δh) are dimensionless ratios.

Utility Specifications Form

	Name	Duty Type	Fixed Duty	Segments	Ts	Δh1	Tt1	Δh2	Tt2	Δh3	Tt3	ΔTmin Contribution
Hot Utilities:												
h1												
h2												
h3												
h4												
h5												
h6	Fired Heater	Calculate		1	427		426					20
Cold Utilities												
c1	17 barg Steam Gen	Calculate		2	107	442	208	1906	208.1			-5
c2	Air	Calculate		1	43		44					0
c3	Cooling Water	Calculate		1	27		28					-5
c4												
c5												
c6												

Example Problem

This example was initially developed for Natural Resources Canada[5]. The process under examination is the feed and fractionation section of an existing 35,000 BPD FCC Unit, together with the associated preheat train, pumparounds, product streams and heat recovery network (see simplified process flow diagram [PFD], Figure 1). There are two feed streams, one coming hot from a crude unit and the other coming cold from storage. The feed is preheated and then passes through a feed furnace (H1) before going to the reactor for cracking. The reactor effluent then goes to the main fractionator for separation into product streams. Excess heat is removed to generate 17 barg steam or rejected in air or cooling water.

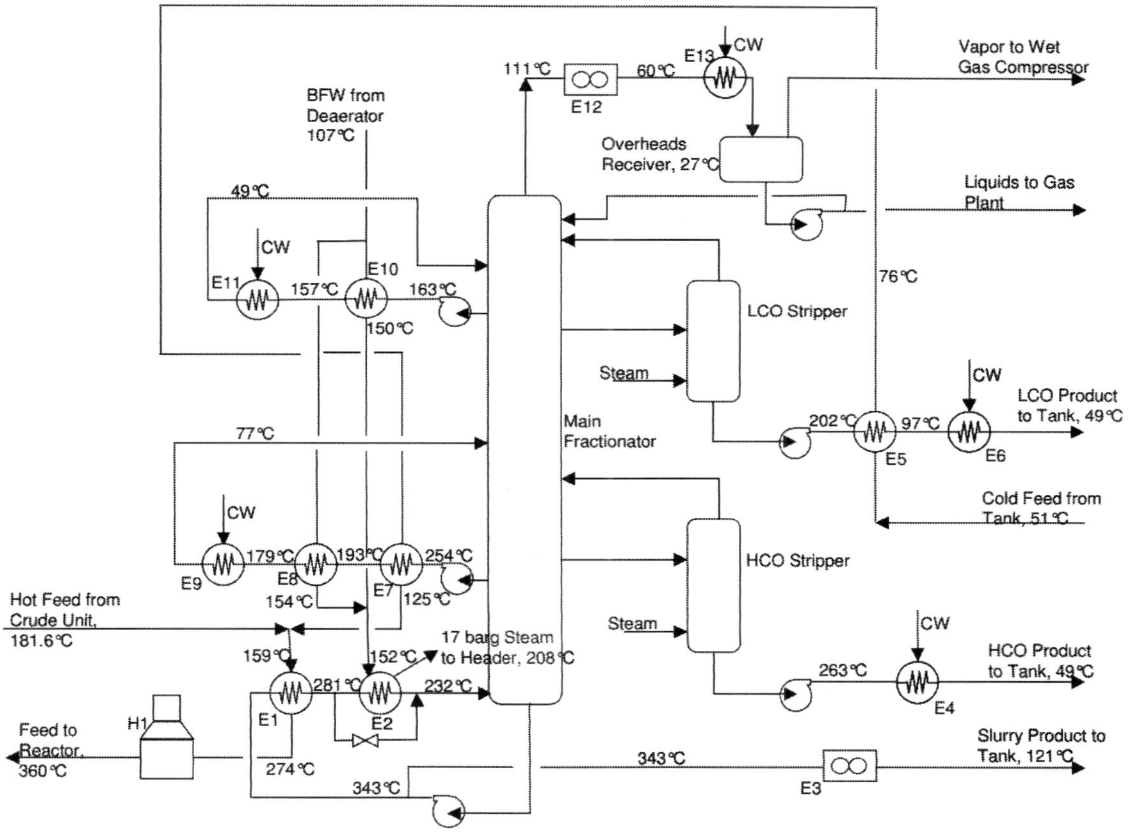

Figure 1: Simplified Flowsheet for FCC Process

Issues in Software Design and Implementation

The basic energy targets, composite curves (Figure 2), andgrand composite curve (Figure 3) were implemented using established algorithms[1]. The graphical outputs are readily generated using the curve plotting capabilities of Excel.

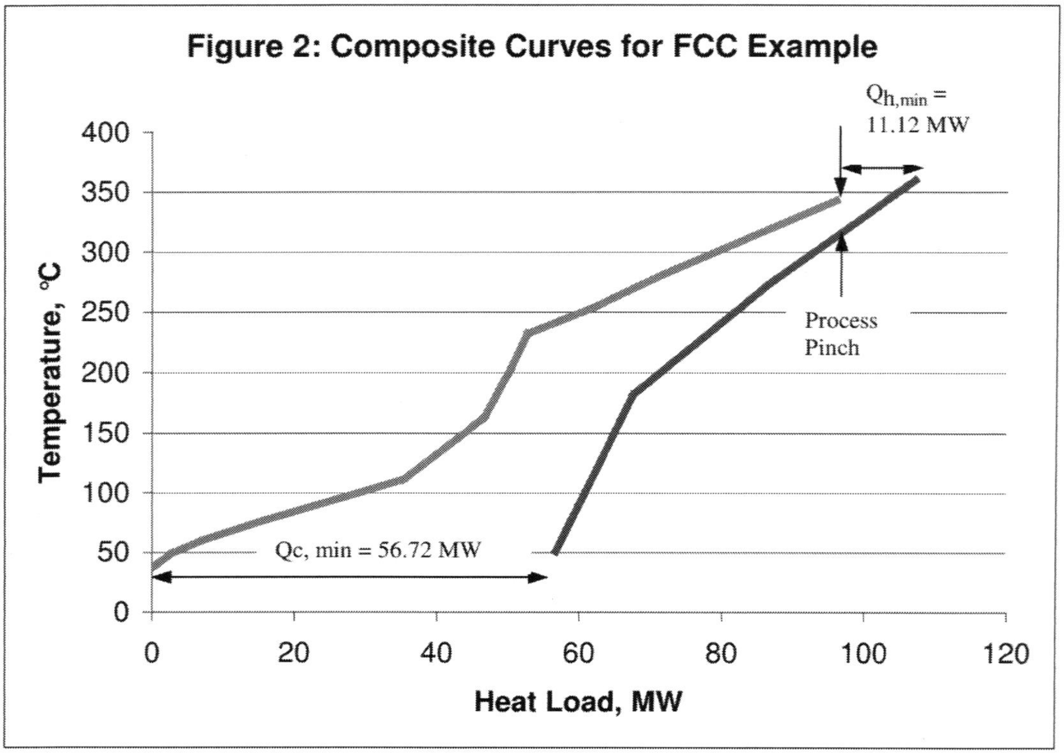

Figure 2: Composite Curves for FCC Example
Qh,min =
11.12 MW
Process
Pinch
Qc, min = 56.72 MW
Temperature, °C
Heat Load, MW
400
350
300
250
200
150
100
50
0
0
20
40
60
80
100
120

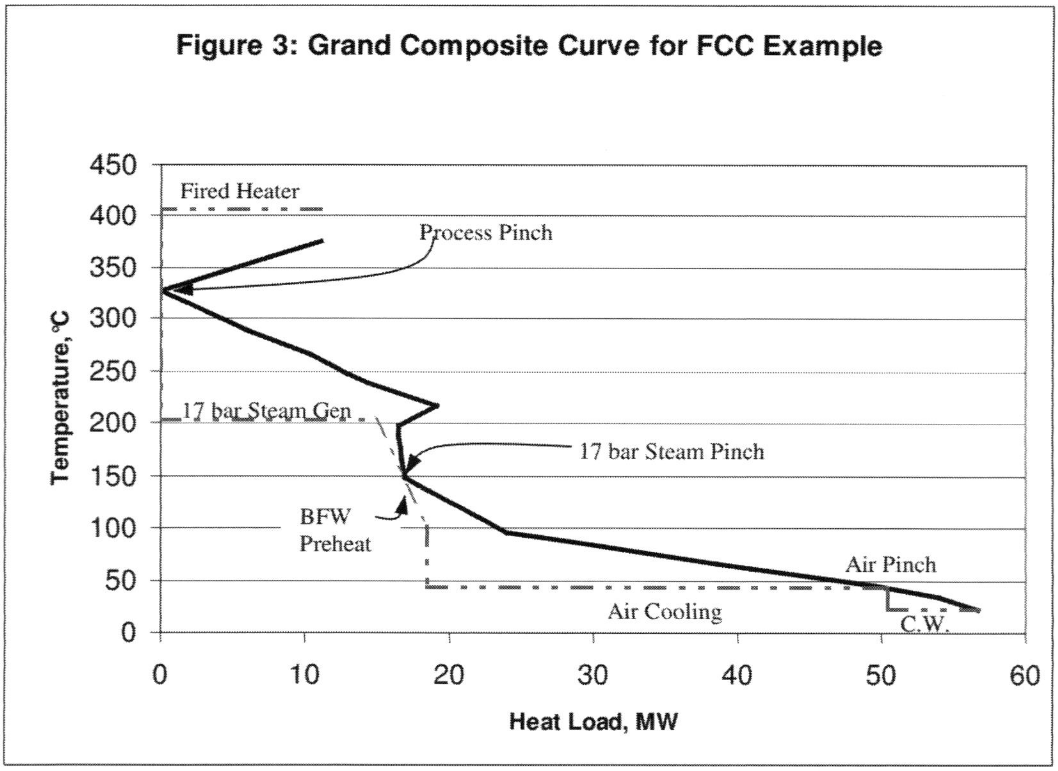

Figure 3: Grand Composite Curve for FCC Example
Fired Heater
Process Pinch
17 bar Steam Gen
17 bar Steam Pinch
BFW
Preheat
Air Pinch
Air Cooling
C.W.
Temperature, °C
Heat Load, MW
450
400
350
300
250
200
150
100
50
0
0
10
20
30
40
50
60

New and unpublished algorithms were used for multiple utility targeting, pinch identification (also on Figure 3) and heat crossing pinches. The algorithm for multiple utility targeting is perhaps of greatest interest, as it incorporates a technique that overcomes a shortcoming noted in the leading commercial packages when dealing with segmented utilities (such as boiler feed water followed by steam generation). This is discussed in more detail later.

However, probably the most challenging part of the spreadsheet was the grid diagram. The specific issues in this part of the program are discussed below.

The Grid Diagram

The essential problem of the grid diagram is to convert the textual and numerical data from the heat exchanger data entry form (Table 1) into a graphical display, showing process streams and heat exchangers in the correct location relative to the process and utility pinches in the process. This is primarily a matter of data management.

The first part of the solution lies in the structure of the heat exchanger data entry form itself. Here the user has to list the names of the hot and cold process streams first (at the extreme left-hand side of the form, see Table 1) before providing details of the heat exchangers themselves. The names of the process streams are now made available to the user in drop down menus within the heat exchanger data entry form. This avoids possible inconsistencies when naming process streams (e.g. "BPA" or "Bottom Pumparound" could both be used as names for the same stream in different heat exchangers). The grid diagram algorithm then identifies and links the segments of each stream and places heat exchangers on each stream in the correct order.

Pinch locations (identified by the utility targeting algorithm) are transferred to the grid diagram, and the stream and heat exchanger information is then displayed contiguously and in correct locations with respect to the pinches. Heat loads and data on heat crossing the pinch are also transferred to the grid diagram from other calculations in the spreadsheet. These tasks are accomplished primarily by using Excel's lookup and reference functions.

The utilities must also be named and their temperatures specified (see Table 2) before completing the heat exchanger data. These names are made available to the user in drop down menus in the heat exchanger data entry form (Table 1). This ensures that only pre-defined utilities are selected when heat exchangers are specified. The utilities are then displayed on the grid diagram in the correct locations with respect to the pinches.

Streams and utilities are represented as horizontal lines on the grid diagram, and a heat exchanger is shown as a circle on a stream. This is easily accomplished by using graphical elements, which are placed by the program's logic to construct the diagram. Each heat exchanger (circle) is labeled with the heat exchanger number (Figure 4).

Figure 4: Grid Diagram Before Manual Completion of Dumbbells

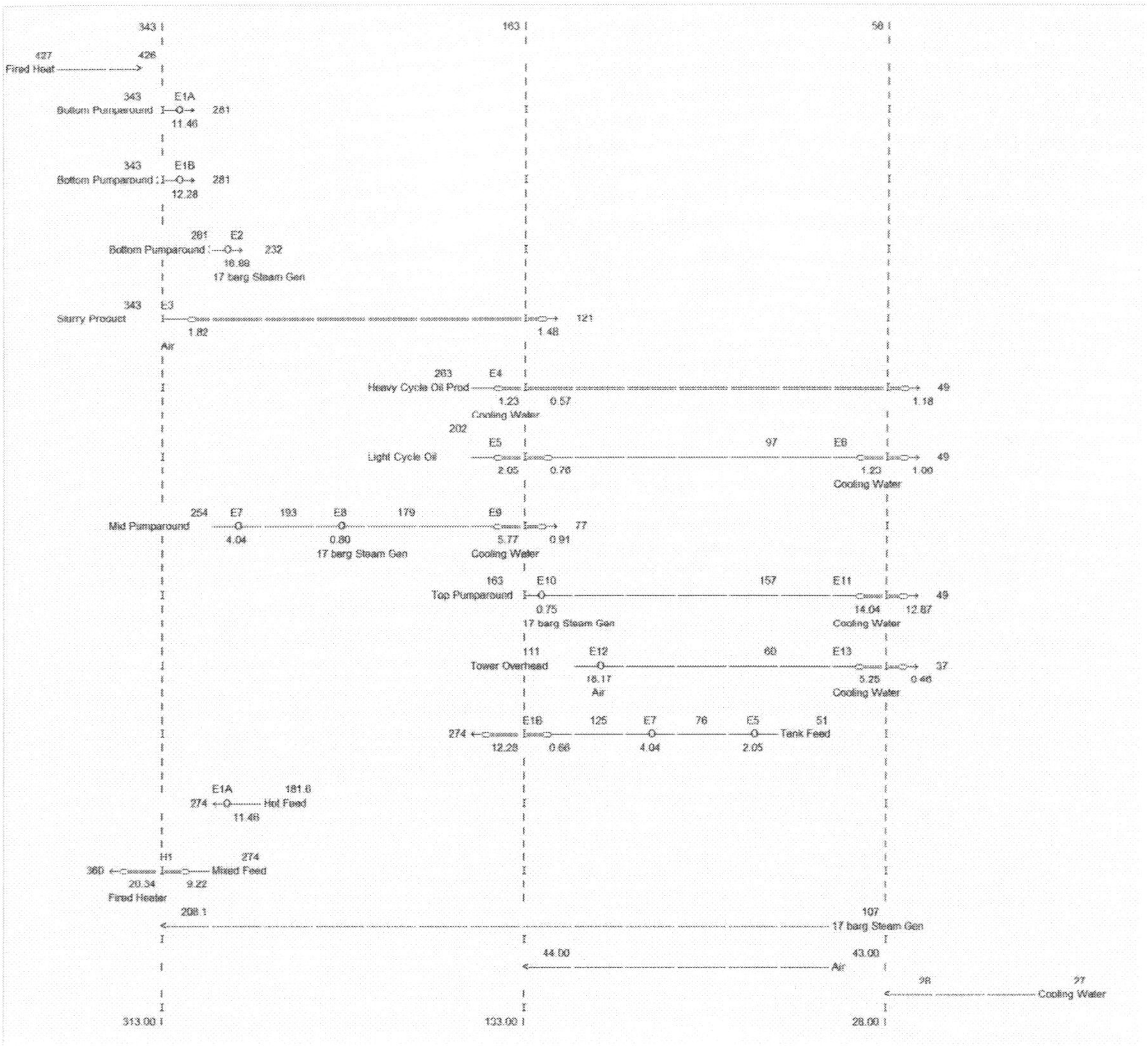

Temperatures are in °C; Duties are in MW
Where part of the duty in a heat exchanger crosses a pinch, the total duty is shown to the left of the pinch and cross-pinch duty to the right.

In order to complete the diagram we must now link the circles on the hot streams with the corresponding circles on cold streams to create "dumbbells" representing the complete heat exchangers. It would be possible to do this using Excel, but the logic required to accomplish this was considered excessively complex. The final connecting of the circles is therefore left to the user. In another simplification, the program does not explicitly handle stream splits in the network; it is left to the user add mixes and splits on the grid diagram (see Figure 5).

Although it requires a little effort to finalize the grid diagram, the use of Excel gives the user the opportunity to customize its appearance. For example, we can show the mixing of the tank feed and the hot feed to create the mixed feed (Figure 5) – something that could not ready be done in most commercial pinch programs. Also, utility exchangers can be shown as circles rather than dumbbells, to minimize clutter in the final diagram.

Figure 5: Grid Diagram After Manual Completion of Dumbbells

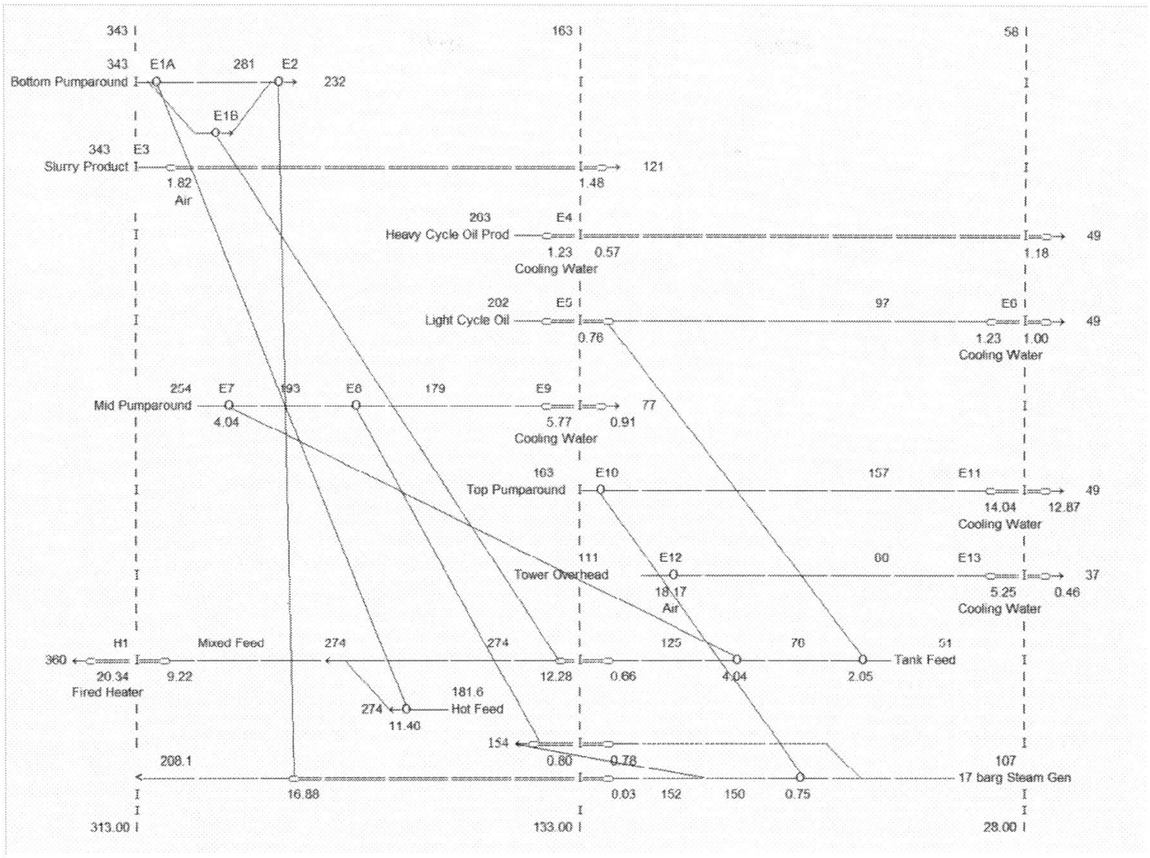

Temperatures are in °C; Duties are in MW
Where part of the duty in a heat exchanger crosses a pinch, the total duty is shown to the left of the pinch and cross-pinch duty to the right.

Segmented Utilities

Many of the opportunities identified in pinch studies are associated with segmented utilities – i.e., utilities with two or more distinct segments. The most common situations of this type arise with steam, where we can consider boiler feed preheat (or condensate cooling), vaporization and superheat as three separate parts of the utility. In the FCC example that we are considering, we generate saturated 17 barg steam from boiler feed water at 107°C. The corresponding temperature-specific enthalpy plot is shown in Figure 6.

Figure 6: Temperature-Enthalpy Profile for Steam Generation

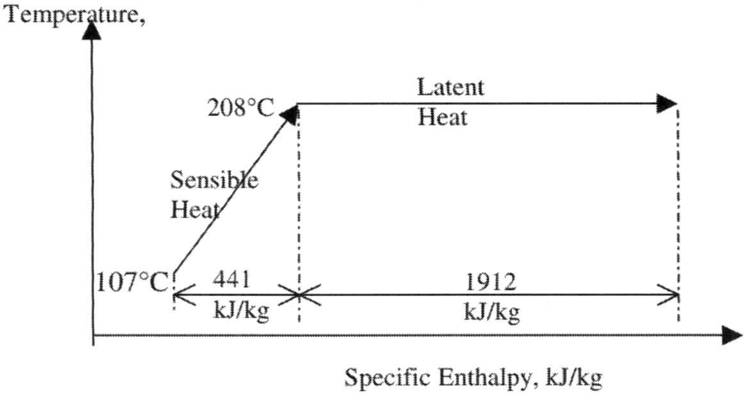

In the existing design, boiler feed water is preheated in two parallel heat exchangers (E-8 and E-10), ahead of the steam generator itself, E-2. In the main commercial pinch programs, if we specified 17 barg steam generation as a segmented utility and placed E-8 and E-10 on that utility, the program would assume that we were heating boiler feed water from 107°C to the saturation temperature, and then generating steam, within these heat exchangers. However, as the heat sources in both cases are below 208°C (see Table 1) the main commercial programs would report these heat exchangers as infeasible.

In PinchExcel, the user has the option of specifying the temperature range over which a utility is used. In Table 1, E-8, E-10 and E-2 are all shown as with 17 barg steam as the cold utility (see Table 1). However, E-8 and E-10 are shown with temperature ranges of 107°C to 154°C and 107°C to 150°C, respectively, identifying them as low-temperature boiler feed water preheaters. E-2 shows a temperature range of 152°C to 208.1°C, identifying it as a high-temperature boiler feed water preheater and steam generator. For all other utility heat exchangers the temperature fields are left blank in Table 1. The program interprets this as meaning that the full temperature range of the utility is to be used.

With this appropriate allocation of temperature ranges, the cross-pinch heat flows are correctly computed (see Table 3), and none of the heat exchangers are infeasible.

Table 3: Cross-Pinch Table

ID	Name	Hot Stream	Cold Stream	Pinch		
				Process	17 barg Stm Gen	Air
				328°C	148°C	43°C
E1A	Mixed Feed Prehtr A	Bottom Pumparound	Hot Feed			
E1B	Mixed Feed Prehtr B	Bottom Pumparound	Tank Feed		0.66	
H1	Feed Heater	Fired Heater	Mixed Feed	9.22		
E2	17 barg Steam Generator	Bottom Pumparound	17 Bar Steam Gen		0.03	
E3	Slurry Product Cooler	Slurry Product	Air		1.48	
E4	HCO Cooler	Heavy Cycle Oil Product	Cooling Water		0.57	1.18
E5	Tank Feed Prehtr 1	Light Cycle Oil	Tank Feed		0.76	
E6	LCO Cooler	Light Cycle Oil	Cooling Water			1.00
E7	Tank Feed Prehtr 2	Mid Pumparound	Tank Feed		4.04	
E8	BFW Preheater 1	Mid Pumparound	BFW (17 Bar Steam Gen)		0.78	
E9	MPA Cooler	Mid Pumparound	Cooling Water		0.91	5.77
E10	BFW Preheater 2	Top Pumparound	BFW (17 Bar Steam Gen)			
E11	TPA Cooler	Top Pumparound	Cooling Water			12.87
E12	Tower Ovhd Cooler 1	Tower Overhead	Air			
E13	Tower Ovhd Cooler 2	Tower Overhead	Cooling Water			0.46
	Total			9.22	9.23	21.27

Study Results

The analysis of this FCC Unit is discussed in detail elsewhere(5). The main results are given below:

The basic targets are presented in Table 4. There is potentially scope to reduce the fired heater duty (heat absorbed) by 9.22 MW, with a very small (0.17 MW) reduction in 17 barg steam generation. The target savings are worth about $1,242,000/year.

Table 4: Targets for Energy and Utilities

Process ΔT_{min} = 30°C; Steam and CW ΔT_{min} = 10°C; Air ΔT_{min} = 15°C

	Existing MW	Target MW	Scope MW
Total Hot	20.34	11.12	9.22
Total Cold	65.94	56.72	9.22
Hot Utilities:			
Fired Heater	20.34	11.12	9.22
Cold Utilities:			
17 barg Steam Gen	18.43	18.26	0.17
Air	19.99	32.21	-12.22
Cooling Water	27.52	6.25	21.27

The final design, after working through a retrofit heat exchanger network analysis, is shown in Figure 7. The changes incorporated in the process are as follows:

Figure 7: FCC Unit with Proposed Changes

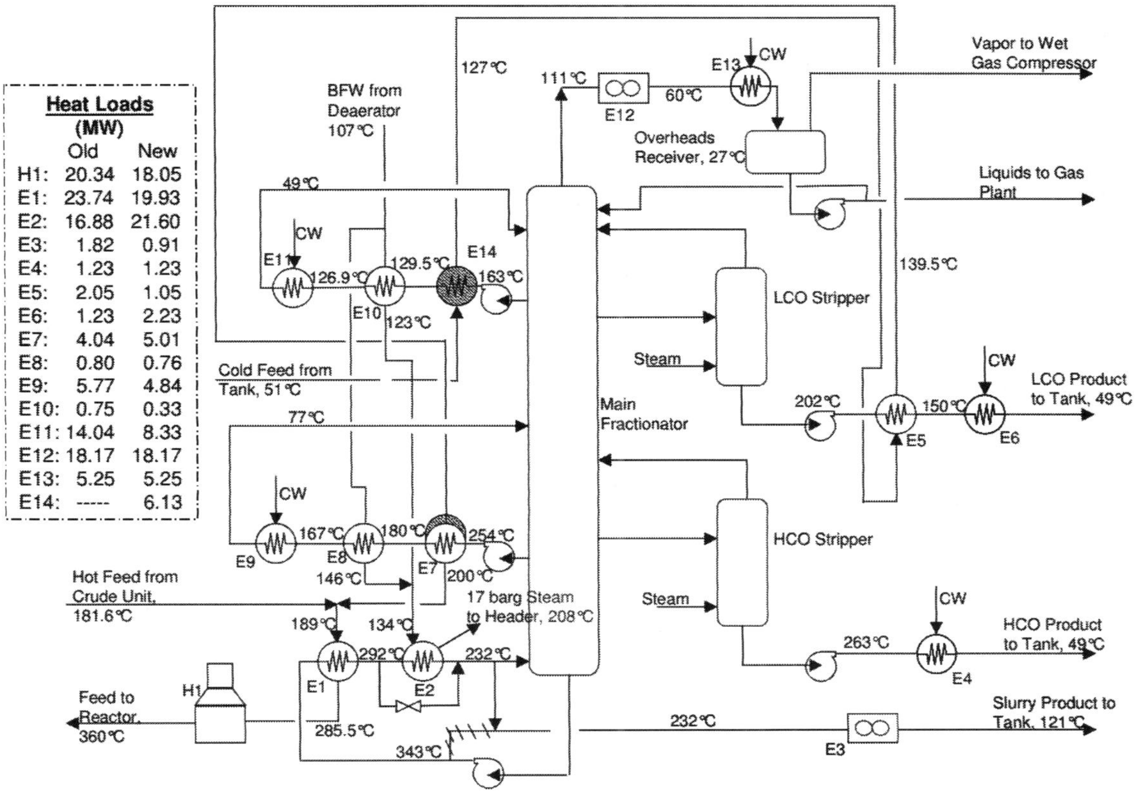

a. Add a new service E14, which matches TPA against the cold tank feed ahead of the existing service E5 (light cycle oil/cold tank feed).

b. Add a shell to E7, the existing MPA/cold tank feed service.

c. Re-pipe the slurry product draw to a point downstream of steam generator E2.

Overall the changes shown in Figure 7 save 2.29 MW in heat absorbed in furnace H1 and increase total heat to steam generation by 4.26 MW. The combined savings are 6.55 MW, worth roughly $828,000/year. These savings amount to 72% of the target energy savings and 67% of the target monetary savings. It was found to be uneconomic in this case to approach closer to the "ideal" pinch targets.

Conclusion

A spreadsheet has been developed to carry out the most common calculations used in pinch studies. The spreadsheet has most of the functionality of mainstream commercial pinch software, and it has been demonstrated in an example based on a typical FCC Unit.

References

1. B. Linnhoff, D. W. Townsend, D. Boland, G. F. Hewitt, B. E. A. Thomas, A. R. Guy, R. H. Marsland, *A User Guide on Process Integration for the Efficient Use of Energy, Revised First Edition,* I.Chem.E., Rugby, England, 1994.

2. Alan P. Rossiter and Jimmy D. Kumana, "Pollution Prevention and Process Integration – Two Complementary Philosophies", in *Waste Minimization through Process Design*, A. P. Rossiter (ed.), McGraw-Hill, New York, 1995, pp. 43-49.

3. A. P. Rossiter, M. A. Rutkowski and A. S. McMullan, "Pinch Technology Identifies Process Improvements," *Hydroc. Proc.*, 70(1), pp. 63-66, January 1991.

4. A. P. Rossiter, "Succeed at Process Integration," *Chem. Eng. Prog.*, 100(1), pp. 58-62, January 2004.

5. CANMET Energy Technology Centre of Natural Resources Canada - Government of Canada, http://cetc-varennes.nrcan.gc.ca/